QUANTITATIVE REMOTE SENSING OF LAND SURFACES

WILEY SERIES IN REMOTE SENSING

Jin Au Kong, Editor

Quantitative Remote Sensing of Land Surfaces

Shunlin Liang

A JOHN WILEY & SONS, INC., PUBLICATION

Library of Congress Cataloging-in-Publication Data

Liang, Shunlin.
   Quantitative remote sensing of land surfaces / Shunlin Liang.
      p. cm.
   Includes bibliographical references and index.
   ISBN 0-471-28166-2
   1. Earth sciences--Remote sensing. 2. Environmental sciences--Remote sensing. 3.
Remote sensing. I. Title.

   QE33.2.R4L53 2004
   550'.28'7--dc21

                                                                    2002044603

Printed in the United States of America

10 9 8 7 6 5 4 3 2 1

Contents

Preface

Remote sensing has entered a new era in the twenty-first century with a series of operating satellites from the NASA Earth Observing System (EOS) program, other international programs, and commercial programs. Since November 2000 the first civilian spaceborne hyperspectral sensor, Hyperion, has observed land surfaces on the NASA Earth Observer-1 (EO-1) platform. The first hyperspatial commercial satellite, IKONOS, produces global imagery at the spatial resolutions of 1 m (panchromatic band) and 4 m (multispectral bands) operationally. An increasing number of people are processing and analyzing vast amounts of optical remotely sensed observations for monitoring land surface processes and other applications on both local and global scales.

One basic characteristic of optical remote sensing in the twenty-first century is the extensive use of quantitative algorithms for estimating Earth surface variables. New sensors have much higher measurement precision with new technology and are far better calibrated. Empirical statistical models are being replaced by physically based models developed mostly since the early 1980s. In the NASA Earth Vision by 2020, environmental prediction is the primary goal. The reliable prediction relies on the physical dynamic models of land surface processes whose variables must be quantitatively estimated from Earth observing data.

This book emphasizes both the basic principles of optical remote sensing (0.4–14 μm) and practical algorithms for estimating land surface variables quantitatively from remotely sensed observations. It presents the current physical understanding of remote sensing as a system with a focus on radiative transfer modeling of the atmosphere, canopy, soil, and snow, and also the state-of-the-art quantitative algorithms for sensor calibration, atmospheric and topographic correction, estimation of a variety of biophysical variables (e.g., leaf area index, fraction of photosynthetically active radiation absorbed by vegetation) and geophysical variables (e.g., broadband albedo, emissivity, and skin temperature), and four-dimensional data assimilation.

This book evolved from the lecture notes for a graduate-level remote sensing course that I offered at the University of Maryland, but it can serve as a valuable reference book for advanced undergraduates and professionals who are involved in remotely sensed data in different scientific disciplines. This book does not assume a special mathematical background beyond a good working knowledge of statistics, calculus, and linear algebra on an undergraduate level. Although some detailed formulas are given in certain sections as important references, the derivations of these formulas are not provided. To aid the reader, each chapter starts with a brief overview of its content and concludes with a summary containing an overview of open research issues and pointers to relevant references. Extensive verbal "explanations" of what is being addressed in the corresponding formulas are also given.

This book should be studied in sequential order, but readers who are interested primarily in practical algorithms might follow the shorter sequence of Chapters 1, and 5–13. Book chapters are conceptually organized into two parts—Chapters 2–4 and 5–12—plus an introductory chapter (Chapter 1) and a final chapter (Chapter 13) on applications. Chapter 1 introduces some basic concepts and presents an overview of an optical remote sensing modeling system that links surface environmental variables to remotely sensed data. Major components of the systems are outlined as a background to the following chapters. Chapter 13 describes many different examples demonstrating how quantitative remote sensing techniques can be used to address scientific issues and solve practical societal problems.

Quantitative estimation of land surface variables relies greatly on physical understanding of the remotely sensed data and their relation to these variables. In Chapters 2–4, I summarize major physical modeling methods in atmosphere, vegetation canopy, soil, and snow media in the reflective spectrum with an emphasis on radiative transfer modeling. Radiative transfer modeling in the thermal infrared spectrum in conjunction with the estimation of land surface skin temperature is discussed in Chapter 10.

Chapter 2 introduces the basic principles of radiative transfer theory and its connections with atmospheric properties with numerical and approximate solutions to the atmospheric radiative transfer equation. Chapter 3 presents three types of canopy reflectance models: radiative transfer models, geometric optical models, and computer simulation models. In *radiative transfer modeling*, different leaf models are also introduced, different methods for solving one-dimensional (1D) canopy radiative transfer equation are presented, and methods for handling spatial heterogeneity of the canopy field are outlined. In *geometric optical modeling*, major models are outlined. A simply but widely mentioned Li–Strahler model is presented in greater detail. In *computer simulation models*, the basic principles of both Monte Carlo ray tracing and radiosity modeling methods are presented. Chapter 4 presents the modeling methods for soil and snow media. The emphasis is on radiative transfer modeling, but geometric optical modeling methods are also outlined.

Chapters 5–12 present the practical algorithms for estimating land surface variables from optical remotely sensed data. It starts with sensor calibration and ends with product validation. Chapter 5 discusses the need for sensor radiometric calibration and various postlaunch vicarious calibration methods. The calibration coefficients for land remote sensing satellite thematic mapping (Landsat TM) and National Oceanic and Atmospheric Administration advanced very high-resolution radiometer (NOAA AVHRR) data are provided for easy reference. Chapter 6 presents the basic principles and practical methods of atmospheric correction that convert top-of-atmosphere (TOA) radiance to surface reflectance. It focuses mainly on estimating aerosol optical properties from single-viewing remotely sensed data, but multiangle imaging spectroradiometric (MISR) aerosol estimation methods representing multiangular remote sensing and those for estimating total column water vapor content of the atmosphere are also discussed. Chapter 7 discusses various topographic correction algorithms that remove the disturbance caused by the variable land surface topography. The current status of digital elevation model (DEM) data is also evaluated. Chapter 8 presents various methods for estimating land surface variables, such as leaf area index, fraction of the photosynthetically active radiation (FAPAR) absorbed by vegetation canopies, the fraction of vegetation coverage, and biochemical concentrations. Statistical algorithms (based mainly on vegetation indices from multispectral and hyperspectral remotely sensed data) and physical inversion methods (based mainly on process models introduced in the first few chapters of the book) are presented.

Chapters 9 and 10 discuss various methods for estimating land surface radiation budget components, such as broadband albedo, broadband emissivity, and skin temperature. In calculating broadband albedo, we focus mainly on the conversion algorithms from narrowband to broadband. In estimating skin temperature, split-window algorithms as well as other temperature and emissivity separation algorithms are presented. Chapter 11 presents four-dimensional data assimilation algorithms and applications. This is a quantitative, objective method for inferring the state of the dynamic system from heterogeneous, irregularly distributed, and temporally inconsistent observational data with differing accuracies. One unique feature is the ability to estimate environmental variables that are not related to any remotely sensed radiometric quantities. The basic principles and the typical algorithms are introduced, and the applications to hydrology and crop growth are demonstrated. Chapter 12 discusses the validation of the estimated land surface variables and various methods of spatial scaling. Since land surface variables drive various models related to land surface processes, the accuracy of the remotely sensed products will largely affect the model outputs and the final conclusions. The validation rationale, NASA EOS validation programs, and some validation methodologies are presented. Because of its importance in remote sensing besides validation, spatial scaling is presented separately. The emphasis is on downscaling methods, including linear unmixing, generating continuous fields, decomposition of normalized difference vegetative index (NDVI) temporal profiles, multiresolution data fusion, and statistical downscaling global (or general) circulation model (GCM) outputs.

This book builds on many journal papers and books cited in the references at the end of each chapter, which is at best representative but by no means exhaustive. Many references are made in the text to websites that serve as references and as sources of complimentary information. As most users of the Internet know, uniform resource locators (URLs) are often changed, while some are even dropped and removed. We have strived to make these as current as possible, but as time progresses some will disappear or otherwise be altered without our knowledge.

This book cites more than 1300 references. Some useful computer program codes and valuable datasets are also included in the CD-ROM for easy reference.

ACKNOWLEDGMENTS

Many people substantially contributed to this book, both directly and indirectly. First and foremost, I am greatly indebted to Professors Alan Strahler at Boston University and John Townshend at the University of Maryland for their consistent encouragement and guidance. Parts of the book were written during my sabbatical leave while visiting the Center for Remote Sensing and Geographic Information Science, Beijing Normal University (BNU), China; the Earth Observation Research Center, National Space Development Agency of Japan (NASDA); CSIRO Land and Water at Canberra, Australia; and USDA Beltsville Agricultural Research Service (BARC), Maryland. I greatly appreciate all the support from Professors Xiaowen Li, Qijiang Zhu, and Jindi Wang at BNU; Dr. Tamotsu Igarashi at NASDA; Dr. Tim McVicar at CSIRO; and Drs. Craig Daughtry and Charlie Walthall at BARC. I'd like to thank Drs. Mat Disney, Philip Lewis, and David Jupp for their kindly providing me with some of the illustrations used in this book, and my co-authors of the journal papers that largely build this book. I am indebted to my students who made great contributions to this book, particularly Chad Shuey, Hongliang Fang, and Lynn Thorp for their help in preparing the manuscript. I gratefully acknowledge the assistance of George Telecki (Editor), Rosalyn Farkas (Production Editor), and other staff at Wiley. I would also like to acknowledge the research funding support from several agencies, particularly from NASA through its Terrestrial Ecology, EOS, EO-1 and other programs. Finally, I would like to express my most heartfelt thanks to my lovely wife, Jie, for her endless support and encouragement. Without her sacrifice, this project would never come to the end.

SHUNLIN LIANG

College Park, Maryland, USA
October 2002

Acronyms

1D	One-dimensional
2D	Two-dimensional
3D	Three-dimensional
4D	Four-dimensional
4DDA	Four-dimensional data assimilation
6S	Second simulation of satellite signals in the solar spectrum
AAC	Autonomous atmospheric concentration
AATSR	Advanced along-track scanning radiometer
ADE	Alpha-derived emissivity
ADEOS	Advanced Earth Observing System
ADM	Angular distribution model
AERONET	Aerosol robotic network
AFV1	Aerosol-free vegetation index
AFWA	(U.S.) Air Force Weather Agency
AGCM	Atmospheric general circulation models
AgRISTARS	Agriculture and Resources Inventory Surveys through Aerospace Remote Sensing
AIRS	Atmospheric Infrared Sounder
AIS	Airborne imaging spectrometer
AISA	Airborne imaging spectroradiometer
ALA	Average leaf (inclination) angle
ALI	Advanced land imager
ALOS	Advanced land observation satellite
AMBRALS	Algorithm for MODIS bidirectional reflectance anisotropy of the land surface
AMSR	Advanced microwave scanning radiometer
AMSU	Advanced microwave sounding unit
ANN	Artificial neural network
APAR	Absorbed (also absorption of) photosynthetically active radiation
ARM	Atmospheric radiation measurement

ARV1	Atmospherically resistant vegetation index
ASAR	Advanced synthetic aperture radar
ASAS	Advanced solid-state array spectroradiometer
ASD	Analytic spectral device
ASTER	Advanced spaceborne thermal emission and reflection radiometer
ATBD	Algorithm theoretical basis document
ATSR	Along-track scanning radiometer
AVHRR	Advanced very high-resolution radiometer
AVIRIS	Airborne visible–infrared imaging spectrometer
BARC	Beltsville (MD) Agricultural Research Service (USDA)
BATS	Biosphere-Atmosphere Transfer Model
BDGP	Bidirectional gap probability
BEPS	Boreal ecosystem productivity simulator
BFGS	Broyden–Fletcher–Goldfarb–Shanno
BGC	Biogeochemical cycle
BHR	Bidirectional hemispherical reflectance
BNA	Band depth normalized to area of absorption feature
BNC	Band depth normalized to center of absorption feature
BOREAS	Boreal ecosystems–atmosphere study
BPMS	Botanical plant modeling system
BRDF	Bidirectional reflectance distribution function
BRF	Bidirectional reflectance factor
BSRN	Baseline surface radiation network
CACI	Chlorophyll absorption continuum index
CAI	Cellulose absorption index
CARI	Chlorophyll absorption ratio index
CART	Cloud and radiation testbed
CASI	Compact airborne spectrographic imager
CCDRE	Crop condition data retrieval and evaluation
CCRS	Canadian Center for Remote Sensing
CEC	Cation-exchange capacity
CEOS	Committee on Earth Observation Satellites
CERES	Clouds and Earth's radiant energy system
CG	Computer graphics
CGMS	Crop growth monitoring system
CIBR	Continuum interpolation band ratio
CIGSN	Continental Integrated Ground-truth Site Network (Australia)
CLM	Climate land model (common)
CWSI	Crop water stress index
DAAC	Distributed Active Archive Center (NASA)
DAO	Data Assimilation Office
DART	Discrete anisotropic radiative transfer (canopy model)

DAS	Data assimilation system
DEM	Digital elevation model
DFP	Davidon–Fletcher–Powell
DHR	Directional hemispherical reflectance
DISORT	Discrete ordinate radiative transfer
DMSP	Defense Meteorological Satellite Program
DN	Digital number
DSS	Decision support system
DU	Dobson unit
ECMWF	European Centre for Medium-Range Weather Forecasts
EDC	EROS Data Center
EF	Evaporation fraction
EIFOV	Effective IFOV
EKF	Extended Kalman filter
ELDAS	European land data assimilation system
ENVISAT	Environmental satellite, ESA
EOI	Earth Observer I
EOF	Empirical orthogonal function
EOS	Earth Observing System (NASA program)
EOSDIS	EOS Data and Information System
EPIC	Erosion–productivity impact calculat(or)(ion)
ER-2	NASA Research Aircraft
ERB	Earth radiation budget
ERBE	Earth Radiation Budget Experiment
ERBS	Earth Radiation Budget Satellite
erfc	Complementary error function
EROS	Earth Resources Observation System
ESA	European Space Agency
ET	Evapotranspiration
ETM +	Enhanced thematic mapper plus
EVI	Enhanced vegetation index
FAO	Food and Agriculture Organization
FAPAR	Fraction of absorbed photosynthetically active radiation
FAS	Foreign Agricukture Service (USDA)
FASCODE	Fast atmospheric signature code
FEM	Finite element method
FEWSNET	Famine Early Warning System Network (USGS)
FFT	Fast Fourier transform(ation)
FIFE	First ISLSCP Field Experiment
FLUXNET	Flux network
FPAR	Fraction of photosynthetically active radiation
FTP	File transfer protocol
FWHM	Full width at half maximum
GA	Genetic algorithm
GAC	Global area coverage

GADS	Global aerosol dataset
GCIP	GEWEX Continental-Scale International Project
GCM	General circulation model (*also* global climate model)
GCOS	Global climate observing system
GCTE	Global change in terrestrial ecosystems
GEBA	Global energy balance archive
GEMI	Global environment monitoring index
GEOS	Goddard Earth Observing System
GERB	Geostationary Earth Radiation Budget
GEWEX	Global energy and water cycle experiment
GIS	Geographic information system
GLDAS	Global land data assimilation system
GLI	Global land imager
GloPEM	Global production efficiency model
GMRF	Gaussian–Markovian random field
GMS	Geostationary meteorological satellite
GO	Geometric optical
GOES	Geostationary operational environmental satellite
GPP	Gross primary productivity
GPS	Global positioning system
GRAMI	Ground radiometric measurement index
GSFC	Goddard Space Flight Center
GTOS	Global terrestrial observing system
GVI	Global vegetation index
HAPEX	Hydrological–atmospheric pilot experiment
HCMM	Heat Capacity Mapping Mission
HDRF	Hemispherically directional reflectance factor
HIRS	High-resolution infrared radiation sounder
HIS	Hue–intensity–saturation
HITRAN	High-resolution transmission (molecular absorption database)
HWHM	Half-angular width at half-maximum magnitude
ICESat	Ice, clouds, and land elevation satellite
IFOV	Instantaneous field of view
IGBP	International Geosphere–Biosphere Programme
IGOSP	Integrated global observing strategy partnership
ILTER	International long-term ecological research
InTEC	Integrated terrestrial ecosystem C-budget (model)
IPCC	Intergovernmental Panel on Climate Change
IPP	Image processing package
IPW	Image processing workbench
IR	Infrared
ISIS	Integrated surface irradiance study
ISLSCP	International Satellite Land Surface Climatology Project
KF	Kalman filter

KM	Kubelka−Munk (theory)
LACIE	Large area crop inventory experiment
LAD	Leaf angle distribution
LAI	Leaf area index
Landsat	Land remote sensing satellite
LBA	Large-scale biosphere−atmosphere experiment in Amazonia
LEO	Low Earth orbiting
LIBERTY	Leaf incorporating biochemistry exhibiting reflectance and transmittance yields
LiDAR	Light detection and ranging
LSF	Line spread function
LSM	Land surface model
LSP	Land surface parameterizations
LST	Land surface temperature
LTER	Long-term ecological research
LW	longwave
MARS	Monitoring agriculture through remote sensing
MC	Monte Carlo
MCRT	Monte Carlo ray tracing
MERIS	Medium-resolution imaging spectrometer
MISR	Multiangle imaging spectroradiometer
MLRA	Major land resource area
MMD	Maximum−minimum difference
MODIS	Moderate-resolution imaging spectroradiometer
MODLAND	MODLIS land
MOS	Modular optoelectronic scanner
MODTRAN	MODerate resolution TRANSmittance
MSAVI	Modified SAVI
MSG	Meteosat second generation
MSR	Modified simple ratio
MSS	Multispectral scanner
MSU	Microwave sounding unit
MTF	Modulation transfer function
NASA	National Aeronautics and Space Administration
NASDA	National Space Development Agency (of Japan)
NASS	National Agricultural Statistics Survey
NBP	Net biome productivity
NCAR	National (U.S.) Center for Atmospheric Research
NDEP	National Digital Elevation Program
NDTI	Normalized difference temperature index
NDVI	Normalized difference vegetation index
NEM	Normalized emissivity method
NEP	Net ecosystem productivity
NIMA	National (U.S.) Imagery and Mapping Agency

NIR	Near infrared
NLDAS	North American land data assimilation system
NOAA	National Oceanic and Atmospheric Administration
NPOESS	National Polar-orbiting Operational Environmental Satellite System
NPP	Net primary productivity
NSF	National Science Foundation
NWF	Numerical weather forecast
NWP	Numerical weather prediction
OLR	Outgoing longwave radiation
OPAC	Optical properties of aerosols and clouds
OTF	Optical transfer function
OTTER	Oregon Transect Terrestrial Ecosystem Research
PAR	Photosynthetically active radiation
PCA	Principal-component analysis
PECAD	Production Estimates and Crop Assessment Division (USDA)
PEM	Production efficiency model
PILPS	Project for Intercomparison of Land Surface Parameterization Schemes
POLDER	Polarization and directionality of Earth's reflectance
ppm	Parts per million
ppmv	Parts per million by volume
ppmw	Parts per million by weight
PPR	Projection–pursuit regression
PRI	Photochemical reflectance index
PSAS	Physically spaced statistical analysis scheme
PSF	Point spread function
PVI	Perpendicular vegetation index
PW	Precipitable water
PWL	Piecewise linear
QA	Quality assessment
RAM	Random access memory
REIP	Red edge inflection point
ReSeDA	Remote sensing data assimilation
RGB	Red-green-blue
RMSE	Root mean square error
RP	Recursive partitioning
RSE	Residual standard error
RT	Radiative transfer
RUE	Radiation use efficiency
SAFARI	South African Fire–Atmospheric Research Initiative
SAGE	Stratospheric aerosol and gas experiment
SAIL	(Light) scattering by arbitrarily inclined leaves (canopy model)

SALSA	Semiarid land surface–atmosphere
SAR	Synthetic aperture radar
SARVI	Soil and atmospherically resistant vegatation index
SAVI	Soil-adjusted vegetation index
SBDART	Santa Barbara DISORT atmospheric radiative transfer
SCA	Spatial coordinate apparatus
ScaRab	Scanner for Earth Radiation Budget
SCM	Successive correction method
SeaWiFS	Sea-viewing wide-field-of-view sensor
SEB	Surface energy balance
SEBASS	Spatially enhanced broadband array
SiB	Simple biosphere model
SIPI	Structure-independent pigment index
SSM/I	Special sensor microwave/imager
SMOS	Soil Moisture/Ocean Salinity (mission)
SMMR	Scanning multichannel microwave radiometer
SNR	Signal-to-noise ratio
SOSA	Successive orders of scattering approximation
SPOT	Systeme pour l'Observation de la Terre
SPRINT	Spreading of photons for radiation interception
SR	Simple ratio
SRB	Surface radiation budget
SS	Sum of squares
SST	Sea surface temperature
SSU	Stratospheric sounding unit
STP	Standard temperature and pressure
SURFRAD	Surface radiation budget network
SVAT	Soil–vegetation–atmosphere transfer (model)
SVI	Spectral vegetation index
SW	Shortwave
SWIR	Shortwave infrared
SZA	Solar zenith angle
TCI	Temperature condition index
TCT	"Tasseled cap" transformation
TIMS	Thermal infrared multispectral scanner
TIR	Thermal infrared
TIROS	Television infrared observation (also *operational*) satellite
TISI	Temperature-independent spectral index
TM	Thematic mapper
TOA	Top of (the) atmosphere
TOMS	Total ozone mapping spectrometer
TOVS	TIROS operational vertical sounder
TRMM	Tropical Rainfall Measuring Mission
TSAVI	Transformed SAVI
TSI	Total solar irradiance

TVI	Triangular vegetation index
TVX	Temperature/vegetation index
UHI	Urban heat island
USDA	United States Department of Agriculture
USGS	United States Geological Survey
UTM	Universal transverse mercator
UV	Ultraviolet
UVB	Ultraviolet B
VCI	Vegetation condition index
VI	Vegetation index
VIRS	Visible infrared scanner
VIS	Visible (spectrum)
VNIR	Visible and near infrared
VPD	Vapor pressure deficit
WASDE	World Agricultural Supply and Demand Estimates (report)
WDI	Water deficit index
WCRP	World Climate Research Programme
WGCV	Working Group on Calibration and Validation
WMO	World Meteorological Organization
WDVI	Weighted difference vegetation index
WI	Water index
www	World wide web

1

Introduction

This chapter introduces the background information and basic concepts of optical remote sensing. *Optical remote sensing* refers to wavelengths ranging from the visible spectrum to the thermal infrared (IR) spectrum (0.4–14 μm). This chapter basically consists of two parts. The first part (Sections 1.1 and 1.2) discuss various radiometric variables that will be extensively used in the book, and the second part (Section 1.3) briefly describes the major components of a remote sensing modeling system that links remotely sensed data with land surface variables, which serves as the pointers to various chapters and sections in the rest of the book. For some topics that are not discussed in detail later, more details will be presented in this chapter.

  Section 1.1 classifies all quantitative models for retrieving land surface variables from optical remotely sensed data into three major categories: statistical, physical, and hybrid. Their major characteristics are briefly discussed. Section 1.2 defines some basic physical concepts and illustrates the conversion from digital numbers to a series of physical variables that will be discussed in the following chapters in more detail. Section 1.3 discusses the major components of a remote sensing modeling system on which physical models are based. The concepts of forward modeling and inversion methods are also illustrated.

1.1 QUANTITATIVE MODELS IN OPTICAL REMOTE SENSING

All models in optical remote sensing are traditionally grouped into two major categories: statistical and physical. *Statistical models* are based on correlation relationships of land surface variables and remotely sensed data. They are easy to develop and effective for summarizing local data; however, the

Quantitative Remote Sensing of Land Surfaces.   By Shunlin Liang
ISBN 0-471-28166-2   Copyright © 2004 John Wiley & Sons, Inc.

developed models are usually site-specific. They also cannot account for cause–effect relationships.

On the other hand, physically based models follow the physical laws of the remote sensing system. They also establish cause and effect relationships. If the initial models do not perform well, we know where to improve by incorporating the latest knowledge and information. However, there is a long curve to develop and learn these physical models. Any models represent the abstract of the reality; thus a realistic model could potentially be very complex with a large number of variables.

There is a new trend in remote sensing to develop hybrid models that combine both statistical and physical models, that may take advantage of the unique features of each and overcome their shortcomings.

All three categories of quantitative models for estimating land surface variables will be presented systematically in Chapter 8 and also individually in several other chapters. All quantitative models in remote sensing utilize five signatures: spectral, spatial, temporal, angular, and polarization. Because of the scope of the book, we will not present any models based on polarization signatures.

Land surfaces can be characterized by *continuous variables* (e.g., leaf area index, albedo) and *categorical variables* (e.g., land cover) through various quantitative models. Because of the scope, we will address only continuous variables; thus, we have excluded a large group of image classification algorithms and applications (e.g., land cover and use and change mapping) out of this book.

1.2 BASIC CONCEPTS

1.2.1 Digital Numbers

Early statistical models in remote sensing often employed *digital numbers* (DNs) to estimate surface characteristic variables directly. DNs are what we get after purchasing data from the data providers. But DNs are the scaled integers from *quantization*, which is not a physical quantity. Although it might be desirable to use nonlinear quantization for low-reflectance cases such as oceanography, most quantization systems in remote sensing are linear, typically 6–12 bits. The DN can be any integer in the range

$$DN \in [1, 2^Q] \tag{1.1}$$

where Q is an integer representing the bits. For example, an 8-bit ($Q = 8$ and $2^8 = 256$) linear quantization system equally divides the dynamic range of response of the sensor into 255 steps, from 1 to 256. The same response of the sensor will produce totally different DNs if a 10- or 12-bit quantization system is used. It is obvious that a larger Q leads to a higher radiometric precision. Table 1.1 shows the quantization levels for several common satel-

TABLE 1.1   Quantization Levels of Some Common Sensor Systems for Land Applications

Sensor	TM	SPOT	AVHRR	IKONOS	MODIS/MERIS
Q	8	8	10	11	12

lite sensor systems, the thematic mapper (TM) in the U.S. Landsat system, the French SPOT (Systeme pour l'Observation de la Terre) satellite system, the advanced very high-resolution radiometer (AVHRR), the private IKONOS satellite, moderate-resolution imaging spectroradiometer (MODIS), and medium-resolution imaging spectrometer (MERIS).

1.2.2   Radiance

We now realize that DNs should be converted to physical quantities for estimating land surface variables such as *radiance* (sometimes called *intensity*), measured in the energy per area per solid angle. The solid angle is explained in Section 1.2.3. Spectral (monochromatic) radiance represents the energy per area per solid angle per unit wavelength. A typical unit could be W cm^{-2} Sr^{-1} μm^{-1}, where sr (steradian) is the unit for measuring solid angles, and μm (micrometer) measures wavelength. Note that different units may be used in the literature and it is important to know their conversions. In particular, wavelength (λ: μm) in thermal infrared (IR) remote sensing is customarily specified by *wavenumber* ν, which is the reciprocal of the wavelength ($1/\lambda$). Traditionally, wavenumber is expressed in inverse centimeters, which is numerically equivalent to $10^4/\lambda$, where λ is in μm (1 cm = 10^4 μm). For example, the wavelength at 10 μm has a wavenumber of 1000 cm^{-1}. To convert radiance in (W cm^{-2} sr^{-1})/cm^{-1} to (W cm^{-2} sr^{-1})/μm, one must multiply by $\nu^2/10^4$.

Normally, DNs are linearly related to radiance, and most remote sensing data providers produce the conversion coefficients for the users. Sensors receive radiance at the range of wavelengths (waveband), but these conversion coefficients are typically for generating spectral radiance to avoid the difference of waveband widths. These coefficients are usually included in the image data header file (or metadata). The procedure that determines these conversion coefficients is called *sensor calibration*. It is an important procedure in remote sensing since many sensors deteriorate in space after the satellite launch and the preflight conversion coefficients seldom remain valid. A detailed description of sensor calibration will be discussed in Chapter 5.

1.2.3   Solid Angle

The directional dependence of radiance is taken into account by employing the *solid angle* (Ω), which is an extension of two-dimensional angle measure-

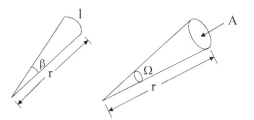

Figure 1.1 Planar angle (β) and solid angle (Ω).

ment. Understanding the angular dependence is very important since most sensors are targeting the Earth surface in a specific direction. As illustrated in Fig. 1.1, the angle β between two radii of a circle of radius r is

$$\beta = \frac{1}{r} \quad \text{(rad)} \tag{1.2}$$

The arc length of the full circle is $2\pi r$, so the angular measure of the full circle is 2π rad. The solid angle Ω is defined as the ratio of the area A of a spherical surface intercepted by the cone to the square of the radius

$$\Omega = \frac{A}{r^2} \quad \text{(sr)} \tag{1.3}$$

For a sphere whose surface area is $4\pi r^2$, its solid angle is 4π sr (steradians). Thus, the solid angle of the upper or lower hemisphere is 2π sr.

A solid angle is often represented by the zenith and azimuth angles in polar coordinates. If θ represents the zenith angle (the angle measured from the vertical or from the horizontal to a surface), ϕ represents the azimuth angle (Fig. 1.2), then a differential element of solid angle is mathematically given by

$$d\Omega = \frac{dA}{r^2} = \frac{(r\,d\theta)(r\sin\theta\,d\phi)}{r^2} = \sin\,\theta\,d\theta\,d\phi = d\mu\,d\phi \tag{1.4}$$

where $\mu = \cos\theta$. Note that the zenith angle θ ranges from 0 to 180°. In the literature, the range of 0–90° (i.e., $0 \le \mu \le 1$) usually represents the up-welling hemisphere and 90–180° (i.e., $-1 \le \mu \le 0$) the downward hemisphere, which will be used throughout this book. But this is arbitrarily defined, we can also find the exactly opposite definition in the literature. The azimuth angle ϕ ranges from 0 to 360°, that is, $0 \le \phi \le 2\pi$.

It is worth pointing out that the viewing zenith angle is not equivalent to the sensor scan angle that has been frequently used in the remote sensing literature. The difference can be observed in Fig. 1.3. If Earth is assumed to be a sphere, the relationship between the two angles can be easily expressed

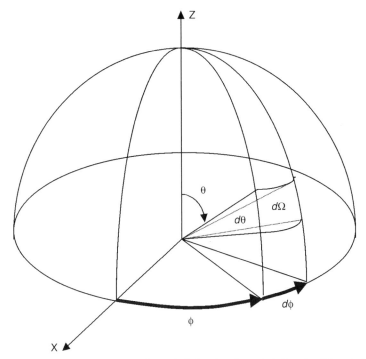

Figure 1.2 Illustration of the zenith angle (θ), azimuth angle (φ), and solid angle (Ω).

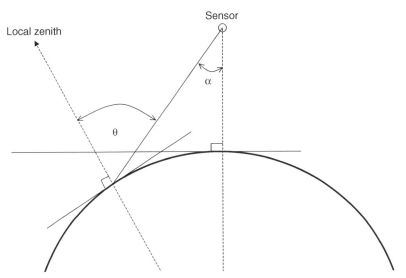

Figure 1.3 Relationship between the sensor scanning angle (α) and the viewing zenith angle (θ).

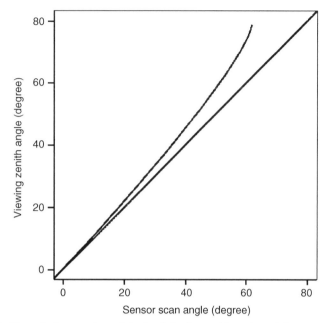

Figure 1.4   Numerical comparison of the local viewing zenith angle and the sensor scan angle.

as

$$\sin \theta = \frac{R_0 + h}{R_0} \sin \alpha \qquad (1.5)$$

where the Earth radius is $R_0 = 6378$ km and h is the orbital height of the satellite above Earth. The typical height of the remote sensing satellites varies from 600 to 900 km (e.g., $h = 705$ km for Landsat-4 and -5, $h = 832$ for the first three SPOT (Systeme pour l'Observation de la Terre) satellites, $h = 681$ km for IKONOS, $h = 705$ km for Aqua). For $h = 700$ km, the relationship between the local viewing zenith angle and the sensor scanning angle is shown in Fig. 1.4. It is clear that the viewing zenith angle θ at a specific location of the Earth surface is always larger than the sensor's scan angle α. The larger the sensor scan angle, the larger their difference.

1.2.4   Irradiance

By definition, *irradiance* (E) is the integration of radiance (L) over the entire solid angle of a hemisphere consisting of the zenith angle θ and the azimuth angle ϕ:

$$E = \int_0^{2\pi} \int_0^{\pi/2} L(\theta, \phi) \cos \theta \sin \theta \, d\theta d\phi = \int_0^{2\pi} \int_0^1 L(\mu, \phi) \mu \, d\mu d\phi \quad (1.6)$$

If radiance is independent of the direction (isotropic), Eq. (1.6) becomes $E = \pi L$. Note that irradiance is often called *flux density* or simply *flux* in the remote sensing literature.

1.2.5 Bidirectional Reflectances and Albedos

Upwelling radiance received by the Earth-viewing sensors depends on the incoming solar radiation. To normalize the variation of the incoming solar radiation, the top-of-atmosphere (TOA) radiance $I(\theta_v, \phi_v)$ converted from DN at the specific viewing direction (θ_v, ϕ_v) is often further converted into reflectance

$$R(\theta_i, \phi_i, \theta_v, \phi_v) = \frac{\pi I(\theta_i, \phi_i, \theta_v, \phi_v)}{\cos \theta_i E_0} \tag{1.7}$$

where θ_i is the solar zenith angle and E_0 is the incoming TOA irradiance. It is not difficult to understand that reflectance obviously depends on both the solar incoming direction and the sensor viewing direction, which is usually denoted as *bidirectional reflectance factor*. The same concept is applied to land surface (Martonchik et al. 2000).

A complete description of the directional reflectance properties of a surface is provided by the *spectral bidirectional reflectance distribution function* (BRDF) $f(\theta_i, \phi_i, \theta_v, \phi_v)$, which is defined as the ratio of the reflected spectral radiance from the surface in the direction θ_v, ϕ_v to the directional spectral irradiance on the surface in the direction θ_i, ϕ_i:

$$f(\theta_i, \phi_i, \theta_v, \phi_v) = \frac{dL(\theta_i, \phi_i, \theta_v, \phi_v)}{dE(\theta_i, \phi_i)} \tag{1.8}$$

and has the units of reciprocal steradians. This concept has been extensively used in multiangle remote sensing, but in practice people tend to use the dimensionless *bidirectional reflectance factor* (BRF) EMBED $R(\theta_i, \phi_i, \theta_v, \phi_v)$, which is numerically equivalent to BRDF multiplied by π:

$$R(\theta_i, \phi_i, \theta_v, \phi_v) = \pi f(\theta_i, \phi_i, \theta_v, \phi_v) \tag{1.9}$$

Different statistical BRDF/BRF models are discussed in Section 2.2. Chapters 2-5 are devoted to introducing various physically based methods for calculating land surface BRDF/BRF. The in-depth reviews on this subject are available in a special issue of *Remote Sensing Reviews* (Liang and Strahler 2000).

For studies of surface shortwave energy balance (see Chapter 9) and other applications, we need surface *directional hemispherical reflectance* (DHR), which can be integrated from BRF over all reflected directions:

$$r(-\mu_i, \phi_i) = \frac{1}{\pi} \int_0^{2\pi} \int_0^1 R(\mu_i, \phi_i, \mu, \phi) \mu \, d\mu \, d\phi \qquad (1.10)$$

DHR is often called *local* or *planar albedo*, but is called *black-sky albedo* in the MODIS products of the NASA Earth Observing System (EOS) program. *Bihemispherical reflectance* (BHR) is a further integration of DHR over all illumination directions

$$r_0 = 2 \int_0^1 r(-\mu_i, \phi_i) \mu_i \, d\mu_i \qquad (1.11)$$

which is also called *global* or *spherical albedo*, or *bright-sky albedo* in the MODIS products.

In the thermal IR spectrum, emissivity (ε) is much more often used than reflectance. $\varepsilon = 1 - r_0$ according to the Kirchhoff's law. More details are available in Chapter 10. Note that all these variables are the function of wavelength.

1.2.6 Extraterrestrial Solar Irradiance

Incoming TOA irradiance depends on the astronomical distance between the Sun and Earth (D). Given the solar irradiance $\overline{E}_0$ at the average Earth–Sun distance ($D = 1$), a simple expression can be used to approximate the solar irradiance on any day

$$E_0 = \frac{\overline{E}_0}{D^2} = \overline{E}_0 \left[1 + 0.033 \cos\left(\frac{2\pi d_n}{365} \right) \right] \qquad (1.12)$$

where d_n is the day number of the year, ranging from 1 on January 1 to 365 on December 31; February is always assumed to have 28 days. A more accurate formula is

$$E_0 = \overline{E}_0 [1.00011 + 0.034221 \cos \chi + 0.00128 \sin \chi$$
$$+ 0.000719 \cos 2\chi + 0.000077 \sin 2\chi] \qquad (1.13)$$

where $\chi = 2\pi(d_n - 1)/365$. For most applications, Eqs. (1.12) and (1.13) will not make any significant difference.

There have been many efforts to compile the solar irradiance curves (Gao and Green 1995) with variable accuracies. The total irradiance integrated over all wavelengths is called the *solar constant*:

$$S_0 = \int_0^\infty \overline{E}_0(\lambda)\, d\lambda \tag{1.14}$$

The results of modern solar constant monitoring during the 1980s reveal that the average S_0 is around 1369 W/m^2 with an uncertainty of $\pm 0.25\%$ (Hartmann et al. 1999). It also has solar cycle variations. The new generation of sensors will provide more accurate observations.

There are four data sets available in MODTRAN4 compiled from different sources. MODTRAN is an atmospheric radiative transfer software package that has been widely used in various remote sensing applications, and more details on this package will be provided in this book from time to time. They are included in the CD-ROM (newkur.dat, chkur.dat, cebchkur.dat, and thkur.dat). The corresponding solar constants are 1362.12, 1359.75, 1368.00, and 1376.23 W/m^2. Figure 1.5a shows the spectral solar irradiance curve (W m^{-2} μm^{-1}) of thkur.dat from 0.2–3.5 μm, Fig. 1.5b–Fig. 1.5d show the difference between thkur.dat and three other datasets in the same unit from 0.2 to 1.0 μm. It is surprising to see that the differences are so significant, particularly over the visible spectrum.

We need to be very careful to select the right data source of the TOA solar irradiance when we analyze hyperspectral data, although this may not be a serious problem when we work with the multispectral remotely sensed data. For different sensors whose spectral response functions (see Section 1.3.5.1) are known, we can easily calculate $\overline{E}_0$ for each waveband by integrating TOA spectral solar irradiance with the sensor spectral response functions. For ease of reference, $\overline{E}_0$ of the six reflective bands of Landsat-4/5 thematic mapper and Landsat-7 enhanced thematic mapper plus (ETM $+$) and the two reflective bands of NOAA advanced very high-resolution radiometer (AVHRR) are listed in Tables 1.2 and 1.3. In fact, many sensor spectral response functions and these four TOA spectral irradiance datasets displayed in Fig. 1.5 are provided in the accompanying CD-ROM, and readers can easily generate similar numbers. Do not be surprised if you see different numbers for TM or ETM $+$ in the literature since different TOA spectral solar irradiance datasets may be used.

TABLE 1.2 TOA Solar Irradiance of Landsat TM / ETM $+$ Reflective Bands (W m^{-2} μm^{-1})

Bands	1	2	3	4	5	7	Pan[a]
ETM $+$	1970	1843	1555	1047	227.1	80.53	1368
TM	1954	1826	1558	1047	217.2	80.29	N/A

[a] Panchromatic band.

TABLE 1.3 TOA Solar Irradiance (W m^{-2} μm^{-1}) of NOAA AVHRR Band 1 (0.58 0.68 μm) and Band 2 (0.72–1.1 μm)

Bands	NOAA7	NOAA9	NOAA11	NOAA12	NOAA14	NOAA16
1	1643.5	1635	1629.2	1613.7	1628.1	1644
2	1051.8	1053.6	1052.8	1049.8	1029.8	1034

Source: Data derived from Neckel and Labs (1984).

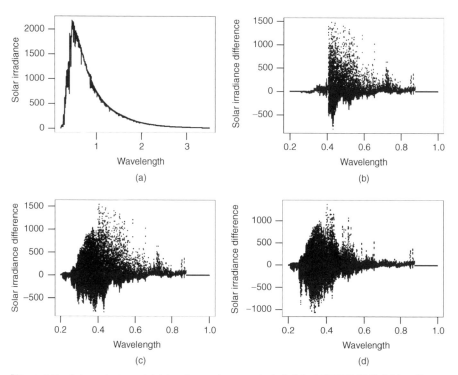

Figure 1.5 Solar extraterrestrial irradiance data sets included in MODTRAN4: (a) irradiance spectra data of thkur.dat; irradiance difference spectra are shown between (thkur.dat) and (cebchkur) (b), between (thkur.dat) and (chkur) (c), and between (thkur.dat) and (newkur) (d). The unit is W m^{-2} μm^{-1}.

1.3 REMOTE SENSING MODELING SYSTEM

An optical remote sensing system can be physically modeled with several subsystems: scene generation, scene radiation modeling, atmospheric radiative transfer modeling, navigation modeling, and mapping and binning. *A forward modeling scheme* predicts what remotely sensed data will be under a set of environmental and sensing conditions. *A physically based inversion scheme* determines various land surface geophysical and biophysical variables from remotely sensed data. Both schemes rely on a good physical understand-

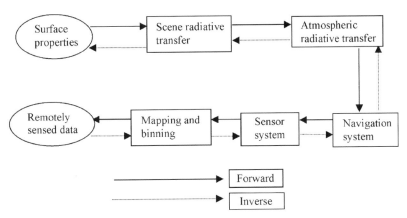

Figure 1.6   Illustration of a remote sensing modeling system (both forward and inversion schemes).

ing of different components in a remote sensing system (Fig. 1.6). Each individual component is briefly outlined below, and most of the components are discussed in more detail in later chapters. For further reading, readers are referred to some related textbooks for some of the components (e.g., Schott 1997, Schowengerdt 1997, Milman 1999)

1.3.1   Scene Generation

A scene generation model quantifies the relationships among the type, number, and spatial distribution of objects and the background in the scene of the land surface. It is a quantitative description of our understanding of the landscape. Strahler et al. (1986) identify two different scene models in remote sensing: H- and L-resolution (high- and low-resolution) models. The H-resolution models are applicable where the elements of the scene are larger than the pixel size; with the L-resolution model, the converse is true. L-resolution models can be regarded as a type of (continuous) mixture model where the proportions are functions of the sizes and shapes of the elements in the scene model and their relative densities within the pixel. H-resolution models are highly related to computer graphics. These techniques are relatively mature now, and motion pictures have been generated using similar techniques without any human actors. Many such algorithms are based on these two mathematical theories: fractals and L system.

Fractals are complex, detailed geometric patterns found throughout the natural world. They can be created using mathematical formulas and are infinite in their ability to be viewed in ever increasing detail. They are recursively defined and small sections of them are similar to large ones. One way to think of fractals for a function $f(x)$ is to consider x, $f(x)$, $f(f(x))$,

$f(f(f(x)))$, $f(f(f(f(x))))$, and so on. More details are available in the literature (e.g., Mandelbrot 1983, Fisher 1995).

Lindenmayer systems, or L systems for short, are a particular type of symbolic dynamical system with the added feature of a geometric interpretation of the evolution of the system. An L system is a formal grammar for generating strings. By recursively applying the rules of the L systems to an initial string, one can create a string with fractal structure. Interpreting this string as a set of graphical commands allows the fractal to be displayed. L systems are very useful for generating realistic plant structures (Prusinkiewicz and Lindenmayer 1990). L systems have been used for generating landscapes in computer simulation models (e.g., Monte Carlo ray tracing and radiosity), discussed in Sections 3.5.1 and 3.5.2.

Suppose that we want to generate a H-resolution forest scene. The first thing we need to know is the spatial distribution of the trees: regular or random. For regularly distributed canopies, we need to specify the row intervals. If they are randomly distributed, we need to specify the spatial distribution function (e.g., Poisson) and its characteristic parameters. We also need to specify the geometric parameters of the tree, such as the densities and dimensions of the leaves, trunk, and branches, and their space and inclination angles. Sometimes, we also need to specify the cluster density of the leaves and the locations of their central points. These parameters can be measured in the field. Some examples are given in Section 3.2.3.

In generating an L-resolution scene, it is not necessary to specify the shape and size of the individual elements since they cannot be distinguished within a pixel. There are two common numerical generation techniques. The first is based on the geographic information system (GIS) concept. The spatial distributions of different elements are expressed by using different data layers in a GIS environment. Overlapping these layers is one of the basic functions in any GIS software program that can generate an L-resolution scene. The second is based on stochastic field theory. The brightness of the scene is described by a random function. The common stochastic field functions used for generating remote sensing imagery include Gibbs random field and Markov random field. A discrete Gibbs random field is specified by a probability mass function of the image as follows:

$$P(x) = \frac{1}{Z} e^{[H(x)/T]} \tag{1.15}$$

where Z is the normalization factor, $H(x)$ is an energy function, and T represents temperature, $T = \Sigma H(x)$, over all G^n images; G is the number of pixel value levels, and the image is of size $\sqrt{n} \times \sqrt{n}$. Except in very special circumstances, it is not feasible to compute T. A relaxation-type algorithm described in the literature (e.g., Cross and Jain 1983, Dubes and Jain 1989) simulates a Markov chain through an iterative procedure that readjusts the gray levels at pixel locations during each iteration. This algorithm sequentially initializes the value of each pixel using a uniform distribution. Then a single pixel location is selected at random, and using the conditional distribution that describes the Markov chain, the new pixel value at that location is selected, dependent only on the pixel values of the pixels in its local neighborhood. The sequential algorithm terminates after a given number of iterations.

The Gaussian–Markovian random field (GMRF) models assume that pixel values have joint Gaussian distributions and correlations controlled by a number of parameters representing the statistical dependence of a pixel value on the pixel values in a symmetric neighborhood. There are two basic schemes for generating a GMRF image model: direct or iterative. More details can be found in the literature (e.g., Chellappa 1985, Jeng et al. 1993).

In geostatistics, a conditional simulation is an effective technique to generate the random field. Sequential Gaussian simulations, Gaussian simulations, nonparametric simulation techniques such as sequential indicator simulations, and probability field simulations are becoming more and more popular. Some of these algorithms are available in the public domain software packages, such as GSLIB (Deutsch and Journel 1992). The interested reader is referred to the textbooks on this subject (e.g., Journel and Huijbregts 1978, Isaaks and Srivastava 1989, Cressie, 1993).

1.3.2   Scene Radiation Modeling

There are roughly three types of models that characterize the radiation field of the scene and are mostly used in optical remote sensing: geometric optical models, turbid-medium radiative transfer models, and computer simulation models. In *geometric optical models*, canopy or soil is assumed to consist of geometric protrusions with prescribed shapes (e.g., cylinder, sphere, cones, ellipsoid, spheroid), dimensions, and optical properties that are distributed on a background surface in a defined manner (regularly or randomly distributed). The total pixel value is the weighted average of sunlit and shadowed components. The turbid-medium radiative transfer models treat surface elements (leaf or soil particle) as small absorbing and scattering particles with given optical properties, distributed randomly in the scene and oriented in given directions. The further development of geometric optical models has incorporated radiative transfer theory in calculating the individual sunlit/shadow components; the resulting models are often called *hybrid models*. In computer simulation models, the arrangement and orientation of scene elements are simulated on a computer and the radiation properties are determined on the basis of *radiosity* equations and/or *Monte Carlo ray tracing* methods. The details on these modeling methods for vegetation canopies will be given in Chapter 3 and for soil and snow in Chapter 4.

1.3.3 Atmospheric Radiative Transfer Modeling

The Earth's Atmosphere affects remote sensing imagery significantly. The atmospheric gases, aerosols, and clouds scatter and absorb the incoming solar radiation and the reflected and/or emitted radiation from the surface. As a result, it greatly modulates the spectral dependence and spatial distribution of the surface radiation. Therefore, understanding the atmospheric radiative transfer and correcting atmospheric effects from remotely sensed imagery are very critical for land surface characterization. These two topics are discussed in Chapters 2, 6, and 10 in detail.

1.3.4 Navigation Modeling

For a spaceborne remote sensing system, we must be able to calculate the position of the satellite in space, to track it from Earth and to know where the instruments are pointing. Despite their importance, positioning in space, tracking, and navigation are more or less the scientific and technical issues that engineers are mainly concerned with. Within the scope of this book, we will not provide with any more details on this subject. The interested reader is referred to other publications (e.g., Emery et al. 1989).

1.3.5 Sensor Modeling

The upwelling TOA radiance is recorded and modulated by a sensor. Sensor models characterize the process converting the spectral radiance of the land surface–atmosphere system into digital numbers that users obtain when purchasing from the data distribution centers. Figure 1.7 shows the block diagram of the sensor model. Sensor spectral response functions are integrated with scene spectral radiance to generate the band values, and spatial response functions define the pixel values with a discrete convolution of separable linespread functions in the two spatial dimensions. Noise can be modeled explicitly before the signals are converted to digital numbers by scaling each band. Because of their importance, we will focus on the sensor spectral response functions and spatial responses below.

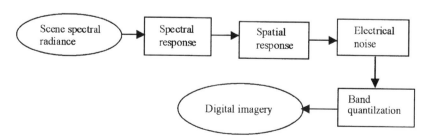

Figure 1.7 Flowchart of a sensor modeling system.

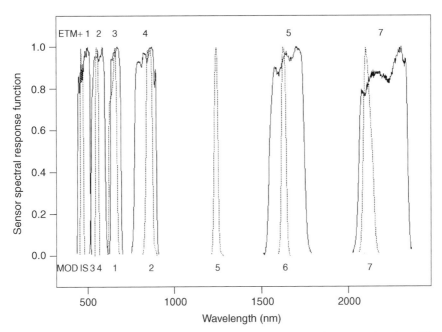

Figure 1.8   ETM + and MODIS visible and near-IR sensor spectral response functions. The solid lines correspond to ETM + and the dashed lines, to MODIS.

1.3.5.1   *Spectral Response*

Letting $L_{\lambda,\,\text{sensor}}$ be the spectral radiance of the atmosphere-land surface scene reaching the sensor detectors, the resulting array, after the spectral response function $f(\lambda)$ is applied, is computed as follows:

$$L_b(x,y) = \frac{\int_{\lambda_{\min}}^{\lambda_{\max}} L_{\lambda,\,\text{sensor}}(x,y)f(\lambda)\,d\lambda}{\int_{\lambda_{\min}}^{\lambda_{\max}} f(\lambda)\,d\lambda} \qquad (1.16)$$

where x, y are pixel locations. Few radiative transfer solvers incorporate sensor spectral response functions in their software packages; users can easily calculate spectral radiance using Eq. (1.15). The sensor spectral response functions of the first few bands of ETM + and MODIS for land applications are shown in Fig. 1.8. The reader can make such a figure for many sensors whose spectral response functions are included in the CD-ROM accompanying this book.

Note that L_b is the effective (average) spectral radiance of a specific band. The band integrated radiance equals L_b multiplied by the effective *bandwidth*. Bandwidths of spectral responses are usually expressed as the full

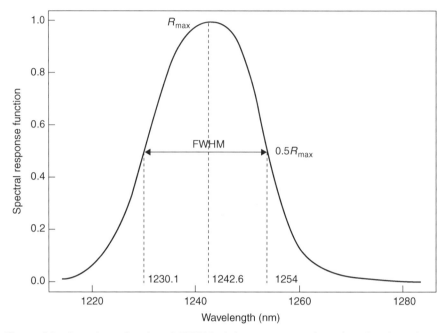

Figure 1.9 Central wavelength and FWHM of the sensor spectral response function using a simple traditional method.

width at half maximum (FWHM), which is shown in Fig. 1.9. This is the sensor response function of MODIS band 5. The central wavelength of a spectral band responds to the maximum value of the response function. This approach is very empirical and simple, but may not be appropriate if the response function is not skew and irregular.

Palmer (1984) suggested the moment method for determining the central wavelength λ_c and effective bandwidth $\Delta\lambda$. The pertinent equations are

$$\lambda_c = \frac{\int_{\lambda_{\min}}^{\lambda_{\max}} f(\lambda)\,\lambda\,d\lambda}{\int_{\lambda_{\min}}^{\lambda_{\max}} f(\lambda)\,d\lambda} \qquad (1.17)$$

$$\sigma^2 = \frac{\int_{\lambda_{\min}}^{\lambda_{\max}} f(\lambda)\,\lambda^2\,d\lambda}{\int_{\lambda_{\min}}^{\lambda_{\max}} f(\lambda)\,d\lambda} - \lambda_c^2 \qquad (1.18)$$

$$\lambda_1 = \lambda_c - \sqrt{3}\,\sigma \qquad (1.19)$$

$$\lambda_2 = \lambda_c + \sqrt{3}\,\sigma \qquad\qquad (1.20)$$

$$\Delta\lambda = \lambda_2 - \lambda_1 = 2\sqrt{3}\,\sigma \qquad\qquad (1.21)$$

where λ_{min} and λ_{max} are the minimum and maximum wavelengths of the waveband beyond which the sensor spectral response function is zero. The spectral response functions of ETM+ and MODIS visible and near-IR (VNIS) sensors are displayed in Fig. 1.8. The digital files of these two sensors as well as others are included in the CD-ROM for ease of reference. For MODIS band 5, the calculated central wavelength is 1242.1 nm with $\lambda_1 =$ 1226.47 nm and $\lambda_2 = 1257.79$ nm, and the effective bandwidth is 31.3 nm. If we use the simple method as illustrated in Fig. 1.9, FWHM $= 23.9$ nm. The difference is significant.

The number of multispectral bands varies dramatically from one sensor to another. Figure 1.10 illustrates the spectral bands and their spectral ranges of some common spaceborne multispectral sensors in the visible and near-IR spectrum for land applications. Hyperspectral sensors are discussed in Section 8.1.2.

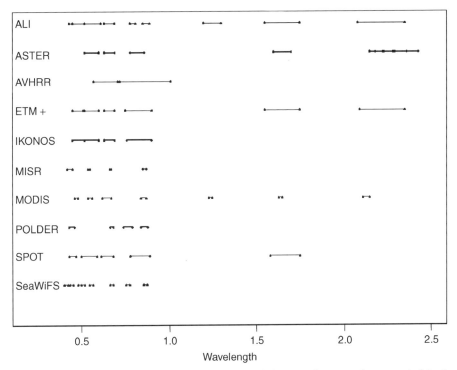

Figure 1.10 Illustration of multispectral bands and their spectral ranges of some typical land sensors.

In Fig. 1.10, SPOT bands actually represent SPOT VEGTATION. SPOT1, -2, and -3 HRV has three multispectral bands: green, red, and near IR. SPOT4 and -5 HRV has one extra band in middle IR. Those bands have almost identical spectral coverage with the VEGETATION bands.

1.3.5.2 *Spatial Response*

The spatial response of the sensor is often characterized by a parameter called *modulation transfer function* (MTF), a precise measurement of details and contrast made in the frequency domain. Each component of the sensor imaging system has a characteristic frequency response. The beauty of working in the frequency domain is that the response of the entire system (or group of components) can be calculated by multiplying the responses of each component. The Fourier transform of the MTF, which is fast and easy to perform on modern computers using the fast Fourier transform (FFT) algorithm, is called *point spread function* (PSF) in the spatial domain. In fact, the Fourier transform of PSF is called the *optical transfer function* (OTF), and the normalized magnitude of the OTF is MTF. The sensor spatial response function is often modeled as a Gaussian PSF:

$$h(x, y) = \frac{1}{2\pi\sigma^2} \exp\left(-\frac{x^2 + y^2}{2\sigma^2}\right) \tag{1.22}$$

where σ is one-half the number of scene cells (on a side) in the sensor ground instantaneous field of view (IFOV). If we can assume that PSF is a separable function with respect to the (x, y) coordinate system, then $\text{PSF}(x, y) = \text{LSF}_x(x) \cdot \text{LSF}_y(y)$ where PSF and LSF are denoted as point spread and linespread functions.

There are many different techniques for estimating the MTF of a sensor system in a laboratory, such as the impulse input method, sinusoidal input method, knife-edge input method, and pulse input method. However, only some of them can be used for estimating the sensor MTF from remote sensing imagery. For example, the pulse input method was used to measure Landsat TM MTF in which the source for the pulse is the bridge over the San Francisco Bay (Schowengerdtet al. 1985). Both the pulse and edge methods were used to estimate IKONOS sensor MTF (Helder and Choi 2001). Ruiz and Lopez (2002) derived the normalized PSF matrix of the SPOT imagery using the edge method, which is listed in Table 1.4. The TM PSF is shown in Fig. 1.11.

From MTF we can define the effective instantaneous field of view (EIFOV), which is the spatial dimension equivalent to half a cycle of the spatial frequency where the MTF falls to 0.5 (NASA 1973). The measured PSF of the overall TM system is shown in Fig. 1.10. It is not an ideal square shape with a width of 30 m, which implies that signals beyond the 30×30-m grid also contribute to the current pixel value. In other words, the current pixel value depends on the values of its neighboring pixels. In fact, the

TABLE 1.4   SPOT Sensor Point Spread Function (PSF) Matrix

0.0017	0.0018	0.0020	0.0022	0.0024	0.0022	0.0020	0.0018	0.0017
0.0018	0.0050	0.0057	0.0064	0.0071	0.0064	0.0057	0.0050	0.0018
0.0020	0.0057	0.0167	0.0201	0.0236	0.0201	0.0167	0.0057	0.0020
0.0022	0.0064	0.0201	0.0317	0.0448	0.0317	0.0201	0.0064	0.0022
0.0024	0.0071	0.0236	0.0448	0.6513	0.0448	0.0236	0.0071	0.0024
0.0022	0.0064	0.0201	0.0317	0.0448	0.0317	0.0201	0.0064	0.0022
0.0020	0.0057	0.0167	0.0201	0.0236	0.0201	0.0167	0.0057	0.0020
0.0018	0.0050	0.0057	0.0064	0.0071	0.0064	0.0057	0.0050	0.0018
0.0017	0.0018	0.0020	0.0022	0.0024	0.0022	0.0020	0.0018	0.0017

Source: Ruiz and Lopez (2002).

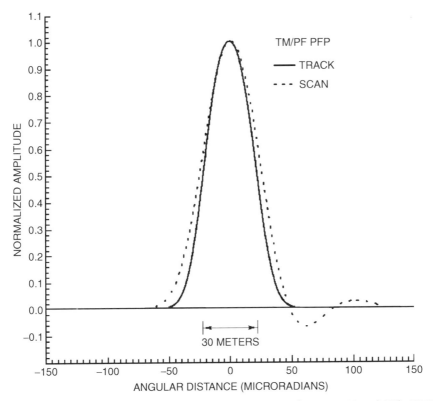

Figure 1.11   TM PSF function at both track and scan directions. [From Markham (1985), *IEEE Trans. Geosci. Remote Sens.* Copyright © 1985 with permission from IEEE.]

TABLE 1.5 Spatial Resolution (in meters) of Some Typical Multispectral Remote Sensors

Sensors	Spectral Bands	
	Multispectral	Panchromatic
ALI	All bands: 30	10
ASTER	Bands 1–3: 15	
	Bands 4–9: 30	
AVHRR	All bands: 1000	
ETM +	Bands 1–5 and 7: 30	15
	Band 6: 60	
GLI	Bands 1–6: 250	
IKONOS	4	1
MERIS	300	
MISR	1000	
MODIS	Bands 1–2: 250	
	Bands 3–7: 500	
	Other bands: 1000	
POLDER	6000 × 8000	
SPOT 1–4	All bands: 20	10
SPOT-5	Bands 1–3: 10	5/2.5
	Band 4: 20	
SPOT-VEGETATION	All bands: 1000	
SeaWiFS	All bands: 1100	

EIFOVs for TM 1–4 are 35.9 and 32.1 m for TM bands 1–4 in the scan and track directions, respectively. For bands 5 and 7, the EIFOVs are 35.7 and 33.3 m in the scan and track directions, respectively (Markham 1985).

The impacts of sensor PSF on land cover classifications are evaluated by Townshend et al. (2000) and Huang et al. (2002).

For ease of reference, Table 1.5 lists the spatial resolutions of some typical land sensors. Note that the spatial resolution of some off-nadir viewing sensors is the same as that at nadir. The acronyms of these sensors are listed in the Frontmatter of the book.

1.3.6 Mapping and Binning

After navigation of the sensor output, the geolocation of the central point and other parameters of each measurement need to be determined. Since each measurement corresponds to a surface region with different shapes and sizes, an important process in the image generation is to bin these measurements into a regular two-dimensional array with one particular map projection.

Selection of a map projection becomes very important when we are concerned with the global monitoring because any map projection that

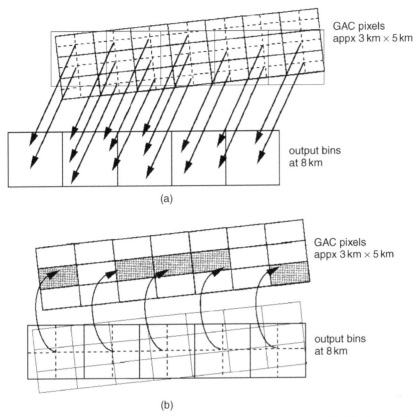

Figure 1.12 Illustration of both forward (a) binning and inverse (b) binning methods. [From James and Kalluri (1994), *Int. J. Remote Sens*. Reproduced by permission of Taylor & Francis, Ltd.]

transforms the curve surface to a flat plane introduces errors. Many equal-area map projections have been proposed and used in the remote sensing product generation, such as the space oblique mercator or the Universal Transverse Mercator (UTM) projection for TM/ETM+ imagery (Snyder 1981), the interrupted Goode homolosine projection for AVHRR (Steinwand 1994), the intergerized sinusoidal projection and Lambert azimuthal projection for MODIS (Masuoka et al. 1998), and space oblique mercator projection for MISR (Diner et al. 1998).

There are two general methods for mapping each measurement to the output bins: *forward binning*, where the location of each measurement is used to locate the output bin; and *inverse binning*, where the location of the output bin is used to locate each measurement to the bin. Figure 1.12 illustrates the two binning methods for generating AVHRR Pathfinder Land data from the Global Area Coverage (GAC) data (James and Kalluri 1994).

No matter which binning method is used, the original observations have to be resampled. There are several resampling methods used in remote sensing, including nearest-neighbor, bilinear interpolation, and cubic convolution (e.g., Schowengerdt 1997). When ordering Landsat ETM + imagery from USGS EDC, the user can specify a specific resampling method.

1.4 SUMMARY

This is an introductory chapter that lays the foundation for the rest of the book. It discusses various radiometric variables and then describes the major components of a remote sensing modeling system. Most of the discussion was very brief, but landscape generation, spectral and spatial sensor characteristics, and mapping and binning were discussed in a bit more detail. Because of the scope of this book, readers should consult the references cited in the text for further details on some components such as navigation modeling and image mapping and binning.

REFERENCES

Anderson, G. P., Berk, A., Acharya, P. K., Matthew, M. W., Bernstein, L. S., Chetwynd, J. H., Dothe, H., Adler-Golden, S. M., Ratkowski, A. J., Felde, G. W., Gardner, J. A., Hoke, M. L., Richtsmeier, S. C., and Jeong, L. S. (2001), MODTRAN4, version 2: Radiative transfer modeling, *Proc. SPIE*, 455–459.

Chellappa, R. (1985), Two-dimensional discrete gaussian markov random field models for image processing, in *Progress in Pattern Recognition*, L. N. Kanal and A. Rosenfeld, eds., Elsevier Science Publishers, pp. 79–112.

Cressie, N. (1993), *Statistics for Spatial Data*, rev. ed., Wiley.

Cross, G. R. and Jain, A. K. (1983), Markov random field texture models, *IEEE Trans. Pattern Anal. Mach. Intel.* **5**: 25–39.

Deutsch, C. V. and Journel, A. G. (1992), *GSLIB: Geostatistical Software Library and User's Guide*, Oxford Univ. Press.

Diner, D., Beckert, J. C., Reilly, T. H., Bruegge, C. J., Conel, J. E., Kahn, R. A., Martonchik, J. V., et al. (1998), Multi-angle imaging spectroradiometer (MISR) instrument descrition and experiment overview, *IEEE Trans. Geosci. Remote Sens.* **36**: 1072–1097.

Dubes, R. C. and Jain, A. K. (1989), Random field models in image analysis, *J. Appl. Stat.* **16**: 131–164.

Emery, W. J., Brown, J., and Nowak, Z. P. (1989), AVHRR image navigation: Summary and review. *Photogram. Eng. Remote Sens.* **55**: 1175–1183.

Fisher, Y., ed. (1995), *Fractal Image Compression: Theory and Application to Digital Images*, Springer-Verlag.

Gao, B. C. and Green, R. O. (1995), Presence of terrestrial atmospheric gas absorption bands in standard extraterrestrial solar irradiance curves in the near-infrared spectral region, *Appl. Opti.* **34**: 6263–6268.

Hartmann, D. L., Bretherrton, C. S., Charlock, T. P., Chou, M. D., Del Genio, A., Dickinson, R. E., Fu, R., Houze, R. A., King, M. D., Lau, K. M., Leovy, C. B., Sorooshian, S., Washburne, J., Wielicki, B., and Willson, R. C. (1999), Radiation, clouds, water vapor, precipitation, and atmospheric circulation, in *EOS Science Plan*, NASA GSFC, pp. 39–114.

Helder, D. and Choi, J. (2001), MTF modeling and error analysis for high spatial resolution satellite sensors.

Huang, C., Townshend, J. R. G., Liang, S., Kalluri, S. N. V., and DeFries, R. S. (2002), Impact of sensor's point spread function on land cover characterization: Assessment and deconvolution, *Remote Sens. Environ.* **80**: 203–212.

Isaaks, E. and Srivastava, R. M. (1989), *An Introduction to Applied Geostatistics*, Oxford Univ. Press.

James, M. E. and Kalluri, S. N. V. (1994), The pathfinder AVHRR land data set: An improved coarse resolution data set for terrestrial monitoring, *Int. J. Remote Sens.* **15**: 3347–3363.

Jeng, F. C., Woods, J. W. and Rastogi, S. (1993), Compound Gauss-Markov random fields for parallel image processing, in *Markov Random Fields: Theory and Application*, R. Chellappa and A. K. Jain, eds., Academic Press, pp. 11–38.

Journel, A. G. and Huijbregts, C. J. (1978), *Mining Geostatistics*, Academic Press.

Kerekes, J. P., and Landgrebe, D. A. (1991), An analytical model of Earth-observational remote sensing systems. *IEEE Trans. Geosci. Remote Sens.* **21**: 125–133.

Liang, S. and Strahler, A. eds. (2000), Land surface bidirectional reflectance distribution function (BRDF): Recent advances and future prospects, *Remote Sens Rev* **18**: 83–551.

Mandelbrot, B. (1983), *The Fractal Geometry of Nature*, Freeman.

Markham, B. L. (1985), The Landsat sensors' spatial responses, *IEEE Trans. Geosci. Remote Sens.* **23**: 864–875.

Martonchik, J. V., Bruegge, C. J., and Strahler, A. H. (2000), A review of reflectance nomenclature used in remote sensing, *Remote Sensing Rev.* **19**: 9–20.

Masuoka, E., Fleig, A., Wolfe, R., and Patt, F. (1998), Key characteristics of MODIS data products. *IEEE Trans. Geosci. Remote Sens.* **36**: 1313–1323.

Milman, A. S. (1999), *Mathematical Principles of Remote Sensing*: *Making Inferences from Noisy Data*, Sleeping Bear Press.

NASA (1973), Advanced scanners and imaging systems for earth observations, *NASA Tech. Report* SP-335.

Palmer, J. M. (1984), Effective bandwidths for Landsat-4 and Landsat-d' multispectral scanner and thematic mapper subsystems, *IEEE Trans. Geosci. Remote Sens.* **22**: 336–338.

Prusinkiewicz, P. and Lindenmayer, A. (1990), *The Algorithmic Beauty of Plants*, Springer-Verlag.

Ruiz, C. P. and Lopez, F. J. A. (2002), Restoring SPOT images using PSF-derived deconvolution filters, *Int. J. Remote Sens.* **23**: 2379–2391.

Schott, J. R. (1997), *Remote Sensing: The Image Chain Approach*, Oxford Univ. Press.

Schowengerdt, R., Archwamety, C., and Wrigley, R. C. (1985), Landsat Thematic Mapper image derived mtf, *Photogram. Eng. Remote Sens.* **51**: 1395–1406.

Schowengerdt, R. A. (1997), *Remote Sensing, Models and Methods for Image Processing*, 2nd ed., Academic Press.

Snyder, J. P. (1981), Space oblique mercator projection—mathematical development, *U.S. Geol. Survey Bull.* 1518.

Steinwand, D. R. (1994), Mapping raster imagery to the interrupted good homolosine projection, *Int. J. Remote Sens.* **15**: 3463–3471.

Strahler, A. H., Woodcock, C. E., and Smith, J. A. (1986), On the nature of models in remote sensing. *Remote Sens. Envir.* **20**: 121–139.

Townshend, J. G. R., Huang, C., Kalluri, S., DeFries, D., Liang, S., and Yang, K. (2000), Beware of per-pixel characterization of land cover, *Int. J. Remote Sens.* **21**: 839–843.

2

Atmospheric Shortwave Radiative Transfer Modeling

Earth's atmosphere modulates any surface signals twice. It affects the distribution of incoming solar radiation at the surface that is related to the surface reflectance responses. The solar radiation reflected by the surface is further scattered and absorbed by the atmosphere before reaching the sensor. Since the sensors are located at the top or the middle of the atmosphere, the radiance received by a sensor contains information of both the atmosphere and land surface. The main theme of this book is the estimation of land surface variables from remote sensing observations, and removal of atmospheric effects is a necessary step. We must introduce radiation transfer in the atmosphere first before discussing atmospheric correction.

Radiative transfer theory has been recognized as the principal modeling method that accounts for the solar radiation in the atmosphere. It is also widely used for land surface modeling (e.g., vegetation canopy, snow, and soil) as discussed in Chapters 3 and 4. This chapter introduces shortwave atmospheric radiative transfer models and their solutions, and the longwave radiative transfer in the atmosphere will be presented in Chapter 10. In this book we are primarily concerned with the scalar approximation of the full vector description of the theory, and thus polarization is not discussed within this book. Readers who are interested in vector radiative transfer theory should consult related books (e.g., Fung 1994). The introduction of radiative transfer theory in this chapter will be fundamental to understanding subsequent chapters since some of the important concepts are not repeated.

After radiative transfer formulation and its linkage with atmospheric optical properties are introduced, both numerical and approximate solutions to atmospheric radiative transfer equations are discussed. Some typical radiative transfer solvers (computer software packages) are also summarized.

Quantitative Remote Sensing of Land Surfaces. By Shunlin Liang
ISBN 0-471-28166-2 Copyright © 2004 John Wiley & Sons, Inc.

Section 2.1 defines the atmospheric radiative transfer equation in the solar spectrum. Radiative transfer in the thermal infrared spectrum is presented in Chapter 10. The derivation of the radiative transfer theory is not given because of space limitations in the book; only some general ideas and key formulas are described. Section 2.2 presents some analytic BRDF models of land surfaces to specify the lower boundary conditions of the atmospheric radiative transfer equation. To obtain unique solutions, boundary conditions must be specified before solving the radiative transfer equation. Because of its importance in estimating land surface variables, the details of some representative models are described. Section 2.3 links the quantities in radiative transfer equations with atmospheric properties. The algorithms for calculating the optical properties of a single particle are introduced. The aerosol size distribution, refractive indices, and aerosol climatology are also presented. Section 2.4 describes various numerical and approximate solutions to the atmospheric radiative transfer equation, including two-stream approximations, successive orders of scattering numerical method, and the method of discrete ordinates. Some typical radiative transfer solvers in the public domain are also summarized. Section 2.5 outlines some approximate formulations of radiative interactions between atmosphere and surface. For a Lambertian surface, formulating this interaction is very straightforward. For a surface that reflects anisotropically, a numerical scheme is needed. For quick calculations in quantitative land remote sensing, approximate but accurate formulations are necessary.

2.1 RADIATIVE TRANSFER EQUATION

If polarization effects are ignored, the radiative transfer equation can be expressed as

$$\frac{dI(\mathbf{s})}{d\mathbf{s}} = -K(I - J) \tag{2.1}$$

This is the differential integral equation regarding radiance (I) at a specific direction ($\mathbf{s}$); K is the volume extinction coefficient, J usually called the *source function*. The viewing direction ($\mathbf{s}$) is characterized by the vector of five elements, (x, y, z, μ, ϕ); (x, y, z) are the coordinates in the three-dimensional space. Conventionally, (x, y) denote the two-dimensional horizontal plane and z the vertical direction. Coordinates (μ, ϕ) are the angular coordinates, $\mu = \cos(\theta)$. If polarization is considered, radiance I itself is a vector of four elements, but the basic form of the radiative transfer equation is the same.

Assuming that the atmosphere is homogeneous in the horizontal plane and variable only in the vertical direction, the basic radiative transfer equation for radiance at any direction $I(\mu, \phi)$ in the solar reflective region can be written

$$\mu \frac{dI(z, \mu, \phi)}{dz} = \sigma_e I(z, \mu, \phi)$$

$$- \int_0^{2\pi} \int_{-1}^{1} I(z, \mu_i, \phi_i) \sigma_s(\mu, \phi, \mu_i, \phi_i) \, d\mu_i \, d\phi_i - J_0 \quad (2.2)$$

where σ_e is the volume extinction coefficient: σ_s is the differential scattering coefficient, z the vertical coordinate (increasing from the bottom of the atmosphere to the top of the atmosphere or vice versa); and J_0 is the source function of the medium. This equation is a typical plane-parallel, one-dimensional (1D) radiative transfer equation.

If either atmosphere or surface is not horizontally homogeneous, the atmospheric radiative transfer has to be a multidimensional problem. We consider only 1D atmospheric radiative transfer problems in this book, but the effect of surface heterogenity is discussed in Section 2.5.

Equation (2.2) is a differential integral equation and can be simply explained in this way. The left side of the equation is about the change rate of radiance at the direction (μ, ϕ). The first term on the right side represents the loss (attenuation) of radiance in the direction (μ, ϕ) due to both absorption and scattering, while the second term indicates the gain of radiance due to the scattering from all other directions (μ_i, ϕ_i) into the direction (μ, ϕ).

If the particles can be assumed to be isotropic (e.g., molecular particles), the 1D equation can be rewritten as

$$\mu \frac{dI(\tau, \mu, \phi)}{d\tau} = I(\tau, \mu, \phi)$$

$$- \frac{\omega}{4\pi} \int_0^{2\pi} \int_{-1}^{1} I(\tau, \mu_i, \phi_i) P(\mu, \phi, \mu_i, \phi_i) \, d\mu_i \, d\phi_i - J_0 \quad (2.3)$$

where τ, called *optical depth* or *optical thickness*, is proportional to z but also depends on the extinction coefficient $\tau = \int_0^z \sigma_e(z) dz$. ω is the single scattering albedo, representing the scattering probability when photons hit the particles in the medium. For example, if $\omega = 0.92$, the photon, when hit by the particle, will be scattered in another direction with a probability of 92%. It might be absorbed by the particle with a probability of 8%. $P(\cdot)$ is the phase function, characterizing the probability of photon scattered from one direction to another. It is usually assumed in the atmosphere that $P(\cdot)$ depends only on the angle between the incident and emergent angles, and not on the absolute angles themselves. In the visible and near-IR spectrum, $J_0 = 0$ if the medium does not have any extra light source.

The radiance distribution of the atmosphere depends on both optical properties and surface reflectivity. Atmospheric properties determine the key parameters in the radiative transfer equation, such as extinction coefficient, single scattering albedo, and phase function, and are discussed in Section 2.3. The surface influence is taken into account through the lower boundary condition.

In fact, the radiative transfer equation cannot be solved unless the boundary conditions are specified. For the 1D atmosphere, two boundary conditions are needed. At the top of the atmosphere

$$I(\tau_t, -\mu, \phi) = \delta(\mu - \mu_0)\delta(\phi - \phi_0)\pi E_s \qquad (2.4)$$

where δ is the Dirac delta function, indicating that the incoming radiation is from only one single direction at the specific solar zenith angle and azimuth angle (μ_0, ϕ_0); πE_s denotes the incident solar flux density perpendicular to the incident beam, and $E_0 = \pi E_s$ has been discussed in Section 1.2; τ_t is the total optical depth of the atmosphere.

The lower boundary condition at the Earth surface is

$$I(0, \mu, \phi) = \frac{1}{\pi} \int_0^{2\pi} \int_{-1}^0 R(\mu, \phi, \mu', \phi') I(\mu', \phi') \mu' d\mu' d\phi' \qquad (2.5)$$

where $R(\mu, \phi, \mu', \phi')$ is the surface BRF (see Section 1.2.5). If the Earth surface is assumed to be Lambertian (isotropic), then

$$R(\mu, \phi, \mu', \phi') = r_s \qquad (2.6)$$

Thus, surface reflectance is a constant value r_s in all directions. The atmospheric radiative transfer and the boundary conditions are illustrated in Fig. 2.1, where θ_0 is the solar zenith angle and θ_v is the viewing zenith angle.

In the next section, we specifically discuss various analytic forms of surface directional reflectance in terms of BRDF since almost all surfaces reflect solar radiation anisotropically. When the atmosphere is very clear, the surface contribution dominates the total upwelling radiance that can be detected by remote sensors, and thus the realistic specification of the lower boundary condition using simple functions is very important in determining the atmospheric radiation field. Note that estimation of various surface bio/geophysical variables is the main theme of this book. The common practice is to retrieve surface directional reflectance first through atmospheric correction and/or other techniques and then to estimate those surface variables from the retrieved surface reflectance. The related details are discussed in other chapters of the book.

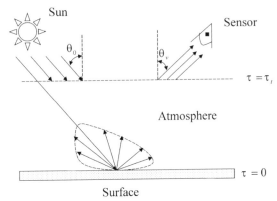

Figure 2.1   Illustration of atmospheric radiative transfer and the boundary conditions. The solar and viewing zenith angles are θ_0 and θ_v, respectively. The surface is non-Lambertian.

2.2   SURFACE STATISTICAL BRDF MODELS

The bidirectional reflectance distribution function (BRDF) was introduced in Section 1.2.5. It is convenient to describe surface directional reflectance using a simple function with a small number of adjustable parameters. Certain so-called kernel BRDF parametric models have been suggested in the literature and used in satellite data processing. They are usually denoted as empirical or semiempirical statistical models. A set of representative models is given below. Note that they are organized under BRDF models in this book; some of them might be called BRF models in the literature. We do not intend to distinguish them since they are exchangeable with a constant π [see Eq. (1.9)].

2.2.1   Minnaert Function

One of the earliest formulas that has been widely used in planetary astronomy is

$$R(\theta_i, \theta_v) = \frac{\rho_0}{(\mu_i \mu_v)^{1-k}} = \frac{\rho_0}{(\cos \theta_i \cos \theta_v)^{1-k}} \qquad (2.7)$$

where ρ_0 is a coefficient, and k is a dimensionless parameter from 0 to 1. When $k = 1$, we obtain the Lambertian reflectance formula. For dark surfaces, k is about 0.5. For brighter surfaces, k increases. It is close to 1 when the surface is very bright. When the solar zenith angle θ_i or viewing zenith angle θ_v increases (smaller μ), reflectance increases. This is illustrated in Fig. 2.2. This formula does not consider azimuthal dependence; thus is not sufficient to describe the directional reflectance of structured surfaces, such as forests.

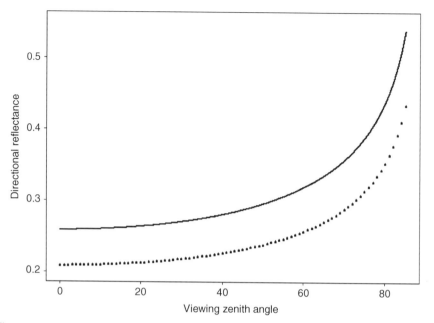

Figure 2.2   Directional reflectance of Minnaert model with two solar illumination points (30°, dotted line; 65°, solid line) and different viewing zenith angles. Other parameters are $k = 0.7$ and $\rho_0 = 0.2$.

2.2.2 Lommel–Seeliger Function

The Lommel–Seeliger function also originates from the planetary astronomy community:

$$R(\theta_i, \theta_v) = \frac{2\rho_0}{\mu_v + \mu_i} \tag{2.8}$$

where all the variables are the same as those in Eq. (2.7). It is similar to the Minnaert model and also does not consider azimuthal dependence but describes the directional reflectance of dark objects very well. For a given solar zenith angle, reflectance increases as the viewing zenith angle becomes larger (with smaller μ_v). At the same viewing zenith angle, this model predicts larger reflectances when the solar zenith angle increases from 0° to 90°.

2.2.3 Walthall Function

Walthall et al. (1985) proposed this empirical statistical model for the soil based on extensive field experimental data

$$R_{soil}(\theta_i, \theta_v, \phi) = \frac{R_{H,soil}}{0.13} \left[a\theta_v^2 + b\theta_v \cos(\phi) + c \right] \tag{2.9}$$

where $R_{H,\,\text{soil}}$ is the soil bihemispherical reflectance; θ_v and ϕ are the viewing zenith angle and relative azimuth angle between the solar illumination direction and the viewing direction, respectively; and a, b, c are coefficients dependent on solar zenith angle $\theta_i (< \pi/2)$:

$$a = -4.988 \times 10^{-4} - 2.953 \times 10^{-5}\theta_i + 8.920 \times 10^{-7}\theta_i^2 \qquad (2.10)$$

$$b = 6.988 \times 10^{-4} + 2.243 \times 10^{-3}\theta_i \qquad (2.11)$$

$$c = 14.46 + 3.216 \times 10^{-2}\theta_i - 1.7373 \times 10^{-3}\theta_i \qquad (2.12)$$

Nilson and Kuusk (1989) revised this formula to be reciprocal. In the literature, the Walthall function is generally expressed by the following simple equation:

$$R(\theta_i, \theta_v, \phi) = a\left(\theta_i^2 + \theta_v^2\right) + b\theta_i^2\theta_v^2 + c\theta_i\theta_v \cos\phi + d \qquad (2.13)$$

where a, b, c, d are the coefficients. Note that both Minnaert and Lommel–Seeliger functions are also reciprocal. However, several studies question the reciprocity of the surface BRDF (Kriebel 1996, Li et al. 1999, Leroy 2001, Snyder 2002).

This model was developed for soils, but had been used as a general statistical BRDF model. In fact, it was found in the original paper (Walthall et al. 1985) that this model can fit soybean directional reflectance very well. It is as simple as the previous two models, but it considers the azimuthal dependence.

Like the previous two functions, the Walthall function does not consider the hotspot effect, a reflectance enhancement when the viewing direction is exactly opposite the solar illumination direction. Liang and Strahler (1994a) therefore proposed adding a hotspot kernel to the regular Walthall function (2.13) with additional two coefficients c_1 and c_2:

$$c_1 e^{-c_2 \tan(\pi - \xi)} \qquad (2.14)$$

where ξ is the phase angle: $\cos\xi = \cos\theta_i \cos\theta_v + \sin\theta_i \sin\theta_v \cos\phi$.

2.2.4 Staylor–Suttles Function

For desert and other arid regions, Staylor and Suttles (1986) developed an analytic model as follows from Nimbus 7 Earth Radiation Budget (ERB) scanner measurements:

$$R(\theta_i, \theta_v, \phi) = B(\mu_v, -\mu_i)\frac{1 + c_3(\mu_v\mu_i - \sin\theta_i \sin\theta_v \cos\phi)^2}{1 + c_3\left[(\mu_v\mu_i)^2 + 0.5(\sin\theta_i \sin\theta_v)^2\right]} \qquad (2.15)$$

TABLE 2.1 Fitted Parameters of Eq. (2.15)

Desert Site	c_1	c_2	c_3	N
Sahara–Arabian (Rub al Khali)	0.011	0.920	0.33	1.764
Gibson	0.009	0.623	0.60	1.786
Saudi	0.008	1.088	0.18	1.678

where

$$B(-\mu_i, \mu_v) = \frac{1}{\mu_v \mu_i}\left[c_1 + c_2\left(\frac{\mu_v \mu_i}{\mu_v + \mu_i}\right)^N\right] \qquad (2.16)$$

where c_1, c_2, c_3, and N are free parameters, and the angular variables are defined in Sections 2.2.1–2.2.3. The local albedo can be expressed as

$$A(-\mu_i) = 2\int_0^1 B(-\mu_i, \mu)\mu\, d\mu = \frac{2c_1}{\mu_i} + 2c_2\,\mu_i^{N-1}\int_0^1\left[\frac{\mu}{\mu + \mu_i}\right]^N d\mu \qquad (2.17)$$

On the basis of the direction reflectance data observed from several desert sites, they fitted Eq. (2.15) with the parameters listed in Table 2.1. Note that the ERB scanner is a sensor for measuring broadband shortwave and longwave radiation. This model therefore was designed for broadband directional reflectance. Since it was also developed for the nonvegetated surfaces, its applicability to vegetation surfaces and other land cover types has not been well tested.

2.2.5 Rahman Function

To describe the reflectance angular pattern of the coupled atmosphere and surface media, Rahman et al. (1993) suggested a parametric model:

$$R(\theta_i, \theta_v, \phi) = \rho_0 M(\theta_i, \theta_v) F(\theta_i, \theta_v, \phi)[1 + G(\theta_i, \theta_v, \phi)] \qquad (2.18)$$

where

$$M(\theta_i, \theta_v) = \frac{\cos^{k-1}\theta_v \cos^{k-1}\theta_i}{(\cos\theta_v + \cos\theta_i)^{1-k}} \qquad (2.19)$$

$$F(\theta_i, \theta_v, \phi) = \frac{1 - b^2}{[1 + b^2 - 2b\cos(\pi - g)]^{3/2}} \qquad (2.20)$$

$$\cos g = \cos\theta_i \cos\theta_v + \sin\theta_i \sin\theta_v \cos\phi \qquad (2.21)$$

$$G(\theta_i, \theta_v, \phi) = \frac{1 - a}{1 + \sqrt{\tan^2\theta_i + \tan^2\theta_v - 2\tan\theta_i \tan\theta_v \cos\phi}} \qquad (2.22)$$

in which ρ_0, b, and k are the coefficients, and the angular variables are defined in Sections 2.2.1–2.2.3. The M term on the right side of the function is actually a combined form of both Minnaert and Lommel–Seeliger functions. The F term is the single-term Henry–Greenstein function used to describe the phase function of aerosol scattering. The G term is used to describe the hotspot feature (a high reflectance peak at the antisolar illumination direction) of the bidirectional reflectance function.

To facilitate linearization of the reflectance function for parameter retrieval, Martonchik et al. (1998) further modified this model by using an exponential function to replace the phase function:

$$F(\theta_i, \theta_v, \phi) = \exp(-b) \cos(g) \qquad (2.23)$$

This model has been used in producing the MISR land surface products, discussed further in Section 6.3.2. Although this model was developed for the coupled atmosphere–surface system, most people have used it for characterizing land surface reflectance.

2.2.6 Kernel Functions

A kernel function typically consists of two components (kernels): (1) the reflectance kernel from objects on the surface and (2) the volume scattering kernel from the infinite surface body. Roujean et al. (1992) modeled the surface objects as vertical opaque protrusions that reflect as Lambertian and also cast shadows. They represent irregularities and roughness of bare soil surfaces, or structured features of canopies with low transmittance. The geometric kernel (k_{geo}) was determined based on a large number of identical protrusions characterized by the height, width, and length (see Fig. 2.3). The second component may represent a collection of randomly located facets

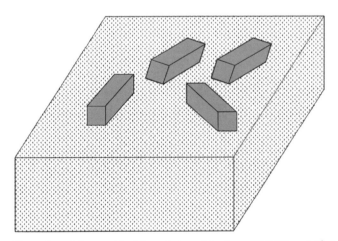

Figure 2.3 Illustration of the empirical formulation of the kernel BRDF model [after Roujean et al. (1992)].

absorbing and scattering radiation. This infinite surface body may be soil or vegetation canopy. A simple radiative transfer model was used to represent this component (k_{vol}).

Mathematically, the kernel model can be expressed as

$$R(\theta_i, \theta_v, \phi) = \alpha_0 + \alpha_1 K_{vol}(\theta_i, \theta_v, \phi) + \alpha_2 K_{geo}(\theta_i, \theta_v, \phi) \quad (2.24)$$

where α_0, α_1, and α_2 are coefficients, and the angular variables are defined in Sections 2.2.1–2.2.3. The volume scattering kernel is given by (Roujean et al. 1992)

$$K_{vol}(\theta_i, \theta_v, \phi) = \frac{(0.5\pi - g)\cos g + \sin g}{\cos \theta_i + \cos \theta_v} - \frac{\pi}{4} \quad (2.25)$$

where g is the phase angle defined by Eq. (2.21). It is a single-scattering solution to the classic canopy radiative transfer equation for plane-parallel dense vegetation canopy with uniform leaf angle distribution, and equal leaf reflectance and transmittance. It also does not account for the hotspot.

The surface reflectance kernel can be determined on the basis of geometric optical principles. Roujean et al. (1992) gave this kernel based on randomly distributed thin rectangular protrusions:

$$K_{geo}(\theta_i, \theta_v, \phi) = \frac{(\pi - \phi)\cos \phi + \sin \phi}{2\pi} \tan \theta_i \tan \theta_v$$

$$- \frac{1}{2}\left[\tan \theta_i + \tan \theta_v + \sqrt{\tan^2\theta_i + \tan^2\theta_v - 2\tan \theta_i \tan \theta_v \cos \phi}\right] \quad (2.26)$$

The LiSparse kernel of the MODIS BRDF algorithm (Wanner et al. 1995) is based on the geometric optical mutual shadowing BRDF model developed by Li and Strahler (1992). It is given by the proportions of sunlit and shaded scene components in a scene consisting of randomly placed spheroids of height to center of crown h and crown vertical to horizontal radius ratio b/r:

$$K_{geo}(\theta_i, \theta_v, \phi) = O(\theta_i, \theta_v, \phi) - \sec \theta_i' - \sec \theta_v'$$

$$+ \tfrac{1}{2}(1 + \cos g') \sec \theta_i' \sec \theta_v' \quad (2.27)$$

where

$$O = \frac{1}{\pi}\left(\arccos X - X\sqrt{1 - X^2}\right)(\sec \theta_i' + \sec \theta_v') \quad (2.28)$$

$$X = \frac{h}{b}\frac{\sqrt{D^2 + (\tan \theta_i' \tan \theta_v' \sin \phi)^2}}{\sec \theta_i' + \sec \theta_v'} \quad (2.29)$$

X has to be constrained to the range $(-1, 1)$, and

$$D = \sqrt{\tan^2\theta_i' \tan^2\theta_v' - 2\tan\theta_i' \tan\theta_v' \cos\phi} \qquad (2.30)$$

$$\cos g' = \cos\theta_i' \cos\theta_v' + \sin\theta_i' \sin\theta_v' \cos\phi \qquad (2.31)$$

and

$$\theta' = \tan^{-1}\left(\frac{b}{r}\tan\theta\right) \qquad (2.32)$$

In the MODIS BRDF algorithm, $h/b = 2$, and $b/r = 1$. This model has also been used for mapping surface albedo from airborne multiple observations and other data (e.g., Barnsley et al. 1997, Lewis et al. 1999).

We do not intend to discuss every model published in the literature. Readers are referred to additional papers for details (e.g., Goel and Reynolds 1989, Ba et al. 1995, Liang and Townshend 1997).

2.3   ATMOSPHERIC OPTICAL PROPERTIES

Before discussing how to solve the radiative transfer equation, it is necessary to introduce the optical properties of the atmosphere that are associated with the radiative transfer equation.

The optical properties (e.g., optical depth, single scattering albedo, phase function) of the medium are determined by the particles that compose the medium and their properties. If the molecular particles in the atmosphere are far smaller than the wavelength, its scattering pattern can be calculated by the Rayleigh scattering law. For spherical particles, their scattering behaviors depend on the *refractive index* and the *size parameter* defined as

$$\chi = \frac{2\pi r}{\lambda} \qquad (2.33)$$

where r is the radius of the sphere.

In the following two sections, we discuss scattering theory for both small particles (Rayleigh scattering) and large particles (Mie scattering).

2.3.1   Rayleigh Scattering

If χ is smaller than 0.01, then the *Rayleigh scattering* formulas are valid. The phase function for Rayleigh particles is

$$P(\mu) = \frac{3}{16\pi}\frac{2}{2+\delta}\left[(1+\delta) + (1-\delta)(1+\mu^2)\right] \qquad (2.34)$$

where δ is the depolarization factor that gives the correction for the depolarization effect of scattering from anisotropic molecules. When $\delta = 0$ for symmetric molecules (no depolarization), then (2.34) becomes

$$P(\mu) = \frac{3}{16\pi}(1 + \mu^2) \tag{2.35}$$

Young (1980) suggested that $\delta = 0.0279$ for dry air; thus Eq. (2.34) can be written as

$$P(\mu) = 0.06055 + 0.05708\,\mu^2 \tag{2.36}$$

This does not depend on the incoming direction (θ_i, ϕ_i) and outgoing direction (θ_v, ϕ_v), but only their relative angle, usually called the *scattering phase angle* θ:

$$\mu = \cos\theta = \cos\theta_i \cos\theta_v + \sin\theta_i \sin\theta_v \cos(\phi_i - \phi_v) \tag{2.37}$$

The single scattering albedo ω for Rayleigh particles is one.

Rayleigh scattering is easy to handle in remote sensing because its principles are easily understood. The only variable is the optical depth, which is quite stable globally. It depends mainly on the surface elevation (Russell et al. 1993) and can be calculated by

$$\tau_R = \frac{P}{P_0}(0.00864 + 6.5 \cdot 10^{-6} \cdot z)\lambda^{-(3.916 + 0.074\lambda + (0.05/\lambda))} \tag{2.38}$$

where P is the ambient pressure in millibars, $P_0 = 1013.25\,\mathrm{mbar}$, z the height above sea level in kilometers, and λ the wavelength in micrometers. For most applications, it may not affect the accuracy significantly if we drop the dependence of surface elevation and pressure. If so, the equation becomes

$$\tau_R = 0.00864\lambda^{-(3.916 + 0.074\lambda + (0.05/\lambda))} \tag{2.39}$$

Fig. 2.4 illustrates Eq. (2.39), from which we can see that the optical depth decreases quickly as wavelength increases. Taking Rayleigh scattering into account is meaningful only in the shorter wavelengths (i.e., the blue-green-red visible spectrum).

2.3.2 Mie Scattering

If the particle size is very close to the length of the wavelength (i.e., $0.1 < \chi < 50$), such as the most aerosol particles in the atmosphere, their scattering behavior can be characterized by *Mie theory*. Atmospheric aerosols originate from many different sources and mainly from two processes. The

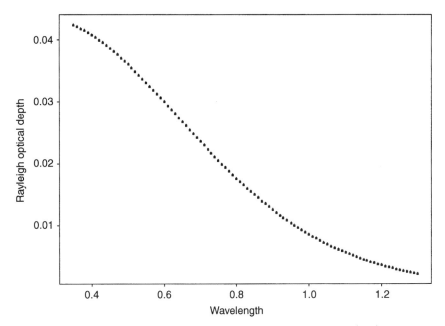

Figure 2.4 Rayleigh optical depth as a function of wavelength (μm).

primary source includes dispersion of materials from the Earth surface, and the secondary source may also be generated from atmospheric chemical reactions or condensation or coagulation processes. Aerosols may include the sea-salt particles from the ocean, wind-blown mineral particles (e.g., desert dust, sulfate, and nitrate aerosols resulting from gas–particle conversion), organic materials, carbonaceous substance from biomass burning, and industrial combustions. These particles are generally produced at the Earth surface and remain in the lower boundary layer, but could be transported to higher altitudes. For example, Saharan desert dust has been observed 5–6 km above the Atlantic Ocean, and smoke from large fires has been observed at 3–4 km in height (Kaufman et al. 1997).

A complete discussion of Mie scattering is outside of the scope of this book. The detailed derivations of absorption and scattering cross sections and angle-dependent scattering functions from Mie theory are provided by Bohren and Huffman (1983) and many other textbooks and references. Numerical code for calculating these quantities is on the CD-ROM included with this book. For solving the atmospheric radiative transfer equation, we need only the phase function $P(\mu)$ and the single scattering albedo ω, which are the outputs of the Mie code. To help understand and use the inputs and outputs of the Mie code that we include in the CD-ROM, the resulting formulas are given below.

The exact expression for the element of the amplitude scattering matrix (2×2) for isotropic spherical particles illuminated by a linearly polarized plane electromagnetic wave are (van de Hulst 1981)

$$S_{11}(\theta) = \sum_{n=1}^{\infty} \frac{2n+1}{n(n+1)} \{a_n \tau_n(\cos\theta) + b_n \pi_n(\cos\theta)\} \qquad (2.40)$$

$$S_{22}(\theta) = \sum_{n=1}^{\infty} \frac{2n+1}{n(n+1)} \{a_n \pi_n(\cos\theta) + b_n \tau_n(\cos\theta)\} \qquad (2.41)$$

$$S_{12}(\theta) = S_{21}(\theta) = 0 \qquad (2.42)$$

where

$$a_n = \frac{\psi_n'(m\chi)\psi_n(\chi) - m\psi_n(m\chi)\psi_n'(\chi)}{\psi_n'(m\chi)\xi_n(\chi) - m\psi_n(m\chi)\xi_n'(\chi)} \qquad (2.43)$$

$$b_n = \frac{m\psi_n'(m\chi)\psi_n(\chi) - \psi_n(m\chi)\psi_n'(\chi)}{m\psi_n'(m\chi)\xi_n(\chi) - \psi_n(m\chi)\xi_n'(\chi)} \qquad (2.44)$$

where $m = n - i\upsilon$ is the relative refractive index of the spherical particle with radius r, χ is defined by (2.33)

$$\psi_n(\chi) = \sqrt{\pi\chi/2}\, J_{n+1/2}(\chi) \qquad (2.45)$$

$$\xi_n = \sqrt{\pi\chi/2}\, H_{n+1/2}(\chi) \qquad (2.46)$$

where $J_{n+1/2}(\chi)$ and $H_{n+1/2}(\chi)$ are the Bessel and Hankel functions. The angular functions are determined by the following formulas

$$\pi_n(\cos\theta) = \frac{P_n^1(\cos\theta)}{\sin\theta} \qquad (2.47)$$

$$\tau_n(\cos\theta) = \frac{dP_n^1(\cos\theta)}{d\theta} \qquad (2.48)$$

where $P_n^1(\cos\theta)$ is the associated Legendre polynomial, defined in (2.95), and θ is the scattering angle. The Mie codes calculate the following variables:

Extinction efficiency factor:

$$Q_{ext} = \frac{2}{\chi^2} \sum_{n=1}^{\infty} (2n+1)\mathrm{Re}(a_n + b_n) \qquad (2.49)$$

Scattering efficiency factor:

$$Q_{sca} = \frac{2}{\chi^2} \sum_{n=1}^{\infty} (2n+1)\left[|a_n|^2 + |b_n|^2\right] \qquad (2.50)$$

Phase function:

$$P(\theta) = \frac{2(|S_{11}|^2 + |S_{22}|^2)}{k^2 r^2 Q_{sca}} \tag{2.51}$$

where $k = 2\pi/\lambda$.

Asymmetry parameter:

$$g = \frac{4}{\chi^2 Q_{sca}} \sum_{n=1}^{\infty} \left[\frac{n(n+2)}{n+1} \text{Re}(a_n a_{n+1}^* + b_n b_{n+1}^*) + \frac{2n+1}{n(n+1)} \text{Re}(a_n b_n^*) \right] \tag{2.52}$$

For particles with a size distribution $f(r)$, we can calculate the extinction and scattering coefficients σ_{ext}, σ_{sca}, phase function $p(\theta)$, single scattering albedo ω, and the asymmetry parameter g:

$$\sigma_{ext} = N \int_0^\infty \pi r^2 Q_{ext} f(r) \, dr \tag{2.53}$$

$$\sigma_{sca} = N \int_0^\infty \pi r^2 Q_{sca} f(r) \, dr \tag{2.54}$$

$$P(\theta) = \frac{2\pi N \int_0^\infty (|S_{11}|^2 + |S_{22}|^2) f(r) \, dr}{k^2 \sigma_{sca}} \tag{2.55}$$

$$g = \frac{\int_0^\infty r^2 Q_{sca} g(r) f(r) \, dr}{\int_0^\infty r^2 Q_{sca} f(r) \, dr} \tag{2.56}$$

$$\omega = \frac{\sigma_{sca}}{\sigma_{ext}} \tag{2.57}$$

We just discussed the scattering behavior of a large particle and a group of particles using Mie theory. For a group of particles, we need to specify the particle size distribution function $f(r)$, which is the subject of the following section. Some examples are also given below.

2.3.3 Aerosol Particle Size Distributions

In general, aerosol particle sizes are not identical. Their radii can be represented by many different functions, such as the power-law function, the modified gamma distribution function, and the lognormal distribution function. Letting $n(r)dr$ be the number of particles per unit volume in the size range r to $r + dr$, this distribution can be expressed as follows:

- The power-law (Junge) distribution:

$$n(r) = \begin{cases} Cr^{-\alpha} & r_1 \leq r \leq r_2 \\ 0 & \text{otherwise} \end{cases} \qquad (2.58)$$

where α is a parameter between 2.5 and 4.0 for the natural aerosols.
- The modified power-law distribution:

$$n(r) = \begin{cases} C\left(\dfrac{r}{r_1}\right)^{-\alpha} & r_1 \leq r \leq r_2 \\ 0 & \text{otherwise} \end{cases} \qquad (2.59)$$

- The gamma distribution:

$$n(r) = Cr^{\alpha}e^{-br} \qquad (2.60)$$

where α and b are parameters.
- The modified gamma distribution:

$$n(r) = Cr^{(1-3b)/b}e^{-r/ab} \qquad (2.61)$$

- The lognormal distribution:

$$n(r) = Cr^{-1}\exp\left[\frac{-(\ln r - \ln r_g)^2}{2\ln^2\sigma_g}\right] \qquad (2.62)$$

where r_g and σ_g are the parameters indicating the mean particle size and its variation.

In the above equations, the constant C for each size distribution is chosen such that the size distribution satisfies the standard normalization

$$\int_{r_{\min}}^{r_{\max}} n(r)\, dr = 1 \qquad (2.63)$$

Mathematically, particle radii in the modified gamma, lognormal, and gamma distributions may extend to infinity. However, a finite $r_{\max}$ must be chosen in actual computer calculations. Most aerosol particles are characterized by the lognormal distribution. Surface fog and mineral dust usually follow a power-law size distribution.

After determining the normalization constant, the lognormal distribution can be written as

$$n(r) = \frac{N}{\sqrt{2\pi}\,\ln\sigma_g}\exp\left[\frac{-\left(\ln r - \ln r_g\right)^2}{2\ln^2\sigma_g}\right] \tag{2.64}$$

where N is the total number of particles per unit volume, given by

$$N = \int_{r_1}^{r_2} n(r)\,dr \tag{2.65}$$

and the characteristic radius (r_g) and characteristic width (σ_g) are defined as

$$\ln r_g = \frac{1}{N}\int_{r_1}^{r_2}\ln r\cdot n(r)\,dr \tag{2.66}$$

$$\left(\ln\sigma_g\right)^2 = \frac{1}{N}\int_{r_1}^{r_2}\left(\ln r - \ln r_g\right)^2\cdot n(r)\,dr \tag{2.67}$$

Note that the actual aerosols may be distributed in several modes. For example, the tropospheric aerosol model is usually characterized by a bimodal lognormal distribution (sum of two lognormals). Whitby (1978) showed that the aerosol distribution may have three modes, a nuclei mode generated by spontaneous nucleation of the gaseous material from particles smaller than 0.04 μm in diameter, the accumulation mode with particles between 0.04 and 0.5 μm resulting from coagulation and in-cloud processes, and the coarse mode with particles larger than 1.0 μm in diameter originated from the Earth's surface. Higurashi and Nakajima (1999) compiled r_g and σ_g values from a series of published papers for one mode, two modes, or three modes (see Table 2.2). After averaging the tabulated values, they calculated the parameters for two modes: $r_{g1} = 0.17$ μm, $\sigma_{g1} = 1.96$, $r_{g2} = 3.44$ μm, and $\sigma_{g1} = 2.37$ μm. If we assume the equal weighting factor for these two modes, the aerosol size distribution is as shown in Fig. 2.5.

If the particles are much larger than the wavelength (i.e., $\chi > 50$), Mie calculation is very time-consuming. The ray tracing method from *geometric optics* is appealing. It is simple and its calculation much faster. The solution is very close to the exact numerical one for large particles. It is also suitable for nonspherical particles for which there are no exact theories. The general idea is illustrated in Fig. 2.6. It is assumed that the incident plane wave can be subdivided into a large number of light rays whose interface behavior is governed by the Fresnel equation and Snell's law (Bohren and Huffman, 1983). At point 1 on the surface of the spherical particle, the incident ray is divided into externally reflected and internally transmitted rays. At point 2 the transmitted ray is partially reflected and partially transmitted. At each

TABLE 2.2   Parameters for Lognormal Distribution with One to Three Modes

Data Set	r_{g1}	σ_{g1}	r_{g1}	σ_{g1}	r_{g1}	σ_{g1}	Notes
1	0.14	1.56	2.29	2.11	—	—	Light aerosol loading
2	—	—	2.84	1.90	35.5	1.38	Moderate aerosol loading
3	—	—	3.08	2.20	35.1	1.37	Heavy aerosol loading
4	0.41	1.36	2.3	1.65	—	—	Winter, Morioka
5	0.36	1.58	—	—	—	—	Oklahoma
6	0.17	1.61	—	—	—	—	Continental background
7	0.19	1.64	—	—	—	—	Urban pollutant
8	0.14	2.6	—	—	—	—	Year average
9	0.13	1.8	—	—	10.0	2.6	
10	0.16	1.79	4.48	3.47	—	—	Winter in Sendai
11	0.17	1.69	2.78	3.06	—	—	Spring in Sendai
12	0.21	1.97	2.98	2.17	—	—	Summer is Sendai
13	0.15	1.96	3.96	3.12	—	—	Autumn in Sendai
14	0.17	2.05	4.375	2.33	—	—	
15	0.175	2.34	3.41	2.23			
16	0.205	2.23	4.09	2.23	—	—	
17	0.189	2.12	3.41	2.23	—	—	
18	0.174	2.01	4.09	2.23	—	—	
19	0.196	2.01	4.09	2.23	—	—	

Source: Detailed information for each dataset is available in the paper by Higurashi and Nakajima (1999).

point where a ray encounters a boundary it is partially reflected and partially transmitted into the surrounding space. By tracing all these rays, we can calculate the single scattering albedo and the phase function.

Mie theory has found extensive applications since spherically shaped particles are frequent in nature, and some atmospheric aerosols are attached with a thick layer of water on their surfaces to end up as a sphere even though their original configurations may not be spherical. Many types of the atmospheric aerosols, however, retain their nonspherical form for which Mie theory is not valid. Nonspherical particles can be classified into three categories: polyhedral solids, stochastically rough particles, and stochastic aggregates (Lumme and Rahola, 1994). For example, soot particles are highly nonspherical and result mainly from biomass burning and human activities. There are two common approaches to calculating their scattering properties.

1. The first approach is to treat a nonspherical particle as a sphere with an equivalent diameter. Thus, Mie theory still can be used. Different methods have been used to determine the equivalent diameter, for example

    Volume diameter—the spherical volume V equals that of the nonspherical particle, $d = \sqrt[3]{6V/\pi}$.

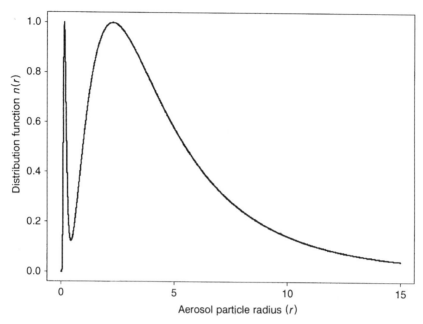

Figure 2.5   Aerosol size distribution characterized by a two-mode lognormal distribution function with the parameters averaged from multiple sources. The mean values of the first and the second mode are 0.17 and 3.44 μm, respectively.

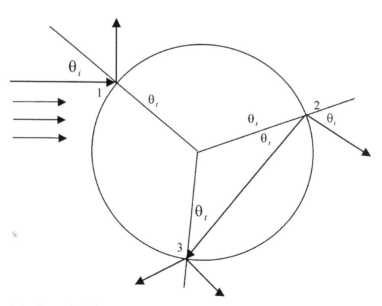

Figure 2.6   Illustration of ray tracing for single spherical particle based on geometric optical principle.

TABLE 2.3 Aerosol Types and Associated Parameters of Lognormal Distribution

Aerosol	r_1 (μm)	r_2 (μm)	r_g (μm)	σ_g	n_r	n_i
Sulfate/nitrate 1 (accumulated)	0.007	0.7	0.07	1.86	1.530	0
Sulfate/nitrate 2 (accumulated)	0.05	2.0	0.45	1.30	1.43	0
Sea salt (accumulated)	0.1	1.0	0.35	2.51	1.50	0
Sea salt (coarse)	1.0	20.0	3.30	2.03	1.50	0
Urban soot	0.005	20.0	0.012	2.0	1.75	0.43–0.55
Biomass burning	0.007	2.0	0.40	1.8	1.43	0.0035

Source: Diner et al. (1999).

> *Surface diameter*—the spherical surface area A equals that of the nonspherical particle, $d = \sqrt{A/\pi}$.
>
> *Surface volume diameter*—the sphere has the same ratio of external surface area A to the volume V as the spherical particle, $d = 6V/A$.

This approach is easy to calculate but is only approximate.

2. The other approach is to develop more rigorous computational methods. For small particles, the T-matrix algorithm (Mishchenko 1993) is a good choice. For large particles, such as ice crystals, ray tracing methods are better (Takano and Liou 1989a, 1989b, 1995).

Besides the particle size distribution, we also need to know the refractive index for each aerosol type. Unfortunately, the refractive index is quite variable in both space and time (d'Almeida et al. 1991). EOS sensors, such as MODIS and MISR, might provide us with some answers. The common practice right now is to assign certain values to each individual aerosol type based on airborne observations and other measurements.

For simulation, the particles listed in Table 2.3 can be assumed to follow the lognormal distribution patterns. Their parameters and the refractive index ($n_r + in_i$) are given in the table (Diner et al. 1999).

The particles listed in Table 2.4 are assumed to follow the power-law distribution. Dusts are seldom spherical and are often modeled as randomly

TABLE 2.4 Aerosol Types and the Associated Parameters of Power-Law Distribution

Aerosol	r_1 (μm)	r_2 (μm)	α	n_r	n_i
Mineral dust (accumulated)	0.05	2.0	3.0	1.53	0.0012–0.0085
Mineral dust (coarse)	0.5	15.0	3.0	1.53	0.0012–0.0085
Near-surface fog	0.5	50.0	2.5	1.33	0

Source: Diner et al. (1999).

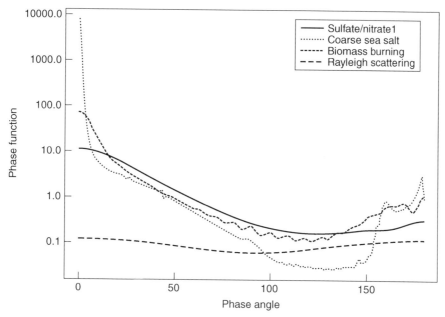

Figure 2.7 Phase function of three aerosol types calculated by Mie code. The Rayleigh phase function is also plotted for comparison.

oriented prolate and oblate spheroids with a uniform distribution of aspect ratios between 1.2 and 1.4. The parameter and the refractive index are given in Table 2.4 (Diner et al. 1999).

Figure 2.7 displays the phase function of three aerosol types: sulfate/nitrate 1, coarse sea salt, and biomass burning. The Rayleigh scattering phase function is also included for comparison. The parameters of their lognormal distribution functions and refractive indices have been given in Table 2.3. It is very evident that the coarser the aerosol particles are, the greater will be the peak of the phase function near the zero phase angle that represents very strong forward scattering.

As mentioned above, the Mie code outputs the single scattering albedo and phase function required by the radiative transfer calculations. The phase function is often expanded by the *Legendre polynomials*:

$$P(\theta) = \sum_{0}^{\infty} (2l+1)a_l P_l(\mu) \tag{2.68}$$

where $\mu = \cos\theta$ and $P_l(\mu)$ is the Legendre polynomial of order l. The first five Legendre polynomials are

$$P_0(\mu) = 1$$
$$P_1(\mu) = \mu$$
$$P_2(\mu) = \frac{3\mu^2 - 1}{2} \qquad (2.69)$$
$$P_3(\mu) = \frac{5\mu^3 - 3\mu}{2}$$
$$P_4(\mu) = \frac{35\mu^4 - 30\mu^2 + 3}{8}$$

The lth expansion coefficient is given by

$$a_l = \frac{1}{2} \int_{-1}^{1} P_l(\mu) P(\mu)\, d\mu \qquad (2.70)$$

a_1 is usually called the *asymmetry parameter* that represents the degree of asymmetry of the angular scattering and denoted by $g \equiv a_1$; $g = 0$ represents isotropic scattering; $g = -1$ represents complete backscattering; and $g = 1$ complete forward scattering. For most aerosol particles, $0 < g < 1$, calculated from (2.52) and (2.56).

The phase function is often approximated by analytic functions. The most widely used is the Henyey–Greenstein function with the asymmetry parameter:

$$P(\mu) = \frac{1 - g^2}{\left(1 + g^2 - 2g\mu\right)^{3/2}} \qquad (2.71)$$

A two-term Henyey–Greenstein function is also often used with the proportional parameter a:

$$P(\mu) = \frac{a\left(1 - g_1^2\right)}{\left(1 + g_1^2 - 2g_1\mu\right)^{3/2}} + \frac{(1 - a)\left(1 - g_2^2\right)}{\left(1 + g_2^2 - 2g_2\mu\right)^{3/2}} \qquad (2.72)$$

2.3.4 Gas Absorption

Atmospheric scattering has been discussed in the previous three sections. Absorption is caused mainly by atmospheric gases, such as water vapor, ozone, and oxygen, as well as aerosols. The aerosol absorption is accounted for by the single scattering albedo ω. If $\omega = 1$, the aerosols are not absorptive. For most multispectral sensors, we are concerned mainly with water vapor and ozone since other gases absorb energy in a very narrow spectral range in the optical region and their concentrations are quite stable. For the hyperspectral sensors, some of these gases have to be taken into account (e.g., oxygen) (Gao et al. 1993, Qu et al. 2003). The absorption in the narrow spectral windows are discussed in Chapter 10.

Most gases are quite stable in both space and time. For example, ozone is distributed mainly in the stratosphere ($\sim 20-50$ km above the surface). Carbon dioxide is usually well mixed with other dry gases, except near sources (e.g., big cities, forest fires). The most variable gas that significantly affects remotely sensed data is water vapor. It is found mostly in the boundary layer (the lowest is 1–2 km). Water vapor content varies between 0.42 g/cm^2 in subartic regions in the winter and 4.12 g/cm^2 in tropical regions. The daily fluctuations from 1.0–4.0 g/cm^2 have been observed (Holben and Eck 1990). Estimation of water vapor from remotely sensed data is discussed in Section 6.4. If the concentrations of these gases are known, radiative transfer codes (e.g., MODTRAN) can be used for calculating the transmittance of these gases.

In the following, we have run MODTRAN to evaluate the impacts of several gases on atmospheric transmittance from 300 to 2500 nm. Figure 2.8 illustrates the impacts of water vapor content. From this figure, we can see that the impacts are mainly in the longer wavelength. Figure 2.9 shows the impacts of ozone absorption where all wavebands from green to near-IR are affected. Figure 2.10 displays the impacts of CO_2 and other gases. It is clear

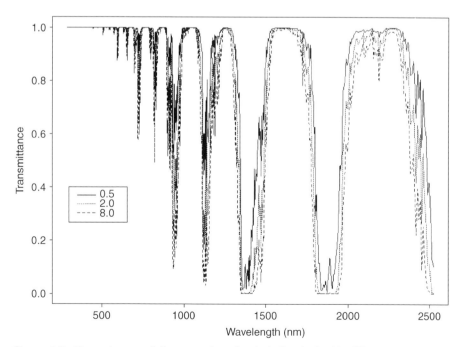

Figure 2.8 Transmittance of the atmosphere (vertical direction) with different water vapor contents (g/cm^2) calculated from MODTRAN.

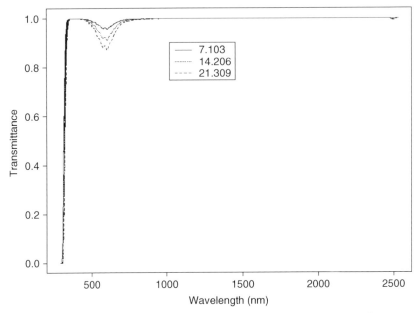

Figure 2.9 Atmospheric transmittance with different ozone concentrations (g/m^2) calculated from MODTRAN.

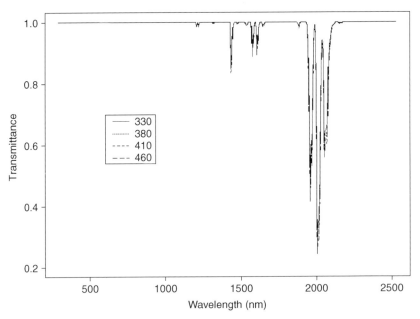

Figure 2.10 Atmospheric transmittance with variable CO_2 concentrations [in parts per million by volume (ppmv)] and other constant gases calculated from MODTRAN.

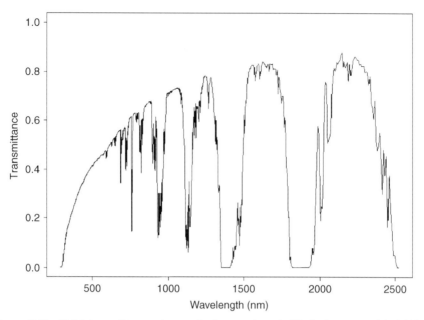

Figure 2.11 Total transmittance of a standard atmosphere (midlatitude, summer) in MOD-TRAN.

their impacts are mainly in the longer wavelength. The total atmospheric transmittance with the default values of the midlatitude summer in MOD-TRAN is shown in Fig. 2.11. It is amazing to see that the maximum transmittance due to gaseous absorption is about 80%.

A simple parametric function has often been used for correcting the effects of the gaseous absorption in both the solar incoming path (sun to surface) and the viewing path (surface to sensor) (Tanré et al. 1992):

$$t_g(-\mu_i, \mu_v, U) = \frac{1}{1 + aU^b} \qquad (2.73)$$

where

$$U = U_0\left(\frac{1}{\mu_i} + \frac{1}{\mu_v}\right) \qquad (2.74)$$

and U_0 is the gaseous content, a and b are the empirical coefficients depending on the wavebands, and μ_i and μ_v are the cosine of the solar zenith angle and the viewing zenith angle, respectively. The coefficients of AVHRR solar bands for water vapor and ozone are shown in Table 2.5 and Table 2.6. Note that ozone absorption in AVHRR band 2 is very small and its coefficients are not given. The coefficients are different for different NOAA

TABLE 2.5   Coefficients *a* and *b* for Ozone Absorption in AVHRR Band 1

Coefficients	NOAA8	NOAA9	NOAA10	NOAA11
a	8.284×10^{-2}	8.137×10^{-2}	8.895×10^{-2}	8.160×10^{-2}
b	1.073	1.070	1.085	1.071

Source: Tarré et al. (1992).

TABLE 2.6   Coefficients *a* and *b* for Water Vapor Absorption in AVHRR Bands 1 and 2

Band	Coefficients	NOAA8	NOAA9	NOAA10	NOAA11
1	*a*	7.561×10^{-3}	5.851×10^{-3}	4.423×10^{-3}	5.587×10^{-3}
	b	0.673	0.747	0.776	0.752
2	*a*	6.701×10^{-2}	7.326×10^{-2}	6.906×10^{-2}	7.008×10^{-2}
	b	0.493	0.490	0.479	0.492

satellites simply because AVHRR sensor spectral response functions are different (see Section 1.3.5.1). In Table 2.6, note that water vapor absorption is much larger in AVHRR band 2 than in band 1.

The French POLDER team (Lafrance et al. 2002) use a similar formula to characterize ozone absorption

$$t_g(-\mu_i, \mu_v, U) = \frac{a_0 + a_1 U}{a_0 + b_1 U + b_2 U^2} \qquad (2.75)$$

where U is defined by (2.74) and U_0 is the column amount of ozone of the atmosphere in cm·atm. The coefficients are listed in Table 2.7. Note that in those land bands, water vapor and oxygen absorption can be neglected.

For other sensors with different spectral responses, we can either develop similar parametric models from model calculations (e.g., MODTRAN) or run the radiative transfer package directly. For most applications, calculation of absorption in the optical region is very fast since we do not need to use the line-by-line codes (see Chapter 10).

TABLE 2.7   Coefficients of the Ozone Transmittance Formula (2.75)

Band	a_0	a_1	b_1	b_2
443	359.759	0	1	0
490	4056.01	1.88253	90.9543	1
670	572.625	8.47641	33.727	1

2.3.5   Aerosol Climatology

It is very useful for us to characterize the average state of the atmospheric scattering and absorption based on extensive ground measurements and inversion from satellite observations. An earlier comprehensive compilation of aerosol optical properties is available in a book by d'Almeida et al. (1991). A complete update and revision leads to the global aerosol dataset (GADS) in conjunction with the software package Optical Properties of Aerosols and Clouds (OPAC) (Hess et al. 1998). For the entire globe, on a grid of 5° longitude and latitude, with seven differentiating height profiles, and for both summer and winter, the aerosol at the gridpoints are composed of the aerosol components. Microphysical and optical properties of six water clouds, three ice clouds, and 10 aerosol components (considered typical cases) are stored in ASCII (American Standard Code for Information Interchange) files. The optical properties are the extinction, scattering, and absorption coefficients; the single scattering albedo; the asymmetry parameter; and the phase function. They are calculated on the basis of the microphysical data (size distribution and spectral refractive index) with the spherical particles of aerosols and cloud droplets and hexagonal columns of cirrus clouds. Data are given for up to 61 wavelengths between 0.25 and 40 μm and up to eight values of the relative humidity. The software package also allows calculation of derived optical properties such as mass extinction coefficients and Angstrom coefficient or Angstrom index, defined by Eq. (6.5). The data sets and software are included in the CD-ROM and were downloaded from `http://www.lrzmuenchen.de/`~`uh234an/www/` `radaer/gads.html`.

The *hi*gh-resolution *trans*mission (HITRAN) molecular absorption database also contains the aerosol optical properties. HITRAN is a compilation of spectroscopic parameters that are used by a variety of computer codes to predict and simulate the transmission and emission of light in the atmosphere. A new edition of the HITRAN molecular spectroscopic database and associated compilation (v11.0) is available on an ftp (File Transfer Protocol) site at `http://www.hitran.com/`. In addition to the megaline HITRAN-2000 database, there are directories containing files of aerosol indices of refraction, UV line-by-line and absorption cross-sectional parameters, and more extensive IR absorption crosssections. For convenience, the aerosol indices of refractive data are also included in the accompanying CD-ROM.

2.4   SOLVING RADIATIVE TRANSFER EQUATIONS

Few readers need to write the codes to solve the radiative transfer equation for general applications since many radiative transfer solvers are available in the public domain (see Section 2.4.4). Nevertheless, it is important to understand some of the basic principles.

2.4.1 Radiation Field Decomposition

The standard practice now is to divide the total radiance into direct and diffuse components ($I = I_I + I_D$). The direct component (i.e., direct sunlight without scattering) in the downward direction is calculated from Beer's law:

$$I_I(\tau, -\mu, \phi) = \pi E_s \delta(\mu - \mu_0) \delta(\phi - \phi_0) \exp\left(-\frac{\tau_t - \tau}{\mu_0}\right) \quad (2.76)$$

If we account for the upwelling reflected component into the diffuse component, the upwelling direct component is zero.

The radiative transfer equation for diffuse radiance $I_D(\tau, \mu, \phi)$ can be written based on Eq. (2.3) as

$$\mu \frac{dI_D(\tau, \mu, \phi)}{d\tau}$$

$$= I_D(\tau, \mu, \phi) - \frac{\omega}{4\pi} \int_0^{2\pi} \int_{-1}^{1} I_D(\tau, \mu_i, \phi_i) P(\mu, \phi, \mu_i, \phi_i) \, d\mu_i \, d\phi_i$$

$$- \frac{\omega}{4} P(\mu, \phi, -\mu_0, \phi_0) E_s e^{-\tau_t - \tau / \mu_0} \quad (2.77)$$

where the incoming radiation at the top of the atmosphere is zero

$$I_D(\tau_t, -\mu, \phi) = 0 \quad (2.78)$$

and the lower boundary condition is

$$I_D(0, \mu, \phi) = \frac{1}{\pi} \int_0^{2\pi} \int_{-1}^{0} R(\mu, \phi, \mu', \phi') I_D(\mu', \phi') \mu' \, d\mu' \, d\phi'$$

$$+ \mu_0 \pi E_s \exp\left(-\frac{\tau_t}{\mu_0}\right) R(\mu, -\mu_0, \phi - \phi_0) \quad (2.79)$$

where $R(\cdot)$ is the surface BRDF discussed in Section 2.2. The first term of the right side of Eq. (2.79) corresponds to diffuse sky radiance and the second term, the reflected direct solar radiation. In the following paragraphs, we will drop the subscript D in the notation and discuss only the diffuse radiation.

Given the optical properties of the medium and their boundary conditions, the next key step is to solve the radiative transfer equation. There are roughly two types of algorithms, approximate solutions, and numerical solutions. Approximate solutions, such as two-stream approximations (Meador and Weaver 1980, Kaufman and Joseph 1982, Liang and Strahler 1994a) and four-stream approximations (Liou 1974, Liang and Strahler 1994b), widely used in earlier radiative transfer calculations, are still used in GCM (global

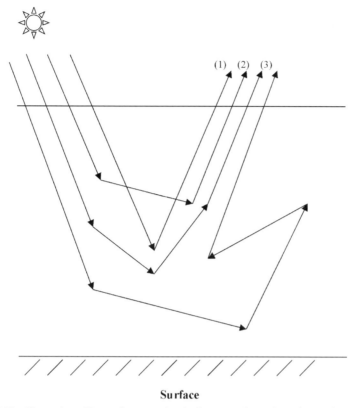

Figure 2.12   Illustration of low-order scattering in the successive orders of scattering where (1), (2), and (3) respectively represent single, double, and triple scattering.

circulation model) and other applications. With the rapid development of computer technology, more people tend to use numerical algorithms, which are computationally expensive but more accurate.

2.4.2 Numerical Solutions

There are many different methods available to calculate the diffuse radiation field (e.g. Lenoble 1985). We will mainly introduce two of them for the radiative transfer problem of an one-dimensional atmosphere: method of successive orders of scattering and method of discrete ordinates.

2.4.2.1 Method of Successive Orders of Scattering

This is one of the oldest and simplest concepts in all the different methods. Various algorithms have been developed, but the basic idea is to calculate the diffused radiance associated with photons scattered once $[L^{(1)}]$, twice $[L^{(2)}]$, 3 times $[L^{(3)}]$, and so on (see Fig. 2.12). The total diffuse radiance is the sum of all orders:

$$I(\tau, \mu, \phi) = \sum_{n=1}^{\infty} L^{(n)}(\tau, \mu, \phi) \tag{2.80}$$

From Eq. (2.77), we can easily write the radiative transfer equation for single scattering:

$$\mu \frac{dL^{(1)}(\tau, \mu, \phi)}{d\tau} = L^{(1)}(\tau, \mu, \phi) - \frac{\omega}{4} P(\mu, \phi, -\mu_0, \phi_0) E_s e^{-\tau_t - \tau/\mu_0} \tag{2.81}$$

This is the ordinary differential equation and we can formulate an analytic solution. Singly scattered radiance thus can be derived and calculated from the following analytic formulas.

In the downward direction:

$$L^{(1)}(\tau, -\mu, \phi) = \frac{\omega \mu_0 E_s}{4(\mu_0 - \mu)} P(-\Omega_0, -\Omega) \left[\exp\left(-\frac{\tau_t - \tau}{\mu_0}\right) - \exp\left(-\frac{\tau_t - \tau}{\mu}\right) \right]$$

$$\text{when} \quad \mu \neq \mu_0 \tag{2.82}$$

$$L^{(1)}(\tau, -\mu, \phi) = \frac{\omega \tau_t - \tau E_s}{4\mu_0} P(-\Omega_0, -\Omega) \exp\left(-\frac{\tau_t - \tau}{\mu_0}\right)$$

$$\text{when} \quad \mu = \mu_0 \tag{2.83}$$

In the upwelling direction:

$$L^{(1)}(\tau, \mu, \phi) = \frac{\omega \mu_0 E_s}{4(\mu + \mu_0)} P(-\Omega_0, \Omega)$$

$$\times \left[\exp\left(-\frac{\tau_t - \tau}{\mu}\right) - \exp\left(-\frac{\tau_t}{\mu_0} - \frac{\tau}{\mu}\right) \right] \tag{2.84}$$

where Ω is denoted as the direction characterized by both μ and ϕ.

The higher order scattered radiance is related to the lower order scattered radiance through the source function J in the following way

$$L^{(n)}(\tau, -\Omega) = \int_\tau^{\tau_t} \frac{d\tau'}{\mu} J_n(\tau', \Omega) \exp\left(\frac{\tau - \tau'}{\mu}\right) \quad \text{downward} \tag{2.85}$$

$$L^{(n)}(\tau, \Omega) = L^{(n)}(0, \Omega) \exp\left(-\frac{\tau}{\mu}\right) + \int_0^\tau \frac{d\tau'}{\mu} J_n(\tau', \Omega) \exp\left(-\frac{\tau' - \tau}{\mu}\right)$$

$$\text{upwelling} \tag{2.86}$$

where the first term in the right side of Eq. (2.86) is the upweling radiance from the surface using Eq. (2.79).
The source function can be updated as follows:

$$J^{n+1}(\tau, \Omega) = \frac{\omega}{4\pi} \int_{4\pi} P(\Omega', \Omega) L^{(n)}(\tau, \Omega') \, d\Omega' \tag{2.87}$$

For double scattering, we can write the explicit formulas, but for higher-order scattering, a numerical scheme is needed. The basic approach is to perform the τ integration by dividing the atmosphere into many thin layers and Ω integrals numerically using *Gauss quadrature*.

Gaussian quadrature points x_k have the property that

$$\int_a^b f(x)\,\omega(x)\,dx \approx \sum_{i=1}^N A_k f(x_k) \qquad (2.88)$$

is *exact* if $f(x)$ is any polynomial of degree $\leq 2N - 1$. The values of the x_k are determined by the form of the weighting function $\omega(x)$ and the interval boundaries (a, b). In fact, the x_k are the roots of the orthogonal polynomial with weighting function $\omega(x)$ over the interval (a, b). For example, in radiative transfer calculations,

$$\int_{-1}^1 I(\tau, \mu)\,d\mu \approx \sum_{i=1}^M w_i I(\tau, \mu_i) \qquad (2.89)$$

where w_i is a quadrature weight and μ_i is the discrete ordinate. The low-order quadrature and weights are given in Table 2.8. For high-order quadrature and weights, some programs are available for easy calculation (e.g., Press et al. 1989).

TABLE 2.8   Low-Order Gaussian Quadrature Discrete Ordinates and Weights

M	i	μ_i	w_i
1	1	0.57735	1.00000
2	1	0.33998	0.65215
	2	0.86112	0.34785
3	1	0.23862	0.46791
	2	0.66121	0.36076
	3	0.93247	0.17132
4	1	0.18343	0.36268
	2	0.52553	0.31371
	3	0.79667	0.22238
	4	0.96029	0.10123
5	1	0.14887	0.29552
	2	0.43340	0.26927
	3	0.67941	0.21909
	4	0.86506	0.14945
	5	0.97391	0.06667
6	1	0.12523	0.24915
	2	0.36783	0.23349
	3	0.58732	0.20317
	4	0.76990	0.16008
	5	0.90412	0.10694
	6	0.98156	0.04718

Note that this approximation is accurate if the radiance function can be approximated by a smooth polynomial. However, the radiance at the top of the atmosphere usually changes rather rapidly when μ approaches zero, and this approximation may cause large errors. To remedy this problem, the double-Gauss method (Sykes 1951) was proposed and has been widely used in current radiative transfer calculations. The idea is to apply the Gaussian formula separately to the half-range $-1 < \mu < 0$ and $0 < \mu < 1$. The main advantage of this double-Gauss scheme is that the quadrature points (in even orders) are distributed symmetrically around $|\mu| = 0.5$ and cluster toward both $|\mu| = 1$ and $\mu = 0$, whereas in the Gaussian scheme for the complete range, $-1 < \mu < 1$, they are clustered toward $\mu = \pm 1$. The clustering toward $\mu = 0$ will give superior results near the boundary where radiance varies rapidly around $\mu = 0$. A half-range scheme is also preferred since radiance is not continuous at the boundaries. Another advantage is that upwelling and downward fluxes and the mean radiance can be calculated immediately without further approximation.

2.4.2.2 *Method of Discrete Ordinates*

The idea of discrete ordinates is to replace the integrals in the radiative transfer equation with quadrature sums and thus transform the pair of coupled integral-differential equations into a system of coupled differential equations that are finally converted into a high-dimensional linear algebra system. Let us start from the elimination of dependence of the azimuth angle.

Radiance $I(\tau, \mu, \phi)$ has three arguments. The typical approach is to expand radiance into the Fourier cosine series first:

$$I(\tau, \mu, \phi) = \sum_{m=0}^{2N-1} I^m(\tau, \mu) \cos m(\phi - \phi_0) \qquad (2.90)$$

The original radiative transfer equation (2.77) is replaced by $2N$ independent equations (one for each Fourier component), and the equation for each of the Fourier components is

$$\mu \frac{dI^m(\tau, \mu)}{d\tau} = I^m(\tau, \mu) - \frac{\omega}{2} \int_{-1}^{1} I^m(\tau, \mu) p^m(\tau, \mu_i, \mu)\, d\mu_i$$

$$- \frac{\omega}{4} p^m(\tau, \mu, -\mu_0) E_s (2 - \delta_{0m}) e^{-\tau_t - \tau/\mu_0} \quad (m = 0, 1, 2, \ldots, 2N-1) \quad (2.91)$$

where δ_{0m} is the Krönecker delta ($\delta_{0m} = 1$ for $m = 0$ and $\delta_{0m} = 0$ for $m \neq 0$) and the phase function is also expanded into Fourier series:

$$P(\tau, \mu, \phi, \mu_i, \phi_i) = \sum_{m=0}^{2N-1} p^m(\tau, \mu, \mu_i) \cos m(\phi - \phi_0) \qquad (2.92)$$

The Fourier component of the phase function therefore is

$$p^m(\tau,\mu,\mu_i) = (2-\delta_{0m})\sum_{l=m}^{2N-1}(2l+1)\chi_l(\tau)\Lambda_l^m(\mu)\Lambda_l^m(\mu_i) \quad (2.93)$$

where $\chi_l(\tau)$ are the coefficients. The normalized associated Legendre polynomial is related to the *associated Legendre polynomial $p_l^m(\mu)$* in the following way:

$$\Lambda_l^m(\mu) = \sqrt{\frac{(l-m)!}{(l+m)!}}\,P_l^m(\mu) \quad (2.94)$$

The first few associated Legendre polynomials are

$$p_1^1(\mu) = \sqrt{1-\mu^2}$$
$$p_1^2(\mu) = 3\mu\sqrt{1-\mu^2}$$
$$p_2^2(\mu) = 3\sqrt{1-\mu^2} \quad (2.95)$$
$$p_3^2(\mu) = 15\mu(1-\mu^2)$$
$$p_3^1(\mu) = \tfrac{3}{2}\sqrt{1-\mu^2}\,(5\mu^2-1)$$

After eliminating the dependence of the azimuth angle, the second step is to convert the continuous differential equation into a form suitable for numerical calculation by computer. If the integrals are replaced by quadratures, Eq. (2.91) becomes

$$\mu_i\frac{dI^m(\tau,\mu_i)}{d\tau} = I^m(\tau,\mu_i) - \sum_{j=-N}^{N} w_j I(\tau,\mu_i)D^m(\tau,\mu_j,\mu_i) - Q^m(\tau,\mu_i)$$

$$(i=\pm1,\pm2,\ldots,\pm N) \quad (2.96)$$

where μ_i and w_i are quadrature points and weights and

$$D^m(\tau,\mu_j,\mu_i) = \frac{\omega(\tau)}{2}\sum_{l=m}^{2N-1}(2l+1)\chi_l(\tau)\Lambda_l^m(\mu_j)\Lambda_l^m(\mu_i) \quad (2.97)$$

$$Q^m(\tau,\mu_i) = \frac{\omega(\tau)}{4}E_s(2-\delta_{0m})e^{-\tau_t-\tau/\mu_0}$$

$$\times \sum_{l=0}^{2N-1}(-1)^{l+m}(2l+1)\chi_l(\tau)\Lambda_l^m(\mu_0)\Lambda_l^m(\mu_i) \quad (2.98)$$

Since the single scattering albedo and the phase function are the function of τ in a vertically inhomogeneous atmosphere, Eq. (2.96) constitutes a system

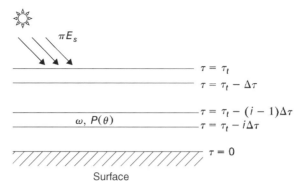

Figure 2.13 Atmospheric stratification for numerical calculation of radiative transfer. Each layer has different single scattering albedos (ω) and aerosol phase functions $P(\theta)$.

of $2N$ coupled differential equations with variable parameters for which analytic solutions do not exist. A common practice is to divide the atmosphere into L homogeneous layers in which the single scattering albedo and the phase function can be approximated as the same (see Fig. 2.13). It is actually in the process of converting the continuous variable τ into a discrete format. For any layer in this figure, we can write Eq. (2.96) in matrix form as

$$
\begin{bmatrix} \dfrac{d\mathbf{I}^+}{d\tau} \\[2ex] \dfrac{d\mathbf{I}^-}{d\tau} \end{bmatrix} = \begin{bmatrix} -\alpha & -\beta \\ \beta & \alpha \end{bmatrix} \begin{bmatrix} \mathbf{I}^+ \\ \mathbf{I}^- \end{bmatrix} + \begin{bmatrix} \tilde{\mathbf{Q}}^+ \\ \tilde{\mathbf{Q}}^- \end{bmatrix} \tag{2.99}
$$

where

$$\mathbf{I}^\pm = \{I^m(\tau, \pm\mu_i)\}, \qquad\qquad i = 1, 2, \ldots, N$$

$$\tilde{\mathbf{Q}}^\pm = \mathbf{M}^{-1}\mathbf{Q}^\pm$$

$$\mathbf{Q}^\pm = \{Q^m(\tau, \pm\mu_i)\} \qquad\qquad i = 1, \ldots, N$$

$$\mathbf{M} = \{\mu_i\delta_{ij}\} \qquad\qquad i, j = 1, \ldots, N$$

$$\alpha = \mathbf{M}^{-1}(\mathbf{D}^+\mathbf{W} - \mathbf{1}) \tag{2.100}$$

$$\beta = \mathbf{M}^{-1}\mathbf{D}^-\mathbf{W}$$

$$\mathbf{W} = \{w_i\delta_{ij}\} \qquad\qquad i, j = 1, \ldots, N$$

$$\mathbf{D}^+ = \{D^m(\mu_i, \mu_j)\} = \{D^m(-\mu_i, -\mu_j)\}, \qquad i, j = 1, \ldots, N$$

$$\mathbf{D}^- = \{D^m(-\mu_i, \mu_j)\} = \{D^m(\mu_i, -\mu_j)\} \qquad i, j = 1, \ldots, N$$

$$\mathbf{1} = \{\delta_{ij}\} \qquad\qquad i, j = 1, \ldots, N$$

Equation (2.99) is a system of $2N$-coupled ordinary differential equations with constant coefficients. Since they are linear, they can be solved using well-known methods of linear algebra. The first step is to seek the solution to the homogeneous version (i.e., $\tilde{\mathbf{Q}} = 0$ of the equation system that is a standard algebraic eigenvalue problem of order $2N \times 2N$ determining the eigenvalues and eigenvectors). The second step is to obtain the particular solution with the source function. The general solution is the sum of the homogeneous solutions and the particular solutions. In a multilayered atmosphere, an integration is needed layer by layer to obtain the expressions for both upwelling and downward radiance at arbitrary τ and μ. The sum of all Fourier components through Eq. (2.90) gives the final solution. The details are omitted here, but interested readers should read the original papers and textbooks (e.g., Liou 1980; Stamnes et al. 1988; Thomas and Stamnes 1999). The discrete ordinate method has been implemented numerically into a FORTRAN code called DISORT (distort ordinate radiative transfer) (Stamnes et al. 1988), which is included in the CD-ROM.

As Stamnes et al. (1988) summarized, this approach has many important features, some of which are:

- It is unconditionally stable for an arbitrarily large number of quadrature angles (streams) and arbitrarily large optical depths.
- It allows for an arbitrary bidirectional reflectivity at the lower boundary.
- It offers rapid computation of albedo and transmissivity when thermal sources are absent.
- Unlike the popular doubling method, computing time for individual layers is independent of optical thickness.

2.4.3 Approximate Solutions: Two-Stream Algorithms

Although numerical radiative transfer solvers are widely available, some approximate methods are still used, primarily for calculating radiant flux (irradiance). In general, they cannot describe the radiance angular distribution accurately because of the nature of approximation.

For Eq. (2.96), if $N = 1$, we have a two-stream solution although it is seldom accurate and some special modifications are needed. If $N = 2$, we can obtain the four-stream solution that can be derived analytically (Liou 1974; Liang and Strahler 1994b). If the optical depth is very large, such as in clouds, the asymptotic fitting algorithm may be very suitable (King and Harshvardhan 1986). The four-stream and asymptotic fitting approximation algorithms are discussed in Section 3.3.1 on canopy radiative transfer modeling. We will briefly discuss various two-stream algorithms as follows.

Meador and Weaver (1980) presented all existing two-stream approximations in identical forms of coupled differential equations for the integrated radiance over hemispheres and a set of common solutions for all approxima-tions. We will first closely follow their notion for deriving the integrated radiance and then present different methods for calculating the angle-depen-dent radiance.

We will consider only the diffuse radiance here. The standard radiative transfer equation has been presented in Eq. (2.77). For the azimuthally integrated radiance

$$I(\tau, \mu) = \int_0^{2\pi} I(\tau, \mu, \phi) \, d\phi \qquad (2.101)$$

the radiative transfer equation, after the azimuthal integration, becomes

$$\mu \frac{dI(\tau, \mu)}{d\tau} = I(\tau, \mu) - \frac{\omega}{2} \int_{-1}^{1} p(\mu, \mu') I(\tau, \mu') \, d\mu'$$

$$- \frac{\omega E_s p(\mu, -\mu_0)}{4} \exp\left(-\frac{\tau_t - \tau}{\mu_0}\right) \qquad (2.102)$$

where

$$p(\mu, \mu') = \frac{1}{2\pi} \int_0^{2\pi} P(\mu, \mu', \phi - \phi_s) \, d\phi = \sum_{l=0}^{\infty} g_l P_l(\mu) P_l(\mu') \quad (2.103)$$

where $g_l = \frac{1}{2} \int_{-1}^{1} P_l(\mu) p(\mu, 1) \, d\mu$ and $P_l(\mu)$ is the Legendre polynomial function of order l.

If we further integrate the azimuth-independent radiance

$$I^{\pm}(\tau) = \int_0^1 I(\tau, \pm \mu) \mu \, d\mu \qquad (2.104)$$

and integrate Eq. (2.102) in a similar way, we can obtain the following pair of equations

$$\frac{dI^+}{d\tau} = \int_0^1 I(\tau, \mu) \, d\mu - \frac{\omega}{2} \int_0^1 \int_{-1}^{1} p(\mu, \mu') I(\tau, \mu') \, d\mu' \, d\mu$$

$$- \pi E_s \omega \beta_0 \exp\left(-\frac{\tau_t - \tau}{\mu_0}\right) \qquad (2.105)$$

and

$$\frac{dI^-}{d\tau} = -\int_0^1 I(\tau, -\mu)\, d\mu + \frac{\omega}{2} \int_0^1 \int_{-1}^1 p(\mu, -\mu')\, I(\tau, \mu')\, d\mu'\, d\mu$$
$$+ \pi E_s \omega (1 - \beta_0) \exp\left(-\frac{\tau_t - \tau}{\mu_0}\right) \qquad (2.106)$$

where

$$\beta_i = \frac{1}{2}\int_0^1 p(\mu_i, -\mu')\, d\mu' = 1 - \frac{1}{2}\int_0^1 p(\mu_i, \mu')\, d\mu' \qquad (2.107)$$

Different approaches have been proposed to simplify these two equations, (2.106) and (2.107), which lead to different two-stream approximations, but all of them can be presented in a unified expression:

$$\frac{dI^+}{d\tau} = \gamma_1 I^+ - \gamma_2 I^- - \gamma_3 \pi E_s \omega \exp\left(-\frac{\tau_t - \tau}{\mu_0}\right) \qquad (2.108)$$

$$\frac{dI^-}{d\tau} = \gamma_2 I^+ - \gamma_1 I^- + \gamma_4 \pi E_s \omega \exp\left(-\frac{\tau_t - \tau}{\mu_0}\right) \qquad (2.109)$$

where γ_i are listed in Table 2.9 corresponding to different two-stream approximations, and $\gamma_3 + \gamma_4 = 1$.

TABLE 2.9   Parameters in Two-Stream Formulas

Algorithm	γ_1	γ_2	γ_3
Eddington	$\frac{1}{4}\{7 - \omega(4 + 3g)\}$	$-\frac{1}{4}\{7 - \omega(4 - 3g)\}$	$\frac{1}{4}(2 - 3g\,\mu_0)$
Modified Eddington	$\frac{1}{4}\{7 - \omega(4 + 3g)\}$	$-\frac{1}{4}\{7 - \omega(4 - 3g)\}$	β_0
Quadrature	$\sqrt{3}/2\{2 - \omega(1 + g)\}$	$\sqrt{3}/2\{2 - \omega(1 + g)\}$	$\frac{1}{4}(2 - 3g\,\mu_0)$
Modified quadrature	$\sqrt{3}\{1 - \omega(1 - \beta_1)\}$	$\sqrt{3}\,\omega\beta_1$	β_0
Hemispheric constant	$2\{1 - \omega(1 - \beta_1)\}$	$2\omega\beta_1$	β_0
Delta function	$\sqrt{3}\{1 - \omega(1 - \beta_1)\}$	$\omega\beta_1$	β_0
Hybrid method	$\dfrac{t_1 + \omega g^2(4\beta + 3g)}{4\left[1 - g^2(1 - \mu_0)\right]}$ $t_1 = 7 - 3g^2 - \omega(4 + 3g)$	$-\dfrac{t_2 - \omega g^2(4\beta + 3g - 4)}{4\left[1 - g^2(1 - \mu_0)\right]}$ $t_2 = 1 - g^2 - \omega(4 - 3g)$	β_0

Source: Meador and Weaver (1980).

Meador and Weaver (1980) solved Eqs. (2.108) and (2.109) for a plane-parallel atmosphere with boundary conditions $I^+(0) = I^-(\tau_t) = 0$ in terms of plane albedo $R = I^+(\tau_t)/(\pi E_s \mu_0)$ and transmittance $T = I^-(0)/(\pi E_s \mu_0) + \exp(-\tau_t/\mu_0)$:

$$R = \frac{\omega[u_1 - u_2 - 2k(\gamma_3 - \alpha_2 \mu_0)\exp(-\tau_t/\mu_0)]}{(1 - k^2\mu_0^2)[(k + \gamma_1)\exp(k\tau_t) + (k - \gamma_1)\exp(-k\tau_t)]} \tag{2.110}$$

$$T = e^{(-\tau_t/\mu_0)}\left\{1 - \frac{\omega[u_3 - u_4 - 2k(\gamma_4 + \alpha_1 \mu_0)\exp(\tau_t/\mu_0)]}{(1 - k^2\mu_0^2)[(k + \gamma_1)\exp(k\tau_t) + (k - \gamma_1)\exp(-k\tau_t)]}\right\} \tag{2.111}$$

where

$$\alpha_1 = \gamma_1\gamma_4 + \gamma_2\gamma_3$$
$$\alpha_2 = \gamma_1\gamma_3 + \gamma_2\gamma_4$$
$$k = \sqrt{\gamma_1^2 - \gamma_2^2}$$
$$u_1 = (1 - k\mu_0)(\alpha_2 + k\gamma_3)\exp(k\tau_t) \tag{2.112}$$
$$u_2 = (1 + k\mu_0)(\alpha_2 - k\gamma_3)\exp(-k\tau_t)$$
$$u_3 = (1 + k\mu_0)(\alpha_1 + k\gamma_4)\exp(k\tau_t)$$
$$u_4 = (1 - k\mu_0)(\alpha_1 - k\gamma_4)\exp(-k\tau_t)$$

Derivation of these formulas was based on zero surface reflectance as the lower boundary condition. Liang and Strahler (1994a) developed specific formulas for the hybrid method whose parameters are listed in Table 2.9 for a reflective surface.

For most aerosol particles, the calculated phase function has a very sharp peak in the forward direction (see Fig. 2.7) and usually requires several hundred terms in a Legendre polynomial expansion [see Eqs. (2.93) and (2.103)]. A scaling approximation is a procedure that assumes that photons scattered within the peak are not scattered at all, and thus the accuracy is not significantly reduced if we scale the single scattering albedo and optical depth

$$\tilde{\omega} = \frac{(1 - f)\omega}{1 - \omega f} \tag{2.113}$$

$$\tilde{\tau} = (1 - \omega f)\tau \tag{2.114}$$

where f is the fraction of the phase function within the forward peak. The value of f is often determined arbitrarily, but $f = g$ is often used where g is the asymmetry parameter of the phase function. The approximation with such a scaling is usually called a δ-approximation.

2.4.4 Representative Radiative Transfer Solvers (Software Packages)

We have just discussed the general principles of two numerical methods and one approximate method for solving radiative transfer equations. Under most circumstances, we do not need to code these algorithms as a user since many radiative transfer solvers (software packages) are available in the public domain. Table 2.10 lists some of the them. Although the www or ftp addresses are given, keep in mind that they may be changed from time to time. No attempt was made to conduct the detailed analysis and comparison of these codes, although some general comments are provided in the table.

2.5 APPROXIMATE REPRESENTATION FOR INCORPORATING SURFACE BRDF

For atmospheric correction and other applications, it is very useful to apply an analytic formula that deals with the interaction between the atmosphere and surface. For a Lambertian surface (isotropic reflectance), the classic formula has been available. However, since most surfaces reflect anisotropically, there are no analytic formulas available. Instead, approximations have to be made. We first discuss the formulation for Lambertian surface, and then discuss various approximations for surfaces characterized by BRDF.

If a land surface is Lambertian with reflectance r_s, then the upwelling radiance $I(\tau, \mu, \phi)$ at the top of the atmosphere and the downward radiance $I(0, -\mu, \phi)$ at the bottom of the atmosphere (sky radiance) can be expressed as

$$I(\tau_t, \mu, \phi) = I_0(\tau_t, \mu, \phi) + \frac{r_s}{1 - r_s \bar{\rho}} \mu_0 E_s \gamma(-\mu_0) \gamma(\mu) \qquad (2.115)$$

$$I(0, -\mu, \phi) = I_0(0, -\mu, \phi) + \frac{r_s}{1 - r_s \bar{\rho}} \mu_0 E_s \gamma(-\mu_0) \gamma(-\mu) \qquad (2.116)$$

where $\bar{\rho}$ is the spherical albedo of the atmosphere; $I_0(\tau_t, \mu, \phi)$ is usually called *path radiance* in remote sensing, which is the upwelling radiance at the top of the atmosphere with a zero-reflectance surface (i.e., black surface); $I_0(0, -\mu, \phi)$ is downward sky radiance over a black surface; and $\gamma(-\mu_0)$ and $\gamma(\mu)$ are the total transmittances in the path from the Sun to the target and in the path from the target to the sensor. The total transmittance is the sum of the direct transmittance $t_i(\mu)$ and the diffuse transmittance $t_d(\mu)$

$$\gamma(\mu) = t_i(\mu) + t_d(\mu) \qquad (2.117)$$

TABLE 2.10 Typical Radiative Transfer Solvers

Name	Author	Type	Description
STREAMER	Key (1998)	Discrete ordinate (atmosphere)	STREAMER is a radiative transfer model that can be used for computing either radiances (intensities) or irradiances (fluxes) for a wide variety of atmospheric and surface conditions (available from `http://stratus.ssec.wisc.edu/streamer/streamer.html`)
HYDROLIGHT	Mobley (1994)	Invariant embedding (ocean)	HYDROLIGHT is a radiative transfer numerical model that computes radiance distributions and derived quantities for natural water bodies, it is based on the invariant imbedding numerical algorithm (available from `http://www.sequoiasci.com/publications2.asp?a_id=9`)
DISORT	Stamnes et al. (1988)	Discrete ordinate	DISORT is one of the most heavily tested radiative transfer code available for N-stream plane-parallel atmospheric cases (available from `ftp://climate.gsfc.nasa.gov/pub/wiscombe/Dis_Ord`) It is also included in the attached CD-ROM
SBDART	Ricchiazzi et al. (1998)	Revised DISORT	SBDART (Santa Barbara DISORT atmospheric radiative transfer) is a FORTRAN computer code designed for the analysis of a wide variety of radiative transfer problems encountered in satellite remote sensing and atmospheric energy budget studies (available from `http://www.icess.ucsb.edu/esrg/pauls_dir/`)
Clirad_sw, clirad_lw	Chou (1990)	Two-stream	Radiative transfer code used in global circulation models and mesoscale models (developed at NASA Goddard; available from `http://climate.gsfc.nasa.gov/~chou/clirad_sw/` or `http://climate.gsfc.nasa.gov/~chou/clirad_lw/`)
FEMRAD	Kisselev et al. (1994)	Finite element	FEMRAD is a FORTRAN code based on the finite element method (FEM) for the solution of the radiative transfer equation (available from `http://atol.ucsd.edu/~`)

Name	Reference	Method	Description
SHDOM	Evans (1998)	Discrete ordinate	This program computes unpolarized monochromatic or spectral band radiative transfer in a one-, two-, or three-dimensional medium for either collimated solar and/or thermal emission sources of radiation (available from `http://nit.colorado.edu/~evans/shdom.html`)
PolRadTran	Evans and Stephens (1991)	Discrete ordinate	PolRadTran is a plane-parallel fully polarized atmospheric radiative transfer model (available from `http://nit.colorado.edu/~evans/polrad.html`)
mc-layer	Macke (1999)	Monte Carlo	Monte Carlo radiative transfer code for multiple scattering in vertically inhomogeneous atmospheres (available from `http://www.ifm.unikiel.de/me/research/Projekte/RemSens/SourceCodes/source.html`)
MCML	Wang and Jacques (1995)	Monte Carlo	MCML is a steady-state Monte Carlo simulation program for multilayered turbid media with an infinitely narrow photon beam as the light source; each layer has its own optical properties of absorption, scattering, anisotropy, and refractive index (available from `http://ee.ogi.edu/omlc/science/software/mc/index.htm`)
DOM VDOM	Haferman (1993)	Discrete ordinate	Three-dimensional (3D) discrete ordinates method (available from `ftp://iihr.uiowa.edu/pub/hml/haferman`
libRadtran	Kylling and Mayer	Discrete ordinate	libRadtran is a collection of C and FORTRAN functions and programs for calculation of solar and thermal radiation in Earth's atmosphere (available from `http://www.libradtran.org`)
MODTRAN	Anderson et al. (2001)	Discrete ordinate	This is the one of the most widely used radiative transfer simulation codes for remote sensing and other applications; many examples in this book were based on MODTRAN simulations UNIX version: `http://www.vsbm.plh.af.mil/soft/modtran4.html` PC version: `www.ontar.com` Parallel machine: `www.hpc.jpl.nasa.gov/PEP/wangp/ParaModtra`
6S	Vermote et al. (1997)	Successive orders of scattering	6S (second simulation of satellite signasl in the solar spectrum) has been widely used in optical remote sensing; many sensor spectral response functions are incorporated; covers wavelength 0.4–2.5 μm. (available at `ftp://loaser.univ-lille1.fr/` or `ftp://kratmos.gsfc.nasa.gov/6S/`)

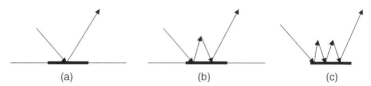

Figure 2.14 Multiple interactions of photons between the atmosphere and land surface.

where the direct transmittance can be calculated by the Beer's law:

$$t_i(\mu) = e^{(-\tau_i/\mu)} \tag{2.118}$$

Without a loss of generality, we will consider only the upwelling radiance at the top of the atmosphere from now on. If we perform a Taylor expansion, Eq. (2.115) becomes

$$I(\tau_t, \mu, \phi) = I_0(\tau_t, \mu, \phi) + \mu_0 E_s \gamma(-\mu_0)$$

$$\times \left[r_s + r_s \bar{\rho} r_s + r_s (\bar{\rho} r_s)^2 + r_s (\bar{\rho} r_s)^3 + \cdots \right] \gamma(\mu) \tag{2.119}$$

It is obvious that the terms in square brackets on the right side of this equation have simple physical interpretations (see Fig. 2.14), with r_s indicating photons that hit the surface once (a), $r_s \bar{\rho} r_s$ indicating photons that hit the surface twice and are reflected by the atmosphere once (b), $r_s(\bar{\rho} r_s)^2$ indicating photons that hit the surface for 3 times and are reflected by the atmosphere twice (c). The value of the term $r_s(\bar{\rho} r_s)^n$ will be very small when n becomes larger, particularly for clear atmosphere (small $\bar{\rho}$) and dark surfaces (small r_s).

If we normalize Eq. (2.119) to reflectance and rearrange the terms in the square brackets, it becomes

$$\rho(\tau_t, \mu, \phi) = \rho_0(\tau_t, \mu, \phi) + \gamma(-\mu_0) \left[r_s + \frac{r_s \bar{\rho} r_s}{1 - r_s \bar{\rho}} \right] \gamma(\mu) \tag{2.120}$$

Thus, the first term in the brackets represents the single interaction between the atmosphere and surface, and the second term represents the multiple interactions.

If we further separate total transmittance into direct and diffuse components, Eq. (2.119) becomes

$$\rho(\tau_t, \mu, \phi) = \rho_0(\tau_t, \mu, \phi) + \left[\exp\left(-\frac{\tau_t}{\mu_0}\right) + t_d(-\mu_0) \right]$$

$$\times r_s \left[\exp\left(-\frac{\tau_t}{\mu}\right) + t_d(\mu) \right] + \gamma(-\mu_0) \frac{r_s \bar{\rho} r_s}{1 - r_s \bar{\rho}} \gamma(\mu) \tag{2.121}$$

Now let us consider a non-Lambertian surface that is characterized by a bidirectional reflectance factor $R(-\mu_0, \phi_0, \mu, \phi)$. Since surfaces reflect the solar radiation dependent on both the solar illumination direction and the sensor viewing direction, we have to consider the bidirectional transmittance of the atmosphere $T(\mu', \phi', \mu, \phi)$. From Eq. (2.121),

$$\rho(\tau_t, \mu, \phi) = \rho_0(\tau_t, \mu, \phi) + \rho_1 + \rho_2 + \rho_3 + \rho_4 \qquad (2.122)$$

where

$$\rho_1 = \exp\left(-\frac{\tau_t}{\mu_0}\right) R(-\mu_0, \phi_0, \mu, \phi) \exp\left(-\frac{\tau_t}{\mu}\right) \qquad (2.123)$$

$$\rho_2 = \exp\left(-\frac{\tau_t}{\mu_0}\right) \cdot \frac{2}{\pi} \int_0^{2\pi} \int_0^1 R(-\mu_0, \phi_0, \mu', \phi') T(\mu', \phi', \mu, \phi) \mu \, d\mu' \, d\phi' \qquad (2.124)$$

$$\rho_3 = \left[\frac{2}{\pi} \int_0^{2\pi} \int_0^1 T(-\mu_0, \phi_0, -\mu', \phi') R(-\mu', \phi', \mu, \phi) \mu \, d\mu' \, d\phi'\right] \exp\left(-\frac{\tau_t}{\mu}\right) \qquad (2.125)$$

$$\rho_4 = \frac{2}{\pi} \int_0^{2\pi} \int_0^1 \left[\frac{2}{\pi} \int_0^{2\pi} \int_0^1 T(-\mu_0, \phi_0, -\mu', \phi') R(-\mu', \phi', \mu'', \phi'') \mu' \, d\mu' \, d\phi'\right]$$
$$\times T(\mu'', \phi'', \mu', \phi') \mu'' \, d\mu'' \, d\phi'' \qquad (2.126)$$

where ρ_i may be interpreted in a simple way (see Fig. 2.15); ρ_0 is the pure contribution from the atmosphere (normalized from the path radiance), ρ_1 is the direct solar illumination component reflected by the surface and then directly transmitted to the sensor, ρ_2 is the direct illumination component reflected by the surface and then diffusely transmitted to the sensor, ρ_3 is the diffuse illumination component reflected by the surface and then directly transmitted to the sensor, and ρ_4 is the diffuse illumination component reflected by the surface and then diffusely transmitted to the sensor.

Equation (2.122) is accurate, but calculating ρ_4 is computationally expensive since it involves double integrations over both the solar illumination direction and the viewing direction. If we compare Eq. (2.121) with Eq. (2.122) item by item and approximate ρ_4 using the corresponding Lambertian term, we obtain

$$\rho_4 = t_d(-\mu_0) r_s t_d(\mu) + \gamma(-\mu_0) \frac{r_s \bar{\rho} r_s}{1 - r_s \bar{\rho}} \gamma(\mu) \qquad (2.127)$$

In 6S (second simulation of satellite signals in the solar system) code, surface BRDF (Tanré et al. 1981, Vermote et al. 1997) is incorporated in the same

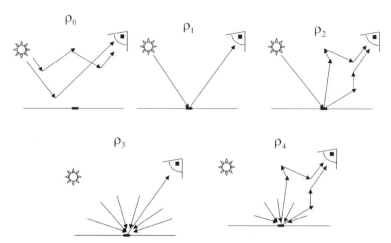

Figure 2.15 Approximate formulation of surface bidirectional reflectance distribution function (BRDF) in calculation of atmospheric radiative transfer.

way as in Eq. 2.122, but the expressions for ρ_2 and ρ_3 are different

$$\rho_2 = \exp\left(-\frac{\tau_t}{\mu_0}\right)\rho^*(-\mu_0, \phi_0, \mu, \phi)t_d(\mu)$$

$$\rho_3 = t_d(-\mu_0)\rho^*(-\mu_0, \phi_0, \mu, \phi)\exp\left(-\frac{\tau_t}{\mu}\right) \qquad (2.128)$$

where

$$\rho^*(-\mu_0, \phi_0, \mu, \phi) = \frac{1}{E_0^d}\int_0^{2\pi}\int_0^1 I_0^d(-\mu', \phi')R(-\mu', \phi', \mu, \phi)\mu'\,d\mu'\,d\phi'$$

$$(2.129)$$

where I_0^d is the sky radiance with zero surface reflectance and E_0^d is the solar diffuse downward irradiance at the zero-reflectance surface

$$E_0^d = \int_0^{2\pi}\int_0^1 I_0^d(-\mu', \phi')\mu'\,d\mu'\,d\phi' \qquad (2.130)$$

Also, the spherical albedo in ρ_4 is integrated from $\rho^*(-\mu_0, \phi_0, \mu, \phi)$.

Qiu (2001) found from extensive numerical calculations that the 6S formulation is reasonably accurate when $\mu_s \approx 1$. However, when the solar zenith angle or the viewing zenith angle increases, the accuracy of the 6S formulation decreases. Even when the surface is Lambertian, the 6S code underestimates the upwelling radiance when the aerosol optical depth is large. Qiu then proposed a correction to ρ_4 for improving its accuracy. Because of the lengthy formulas, no details are provided here.

Qin et al. (2001) suggested a new solution based on an alternative approach. The final formula for TOA bidirectional reflectance $\rho(\Omega_i, \Omega_v)$ is similar to that for the Lambertian surface:

$$\rho(\Omega_i, \Omega_v) = \rho_0(\Omega_i, \Omega_v)$$
$$+ \frac{T(\Omega_i) R(\Omega_i, \Omega_v) T(\Omega_v) - t_{dd}(\Omega_i) t_{dd}(\Omega_v) |R(\Omega_i, \Omega_v)| \bar{\rho}}{1 - r_{hh} \bar{\rho}}$$

$$(2.131)$$

where $\Omega_i \equiv (-\mu_i, \phi_i)$ is the solar incoming direction, $\Omega_v \equiv (\mu_v, \phi_v)$ for the viewing direction. Two groups of coefficients in Eq. (2.131) that are independent of each other: atmosphere-dependent and surface-dependent. These coefficients in each group represent the inherent properties of either atmosphere or surface. This means that we can determine these two groups of coefficients separately.

For the atmosphere, $\rho_0(\Omega_i, \Omega_v)$ is the atmospheric reflectance associated with path radiance (zero surface reflectance), $\bar{\rho}$ is the atmospheric spherical albedo as defined before. The transmittance matrices are defined as

$$T(\Omega_i) = \begin{bmatrix} t_{dd}(\Omega_i) & t_{dh}(\Omega_i) \end{bmatrix} \qquad (2.132)$$

$$T(\Omega_v) = \begin{bmatrix} t_{dd}(\Omega_v) & t_{hd}(\Omega_v) \end{bmatrix}^T \qquad (2.133)$$

where the subscript T denotes transpose and each transmittance has two subscript symbols; d (directional) and h (hemispherical).

t_{dd} is the direct transmittance has the simple analytic expression $t_{dd}(\mu) = \exp(-\tau_t/\mu)$,

t_{dh} is the directional hemispheric transmittance. It defines the fraction of downward diffuse flux generated by atmospheric scattering as the direct beam passes through the atmosphere and can be expressed. It can be calculated as the ratio of the integrated sky radiance at the surface level $L^{\downarrow}(\Omega_i, \Omega_v)$ over the downward hemisphere to the TOA incoming solar radiation:

$$t_{dh}(\Omega_i) = \frac{\int_{2\pi^-} L^{\downarrow}(\Omega_i, \Omega_v) \mu_v \, d\Omega_v}{\mu_i E_0} \qquad (2.134)$$

t_{hd} is the hemispheric directional transmittance defined as the ratio of the integrated upwelling TOA radiance over the upper hemisphere to the upwelling flux at the surface level $F^{\uparrow}$:

$$t_{hd}(\Omega_v) = \frac{\int_{2\pi^+} L^{\uparrow}(\Omega_i, \Omega_v)\mu_i \, d\Omega_i}{F^{\uparrow}} \tag{2.135}$$

Both t_{dh} and t_{hd} have to be calculated numerically. A practical solution is to create tables in advance.

For the surface, the reflectance matrix is defined as

$$R(\Omega_i, \Omega_v) = \begin{bmatrix} r_{dd}(\Omega_i, \Omega_v) & r_{dh}(\Omega_i) \\ r_{hd}(\Omega_v) & r_{hh} \end{bmatrix} \tag{2.136}$$

where $r_{dd}(\Omega_i, \Omega_v)$ is bidirectional reflectance and is actually equivalent to the bidirectional reflectance factor defined by Eq. (1.9); $r_{dh}(\Omega_i)$ is the directional hemispherical reflectance defined as in (1.10); $r_{hd}(\Omega_v)$ is the hemispherical directional reflectance, which can be defined as

$$r_{hd}(\Omega_v) = \frac{1}{\pi} \int_{2\pi^-} r_{dd}(\Omega_i, \Omega_v) \, d\Omega_i \tag{2.137}$$

where the integration is over the lower hemisphere, r_{hh} is the bihemispherical reflectance, as defined by (1.11). The determinant $|R|$ can be easily calculated as

$$|R(\Omega_i, \Omega_v)| = r_{dd}(\Omega_i, \Omega_v)r_{hh} - r_{dh}(\Omega_i)r_{hd}(\Omega_v) \tag{2.138}$$

It is evident that as long as surface BRDF or BRF is known, the surface reflectance matrix can be determined. The authors claim that this approach does not introduce any approximation into the formulation. Their numerical experiments demonstrated that this formulation is very accurate.

2.6 SUMMARY

The atmosphere modulates surface signals twice. It affects the distribution of the incoming solar radiation at surface, which is also related to the surface reflectance. The solar radiation reflected by the surface is further scattered and absorbed by the atmosphere before reaching the sensor. Radiative transfer theory has been recognized as the principal modeling method that accounts for the solar radiation in the atmosphere. The radiative transfer formulation and the linkage with atmospheric optical properties were first introduced in this chapter.

Both numerical and approximate solutions to radiative transfer equations were presented. In the numerical solution category, two representative numerical methods (i.e., successive order of scattering and discrete ordinates) were discussed. More methods (e.g., doubling, invariant embedding) can be found in other books on radiative transfer (e.g., Lenoble 1985). In the approximate solution category, most two-stream algorithms were discussed in a unified format. The Gauss–Seidel algorithm, another numerical method, and several other approximate solutions (e.g., four-stream approximation, asymptotic fitting theory) that can also be applied to solve atmospheric radiative transfer equations will be discussed in Chapter 3 on solving the canopy radiative transfer equation. Few readers will need to code these algorithms since there are many different radiative transfer solvers available in the public domain; these were presented in Section 2.3.3.

One important difference between this chapter and the similar chapters in other books is the extensive discussions of the interactions between atmosphere and land surface (e.g., Sections 2.2 and 2.5). Although many statistical BRDF models have been introduced in Section 2.2, we did not intend to discuss every published statistical model. Readers are referred to some of these papers for details (e.g., Goel and Reynolds 1989, Ba et al. 1995, Liang and Townshend 1997).

Methods described in this chapter will also be valuable to help understand the atmospheric correction algorithms in Chapter 6, and the atmosphere–surface interactions (e.g, broadband albedos) in Chapter 9, for possible application in solving canopy and soil/snow radiative transfer problems (Chapters 3 and 4).

REFERENCES

Anderson, G. P., Berk, A., Acharya, P. K., Matthew, M. W., Bernstein, L. S., Chetwynd, J. H., Dothe, H., Adler-Golden, S. M., Ratkowski, A. J., Felde, G. W., Gardner, J. A., Hoke, M. L., Richtsmeier, S. C., and Jeong, L. S. (2001), MODTRAN4, version 2: Radiative transfer modeling, *Proc. SPIE*, 455–459.

Ba, M. B., Deschamps, P. Y., and Frouin, R. (1995), Error reduction in NOAA satellite monitoring of the land surface vegetation during FIFE, *J. Geophys. Res.* **100**: 25537–25548.

Barnsley, M. J., Lewis, P., Sutherland, M., and Muller, J.-P. (1997), Estimating land surface albedo in the Hapex-Sahel southern super-site: Inversion of two BRDF models against multiple angle ASAS images, *J. Hydrol.*, 188–189; 749–778.

Bohren, C. F. and Huffman, D. R. (1983), *Absorption and Scattering of Light by Small Particles*, Wiley.

Chandrasekhar, S. (1960), *Radiative Transfer*, Dover Publications.

Chou, M. D. (1990), Parametrization for the absorption of solar radiation by O_2 and CO_2 with application to climate studies, *J. Climate* **3**: 209–217.

d'Almeida, G. A., Koepke, P., and Shettle, E. P. (1991), *Atmospheric Aerosols: Global Climatology and Radiative Characteristics*, A. Deepak Publishing.

Diner, D. J., Martonchik, J. V., Borel, C., Gerstl, S. A. W., Gordon, H. R., Knyazikhin, Y., Myneni, R., Pinty, B., and Verstraete, M. M. (1999), *Multi-Angle Imaging Spectro-Radiometer (MISR) Level 2 Surface Retrieval (MISR-10) Algorithm Theoretical Basis*, JPL D-11401, Rev. D.

Evans, K. F. (1998), The spherical harmonics discrete ordinate method for three-dimensional atmospheric radiative transfer, *J. Atmos. Sci.* **55**: 429–464.

Evans, K. F. and Stephens, G. L. (1991), A new polarized atmospheric radiative transfer model, *J. Quant. Spectrosc. Radiat Transfer* **46**: 413–423.

Fung, A. K., (1994), *Microwave Scattering and Emission Models and Their Applications*, Artech House.

Gao, B.-C., Heidebrecht, K. B., and Goetz, A. F. H. (1993), Derivation of scaled surface reflectance from AVIRIS data, *Remote Sens. Envir.* **44**: 165–178.

Goel, N. S. and Reynolds, N. (1989), Bi-directional canopy reflectance and its relationship to vegetation characteristics, *Int. J. Remote Sens.* **10**: 107–132.

Haferman, J. L., Krajewski, W. F., Smith, T. F., and Sanchez, A. (1993), Radiative transfer for a three-dimensional raining cloud, *Appl. Opt.* **32**: 2795–2802.

Hess, M., Koepke, P., and Schult, I. (1998), Optical properties of aerosols and clouds: The software package OPAC, *Bull. Am. Meteorol. Soc.* **79**: 831–844.

Higurashi, A. and Nakajima, T. (1999), Development of a two-channel aerosol retrieval algorithm on a global scale using NOAA AVHRR, *J. Atmos. Sci.* **56**: 924–941.

Holben, B. and Eck, T. F. (1990), Precipitable water in the Sahel measured using sun photometry, *Agric. Forest Meteorol.* **52**: 95–107.

Kaufman, Y. J. and Joseph, J. H. (1982), Determination of surface albedoes and aerosol extinction characteristics from satellite imagery, *J. Geophys. Res.* **20**: 1287–1299.

Kaufman, Y. J., Tanre, D., Gordon, H. R., Nakajima, T., Lenoble, J., Frouin, R., Grassl, H., Herman, B. M., King, M. D., and Teillet, P. M. (1997), Passive remote sensing of tropospheric aerosol and atmospheric correction for the aerosol effect, *J. Geophys. Res.* **102**: 16815–16830.

Key, J. and Schweiger, A. J. (1998), Tools for atmospheric radiative transfer: Streamer and fluxnet, *Computers and Geosciences* **24**: 443–451.

King, M. D. and Harshvardhan, (1986), Comparative accuracy of selected multiple-scattering approximations, *J. Atmos. Sci.* **43**: 784–801.

Kisselev, V. B., Roberti, L., and Perona, G. (1994), An application of the finite-element method to the solution of the radiative transfer equation, *J. Quant Spectrosc. Radiat. Transfer* **51**: 603–614.

Kriebel, K. T. (1996), On the limited validity of reciprocity in measured BRDFs, *Remote Sens. Envir.* **58**: 52–62.

Lafrance, B., Hagolle, O., Bonnel, B., Fouquart, Y., and Brogneiz, G. (2002), Interband calibration over clouds for POLDER space sensor, *IEEE Trans. Geosci. Remote Sens.* **40**: 131–142.

Lenoble, J. (1985), *Radiative Transfer in Scattering and Absorbing Atmospheres*: *Standard Computational Procedures*, A. Deepak Publishing.

Leroy, M. (2001), Deviation from reciprocity in bidirectional reflectance, *J. Geophys. Res.* **106**: 11917–11923.

Lewis, P., Disney, M. I., Barnsley, M. J., and Muller, J.-P. (1999), Deriving albedo maps for Hapex-Sahel from ASAS data using kernel-driven brdf models, *Hydrol. Earth Syst. Sci.* **3**: 1–13.

Li, X. and Strahler, A. H. (1992), Geometric-optical bidirectional reflectance modeling of the discrete crown vegetation canopy: Effect of crown shape and mutual shadowing, *IEEE Trans. Geosci. Remote Sens.* **30**: 276–292.

Li, X. W., Wang, J. D., and Strahler, A. H., (1999), Apparent reciprocity failure in directional reflectance of structured surfaces, *Progress Natural Sci.* **9**: 747–752.

Liang, S. and Strahler, A. H. (1993), An analytic BRDF model of canopy radiative transfer and its inversion, *IEEE Trans. Geosci. Remote Sens.* **31**: 1081–1092.

Liang, S. and Strahler, A. H. (1994a), Retrieval of surface BRDF from multiangle remotely sensed data, *Remote Sens. Envir* **50**: 18–30.

Liang, S. and Strahler, A. H. (1994b), A four-stream solution for atmospheric radiative transfer over an non-lambertian surface, *Appl. Opt.* **33**: 5745–5753.

Liang, S. and Townshend, J. R. G. (1997), Angular signatures of NASA/NOAA pathfinder AVHRR land data and applications to land cover identification, *Proc. IGRASS* Singapore, pp. 1781–1783.

Liou, K. N. (1974), Analytic two-stream and four-stream solutions for radiative transfer. *J. Atmos. Sci.* **31**: 1473–1475.

Liou, K. N. (1980), *An Introduction to Atmospheric Radiation*, Academic Press.

Lumme, K. and Rahola, J. (1994), Light scattering by porous dust particles in the discrete-dipole approximation, *Astrophys. J.* **425**: 653–667.

Macke, A., Mitchell, D. L., and Bremen, L. V. (1999), Monte Carlo radiative transfer calculations for inhomogeneous mixed phase clouds, *Physics and Chemistry of the Earth. B: Hydrology, Oceans and Atmosphere* **24**: 237–241.

Martonich, J. V., Diner, D. J., Kahn, R. A., Ackerman, T. P., Verstraete, M. E., Pinty, B., and Gordon, H. R. (1998), Techniques for the retrieval of aerosol properties over land and ocean using multiangle imaging, *IEEE Trans. Geosci. Remote Sens.* **36**: 1212–1227.

Meador, W. E. and Weaver, W. R. (1980), Two-stream approximations to radiative transfer in planetary atmospheres: A unified description of existing methods and a new improvement, *J. Atmos. Sci.* **37**: 630–643.

Mishchenko, M. I. (1993), Light scattering by size-shape distributions of randomly oriented axially symmetric particles of a size comparable to a wavelength, *Appl. Opt.* **32**: 4652–4666.

Mobley, C. D. (1994), *Light and Water: Radiative Transfer in Natural Waters*, Academic Press, 592 pp.

Nilson, T. and Kuusk, A. (1989), A reflectance model for the homogeneous plant canopy and its inversion, *Remote Sens. Envir.* **27**: 157–167.

Press, W. H., Flannery, B. P., Teukolsky, S. A., and Vetterling, W. T. (1989), *Numerical Recipes. The Art of Scientific Computing* (*Fortran Version*), Cambridge Univ. Press.

Qin, W., Herman, J. R., and Ahman, Z. (2001), A fast, accurate algorithm to account for non-lambertian surface effects on TOA radiance, *J. Geophys. Res.* **106**: 22671–22684.

Qiu, J. (2001), An improved model of surface BRDF-atmospheric coupled radiation, *IEEE Trans. Geosci. Remote Sens.* **39**: 181–187.

Qu, Z., Kindel, B. C., and Goetz, A. F. H. (2003), The high accuracy atmospheric correction for hyperspectral data (HATCH) model, *IEEE Trans. Geosci. Remote Sens.* **41**: 1223–1231.

Rahman, H., Pinty, B., and Verstraete, M. M. (1993), Coupled surface-atmosphere reflectance (CSAR) model, 2, semiempirical surface model usable with NOAA advanced very high resolution radiometer data, *J. Geophys. Res.* **98**: 20791–20801.

Ricchiazzi, P., Yang, S., Gautier, C., and Sowle, D. (1998), SBDART: A research and teaching software tool for plane-parallel radiative transfer in the Earth's atmosphere, *Bull. Amer. Meteor. Soc.* **79**: 2101–2114.

Roujean, J. L., Leroy, M., and Deschamps, P. Y. (1992), A bidirectional reflectance model of the earth's surface for the correction of remote sensing data *J. Geophys. Res.* **97**: 20455–20468.

Russell, P. B., Livingston, J. M., Dutton, E. G., Pueschel, R. F., Reagan, J. A., DeFoor, T. E., Box, M. A., Allen, D., Pilewskie, P., Herman, B. M., Kinne, S. A., and Hoffman, D. J. (1993), Pinatubo and pre-Pinatubo optical depth spectra: Mauna Loa measurements, compositions, inferred particle size distributions, radiative effects, and relationship to lidar data, *J. Geophys. Res.* **98**: 22969–22985.

Snyder, W. C. (2002), Definition and invariance properties of structured surface BRDF, *IEEE Trans. Geosci. Remote Sens.* **40**: 1032–1037.

Stamnes, K., Tsay, S. C., Wiscombe, W., and Jayaweera, K. (1988), Numerically stable algorithm for discrete-ordinate-method radiative transfer in multiple scattering and emitting layered media, *Appl. Opt.* **27**: 2502–2509.

Staylor, W. F. and Suttles, J. T. (1986), Reflection and emission models for deserts derived from Nimbus-7 ERB scanner measurements, *J. Climate Appl. Meteorol.* **25**: 196–202.

Sykes, J. B. (1951), Approximate integration of the equation of transfer, *Monthly Notices Royal Astron. Soc.* **111**: 377–386.

Takano, Y. and Liou, K. N. (1989a), Solar radiative transfer in cirrus clouds. Part I: Single-scattering and optical properties of hexagonal ice crystals, *J. Atmos. Sci.* **46**: 3–19.

Takano, Y. and Liou, K. N. (1989b), Solar radiative transfer in cirrus clouds. Part II: Theory and computation of multiple scattering in an anisotropic medium, *J. Atmos. Sci.* **46**: 20–36.

Takano, Y. and Liou, K. N. (1995), Solar radiative transfer in cirrus clouds. Part III: Light scattering by irregular ice crystals, *J. Atmos. Sci.* **52**: 818–837.

Tanré, D., Herman, M., and Deschamps, P. Y. (1981), Influence of the background contribution upon space measurements of ground reflectance, *Appl. Opt.* **20**: 3673–3684.

Tanré, D., Holben, B. N., and Kaufman, Y. J. (1992), Atmospheric correction algorithm for NOAA-AVHRR products: Theory and application, *IEEE Trans. Geosci. Remote Sens.* **30**: 231–248.

Thomas, G. E. and Stamnes, K. (1999), *Radiative Transfer in the Atmosphere and Ocean*, Cambridge Atmospheric and Space Science Series, Cambridge Univ. Press.

van de Hulst, H. C. (1981), *Light Scattering by Small Particles*, Dover Publications.

Vermote, E., Tanré, D., Deuze, J. L., Herman, M., and Morcette, J. J. (1997), Second simulation of the satellite signal in the solar spectrum: An overview, *IEEE Trans. Geosci. Remote Sens.* **35**: 675–686.

Walthall, C. L., Norman, J. M., Welles, J. M., Campbell, G., and Blad, B. L. (1985), Simple equation to approximate the bi-directional reflectance from vegetation canopies and bare soil surfaces, *Appl. Opt.* **24**: 383–387.

Wang, L., Jacques, S. L., and Zheng, L. (1995), MCML—Monte Carlo modeling of light transport in multi-layered tissues, *Computer Methods and Programs in Biomedicine* **47**: 131–146.

Wanner, W., Li, X., and Strahler, A. (1995), On the derivation of kernels for kernel-driven models of bidirectional reflectance, *J. Geophys. Res.* **100**: 21077–21090.

Whitby, K. Y. (1978), The physical characteristics of sulfur aerosols, *Atmos. Envir.* **12**: 135–159.

Young, A. T. (1980), Revised depolarization corrections for atmospheric extinction, *Appl. Opt.* **19**: 3427–3428.

3

Canopy Reflectance Modeling

Remotely sensed vegetation properties, such as albedo, leaf area index (LAI), fractional photosynthetically active radiation absorbed by canopy (FPAR), surface roughness, and phenology have been considered critical inputs to land surface process models. They have also gained importance in driving biochemical models as well as many other models related to biospheric functions. The mathematical and physical sophistication of the techniques used to estimate these parameters have increased considerably. They have mainly evolved from simple empirical approaches to more physically based approaches that are greatly rooted in our understanding of the radiation regime of the vegetation canopies.

This chapter introduces various physically based canopy reflectance models that link canopy properties with sensor-measured radiance, including radiative transfer models, geometric optical models, and computer simulation models. Because of their extensive applications to all media of atmosphere, canopy, soil, and snow, radiative transfer modeling methods will be discussed in more detail. These models will provide the theoretical foundations for developing practical algorithms to estimate biophysical parameters, discussed in Chapter 8.

Section 3.1 discusses canopy radiative transfer formulations that are typically suitable for dense vegetation canopies. The emphasis is on one-dimensional, plane-parallel formulation, but three-dimensional cases and the approximate methods for handling horizontally heterogeneous canopies are outlined. Hotspot modeling is also briefly discussed.

Section 3.2 introduces various leaf optical models that provide necessary leaf optical properties to most canopy reflectance models. These leaf models can also be used for estimating leaf biochemistry through model inversion. Section 3.3 provides various numerical (i.e., Gauss–Seidel) and approximate solutions (i.e., four-stream, asymptotic fitting) to the canopy radiative transfer equation. These methods can be used for solving radiative transfer equations of other media, such as atmosphere, snow, soil, and water. Section 3.4 introduces the basic principles of geometric optical models that are typically suitable for sparse vegetation canopies. A detailed description of the geometric model developed by Li and Strahler (1985) is also given. Section 3.5 discusses two types of computer simulation models: Monte Carlo ray tracing models and radiosity models. They are typically suitable for very complex simulations of canopy radiation regimes and for validating the simplified canopy reflectance models.

3.1 CANOPY RADIATIVE TRANSFER FORMULATION

Section 3.1.1 will first introduce canopy configurations including leaf area index (LAI) and leaf angle distribution (LAD). The optical properties of the canopy elements are discussed in Section 3.1.2. In Section 3.1.3, we discuss one-dimensional (1D) canopy radiative transfer equations and boundary conditions. Hotspot modeling is discussed in Section 3.1.4. Section 3.1.5 presents extensions to heterogeneous canopies.

3.1.1 Canopy Configuration

The canopy radiation regime depends largely on canopy configuration. Different modeling techniques have represented canopy structures in variable degrees. In radiative transfer models, the vegetation canopy is assumed to be a turbid medium (e.g., atmosphere and water), and the canopy elements are randomly distributed as illustrated in Fig. 3.1. For a 1D case, canopy is horizontally homogeneous and infinite but vertically variable and finite. The medium below the canopy could be soil or another canopy layer (e.g., grass).

Figure 3.1 Illustration of the vegetation leaf canopy as an one-dimensional turbid medium. [From Pinty et al. (2001), *J. Geophys. Res*. Reproduced by permission of American Geophysical Union.]

For simplicity, we consider only the canopy field with leaves. Realistic canopies composed of a mixture of leaf and branch elements can be generalized by averaging effects over their proportionate distributions.

Canopy architecture is described at the leaf level through leaf area index (LAI) and leaf angle distribution (LAD). LAI (L) is defined as one-sided leaf area (m^2) per unit of ground area (m^2), which can be an integration of the leaf area density function $u_l(z)$ over the vertical extent of the canopy (H):

$$L = \int_0^H u_l(z)\, dz \tag{3.1}$$

There are many models of $u_l(z)$. Many canopies tend to have a higher leaf area density toward the top of the canopy. The simplest form assumes that density is constant over canopy height [i.e., $u_l(z) = L/H$].

A geometry function $G(\Omega)$ is usually defined to represent the mean projection of a unit foliage area in the direction Ω characterized by the zenith angle μ and the azimuth angle ϕ

$$G(\Omega) = \frac{1}{2\pi} \int_0^{2\pi} \int_0^1 g_l(\Omega_l) |\Omega_l \Omega| d\Omega_l \tag{3.2}$$

where $1/2\pi \cdot g_l(\Omega_l)$ is the probability density of the distribution of the leaf normals with respect to the upper hemisphere, which is often referred to as *leaf angle distribution* (LAD).

The zenith and azimuthal angles of the distribution of the leaf normals are usually assumed to be independent and the distribution in the azimuth is uniform: $g_l(\Omega_l) = g_l(\mu_l)$ and $\int_0^{\pi/2} g_l(\theta_l) \sin \theta_l\, d\theta_l = 1$. Different theoretical and experimental models for this function have been published, such as elliptical (Campbell 1984), beta, and triangular distributions (Bunnik 1978, Goel and Strebel 1984).

Let us define a new function for convenience: $g_l^*(\theta_l) = g_l(\theta_l) \sin \theta_l$. The triangular formula, based on the data of de Wit (1965), is in a general format

$$g_l^*(\theta_l) = a + b \cos 2\theta_l + c \cos 4\theta_l \tag{3.3}$$

where parameters a, b, c characterize different canopies:

For a planophile canopy, $a = b = 2/\pi$ and $c = 0$:

$$g_l^*(\theta_l) = \frac{2}{\pi(1 + \cos 2\theta_l)} \tag{3.4}$$

For an erectophile canopy, $a = 2/\pi$, $b = -(2/\pi)$, and $c = 0$:

$$g_l^*(\theta_l) = \frac{2}{\pi(1 - \cos 2\theta_l)} \tag{3.5}$$

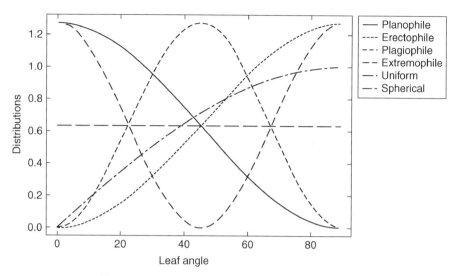

Figure 3.2   Illustration of a few typical LAD functions.

For a plagiophile canopy, $a = 2/\pi$, $b = 0$, and $c = (2/\pi)$:

$$g_i^*(\theta_l) = \frac{2}{\pi(1 - \cos 4\theta_l)} \qquad (3.6)$$

For an extremophile canopy, $a = 2/\pi$, $b = 0$, and $c = 2/\pi$:

$$g_i^*(\theta_l) = \frac{2}{\pi(1 + \cos 4\theta_l)} \qquad (3.7)$$

For a uniform canopy, $a = 2/\pi$, $b = 0$, and $c = 0$:

$$g_i^*(\theta_l) = \frac{2}{\pi} \qquad (3.8)$$

For a spherical canopy, $a = \sin \theta_L$, $b = 0$, and $c = 0$:

$$g_i^*(\theta_l) = \sin \theta_l \qquad (3.9)$$

It is easy to see from Fig. 3.2 that these functions represent some typical cases. For example, the planophile function corresponds to canopies with mainly horizontal leaves; the erectophile function, to canopies with mainly vertical leaves; and the plagiophile function, to leaves around 45°.

A more complicated function is characterized by a two-parameter beta distribution (Goel and Strebel 1984)

$$g_l(\theta_l) = \frac{\pi}{2}\frac{\Gamma(u+v)}{\Gamma(u)\Gamma(v)}\left(1-\frac{\theta_l}{90}\right)^{u-1}\left(\frac{\theta_l}{90}\right)^{v-1} \tag{3.10}$$

where Γ is the gamma function. The two parameters u and v are related to the average leaf (inclination) angle (ALA) and its second moment $\langle\theta_L^2\rangle$ by

$$\text{ALA} = \frac{90v}{u+v} \tag{3.11}$$

and

$$\langle\theta_l^2\rangle = \frac{90^2 v(v+1)}{(u+v)(u+v+1)} \tag{3.12}$$

By changing the values of both parameters (u and v), we can obtain various different distribution functions. This presentation is also extended to characterize the variations of both the zenith and azimuth (Strebel et al. 1985).

Another sophisticated model is based on the elliptical distribution function that has been used by Kuusk (1994, 1995a, 1995b):

$$g_l(\theta_l) = \frac{b}{\sqrt{1-\varepsilon^2\cos^2(\theta_l-\theta_m)}} \tag{3.13}$$

where b can be determined by the normalization condition

$$b = \varepsilon\Big/\left[\cos\theta_m\ln\left(\frac{\cos\eta+\sin\nu}{\cos\nu-\sin\eta}\right)-\sin\theta_m(\eta-\nu)\right] \tag{3.14}$$

where

$$\eta = \sin^{-1}(\varepsilon\cos\theta_m) \tag{3.15}$$

$$\nu = \sin^{-1}(\varepsilon\sin\theta_m) \tag{3.16}$$

ε and θ_m are parameters. ε characterizes the leaf orientation ($0\le\varepsilon\le1$) from $\varepsilon=0$ for the spherical to $\varepsilon=1$ for the fixed inclination angle, θ_m characterizes the actual inclination angle, $\theta_m=0°$ for a planophile canopy, and $\theta_m=90°$ for an erectophile canopy.

Besides LAI and LAD, another canopy feature that needs to be considered in modeling canopy radiative transfer is the leaf dimension, which will be mentioned in Section 3.1.3.

3.1.2   One-Dimensional Radiative Transfer Formulation

The classic radiative transfer equation is for a sparse medium with "point" particles; that is, the particle size is very small compared to the distance between them. The canopy medium is different since the leaf size is much bigger. The classic radiative transfer equation has to be modified to account for the finite dimensions of the canopy elements. We limit our discussion mainly to one-dimensional radiative transfer modeling. The modeling methods for handling heterogeneous vegetation canopies are outlined at the end of the chapter.

Before discussing the detailed radiative transfer formulations, we would like to mention a few early canopy models that may have impacted this field significantly. The first is the Suits (1972) model, in which canopy is assumed to consist of only vertical and horizontal leaves, and the model is parameterized with canopy structure and solar/viewing geometry. By extending the Suits model to allow for the variations of leaf angles, Verhoef (1984) developed the SAIL (light *s*cattering by *a*rbitrarily *i*nclined *l*eaves) canopy model. Gastellu-Etchegorry et al. (1996a) later improved the SAIL model by considering the anisotropic behavior of the soil background.

A classic textbook on canopy radiative transfer modeling was written by Ross (1981). There have also been several reviews on this topic (e.g., Goel 1988; Myneni et al. 1990a, 1990b; Qin and Liang 2000, Strahler 1997). An in-depth review of mathematical modeling development was compiled by Myneni and Ross (1991).

The one-dimensional radiative transfer equation of a flat homogeneous canopy for radiance I at the direction (Ω) is given by

$$-\mu \frac{\partial I(\tau,\Omega)}{\partial \tau} + h(\tau,\Omega)G(\Omega)I(\tau,\Omega) = \frac{1}{\pi}\int_0^{2\pi}\int_{-1}^1 \Gamma(\Omega',\Omega)I(\tau,\Omega')\,d\Omega'$$

$$(3.17)$$

where $h(\tau,\Omega)$ is an empirical correction function to account for the variation of the extinction coefficient and will be discussed in more detail later [see Eq. (3.31)]. The optical depth τ can be defined by the leaf area density $u_l(z)$ as $\tau(z) \equiv \int_0^z u_l(z)\,dz$; $G(\Omega)$ is defined in Eq. (3.2). It is easy to see that the basic form of the equation is very similar to Eq. (2.3) for the atmosphere.

In Eq. (3.17), the area scattering phase function $\Gamma(\Omega',\Omega)$, consisting of both diffuse and specular components, is defined as

$$\Gamma(\Omega',\Omega) = \Gamma_D(\Omega',\Omega) + \Gamma_{sp}(\Omega',\Omega)$$

$$(3.18)$$

We will find that $\Gamma(\Omega',\Omega)$ depends not only on the scattering angle between Ω' and Ω, but also on the absolute value of Ω' and Ω. This is different from the phase function in the atmospheric radiative transfer (Section 2.2), which is dependent only on the scattering angle. If we can assume that the diffuse scattering for the leaves follows the *bi-Lambertian scattering model* (Shultis

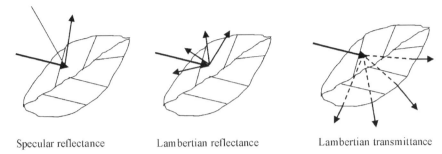

Specular reflectance   Lambertian reflectance   Lambertian transmittance

Figure 3.3 Illustration of bi-Lambertian leaf optical modeling

and Myneni 1988) as illustrated in Fig. 3.3, then

$$\Gamma_D(\Omega', \Omega) = \frac{1}{2\pi} \int_{\Omega^+} g_l(\Omega_l) t_l \alpha' \alpha \, d\Omega_l - \frac{1}{2\pi} \int_{\Omega^-} g_l(\Omega_l) r_l \alpha' \alpha \, d\Omega_l \quad (3.19)$$

Here Ω^- and Ω^+ indicate that the Ω_l integration is over that portion of the $0-2\pi$ range for which the integrand is either positive ($+$) or negative ($-$). In this model, a fraction r_l of the intercepted energy is radiated in a cosine distribution about the leaf normal (i.e., Lambertian reflectance). Similarly, a fraction t_l is transmitted in a Lambertian distribution on the opposite side of the leaf. It is obvious that Ω^+, Ω^-, is a part of the hemisphere for which $\pm \alpha' \alpha > 0$, $\alpha' = \Omega' \Omega_l$, $\alpha = \Omega \Omega_l$.

The area phase function of specular component $\Gamma_{sp}(\Omega', \Omega)$ can be evaluated as

$$\Gamma_{sp}(\Omega', \Omega) = \tfrac{1}{8} g_l(\Omega_l^*) K(\kappa, \Omega' \Omega_l^*) F(n, \Omega' \Omega_l^*) \quad (3.20)$$

where $\Omega_l^* = (\mu_l^*, \phi_l^*)$ defines the direction of the appropriate leaf normal for specular scattering between the incident and the reflected rays. It has been shown that (Card 1987)

$$\mu_l^* = \frac{\mu_i + \mu}{2\mu_\gamma} \quad (3.21)$$

$$\cos \phi_l^* = \frac{\sqrt{1 - \mu_i^2} + \sqrt{1 - \mu^2} \cos(\phi - \phi_i)}{2\mu_\gamma \sqrt{1 - \mu_l^{*2}}} \quad (3.22)$$

where

$$\mu_\gamma = \sqrt{\frac{\mu_i \mu + \sqrt{1 - \mu_i^2} \sqrt{1 - \mu^2} \cos(\phi - \phi_i) + 1}{2}} \quad (3.23)$$

The term $F(n, \alpha')$ is the Fresnel reflectance, indicating the amount of specularly reflected energy for incident unpolarized radiant:

$$F(n, \alpha') = \frac{1}{2} \left[\frac{\sin^2(j-i)}{\sin^2(j+i)} + \frac{\tan^2(j-i)}{\tan^2(j+i)} \right] \tag{3.24}$$

where $j = \cos^{-1}(|\alpha'|)$, $i = \sin^{-1}(\sqrt{1 - \alpha'^2}/n)$ and n is the wax refractive index of canopy leaves (typically 1.3–1.45). If the leaf is optically smooth and flat on a microscale, the Fresnel reflectance law is enough to compute the amounts of specularly reflected radiance, besides considering the number of leaves contributing to specular reflectance $g_l(\Omega_l^*)$. However, a leaf wax surface is rarely optically smooth, and the hair structure on the leaf surface reflects light diffusely, producing incompletely specular reflectance. Therefore, a smoothing factor K is defined to account for this reduction in the amount of specularly reflected light (Vanderbilt and Grant 1985). Nilson and Kuusk (1989) give a form of this factor:

$$K(\kappa, \alpha') = \exp\left[\frac{-2\kappa \tan(\alpha')}{\pi} \right] \tag{3.25}$$

where $\alpha' = \cos^{-1}(\Omega'\Omega_l)$. The argument $0 \le \kappa \le 1$ characterizes the dimension of the hair on the leaf surface.

The area scattering phase function defined above is required to meet the normalization condition (Marshak 1989)

$$\frac{1}{\pi} \int_{4\pi} \frac{\Gamma(\Omega', \Omega)}{\tilde{G}(\Omega')} d\Omega = 1 \tag{3.26}$$

where

$$\tilde{G}(\Omega') = \frac{1}{2\pi} \int_{2\pi+} g_l(\Omega_l) |\alpha'| \left[r_l + t_l + K(\kappa, \alpha')F(n, \alpha') \right] d\Omega_l \tag{3.27}$$

In summary, the interactions of photons with leaves within the canopy are in the following three ways: specular reflectance, Lambertian diffuse reflectance, and Lambertian diffuse transmittance, as shown in Fig. 3.3. The probability that each event occurs is controlled by the canopy configuration and its optical properties.

3.1.3 Boundary Conditions

To solve the canopy radiative transfer equation, the upper (at the bottom of the atmosphere) and lower boundary conditions need to be specified.

The upper boundary condition for the atmosphere is only about direct incoming solar radiation. For canopies, the upper boundary condition needs to be specified by both direct solar radiation i_0 and downward sky diffuse radiance distribution $I_d(-\Omega)$ at the bottom of the atmosphere:

$$I(0, -\Omega) = I_d(-\Omega) + i_0\delta(\mu - \mu_0)\delta(\phi - \phi_0) \qquad (3.28)$$

As we can see from the previous chapter and many other studies (e.g., Liang and Lewis 1996), the sky radiance distribution depends on the reflectance from the canopy. It is actually an iterative process. We cannot specify sky radiance distribution unless we know canopy reflectance; on the other hand, determination of canopy reflectance relies on sky radiance distribution. An ideal solution is to couple both canopy and atmosphere and solve the atmospheric radiative transfer equation and canopy radiative transfer equation simultaneously (Liang and Strahler 1993a). Fortunately, although the upwelling radiance from the canopy is very sensitive to downward sky radiance, the canopy reflectance is less sensitive. A reasonably good approximation of sky radiance distribution is sufficient for calculating canopy reflectance.

The lower boundary condition can be given as

$$I(\tau_t, \Omega) = \int_0^{2\pi}\int_{-1}^0 f_s(\Omega', \Omega)\mu'I(\tau_t, \Omega')\,d\Omega' \qquad (3.29)$$

where $I(\tau_t, \Omega)$ is the downward radiance at surface, and $f_s(\Omega', \Omega)$ is the directional reflectance distribution function (BRDF) of the background (e.g., soil) underneath the canopy. For a Lambertian surface with reflectance r_s this boundary condition can be written as

$$I(\tau_t, \Omega) = \frac{r_s}{\pi}\int_0^{2\pi}\int_{-1}^0 \mu'I(\tau_t, \Omega')\,d\Omega' \qquad (3.30)$$

The basic form of the lower boundary condition for the canopy is exactly the same as that for the atmosphere.

Note that we have changed the direction of the vertical axis to follow the convention in canopy modeling. The optical depth at the top of the canopy is set 0, and the maximum optical depth (e.g., LAI) is at the bottom of the canopy. This is the exact opposite of the atmosphere in Chapter 2.

3.1.4 Hotspot Effects

One of the important features of the directional reflectance patterns of many land surface types (e.g., canopy, soil, snow) is the *hotspot*, a reflectance peak around a viewing direction that is exactly opposite the solar illumination direction. An airphoto of the hotspot is shown in Fig. 3.4. This is the photo of the woodland in Australia, kindly provided by Dr. David Jupp. The shadow of the airplane can be seen from the photo that corresponds to the hotspot direction. The hotspot region appears very bright since no shadows can be seen. Other parts of the image have more shadows and appear darker. Aircraft was sufficiently low, and the ground IFOV of the imagery was about

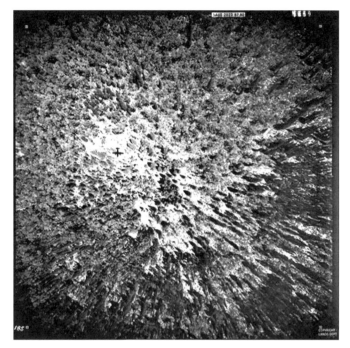

Figure 3.4   An airborne photo of a woodland area to demonstrate the hotspot effects.

5 m. Runs were made at right angles so that the sensor was scanned into and toward the solar principal plane. Figure 3.5 shows that angle-dependent surface reflectance averaged over 1000 scanlines from the previous photo. From the curve of the directional reflectance at the solar principal plane, we can tell the solar zenith angle was about 20° and the hotspot peak is remarkable. In Fig. 3.5(b), the reflectance is largest when the viewing direction was at the nadir because the background is very bright and the proportion of the bare soil at nadir is the largest.

In general, the conceptual explanation of hotspot is based on the shadow hiding theory. When the viewing direction is far away from the solar illumination direction, many shadows are visible in the field of view. When the viewing direction coincides with the solar viewing direction, no shadows can be seen and the observed radiance reaches the local maximum. Classic radiative transfer theory cannot explain this phenomenon since it assumes that all scatters are randomly spaced and infinitely small in relation to the scattering medium, such that no shadows are considered in the formulation. Real canopies show heterogeneity at a variety of scales, and foliage elements have finite size. Shadows cast by these elements are preferentially hidden when the viewing direction and the illumination direction coincide, leading to higher reflectance values. Thus, the classic radiative transfer equation has to be modified according to the canopy hotspot model. It is quite clear if

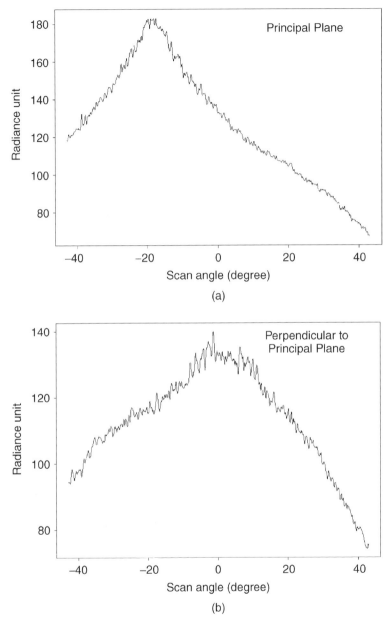

(a)

(b)

Figure 3.5 Reflectance plot as the function of the viewing zenith angle at both the principal plane (a) and the cross-principal plane (b), digitized from the airborne photo shown in Figure 3.4.

we compare the classic radiative transfer equations for the atmosphere [Eq. (2.3)] with that for the canopy [Eq. (3.17)].

The early canopy hotspot models were developed according to the assumption that scatterers are randomly distributed in space and therefore gap probability obeys the Poisson distribution (Kuusk 1985, Jupp and Strahler 1991). The common practice in these models is to estimate the correlation between gap probabilities in both illumination and observing directions when calculating the so-called bidirectional gap probability (BDGP). Through the modified BDGP, the hotspot effect is taken into account in computing the first-order scattering reflectance.

Kuusk (1991) defined a cross-correlation function (Y), and Jupp and Strahler (1991) used a so-called overlap function (S) to account for the correlation. Function Y was approximated with an exponential function of the mean leaf chord length, and function S was specified with a disk model (calculated with the overlap between two disks). However, they are conceptually identical and both can be exactly calculated as a function of the mean area of overlap between the projections of the same scatterers from the two (illumination and observation) directions, regardless of the size and shape of the scatterers (Qin and Xiang 1994). Note that in these two models, only the leaf geometric parameter (i.e., the mean diameter) rather than the gap size, appears explicitly as the driving factor. This is because it is implicitly assumed in these models that, in a statistical sense, the number of gaps is the same as that of leaves, and gap sizes are uniquely determined by the size and amount of leaves. Of course, this assumption is more appropriate for rather dense leaf canopies with elements randomly distributed in space than other cases.

Verstraete et al. (1990) proposed physically rigorous hotspot formalism by modifying the transmission of the outgoing radiation to consider the scatterer-free regions (holes) in the canopy. For a leaf canopy filling with both leaves and finite-depth holes (gaps), they defined two volumes within a hole: one along the illumination direction and the other along observing direction. The two volumes share the same base, which is equal to the circular sunflecks on the leaves. The hotspot effect is quantified by the common part between the two volumes divided by the volume along the observing direction. Unlike the case in previous hotspot models, gap size does appear explicitly in the expression for the sunflecks on the leaves, which depends on leaf size and shape as well. However, properly specifying the mean size of "holes" is a practical challenge in the model application. Pinty et al. (1990) provided a parameterization scheme for the model. Also, with the framework of this model, Gobron et al. (1997) fully considered the hotspot effect in the zero- and first-order scattering for semidiscrete media.

The hotspot kernels used in the models presented above are basically derived under the assumption of horizontally oriented circular leaves. The same is implied for the gap shape between leaves. This is certainly a crude simplification of real vegetation canopies where scatterers (leaves in this case) usually exist in noncircular shapes and nonhorizontal orientations.

Results from computer simulations have demonstrated that even for leaves with the same area but different shapes, the canopy hotspot shapes are different (Qin and Goel 1995). To account for the influences of leaf shape, one has to introduce a second parameter (e.g., mean width) for leaf geometry characterization in addition to the mean leaf dimension (diameter). Qin and Xiang (1994) developed a two-parameter (length and width) hotspot model (rectangle model) that has proved to be much more flexible than disk or exponential model and can be principally applied to any kind of leafs. In later developments (Qin and Goel 1995, Qin and Xiang 1997), a hotspot model was derived to quantify the influence of leaf angle distribution (LAD) on the canopy hotspot. The result showed that the hotspot effect is most significant when the mean leaf angle is around 25°, which is consistent with the simulation results using a computer graphics method (Qin and Goel 1995). These hotspot models can be used as a hotspot kernel in any canopy BRF model.

The approaches described above all deal with the hotspot in the first-scattering problem, although the hotspot effect occurs in any order of scattering. Fortunately, the hotspot effect is smoothed out in higher-order scattering and is not as pronounced as in first-order scattering.

More considerations of the hotspot effects for finite-size media in the radiation transport problem were presented by Myneni et al. (1991) and Knyazikhin et al. 1992). As we discussed in Section 3.1, in their formalism for finite-size scattering centers, the canopy is abstracted as a binary medium of randomly distributed leaf clumps and voids (gaps). Kuusk's hotspot formalism is applied to specify the distribution of voids along the path of photon traveling, and consequently the hotspot effect is implicitly accounted for in the expressions for the total interaction cross section and the differential scattering cross section of the canopy. Theoretically, this approach can handle the hotspot effect in all orders of scattering, and provides a promising direction to completely account for the hotspot effect in the context of radiative transfer problem if a more precise hotspot kernel is incorporated in the future.

To account for the hotspot effect in the context of radiative transfer in homogeneous finite size media, we have to modify the classical extinction coefficient. Marshak (1989) made the first attempt in this direction. His results show that Kuusk's results can be obtained if we appropriately modify the extinction coefficient for first-order scattering in the transport equation. We follow this formulation in the following text. The correction function $h(\tau,\Omega)$ in Eq. (3.17) is used to account for the hotspot phenomenon. Here we use Nilson and Kuusk's (1989) formulation for unscattered (direct) solar radiation and single scattering radiation in the upwelling directions ($\mu > 0$):

$$h(\tau, \Omega) = 1 - \sqrt{\frac{G(\Omega_0)}{G(\Omega)} \frac{\mu}{\mu_0}} \, \exp\left[\frac{\Delta(\Omega_0, \Omega)\tau}{kH}\right] \qquad (3.31)$$

Otherwise, $h(\tau, \Omega) = 1$, where

$$\Delta(\Omega_0, \Omega) = \sqrt{\frac{\mu_0^{-2} + \mu^{-2} + 2(\Omega_0 \Omega)}{\mu_0 \mu}} \qquad (3.32)$$

The parameter k characterizes the dimension of the leaf and is proportional to the ratio of the average diameter of the round leaf to the canopy height H. Notice that in the case of backscattering [i.e., $\Omega = -\Omega_0$, $\Delta(\Omega_0, \Omega) = 0$, and $h(\tau, \Omega) = 0$], the absence of extinction results in the local maximum of reflectance, which is widely termed the *hotspot peak*. The correction function $h(\tau, \Omega)$ becomes smaller than unity only for unscattered sunlit and single scattering components in the upwelling direction. Esposito's numerical calculation (Esposito 1979) shows that including a hotspot effect scattering of a higher order than single-order scattering has a negligible effect.

It is recognized by the land surface BRDF modeling community that the hotspots in BRDFs of vegetated land surfaces are one of the most information-rich subregions within a BRDF space. It has been shown by many model calculations, as well as some more recent measurements with airborne [ASAS (advanced solid-state array spectroradiometer), MAS (MODIS Airborne Simulator), AirPOLDER, and AirMISR] and spaceborne sensors (POLDER on ADEOS) that the hotspot effect is indeed diagnostic for canopy structure and may allow the retrieval of such canopy structural parameters as leaf size and shape, tree crown size, and canopy height for low-LAI stands. For example, Roujean (2000) developed a simple parametric hotspot model for being incorporated into statistical BRDF model for albedo calculations. Lacaze and Roujean (2001) inverted the airborne POLDER data in an attempt to calculate the clumping index of patch vegetation. Qin et al. (2002) used the radiosity model (see Section 3.5.2) to explore the potential of canopy reflectance distributions in the hotspot region for characterizing leaf geometry (leaf size and shape) of grass and crop canopies.

3.1.5   Formulations for Heterogeneous Canopies

In case of horizontally heterogeneous or discontinuous canopies such as row crops, or orchards with isolated tree crowns, the turbid medium analogy is not applicable because foliage enclosures are not finite. Kimes (Kimes and Kirchner 1982, Kimes 1991) first developed a canopy model that handles canopy heterogeneity. The canopy is divided into a rectangular cell matrix, and the radiation transfer is restricted to propagate in a finite number of directions with some simplified assumptions. Gastellu-Etchegorry et al. (1996b) further extended this algorithm to form a model called *discrete anisotropic radiative transfer* (DART) by overcoming some of its drawbacks. This model is illustrated in Fig. 3.6. The scene consists of parallel-piped cells, each of which may contain different components of the landscape (e.g.,

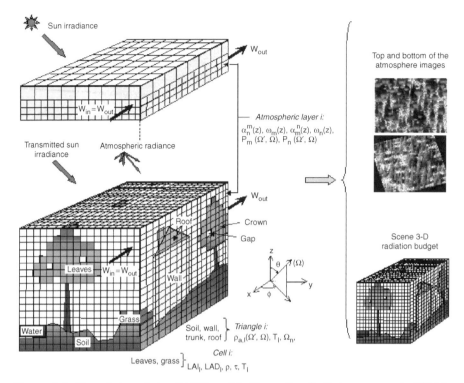

Figure 3.6 Formulation of the DART model. [From Gastellu-Etchegorry et al. (1996b), *Remote Sens. Environ.* Copyright © 1996 with permission from Elsevier.]

leaves, trunks, grass, water, soil). Their optical properties are presented by individual scattering phase functions and structural characteristics of elements within the cell. Radiative transfer is simulated with the exact kernel and discrete ordinate approaches.

Mathematically, the radiative transfer equation for steady state monochromatic radiance $I(\mathbf{r}, \Omega)$ at a position $\mathbf{r}$ along a direction Ω_{ij} in 3D Cartesian geometry is given by

$$\left[\mu_{ij} \frac{d}{dz} + \eta_{ij} \frac{d}{dy} + \zeta_{ij} \frac{d}{dx} \right] I(\mathbf{r}, \Omega_{ij})$$

$$= -\alpha(r, \Omega_{ij}) I(\mathbf{r}, \Omega_{ij}) + Q(\mathbf{r}, \Omega_{ij})$$

$$+ \sum_{u=1}^{U} \sum_{v=1}^{V} C_{uv} \alpha_d(\mathbf{r}, \Omega_{uv} \to \Omega_{ij}) I(\mathbf{r}, \Omega_{uv}) \qquad (3.33)$$

where μ, η, and ζ are directional cosines with respect to the z, y, and x axes, the angular dependence is approximated by discretizing the angular variable Ω into a number of discrete directions Ω_{ij}, α and α_d are the extinction coefficient and differential scattering coefficient, and Q is the first scattering source term. In Eq. (3.33), C_{uv} represents the integration weight (note that the last term is an integration kernel). This equation looks very similar to the 1D radiative transfer equation (3.17). Solving 1D radiative transfer equations numerically has to be carried out from one layer by another, but solving 3D equations has to be carried out from one cell by another.

Myneni et al. (1990a) have explored 3D canopy radiative transfer models and their solutions in a series of studies. Their results have established the relationships between vegetation indices and various biophysical parameters (Myneni and Williams 1994, Myneni et al. 1995).

In principle, 3D canopy radiative transfer models can handle any form of heterogeneous canopies (Myneni et al. 1990a). However, solving a 3D radiative transfer equation is very time-consuming. Another direction is to develop approximate formulations. Nilson and Peterson (1991) developed a new approximate approach for nonhomogeneous canopies. The single scattering component has been extended to the nonhomogenous case because of its significance in forming canopy reflectance, especially its angular distribution and simplicity for this kind of process. It has also been suggested that the multiscattering component of a homogeneous leaf canopy can be treated as an approximation of a real situation. Kuusk has developed a so-called Markov canopy reflectance model (1994, 1995a, 1995b, 1998) following a similar philosophy. The code of the Kuusk model has been included in this CD-ROM of this book.

Goel and Grier (1986a, 1986b) developed a hybrid model for a row-planted canopy (two-dimensional nonhomogeneous canopy). It assumes that the vegetation canopy is composed of subcanopies along rows (see Fig. 3.7). Each subcanopy has an elliptical cross section, and is treated like a turbid medium that is characterized by the SAIL model. This model is also invertible, and the retrieved parameters closely agree with the measured ones. Later, they also extended their model into the three-dimensional case (Goel and Grier 1988).

Gobron et al. (1997) developed a semidiscrete radiative transfer model that takes into account the number, size, and orientation of the leaves and the total height of the canopy. These discrete characterizations of the plant canopy (as opposed to a purely turbid medium) are formulated into the calculation of the first orders of scattering, whereas multiple scattering is approximated from the ordinary radiative transfer formulation for a turbid medium.

Another approach to studying heterogeneous canopies is to follow stochastic theory, as in Anisimov and Menzhulin (1981), Liang and Strahler (1994a) and Shabanov et al. (2000). They treated the leaf canopy as a horizontally homogenous anisotropic turbid medium with local inhomogeneities and used a stochastic method of averaging widely applied in hydrodynamics. These approaches obtain not only mean radiance but also covariance, kurtosis, and

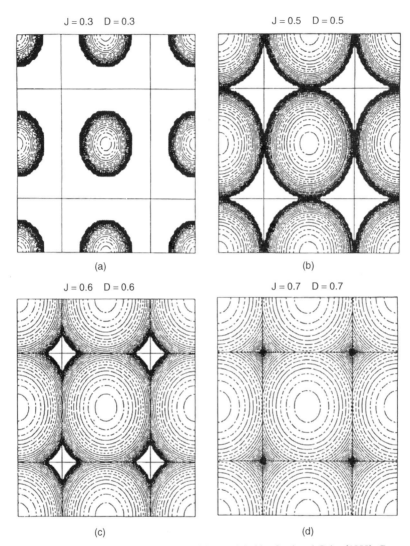

Figure 3.7   Illustration of the 3D canopy field modeled by Goel and Grier (1988). Cross section of each identical ellipsoid canopy crown projected on the ground at four different growth stages (a–d). [From Goel and Grier (1988), *Remote Sens. Environ.* Copyright © 1988 with permission from Elsevier.]

other higher moments that can be useful in evaluating canopy configuration and other properties.

All the abovementioned modeling efforts are based on this turbid medium radiative transfer equation in which the canopy leaf is treated as a particle with no physical size. Although some empirical correlation functions have been developed to account for hotspot effect, the theoretical development of a canopy radiative transfer equation that takes the dimension of the leaf into account (Myneni et al. 1991, Myneni and Ganapol 1991) has never been implemented.

3.2   LEAF OPTICAL MODELS

Major efforts in modeling vegetation reflectance using radiative transfer theory have been at the canopy level, and there are fewer models at the leaf level. But optical properties of scattering elements are needed in a canopy model, which depend on microstructure and material property of leaves. We can either use actual measurements or apply a leaf optical model to calculate the leaf optical properties. Another need for the leaf optical model is to retrieve leaf biochemistry from canopy reflectance observations. Therefore, it is worthwhile to discuss leaf models before discussing various solutions to the canopy radiative transfer equation.

Leaf reflectance illustrates some unique features (Fig. 3.8). In general, leaf pigments have a great impact on visible reflectance, water vapor, protein, and

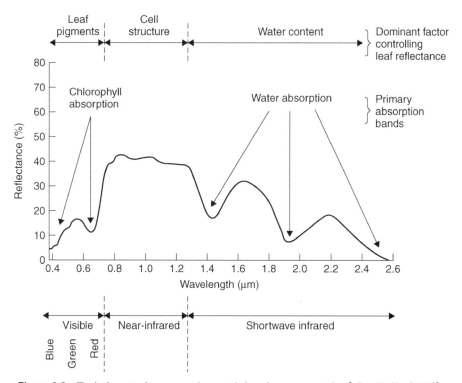

Figure 3.8   Typical spectral response characteristics of green vegetation [after Hoffer (1978)].

other variables, and leaf chemistry affects the reflectance from the near-IR to the middle-IR spectrum. There are several leaf optical models with different modeling techniques available in the literature for both broadleaf and needleleaf. The basic principles of some representative leaf models are introduced below.

3.2.1 "Plate" Models

A well-known leaf reflectance model is the PROSPECT model, first proposed by Jacquemoud and Baret (1990) and modified later (Jacquemoud et al. 1995). PROSPECT simulates leaf reflectance and transmittance from the visible to the middle infrared spectrum as a function of the leaf structure parameter and leaf biochemical parameters. It is based on the so-called "plate model" developed by Allen et al. (1969), who represented a leaf as a uniform plate with rough surfaces. The leaf reflectance and transmittance are determined using geometric optical principles. It is illustrated in Fig. 3.9a. Parameters include the index of refraction and an absorption coefficient. This model was successful in reproducing the reflectance spectra of a compact corn leaf characterized by a few air–cell wall interfaces.

They then extended the model to noncompact leaves by regarding them as piles of N plates separated by $N-1$ airspaces (Allen et al. 1970) (see Fig. 3.9b). The solution of such a system, provided in the last century has been extended to N as a real number, this is the so-called generalized plate model. This additional parameter N actually describes the leaf internal structure and plays a role similar to that of the scattering coefficient s in the Kubelka–Munk model. The original PROSPECT model (Jacquemoud and Baret 1990) requires only three parameters, the structure parameter N, and the chlorophyll and water contents, to calculate the reflectance and transmittance of any fresh leaf over the whole solar domain. Two more parameters, the protein and cellulose + lignin contents, permit the simulation of dry leaf spectra (Jacquemoud et al. 1996).

Mathematically, the total reflectance and transmittance for N layers are given by

$$R_{N,\alpha} = xR_{N,90} + y$$

$$T_{N,\alpha} = xT_{N,90}$$

$$x = \frac{t_{av}(\alpha,n)}{t_{av}(90,n)} \qquad (3.34)$$

$$y = x[t_{av}(90,n) - 1] + 1 - t_{av}(\alpha,n)$$

where n is the refractive index of the leaf plate and α is the maximum incidence angle; $t_{av}(\alpha,n)$ is the transmittance of a dielectric plane surface,

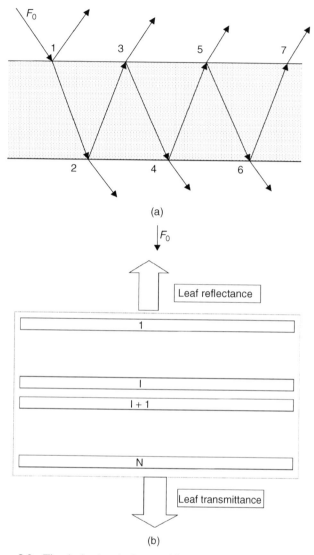

F_0

1   3   5   7

2   4   6

(a)

F_0

Leaf reflectance

1

I

I + 1

N

Leaf transmittance

(b)

Figure 3.9 The single-plate leaf model (a) and the N-plate leaf model (b).

averaged over all directions of incidence and over all polarizations. Its
expression is rather complex, but can be calculated exactly (Stern 1964, Allen
1973). Finally, this model has four parameters: n, N, α, and the transmission
coefficient θ that is related to the absorption coefficient k through the
following equation (Allen et al. 1969):

$$\theta - (1 - k)e^{-k} - k^2 \int_k^\infty x^{-1} e^{-x}\, dx = 0 \tag{3.35}$$

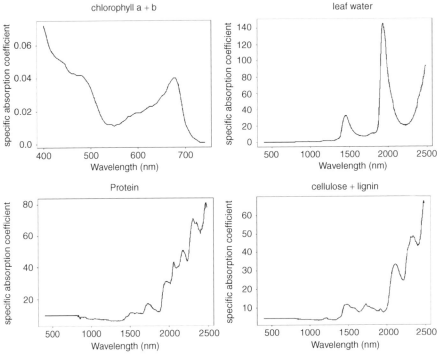

Figure 3.10 Specific absorption coefficients of chlorophyll a + b, leaf water, protein, cellulose, and lignin used in the leaf reflectance model PROSPECT.

The spectral absorption coefficient $k(\lambda)$ can be written in the form

$$k(\lambda) = \sum K_i(\lambda)C_i + k_0 \qquad (3.36)$$

where k_0 is the intercept, $K_i(\lambda)$ is the spectral specific absorption coefficient relative to the leaf component i, and C_i is the leaf component i content per unit leaf area. Figure 3.10 shows the specific absorption coefficients of chlorophyll a + b, water, protein, and cellulose and lignin from PROSPECT. In the visible part, absorption is due to pigments such as chlorophyll a + b, carotenoids, and brown pigments that appear during senescence. In the spectral range 800–2500 nm, absorption by water molecules almost completely masks the effects of cellulose, sugar, proteins, and lignin (Peterson et al. 1988).

The code is included in the accompanying CD-ROM. Figure 3.11 illustrates the impacts of chlorophyll a + b concentration and leaf water content on leaf reflectance, calculated by the PROSPECT model. It is quite clear that chlorophyll concentration mainly affects the reflectance in the visible spectrum, and water content mainly affects the reflectance in the near-IR spectrum. One opinion/interpretation is that chlorophyll a + b and water

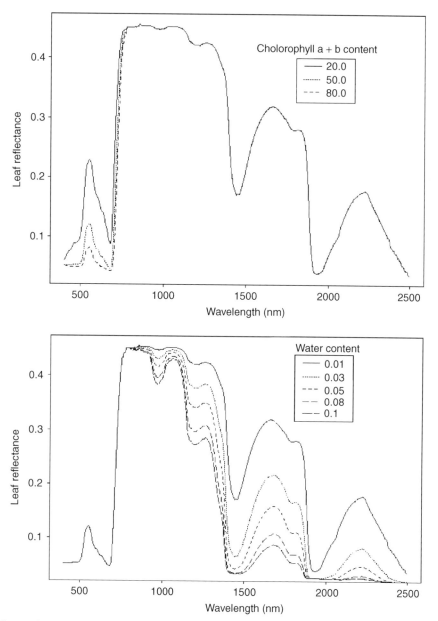

Figure 3.11 Leaf reflectance dependence on chlorophyll a + b concentration and water content from PROSPECT model.

have strong absorptions in several small spectral regions. From this figure, it is very clear that the whole visible or the near-IR reflectance significantly decreases when the concentration of chlorophyll a + b or water content increases.

Through model inversion, Jacquemoud (Jacquemoud 1993, Jacquemoud et al. 1995) succeeded in inferring three biochemical parameters from measured leaf spectra with a good accuracy under certain conditions.

3.2.2 Needleleaf Models

PROSPECT was proposed primarily for broadleaf canopies. LIBERTY (leaf incorporating biochemistry exhibiting reflectance and transmittance yields) was developed to characterize spectral properties of conifer needles (Dawson et al. 1998). Assuming that the leaf cellular structure can be represented by spherical cells, LIBERTY performs a linear summation of the individual absorption coefficients of the major constituent leaf chemicals (chlorophyll, water, cellulose, lignin, protein) according to their content per unit area of leaf. The resultant structural parameters (mean cell diameter, leaf thickness, and an intercellular airgap determinant) provide accurate reflectance and transmittance spectra of both stacked and individual needles. This model has been coupled with a hybrid geometric/radiative transfer BRF model for conifer forests, FLIGHT (forest light) (North 1996, Dawson et al. 1999).

The computer code is also included in the accompanying CD-ROM.

3.2.3 Ray Tracing Models

Another type of leaf reflectance and transmittance model is based on ray tracing (Allen et al. 1973, Govaerts et al. 1996). These models require a detailed description of individual cells and their unique arrangement inside tissues. The optical constants of leaf materials (cell walls, cytoplasm, pigments, air cavities, etc.) also have to be defined. Using the laws of reflection, refraction, and absorption, it is then possible to simulate the propagation of individual photons incident on the leaf surface. Once a sufficient number of rays have been simulated, statistically valid estimates of the radiation transfer in a leaf may be deduced. The technique has been applied with a number of variants. Among various approaches, only ray tracing techniques can account for the complexity of internal leaf structure as it appears in a photomicrograph. Unfortunately, this method requires extensive computation. This approach is suitable for validating simple models or for understanding of the photon process at the leaf level.

3.2.4 Stochastic Models

Stochastic leaf optical models are based on the Markov chain theory (Tucker 1977, Maier et al. 1999). In a Markov chain, the random variable photon

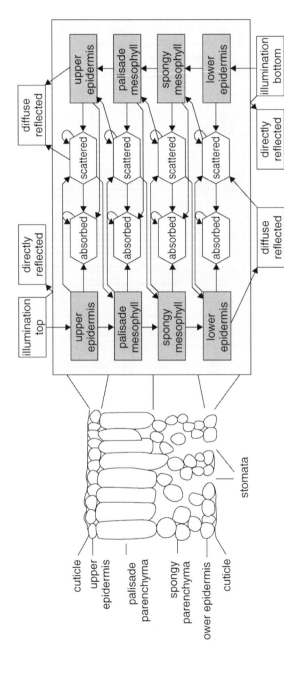

Figure 3.12 Transition matrix between different states in a Markov leaf reflectance model. [From Maier et al. (1999), *Remote Sens. Environ.* Copyright © 1999 with permission from Elsevier.]

state can take only discrete values, like "absorbed in the palisade parenchyma" or "scattered in the spongy parenchyma." The probabilities for the occurrence of these states are described by the elements of a vector, called a *state vector*. In a Markov chain, temporal development occurs in discrete time steps. This is mathematically expressed by matrix multiplications. The elements of the transition matrix are the probabilities for the transitions between the different states. This is illustrated in Fig. 3.12.

3.2.5 Turbid Medium Models

Turbid medium models derived from the Kubelka–Munk theory consider the leaf as a slab of diffusing (scattering coefficient s) and absorbing (absorption coefficient k) material. The N-flux equations are a simplification of the radiative transfer theory—the solution of these equations yields simple analytic formulas for the diffuse reflectance and transmittance. A two-flux model (Allen et al. 1969) and a four-flux model (Fukshansky 1991, Richter and Fukshansky 1996) have been successfully used in the forward mode to calculate the s and k optical parameters of plant leaves. Yamada and Fujimara (1991) later proposed a more sophisticated version in which the leaf was divided into four parallel layers: the upper cuticle, the palisade parenchyma, the spongy mesophyll, and the lower cuticle. The Kubelka–Munk theory is applied with different parameters in each layer, and solutions are coupled with suitable boundary conditions to provide the leaf reflectance and transmittance as a function of the scattering and absorption coefficients. These authors then went even further, interpreting the absorption coefficient determined in the visible region in terms of chlorophyll content. By inversion, their model became a nondestructive method for the measurement of photosynthetic pigments. Finally, a very simple model in terms of the reflectance has been used to estimate the chlorophyll content of sugarbeet leaves (Andrieu et al. 1992). Compared with canopy level, only few models directly use the radiative transfer equation at leaf level (Qin and Liang 2000). The poor information we have on leaf internal structure and biochemical distribution leads to strong simplifications that make such an approach less efficient as compared to more robust equations. Ma et al. (1990) describe the leaf as a slab of water with an irregular surface containing randomly distributed spherical particles. In the LEAFMOD model by Ganapol et al. (1999), a leaf is compared to a homogeneous mixture of biochemicals that scatter and absorb light. Each of these models was able to reproduce a faithful simulation of leaf optical properties.

3.3 SOLVING RADIATIVE TRANSFER EQUATIONS

To calculate canopy directional reflectance, the canopy radiative transfer equation has to be solved. Similar to the atmospheric radiative transfer problem, there are also two types of solutions: numerical and approximate. Numerical solutions are accurate, but the iterative process is computationally expensive. Approximate solutions are fast and easy to implement, but may not be as accurate.

3.3.1   Approximate Solutions

Analytic solutions can be obtained by either approximating the canopy radiative transfer equation with a set of differential equations and solving it for upward and downward fluxes such as Kubelka–Munk (KM) theory, or by decomposing the canopy radiation field into unscattered, single scattering, and multiscattering components, then estimating multiple scattering in such a way that iterative calculations are avoided. In the following, we briefly discuss the radiative transfer models based on KM theory, and models based on accurate calculation of the zero and single scattering but approximate estimation of multiple scattering.

3.3.1.1   Models Based on KM Theory

Most of the early canopy radiative transfer models were based on Kubelka–Munk (KM) theory, which used a linear differential equation set to describe fluxes traveling in the downward (E^-) and upward (E^+) directions in a plane-parallel medium. The KM theory contains two differential equations with two coefficients, and is therefore considered a two-flux theory. Allen et al. (1970) included the direct solar flux (E_s) so that the canopy model consists of three differential equations (called *Duntley*) with five coefficients, which is considered three-flux theory. Suits (1972) further extended their model by including a flux at the sensor viewing direction (E_v). Thus, there are four differential equations with nine coefficients as follows:

$$\frac{dE_s}{dz} = k \cdot E_s \tag{3.37}$$

$$\frac{dE^-}{dz} = a \cdot E^- - \sigma \cdot E^+ - s \cdot E_s \tag{3.38}$$

$$\frac{dE^+}{dz} = -a \cdot E^+ + \sigma \cdot E^- + s' \cdot E_s \tag{3.39}$$

$$\frac{dE_v}{dz} = -K \cdot E_v + \mu \cdot E^+ + v \cdot E^- + w \cdot E_s \tag{3.40}$$

Given these coefficients, solving this differential equation set is not too difficult. The details are omitted here; readers are referred to the literature (Suits 1972, Slater 1980). Suits defined these coefficients in terms of horizontal and vertical leaf area projections. This simplification results in unrealistic canopy reflectance angular patterns. Verhoef (1984) determined these coefficients of the same differential equation set, based on the total canopy LAI and leaf angle distribution. The resulted model is called SAIL model that has been widely used for simulations. One of the SAIL model codes is included in the attached CD-ROM for exercise.

Meanwhile, an independent discrete model based on KM theory was proposed by Idso and de Wit (1970). In this model, the canopy is divided into many layers and radiation at each level is specified in terms of downward and upward fluxes. Using the average transmission theory, equations are derived for both radiation fluxes. Their profiles are obtained by solving the equation set. These models and other discrete models using similar principles have been developed, such as the Goudriaan (1977) model, Cupid model (Norman and Welles 1983), and several others.

There are two major drawbacks of radiative transfer models based on KM theory:

1. Since the differential equation set is established by integrating the original radiative transfer equation over zenith and azimuth angles and then solving for these integrated fluxes (upward and downward), it is, strictly speaking, not suitable for the directional reflectance (zenith and azimuth dependent) calculation, although the coefficients in the equation set are related to solar/viewing angles and canopy structure.

2. The coefficients are complex functions of structure and the phase function of the medium and must be determined empirically or semiempirically for vegetation canopies in general. Therefore, at present, this type of simplification method is utilized mainly in approximating multiple scattering.

3.3.1.2 *Decomposition of the Canopy Radiation Field*

In order to achieve an optimal balance between accuracy and simplicity, a strategy adapted in most of the more recent canopy radiative transfer models is to seek an accurate (and analytic) solution of zero- and first-scattering components and to estimate the multiple scattering contribution. Compared with models that are totally based on KM theory, this strategy has greatly improved the accuracy of the single scattering reflectance by capturing (1) some important phenomena, such as the canopy hotspot, specular reflection from leaf surfaces, and anisotropic reflectance from the soil, which are most pronounced in the first-order scattering radiation field, and (2) the angular characteristics of canopy reflectance that are dominant in the single scattering reflectance component.

The most common practice for achieving this has been to decompose the canopy radiation field into three components: unscattered radiance (I^0), singly scattered radiance (I^1), and multiply scattered radiance (I^M) (e.g., Marshak 1989, Qin and Jupp 1993, Liang and Strahler 1993b). This is illustrated in Fig. 3.13. Equivalently, we can express the canopy reflectance as the sum of three components: single reflection reflectance from the soil,

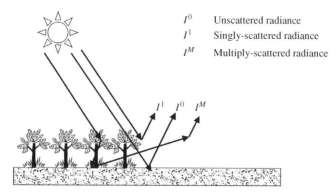

I^0 Unscattered radiance

I^1 Singly-scattered radiance

I^M Multiply-scattered radiance

Figure 3.13   The schema of the radiation field decomposition.

single scattering reflectance from the canopy, and multiscattering reflectance (e.g., Nilson and Kuus, 1989, Nilson and Peterson 1991, Kuusk 1985). The total radiance is the sum of these three components:

$$I = I^0 + I^1 + I^M \tag{3.41}$$

where the downward and upwelling unscattered radiance can be calculated

$$I^0(\tau, -\Omega) = i_0 \exp\left[-\frac{G(\Omega)\tau}{\mu_0}\right] \tag{3.42}$$

$$I^0(\tau, \Omega) = i_0 \exp\left[-\frac{G(\Omega_0)\tau_c}{\mu_0}\right] r_s \mu_0 \exp\left[-\xi(\tau,\Omega)\right] \tag{3.43}$$

where τ_c is the total canopy optical depth, i_0 is the downward solar direct radiance at the top of the canopy. From Eq. (2.76), it follows that

$$i_0 = \pi E_s \exp\left(-\frac{\tau_a}{\mu_0}\right) \tag{3.44}$$

where τ_a is the total optical depth of the atmosphere, and

$$\xi(\tau, \Omega) = G(\Omega)\frac{\tau_c - \tau}{\mu} - \frac{\sqrt{\dfrac{G(\Omega_0)G(\Omega)}{\mu\mu_0}}\, kH}{\Delta(\Omega_0, \Omega)} t_1 \tag{3.45}$$

where

$$t_1 = \exp\left[-\frac{\Delta(\Omega_0,\Omega)\tau}{kH}\right] - \exp\left[-\frac{\Delta(\Omega_0,\Omega)\tau_c}{kH}\right] \tag{3.46}$$

Note that Eq. (3.43) is the reflected radiance by the canopy background but unscattered by the canopy elements. For the single scattering radiance, the solution in the downward direction can be easily written as

$$I^1(\tau, -\Omega) = \frac{i_0 \cdot \mu_0 \Gamma(\Omega_0, \Omega)}{\pi [G(\Omega)\mu_0 - G(\Omega_0)\mu]} \left\{ \exp\left[-\frac{G(\Omega_0)\tau}{\mu_0} \right] - \exp\left[-\frac{G(\Omega)\tau}{\mu} \right] \right\}$$
(3.47)

when $\mu = \mu_0$, then

$$I^1(\tau, -\Omega) = \frac{\tau \cdot i_0 \Gamma(\Omega_0, \Omega)}{\pi \mu_0} \exp\left[-\frac{G(\Omega_0)\tau}{\mu_0} \right]$$
(3.48)

In the upward direction, the solutions are a little complicated because of the hotspot effect:

$$I^1(\tau, \Omega) = \frac{1}{\mu} \int_\tau^{\tau_c} F(\tau', \Omega) \exp\left[-\frac{1}{\mu} \int_\tau^{\tau'} h(\xi, \Omega) G(\Omega) \, d\xi \right] d\tau' \quad (3.49)$$

where the second integration can be explicitly obtained by means of Eq. (3.31) with an alternative integrand range, and

$$F(\tau', \Omega) = \frac{i_0}{\pi} \Gamma(\Omega_0, \Omega) \exp\left[-\frac{G(\Omega_0)\tau'}{\mu_0} \right]$$
(3.50)

Besides the treatment of the canopy hotspot effect, the major differences among different radiative transfer models based on this strategy lie in the algorithms used for calculating multiple scattering. A few representative models are briefly summarized below.

3.3.1.3 *Approximation of Multiple Scattering*
The simplest methods for approximating multiple scattering are based on a two-stream approximation. Some climate process related reflectance models were developed entirely from two-stream approximations (e.g., Dickinson 1983, Sellers 1985) for albedo calculation. However, most of the more recent canopy radiative transfer models used two-stream approximation for estimating multiple scattering reflectance (e.g., Nilson and Peterson 1991, Qin and Jupp 1993, Kuusk 1995a, 1995b). Nilson and Peterson (1991) expressed the reflectance factor of multiple scattering in the form of a symmetric product based on the analytic solution for the two-stream problem in a finite depth medium with a soil boundary condition. Qin and Jupp (1993) improved this solution by deriving the exact expressions for the four coefficients in the equation set for arbitrary leaf angle distribution and including the diffuse sky radiance contribution and anisotropy of the soil. Kuusk (1995a, 1995b) used the results from the SAIL model for diffuse flux components to estimate multiple scattering.

Another example of using two-stream formulation for multiple scattering is based on Hapke's studies on planetary surfaces (Hapke 1981). Hapke derived a completely analytic solution to the physical problem for a homogeneous semi-infinite medium composed of uniformly distributed scatters, as is generally the case for soil surfaces. Dickinson et al. (1990) and Verstraete et al. (1990) adopted Hapke's formulation for vegetation canopies, and the latter fully considered the canopy hotspot effect. However, the assumption of a semi-infinite medium is appropriate for media such as soil and snow, but may not be sufficiently accurate for vegetation canopies, especially those that are optically thin.

In the following, we discuss two approximate methods (four-stream approximation and asymptotic fitting method) that can also be used for solving the radiative transfer equations of other media, such as atmosphere, soil, snow, and water.

Four-Stream Approximation. The popularity of two-stream approximations for radiative flux calculation can be attributed to their simplicity, but when applied to model the angular characteristics of the canopy radiation field, their accuracy is quite limited. On the basis of the same decomposition of the canopy radiation field, Liang and Strahler (1995) developed a four-stream approximation formula for multiple scattering of vegetation canopies. The calculation of the angular radiance distribution for a coupled atmosphere and canopy using the Gauss–Seidel algorithm shows that the multiscattering component is over 50% of the total upwelling canopy radiance in the near-IR region. This spectral region is not as affected by atmospheric scattering as are other wavelengths and is therefore highly useful in the retrieval of biophysical parameters from multiangle remotely sensed imagery.

The four-stream formulation for atmospheric radiative transfer has been discussed by several studies (e.g., Liou 1974, Liang and Strahler 1994b). The general idea is to represent radiance distribution at two zenith angles in both the upper and lower hemispheres. This is illustrated in Fig. 3.14. Obviously, use of more discrete angles will lead to a more accurate representation of the radiance angular distribution, which is what the numerical method of discrete ordinates is about (see Section 2.4.1.2).

To calculate canopy multiple scattering effectively by taking advantage of the existing four-stream formulation, further approximations need to be made. The canopy phase function is not rotationally invariant, but depends on both incident direction and outgoing direction. Numerical results show that the multiple-scattering component is relatively insensitive to variation in azimuthal angle, since when the canopy becomes thicker optically the photons will scatter more times before emerging from the canopy. Thus, we may accept the simplification that multiple scattering is independent of azimuth angle. Also, it seems that the multiple-scattering radiance distribution probably approaches the isotropic case. However, our results in a series of calculations show that the isotropic scattering function will cause large

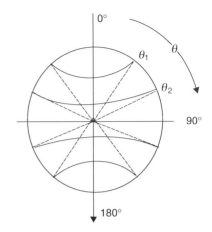

Figure 3.14   Four-stream angular presentation.

errors when there are a number of horizontal or vertical leaves in the canopy, as in the case of a planophile, or erectophile canopy. Instead, a Henyey–Greenstein scattering phase function for multiple scattering and radiance independent of azimuth angles are assumed. Although the Henyey–Greenstein function is an approximation to the real phase function and may result in errors, the single scattering radiance is still evaluated using the exact phase function. As a result, this formulation still can predict accurately the angular dependence of the reflectance.

The Henyey–Greenstein function is empirically related to biophysical parameters as follows. For spherically distributed leaves, the area scattering phase function becomes (Ross 1981):

$$\Gamma(\Omega',\Omega) = \frac{\omega_c}{3\pi}(\sin\beta - \beta\cos\beta) + \frac{t_l}{\pi}\cos\beta \qquad (3.51)$$

where $\beta = \cos^{-1}(\Omega \cdot \Omega')$, the angle between Ω and Ω'. The single scattering albedo of the canopy ω_c is defined as (Myneni 1990b):

$$\omega_c = \omega_D + \omega_{sp} = r_l + t_l + K(\kappa, \Omega' \cdot \Omega_l)F(n, \Omega' \cdot \Omega_l) \qquad (3.52)$$

which is an angular dependent quantity. γ_l and t_l are leaf reflectance and transmittance. In the calculation for multiple scattering, a mean albedo is substituted:

$$\overline{\omega}_c = \frac{1}{2}\int_{-1}^{0}\omega_c(\mu)\mu\,d\mu \qquad (3.53)$$

The phase function is defined by

$$p_c(\Omega',\Omega) = \frac{8\Gamma(\Omega',\Omega)}{\omega_c} = \frac{8}{3\pi}(\sin\beta - \beta\cos\beta) + \frac{8\alpha}{\pi}\cos\beta \quad (3.54)$$

where

$$\alpha = \frac{t_l}{\omega_c} \tag{3.55}$$

In this case, the canopy phase function depends on the phase angle only. From the observation above, we can approximate the above formula using the single term Henyey–Greenstein function. The asymmetry parameter g_c is directly related to the parameter α, and a simple function has been fitted using the least square principle from Eq. (3.54) and the Henyey–Greenstein function:

$$g_c = -0.29 - 0.2478\,\alpha + 2.1653\,\alpha^2 - 1.1248\,\alpha^3 - 0.2059\,\alpha^4 \tag{3.56}$$

For the α ratio varying between 0.3 to 0.5, g_c ranges from 0 to -0.2. The negative value indicates strong backscattering.

For the unscattered radiance and single scattering radiance, the extinction coefficient has been modified to account for the hotspot effect. However, our derivation of the multiscattering component for the canopy is very similar to scattering radiance for the atmosphere. This implies that we do not consider the effect of finite leaf dimension for multiple scattering. The azimuth-independent radiative transfer equation for radiance within canopy I_c is

$$\mu_i \frac{I_c(\tau, \mu_i)}{d\tau} = I_c(\tau, \mu_i) - \frac{\omega_c}{2} \sum_{j=-2}^{2} p_c(\mu_i, \mu_j) I_c(\tau, \mu_j)\, a_j$$

$$- \frac{\omega_c}{4\pi} i_0 p_c(\mu_i, -\mu_0) \exp\left(-\frac{\tau}{\mu_0}\right) \tag{3.57}$$

where $\mu_1 = 0.33998$, $\mu_2 = 0.86112$, $a_1 = 0.65215$, and $a_2 = 0.34785$ (see Table 2.8). The boundary conditions

$$I_c(0, -\mu_i) = I_a(-\mu_i) \tag{3.58}$$

and

$$I_c(\tau_c, \mu_i) = \frac{\pi}{0.52127} \sum_{j=1}^{2} a_j \mu_j r(-\mu_j, \mu_i) I_c(\tau_c, \mu_j)$$

$$+ \mu_0 i_0 r(-\mu_0, \mu_i) \exp\left(-\frac{\tau_c}{\mu_0}\right) \tag{3.59}$$

where subscripts c and a denote the canopy and the atmosphere, respectively, i_0 is the solar direct irradiance arriving at the top of the canopy as attenuated by the atmosphere [Eq. (3.44)], $I_a(-\mu)$ is the sky radiance (atmospheric diffuse radiance) at the top of the canopy, and $r(\cdot)$ is the canopy background (e.g, soil) directional reflectance.

The four-stream discrete ordinate solution to Eq. (3.57) at arbitrary level τ is given by

$$I_c(\tau, x) = \sum_{j=1}^{2} D_j(x) + Z(x) \exp\left(-\frac{\tau}{\mu_0}\right) \tag{3.60}$$

where

$$D_j(x) = L_j W_j(x) \exp(-k_j \tau) + L_{-j} W_j(-x) \exp(k_j \tau) \tag{3.61}$$

and $x = \pm\mu_1, \pm\mu_2$, and functions $W_j(x)$, $Z(x)$, and k_j are known (Liou 1974; Liang and Strahler 1995). $\mathbf{L}$ is a matrix of coefficients to be determined according to the boundary conditions. Detailed expressions are given in the original paper and not repeated here. After calculating radiance at these four directions $(\pm\mu_1, \pm\mu_2)$, we can determine the solutions for any arbitrary direction. For $\mu \neq \mu_0$, we obtain

$$I_c(\tau, \mu) = -\frac{\omega_c}{2} \sum_{l=-2}^{2} a_l p_c(\mu, \mu_l) \left[\sum_{j=1}^{2} E_j(\mu_l) - \frac{Z(\mu_l)\mu_0}{\mu + \mu_0} \exp\left(-\frac{\tau}{\mu_0}\right) \right]$$

$$+ \frac{\omega_c i_0 \mu_0 p_c(\mu, -\mu_0)}{4\pi(\mu + \mu_0)} \exp\left(-\frac{\tau}{\mu_0}\right) + C(\mu) \exp\left(\frac{\tau}{\mu}\right) \tag{3.62}$$

For $\mu = \mu_0$, the radiance can be calculated as follows:

$$I_c(\tau, \mu) = -\frac{\omega_c}{2} \sum_{l=-2}^{2} a_l p_c(\mu, \mu_l) \left[\sum_{j=1}^{2} E_j(\mu_l) - \frac{Z(\mu_l)}{\mu} \tau \right]$$

$$- \frac{\omega_c i_0 p_c(\mu, -\mu_0)\tau}{\mu\pi} + C(\mu) \exp\left(\frac{\tau}{\mu}\right) \tag{3.63}$$

The coefficient $C(\mu)$ needs to be determined according to the boundary condition in Eq. (3.59).

In Eqs. (3.62) and (3.63), $I_c(\tau, \mu)$ is the total scattered radiance. The multiscattering radiance is the difference between total scattering radiance and single scattering radiance, as in Liang and Strahler (1995). Figure 3.15 shows that this formulation is very accurate by comparing it with the Gauss–Seidel numerical solution with the canopy radiative transfer equation (Gauss–Seidel solution will be discussed in Section 3.3.2).

Asymptotic Fitting Method. This method (Liang and Strahler 1993b) extended the concept of semiinfinite media for canopy modeling (Verstraete et al. 1990). It is more appropriate to canopies that are thick but theoretically finite. Since this method was developed in the context of three-component

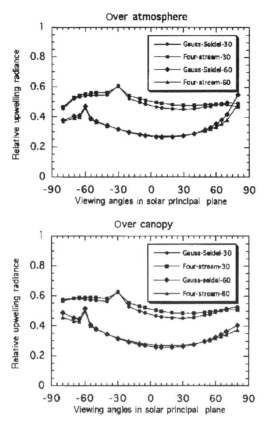

Figure 3.15   Comparison of the four-stream solution with the numerical Gauss–Seidel solution at two solar zenith angles (30° and 60°). The parameters are available in the original paper. [From Liang and Strahler (1995), *J. Geophys. Res.* Reproduced by permission of American Geophysical Union.]

decomposition, it is valid for more general cases. If the canopy is optically very thick, and the multiscattering component dominates, the asymptotic technique is very accurate, so the resulting accuracy is high. If the canopy is optically thin, the multiscattering component is less important although the asymptotic approximation is less accurate, good accuracy is still achieved due to exact zero and single scattering calculations.

The asymptotic solution for a finite-thickness medium with an arbitrary scattering phase function has been derived for many years (van de Hulst 1968). If the canopy is dense with a high optical thickness τ_c and zero soil reflectance is assumed, the multiple scattered upwelling radiance is given by

$$I^M(\mu, -\mu_0) = R^M(\tau_c, \mu, -\mu_0) i_0 \mu_0 + 2 \int_{-1}^{0} R^M(\tau_c, \mu, \mu') I_a(\mu') \mu' \, d\mu'$$

$$(3.64)$$

where the first term on the right side of Eq. (3.64) corresponds to direct sunlight and the second term to the diffuse sky radiance. The directional reflectance from the classic asymptotic theory is expressed as

$$R_0(\tau_c, \mu, -\mu_0) = R_\infty(\mu, -\mu_0) - \frac{mf}{1-f^2} \exp(-\kappa\tau_c) E(\mu) E(-\mu_0) \quad (3.65)$$

where R_∞ is the reflectance function of a semi-infinite medium and $E(\mu)$ is the escape function.

To account for the soil reflectance underneath the finite canopy, the classic reflectance formula in the case of a Lambertian soil is

$$\tilde{R}_0(\tau_c, \mu, -\mu_0) = R_0(\tau_c, \mu, -\mu_0) + \frac{r_s}{r_s A^*} t_0(\tau_c, \mu) t_0(\tau_c, -\mu_0) \quad (3.66)$$

where r_s is the soil Lambertian reflectance, $t_0(\tau_c, \mu)$ is the diffuse transmittance, and A^* is the spherical albedo of the canopy.

With some algebraic manipulation, King (1987) has derived a more simplified formula with the background reflectance on the basis of the preceding relations

$$\tilde{R}_0(\tau_c, \mu, -\mu_0) = R_\infty(\mu, -\mu_0) - \frac{mG(r_s) E(\mu) E(-\mu_0)}{\exp(2\kappa\tau_c) - lG(r_s)} \quad (3.67)$$

where $G(r_s)$ is defined as

$$G(r_s) = l - \frac{mn^2 r_s}{1 - r_s A^*} \quad (3.68)$$

Although this formula originally was for thick clouds, its derivations are also valid for the canopy, and the same results can be derived. If $r_s = 0$, Eq. (3.67) is equivalent to Eq. (3.65). Notice that $f = l \exp(-\kappa\tau_c)$. The parameters κ, l, n, and m have been correlated with the similarity parameter s, defined in terms of the single scattering albedo and the scattering asymmetry factor. Readers should find our original paper (Liang and Strahler 1993b) for details.

The preceding formulas determine the reflectance of all orders of scattering. The multiple scattering reflectance in Eq. (3.67) can be obtained through subtracting the azimuth-independent single-scattering component from Eq. (3.67)

$$R^M(\tau_c, \mu, \mu_0) = \tilde{R}_0(\tau_c, \mu, \mu_0) - \frac{\omega_c p_c(\mu, \mu_0)}{4(\mu + \mu_0)} \left[1 - \exp\left(-\frac{\mu + \mu_0}{\mu\mu_0} \tau_c\right) \right]$$

$$(3.69)$$

where $p_c(\mu, \mu_0)$ is the azimuth-independent part of the Henyey–Greenstein function.

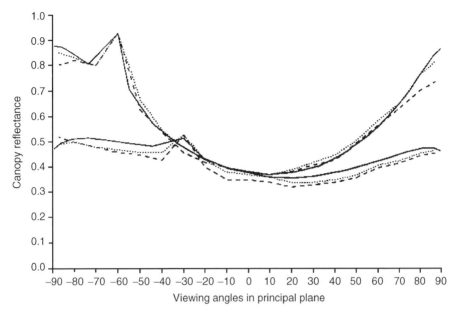

Figure 3.16 Illustration of the accuracy of the asymptotic fitting approximate solution against two Monte Carlo models. The solid curves represent the results of the approximate model; the dotted lines, the Antyufeev–Marshak model; and the dashed curves, the Ross-Marshak model. The detailed parameters for producing these curves are available in the original paper. [From Liang and. Strahler (1993b), *IEEE Trans. Geosci. Remote Sens.* Copyright © 1993 with permission from IEEE.]

This approximate model can also produce reasonably accurate results. Figure 3.16 compares the results of this approximate canopy model with those produced by two Monte Carlo simulation models (Antyufeev and Marshak 1990, Ross and Marshak 1991). There is a good agreement among these three models.

3.3.2 Numerical Solutions: Gauss–Seidel Algorithm

In order to seek accurate solutions for all orders of scattering, numerical methods are usually employed. They are quite flexible and can work on both one-dimensional and three-dimensional geometries. There are several numerical methods to solve the radiative transfer equation for calculating canopy reflectance. The representative methods include the discrete ordinates algorithm (Gerstl and Zardecki 1985, Myneni et al. 1988, Shultis and Myneni 1988), the Gauss–Seidel algorithm (Liang and Strahler 1993a), the integral equation method (Gutschick and Weigel 1984), and successive orders of scattering approximations (Myneni et al. 1987). These methods produce the numerical solutions of the canopy radiative transfer equation, from which canopy reflectance can be calculated. Some of the typical numerical methods have been described in Section 2.4.2, such as the discrete ordinates algorithm, and the successive orders of scattering approximations algorithm. Although they are for solving the atmospheric radiative transfer equation, the basic principles are the same. To illustrate different numerical methods, we will introduce the Gauss–Seidel method that was proposed by Herman and Brown (1965) and then improved by Liang and Strahler (1993a) and used for the coupled canopy-atmosphere radiative transfer modeling.

For the simplicity of our discussions, assume that we are faced with such a general equation for multiple scattered radiance

$$-\mu\frac{\partial I^M(\tau,\Omega)}{\partial\tau} + G(\Omega)I^M(\tau,\Omega) = J(\tau,\Omega) \tag{3.70}$$

where we do not take the hotspot effect into account for multiple scattering, and the source function is

$$J(\tau,\Omega) = \frac{1}{\pi}\int_0^{2\pi}\int_{-1}^1 \Gamma(\Omega',\Omega)\left[I^M(\tau,\Omega') + I^1(\tau,\Omega')\right]d\Omega'$$

$$+ \frac{1}{\pi}\int_0^{2\pi}\int_{-1}^0 \Gamma(\Omega',\Omega)I^0(\tau,\Omega')\,d\Omega' \tag{3.71}$$

If the source function $J(\tau,\Omega)$ is treated independent of multiscattering radiance $I^M(\tau,\Omega)$, Eq. (3.70) becomes an ordinary differential equation and its "formal" solution is

$$\begin{cases} I^M(\tau,-\Omega) = I^M(0,-\Omega)\exp\left[-\frac{G(\Omega)\tau}{\mu}\right] \\ \qquad + \frac{1}{\mu}\int_0^\tau J(\tau',\Omega)\exp\left[-\frac{G(\Omega)(\tau-\tau')}{\mu}\right]d\tau' \quad\text{downward} \\ I^M(\tau,\Omega) = I^M(\tau_c,\Omega)\exp\left[-\frac{G(\Omega)(\tau_c-\tau)}{\mu}\right] \\ \qquad + \frac{1}{\mu}\int_\tau^{\tau_c}J(\tau',\Omega)\exp\left[-\frac{G(\Omega)(\tau'-\tau)}{\mu}\right]d\tau' \quad\text{upwelling} \end{cases}$$

$$\tag{3.72}$$

In fact, the source function $J(\tau,\Omega)$ is related to the multiple scattering radiance $I^M(\tau,\Omega)$ as displayed in Eq. (3.71), and the coupled integral equations [Eq. (3.72)] are not the final solutions. As a result, the coupled medium is split into N contiguous layers, each of thickness $\Delta\tau$. From the spatial interval (τ_i,τ_{i+2}), Herman and Browning (1965) assume that the source function $J(\tau,\Omega)$ may be taken to be independent of τ' and equal to its value at the midpoint of the interval, letting $J(\tau',\Omega) = J(\tau_{i+1},\Omega)$ for all τ' in this interval. But it has been proved (Diner and Martonchik 1984a, 1984b) that for a given choice of $\Delta\tau$, the accuracy of Herman and Browning's scheme could be improved if a linear variation of $J(\tau',\Omega)$ with τ' is assumed inside the integral instead of using a value independent of τ'. For three arbitrary layers $\tau_i \leq \tau' \leq \tau_{i+2}$, the linear relation can be given by

$$
\begin{cases}
J(\tau', \Omega) = J(\tau_i, \Omega) + \dfrac{J(\tau_{i+1}, \Omega) - J(\tau_i, \Omega)}{\Delta\tau}(\tau' - \tau_i) & \text{downward} \\[4mm]
J(\tau', \Omega) = J(\tau_{i+2}, \Omega) + \dfrac{J(\tau_{i+2}, \Omega) - J(\tau_{i+1}, \Omega)}{\Delta\tau}(\tau' - \tau_{i+2}) & \text{upwelling}
\end{cases}
$$

$$(3.73)$$

Replacing the source function here by the corresponding term in Eq. (3.72), we finally obtain after some algebraic manipulation

$$
\begin{cases}
I^M(\tau_{i+2}, -\Omega) = I^M(\tau_i, -\Omega) \exp[-2\Delta\tau'] \\
\qquad + \dfrac{J(\tau_{i+1}, -\Omega)}{G(\Omega)} u_1 - \dfrac{J(\tau_i, -\Omega)}{G(\Omega)} u_2 & \text{downward} \\[4mm]
I^M(\tau_i, \Omega) = I^M(\tau_{i+2}, \Omega) \exp[-2\Delta\tau'] \\
\qquad + \dfrac{J(\tau_{i+1}, \Omega)}{G(\Omega)} u_1 - \dfrac{J(\tau_{i+2}, \Omega)}{G(\Omega)} u_2 & \text{upwelling}
\end{cases}
$$

$$(3.74)$$

where $\Delta\tau'$ is the effective optical depth of each layer in the direction μ defined by

$$\Delta\tau' = \frac{G(\Omega)\,\Delta\tau}{\mu} \tag{3.75}$$

and

$$
\begin{cases}
u_1 = 2 - \dfrac{1 - \exp(-2\Delta\tau')}{\Delta\tau'} \\[4mm]
u_2 = 1 - \dfrac{1}{\Delta\tau'} + (1 + \Delta\tau') \exp(-2\Delta\tau')
\end{cases}.
$$

$$(3.76)$$

For the top and bottom layers, we have

$$
\begin{cases}
I^M(\tau_1, -\Omega) = I^M(\tau_0, -\Omega) \exp[-\Delta\tau'] \\
\qquad + J(\tau_0, -\Omega)[1 - \exp(-\Delta\tau')] & \text{downward} \\[2mm]
I^M(\tau_{N-1}, \Omega) = I^M(\tau_N, \Omega) \exp[-\Delta\tau'] \\
\qquad + J(\tau_N, \Omega)[1 - \exp(-\Delta\tau')] & \text{upwelling}
\end{cases}
$$

$$(3.77)$$

According to Herman and Browning's assumption about the source function within the medium, the iteration formula corresponding to Eq. (3.74) can be written as

$$
\begin{cases}
I^M(\tau_{i+2}, -\Omega) = I^M(\tau_i, -\Omega)\exp[-2\Delta\tau'] \\
\qquad\qquad + \dfrac{J(\tau_{i+1}, -\Omega)}{G(\Omega)}[1 - \exp(-2\Delta\tau')] \quad \text{downward} \\
I^M(\tau_i, \Omega) = I^M(\tau_{i+2}, \Omega)\exp[-2\Delta\tau'] \\
\qquad\qquad + \dfrac{J(\tau_{i+1}, \Omega)}{G(\Omega)}[1 - \exp(-2\Delta\tau')] \quad \text{upwelling}
\end{cases}
\tag{3.78}
$$

Choosing a finite set of M directions using Gauss–Legendre quadrature in the unit sphere, we can discretize the angular variables. The Gauss–Seidel iteration technique works as follows. Start at the top of the canopy using Eq. (3.77) and work downward at successive interfaces using Eq. (3.74) or (3.78) until the lower boundary (soil) is reached. Reflection of the background given at the bottom using Eq. (3.77) again provides the upwelling radiance, and then in similar manner work upward at successive interfaces until the upper boundary is reached. At this state, update the source function based on (3.73). This downward/upward cycling is continued until convergence is obtained.

One thing to note about this algorithm is its parallel property. From Eq. (3.73) and (3.74) or (3.78), it can be shown that for every discrete direction (Ω_{ij}) the iteration formulas are identical and independent of each other. If multiple processors (e.g., a supercomputer) are available, each processor can manipulate the iteration for a specific angle Ω_{ij}, and the calculation time will be dramatically reduced. This is desirable because solving such radiative transfer equations is very time-consuming.

Further, after calculating the downward radiance of the bottom layer, the upward radiances are calculated on the basis of the bottom reflectance function, which can be of any form. Thus, a non-Lambertian boundary condition can be used without extra computational expense. However, if we use the discrete ordinate approach, the non-Lambertian reflectance function usually needs to be expanded in Legendre polynomials (Stamnes et al. 1988). If the number of terms of the Legendre polynomials, which is equal to the number of streams in the discrete ordinate method, is not large enough, the constructed directional function would not represent the actual reflectance well.

Figure 3.17 illustrates the upwelling radiance at the top of both canopy and atmosphere at different LAI levels calculated using the Gauss–Seidel algorithm (Liang and Strahler 1993a). In the near-IR region, when LAI increases, the upwelling radiances increase correspondingly, but the angular dependence remains similar. This observation implies that LAI retrieval is not enhanced by fine resolution in view angles under this particular situation. The upwelling radiance in the visible region above the atmosphere does not vary significantly with increasing LAI if LAI > 2.0. Thus, it will be quite

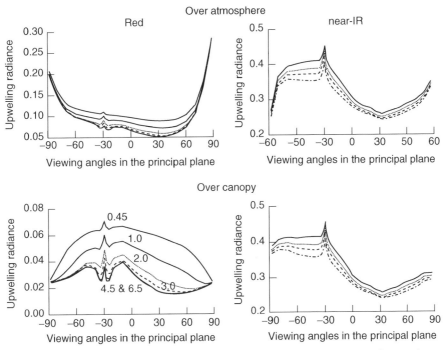

Figure 3.17 Impacts of canopy LAI on upwelling radiance above both canopy and atmosphere. The solar zenith angle is 30°. In the visible region, five curves represent five LAI values: 0.45, 1.0, 2.0, 3.0, 4.5, and 6.5. In the near-IR region, four curves represent four LAI values: 2.0 (dotted dashed line), 3.0 (dashed line), 4.0 (dotted line), and 6.5 (solid line). Other parameters are available in the original paper. [From Liang and Strahler (1993a), *IEEE Trans. Geosci. Remote Sens.* Copyright © 1993 with permission from IEEE.]

difficult to retrieve large LAI values using only visible bands. This also explains why the normalized difference vegetation index (NDVI) saturates when LAI values are high.

We can also observe the atmospheric disturbance from Fig. 3.17. In the visible region, the atmosphere largely masks the hotspot, and the angular distribution of radiance above the atmosphere mainly reflects the atmospheric path radiance characteristics. The forward scattering peaks are also detectable. In the near-IR region, however, the canopy has a higher reflectivity and aerosol optical depth becomes much smaller. Thus the radiances received by sensors above the atmosphere basically depend on the canopy radiation field. Large attenuation occurs in the large viewing angles in the backscattering direction.

3.4 GEOMETRIC OPTICAL MODELS

The geometric optical models assume that the canopy consists of a series of regular geometric shapes (e.g., cylinders, spheres, cones, ellipsoids), placed

Sunlit crown, sunlit/shaded ground/crown, shaded ground/crown

Figure 3.18   Simulated canopy field with the dominating components used in canopy geometric optical modeling.

on the ground surface in a prescribed manner (Fig. 3.18). For sparse vegetation canopies, the reflectance/radiance is the area-weighted sum of different sunlit/shadow components. The fractions of different components are calculated based on the geometric optical principles.

Canopy geometric optical (GO) modeling has a relatively short history, and the earliest models were developed in the mid-70s (e.g., Richardson et al. 1974, Ross and Nilson 1975, Jackson et al. 1979, Norman and Welles 1983). The application of GO modeling to remotely sensed data was simulated by Li and Strahler (1985, 1986), who used simple cone geometry to represent conifer tree crowns. In the later developments, they included the ellipsoidal crown shapes and the effects of mutual shadowing (Li and Strahler 1992, Strahler and Jupp 1990), incorporated radiative transfer processes into GO modeling (Li et al. 1994), and extended them to thermal radiation regimes (Li et al. 1999).

Several other groups in the canopy reflectance modeling community have also developed different types of GO models (e.g. Jupp et al. 1986, 1996; Jupp and Strahler 1991, Nilson and Peterson 1991, Hall et al. 1997, Ni et al. 1997, 1999; Chen 2001). Most of these models and their applications have been evaluated in a review article (Chen et al. 2000). The GO principles have

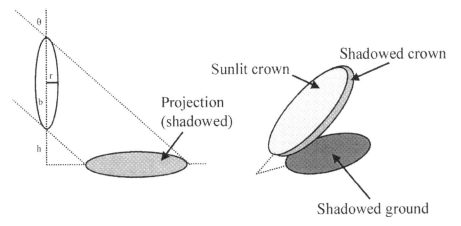

Figure 3.19 An illustration of the ellipsoid crown of the tree in the canopy geometric-optical model [after Strahler and Jupp (1990)].

also been applied in microwave (Sun and Ranson 1995) and LiDAR (light detection and ranging) (Sun and Ranson 2000, Ni-Meister et al. 2001) remote sensing modeling.

To demonstrate the basic principles of GO modeling, we will present the simple GO model developed by Li and Strahler (1985). In this model, a tree is taken as a simple geometric object, an "ellipsoid on a stick" (Fig. 3.19). Tree counts vary from pixel to pixel as a Poisson function. The reflectance of a pixel (S) is an area-weighted sum of the signatures for four components: sunlit crown (C), sunlit background (G), shadowed crown (T), and shadowed background (Z):

$$S = K_g G + (1 - K_g)(K_c C + K_z Z + K_t T) \qquad (3.79)$$

In this expression, K_c, K_z, K_t, are the areal proportions of sunlit crown, shadowed background, and shadowed crown. They are normalized so that $K_c + K_z + K_t = 1$. K_g is simply the proportion of the pixel remaining uncovered by trees or shadows. C, G, T, and Z can be estimated from remotely sensed data.

The three-dimensional geometry of an ellipsoid on a stick casting a shadow on a background, yields the following formulas for the areas in sunlit crown, shadowed crown, and shadowed background:

$$\begin{cases} A_c = \dfrac{\pi}{2}r^2(1 + B) \\[2mm] A_t = \dfrac{\pi}{2}r^2(1 - B) \\[2mm] A_z = r^2(\Gamma - \pi) \end{cases} \qquad (3.80)$$

In these expressions, B is a geometric parameter that describes the proportion of the ellipsoid that is shadowed, Γ is a geometric parameter equal to the area of the sunlit crown, shadowed crown, and shadowed background created by an ellipsoid with a radius of one unit. It is a function of the tree geometry, parameterized by r, b, and h, where r is the horizontal radius of the ellipsoid, b is the vertical radius, and h is the height of the stick; and the solar angles. Γ is derived in Franklin and Strahler (1988) but was reproduced incorrectly:

$$\Gamma = \pi + \frac{\pi}{\cos \theta'} - A_0 \qquad (3.81)$$

where

$$A_0 = \left(\beta - \frac{\sin 2\beta}{2} \right)\left[1 + \frac{1}{\cos \theta'} \right] \qquad (3.82)$$

if $(b + h)\tan \theta > r(1 + 1/\cos \theta')$; otherwise $A_0 = 0$. In this expression,

$$\beta = \cos^{-1}\left[\left(1 + \frac{h}{b} \right)\left(\frac{1 - \cos \theta'}{\sin \theta'} \right) \right] \qquad (3.83)$$

and

$$\theta' = \tan^{-1}\left(\frac{b}{r} \tan \theta \right), \qquad (3.84)$$

The factor B is defined as

$$B = \sqrt{\frac{1 + 16K^2 L^4}{4L^2 + 16K^2 L^4}} \cos \delta \qquad (3.85)$$

where $L = r/(2b)$; $K = \tan(\pi/2 + \theta)$; and $\delta = \tan^{-1}[-1/(4KL^2)]$.

The formulas can be used for calculating K_c, K_z, K_t when the canopy density is low. However, as trees get closer to one another, shadows will be more likely to fall on other tree crowns, increasing the area of shadowed crown and reducing the area of shadowed background. Instead, the following formulas were derived for calculating the proportions of different components:

$$\begin{cases} K_g = e^{-m\Gamma} \\[2mm] K_c = 0.5(1 + \cos \theta')\dfrac{1 - e^{-m\pi}}{1 - e^{-m\Gamma}} \\[2mm] K_z = e^{-m\pi}\dfrac{1 - e^{-m\pi}}{1 - e^{-m\Gamma}} \\[2mm] K_t = 0.5(1 + \cos \theta')\dfrac{1 - e^{-m(\Gamma - \pi)}}{1 - e^{-m\Gamma}} \end{cases} \qquad (3.86)$$

where m characterizes the density of one canopy. These formulas are for the canopies above a flat surface. Topography also needs to be taken into account if the ground is not a perfectly flat plane. If the slope and aspect of the surface are considered, the only modification is to replace the geometric parameter Γ by Γ'

$$\Gamma' = \pi + (\Gamma - \pi)\frac{\cos(\alpha)\cos(\theta_0)}{\cos(\theta - \beta_y)} \qquad (3.87)$$

where α is the slope angle, θ_0 is the solar zenith angle,

$$\beta_y = \mathrm{atan}\left[\tan(\alpha)\cos(\phi)\right] \qquad (3.88)$$

and ϕ is the relative azimuth angle of the slope to the solar azimuth angle.

This simple model has been used for mapping forest tree size and density from both Landsat TM and SPOT imagery (e.g., Franklin and Strahler 1988, Franklin and Turner 1992, Woodcock et al. 1994, 1997). Information on tree size and density, in conjunction with forest types can be used for timber inventory and other resource management.

3.5 COMPUTER SIMULATION MODELS

Both radiative transfer models and geometric optical models treat canopy as a set of statistical ensembles defined over a volume with averaged properties. For accurate computation of the radiation distribution over a complex canopy configuration, computer simulation models are desirable. Given a structural model (one composed of geometric primitives: triangular facets, parametric surfaces, etc.) of an ensemble of plants in a canopy and some method of solving for radiative transfer between surfaces for given viewing and illumination conditions, one can simulate the radiation regime of a canopy for a given set of radiometric attributes of the plant primitives. Two typical methods are Monte Carlo ray tracing (MCRT) and the radiosity method, both of which have been widely used in computer graphics applications.

Both MCRT and radiosity are based on the so-called *light transport* (or *rendering*) equation in one way or another. At any point $\mathbf{r}(x, y, z)$ in the canopy, the exiting radiance $L(\mathbf{r}, \mu, \phi)$ consists of emitted radiance $L_e(\mathbf{r}, \mu, \phi)$ and the all incoming radiance $L_i(\mathbf{r}, \mu, \phi)$ from the surrounding environment reflected by the surface at that point:

$$L(\mathbf{r}, \mu, \phi) = L_e(\mathbf{r}, \mu, \phi)$$
$$+ \int_0^{2\pi}\int_{-1}^{1} L_i(\mathbf{r}, \mu', \phi')R(\mathbf{r}, \mu', \phi', \mu, \phi)\mu'\,d\mu'\,d\phi \qquad (3.89)$$

where $R(\mathbf{r}, \mu', \phi', \mu, \phi)$ is the surface BRDF.

Rendering algorithms are expected to find the solution $L(\mathbf{r}, \mu, \phi)$ of this rendering equation at least for those points $\mathbf{r}$ and directions Ω that are visible from the sensor at direction $-\Omega$. Note that the unknown radiance is present both inside and outside the integral of the equation. These types of integral equations are called *Fredholm integral equations of the second kind*.

3.5.1 Monte Carlo Ray Tracing Models

Ray tracing methods are based on a sampling of photon trajectories within the vegetation canopies. The processing time required for this does not increase dramatically as a scenes complexity increases, and ray tracing can therefore provide a more "scalable" generic solution for radiative transfer. Ray tracing methods are useful for a wide variety of radiation transport problems, and are the only really appropriate methods for applications where path length is specifically required (Disney et al. 2000). Modeling LiDAR (Govaerts et al. 1996, Lewis 1999), and SAR interference effects (Lin and Sarabandi 1999) are examples of this application. The key to effective use of ray tracing methods is the application of effective sampling schemes. The basis for the selection of photon trajectories and various other aspects of ray tracing is the Monte Carlo (MC) method (Disney et al. 2000).

MC methods are most commonly employed in the simulation of canopy reflectance through Monte Carlo ray tracing (MCRT). Intuitively, the intrinsic canopy reflectance, the canopy BRDF, is considered as an averaged probability of a sample set of photons (or rays in the direction of a wavefront, "fired" into a scene) incident on the canopy from a given direction and leaving per unit solid angle around another direction. Alternatively, the bidirectional reflectance factor (BRF) can be considered the probability of photons traveling from a given direction leaving in another, relative to their behavior when scattered by a perfect Lambertian horizontal reflecting surface.

Using the intuitive approach outlined above, the aim is to calculate the required probabilities by simulating the firing of photons into a scene. Within a given constant density medium, a ray, describing a photon trajectory (or alternatively, the direction of propagation of an electromagnetic wave), will travel along a straight line. Consequently, the main computational issue becomes one of testing the intersection of a set of lines (rays) with a defined scene geometric representation. For canopy reflectance modeling, the scene will typically include a lower (ground) boundary, so a ray traveling from a virtual sensor into the scene will intersect with either a vegetation element or a ground element. At this point, the photon is either scattered (reflected or transmitted) or absorbed; that is, the sum of the probabilities of reflectance (P_r), transmittance (P_t), and absorptance (P_a) equal unity. The integral over all conditions can be simulated using the MC approach by generating a random number $\Re$ over the interval $(0, 1)$. If $\Re$ is less than or equal to P_a, then the photon is absorbed. If $\Re$ is between P_a and $P_a + P_r$, the photon is reflected (i.e., scattered from the same side of the object on which the path was initially incident). Otherwise, the photon is transmitted. In the latter two cases, the BRDF of the scattering primitive (e.g., leaf reflectance function) is used as a probability density function to relate another random number to a scattering direction. The trajectory of the scattered photon is then followed until interception by another primitive, or the photon escapes the scene.

In the following, some basic concepts are introduced and then followed by discussions of three different algorithms in Sections 3.5.1.3–3.5.1.5.

3.5.1.1 *Forward and Reverse Ray Tracing*

An important distinction that must be made between MCRT algorithms is that they involve either "forward" or "reverse" ray tracing. In the former, sample photon trajectories are "traced" from illumination sources through to a sensor; in the latter the trajectories traced from the sensor are used to sample the scattering that could have originated at the illumination sources (Disney et al. 2000). Figure 3.20 illustrates the basic principles of both forward and reverse tracing. The forest MCRT that we describe in Section 3.5.1.3 is forward tracing. The botanical plant modeling system (BPMS) developed by Lewis (1999) is based on a reverse tracing algorithm.

The combination of forward tracing and reverse ray tracing is called *bidirectional ray tracing*, because rays are traced in both directions (Glassner 1995).

3.5.1.2 *Canopy Scene Generation*

Scene generation was briefly mentioned in Section 1.3.1. As summarized by Disney et al. (2000), there are also many different methods for generating canopy configurations. Progress is being made toward the type of generic model from various quarters, driven in part by cheaper, faster computers, computer graphics algorithms, and the rapidly emerging field of 3D plant measurement and modeling (Prusinkiewicz 1999). The latter developments lead to the possibility of representing canopy structure as accurately as required, through 3D scanning methods (Room et al. 1996), stereophotogrammetry (Lewis and Boissard 1997), manual measurements (Lewis 1999), algorithmic growth models such as that of Prusinkiewicz and Lindenmayer (1990), or models driven by botanical growth rules (De Reffye and Houllier 1997). Prusinkiewicz (1996), and Mech and Prusinkiewicz (1996) discuss the application of L-system-based models of plant growth in areas such as ecology and epidemiology. Fournier and Andrieu (1999) have developed a model that couples canopy organ growth as a function of temperature and carbon availability to 3D spatial variations of light and temperature within the canopy. Chelle and Andrieu (1999) describe a range of developments in the coupling of numerical solutions of radiation transport within canopy radiation models to physiologically based models for the purpose of characterizing canopy development as a function of incident radiation. The LIGNUM model of Perttunen et al. (1996) is another approach to the

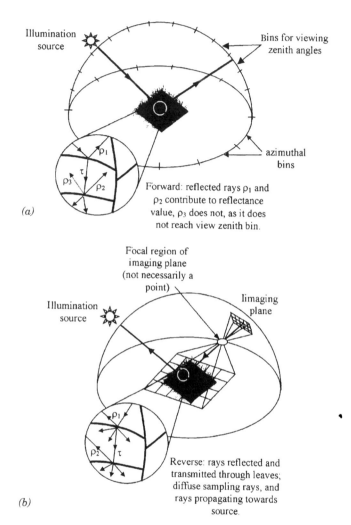

Figure 3.20 Schematic representation of both forward (a) and backward (b) ray tracing of a canopy field [from Disney et al. (2000)].

construction of process models of plant growth (trees in this case). LIGNUM is an attempt to simplify the treatment of metabolic functions controlling growth in the context of structurally detailed 3D plant models.

3.5.1.3 A Forest Ray Tracing Algorithm

North (1996) calculated forest directional reflectance using the MCRT algorithm. To illustrate the basic principle of the MCRT, a detailed description is provided. In this model, forest macrostructure is represented by a set of geometric primitives, positioned in three dimensions above a horizontal plane. The primitives define crown shape (e.g., cone) and size, and may overlap. Crown positions may be directly specified for an individual scene, or estimated from a statistical distribution or growth model. All nonunderstory foliage is considered to lie within the crown boundaries. Within each crown, foliage is approximated by structural parameters of area density, angular distribution and size, and leaf optics (reflectance and transmittance). This algorithm consists of the following steps:

1. Initialize a photon at a position $\mathbf{r}_0$ above the canopy, from source direction Ω_0 and with radiance L^0 set to unity.
2. Simulate the free pathlength s to the next collision. According to radiative transfer theory, the probability of a collision occurring before traveling a distance s through a medium is defined by

$$\int_0^s P(t)\, dt = 1 - \exp(-s\sigma) \qquad (3.90)$$

where $P(t)$ is the collision probability function, governed by the extinction coefficient σ. The free pathlength s can be simulated from a random number R with uniform distribution in the range $(0 < R < 1)$ by the transformation

$$\int_0^s P(t)\, dt = R \qquad (3.91)$$

and thus

$$s = -\frac{\ln(1 - R)}{\sigma} \qquad (3.92)$$

3. Calculate the new photon position from the previous position $\mathbf{r}'$ along the direction Ω'

$$\mathbf{r} = \mathbf{r}' + s\Omega' \qquad (3.93)$$

and new radiance value

$$L = L' \exp(-s\sigma) \qquad (3.94)$$

At each step, the distance s_b to the external boundary of the crown is calculated, and also the distance s_t to the internal trunk, if it lies in the photon path. If the photon path intersects an external boundary of the canopy within distance s, then the photon leaves the crown. Intersection with the internal trunk results in a scattering event. Otherwise, scattering occurs with leaf material.

4. Simulate the new scattering direction and radiance after collision Ω at $\mathbf{r}$. On interception by the medium, the photon may be either absorbed

or scattered in a new direction. If it is absorbed, a new photon is supposed to be launched from the source. This is not an efficient procedure. Instead, aggregate photon radiance is modeled with a continuous probability function. The collision always occurs, and radiance becomes

$$L = L'\Gamma(\mathbf{r}, \Omega' \to \Omega) \tag{3.95}$$

where $\Gamma(\mathbf{r}, \Omega', \Omega)$ denotes the conditional probability of scattering from direction Ω' to a unit solid angle about Ω at $\mathbf{r}$, given that a collision has occurred. It is numerically equivalent to the volume scattering phase function for leaves. For a medium with leaves and branches, the averaged volume scattering function shall be used, which can be approximated by

$$\Gamma(\mathbf{r}, \Omega', \Omega) = p\Gamma^L(\mathbf{r}, \Omega', \Omega) + (1 - p)\Gamma^B(\mathbf{r}, \Omega', \Omega) \tag{3.96}$$

where p is the expected proportion of the area of leaf and the superscripts L and B denote leaf and branch, respectively. The new direction can be simulated by sampling a uniform distribution over a sphere, and the bias in the scattering direction can be corrected by the phase function. Two uniform random numbers ξ and ζ are generated in the range $(0, 1)$, to determine the polar coordinates (θ, ϕ) of the new direction

$$\theta = \cos^{-1}(2\xi - 1) \tag{3.97}$$

and

$$\phi = 2\pi\zeta \tag{3.98}$$

5. Accumulate radiance $L(\Omega)$ in bin corresponding to the viewing angle Ω if the photon has left canopy; otherwise, repeat from step 2, until L has dropped below some threshold.
6. Repeat from step 1, for each new photon until the total number of photons reaches the predefined maximum value.

An additional advantage of the method is that a single photon path can be used to simulate reflectance for any number of wavebands. There will be an additional cost per waveband in processing, in that the cumulative probability/attenuation needs to be updated for each waveband, but the time taken to process this is generally very small compared to the time spent in performing geometric intersection tests, particularly for a complex scene. Lewis (1999) terms this concept a "ray bundle." Dawson et al. (1999) demonstrate the use of this concept in simulating high spectral resolution data over a forest.

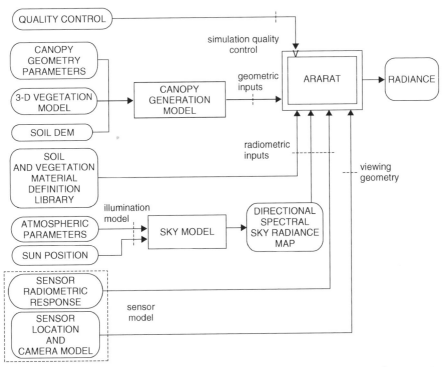

Figure 3.21 Flow chart of the BPMS canopy Monte Carlo ray tracing model [from Lewis (1999)].

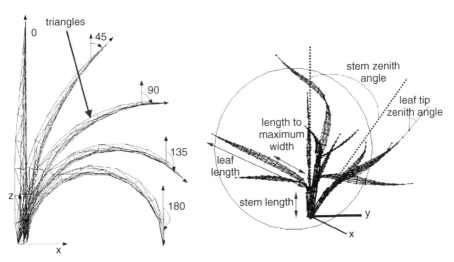

Figure 3.22 Generation of a canopy leaf using the L system [after Lewis (1999) and Disney (2002)].

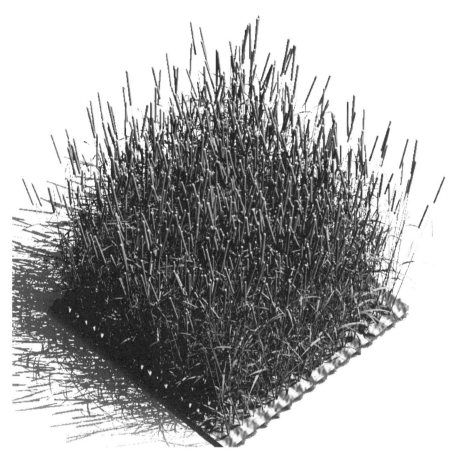

Figure 3.23 A canopy field generated by the L system based on measurements of a barley canopy on June 4, 1997 with LAI = 3.16 [after Disney et al. (2000)].

3.5.1.4 Botanical Plant Modeling System Model

Figure 3.21 is a flowchart of the botanical plant modeling system (BPMS) developed by Lewis (1999). It contains three major components: canopy field generation, sky radiance generation, and a ray tracer called ARARAT (advanced radiometric ray-tracer). For generating the canopy field, the L (Lindenmayer) system is used. The topology string in BPMS can generalize the measured key parameters of the leaf canopy (Fig. 3.22), such as number of leaves, canopy heights, leaf length and angles, and characteristics of canopy variations. The L system can be used to "clone" plants—replications with random rotations (in azimuth). Figure 3.23 is the barley field generated using this approach. It was based on only three or four plants measured at Norfolk, UK and modeled soil reflectance. The rest were cloned and rotated, and replicated in all directions.

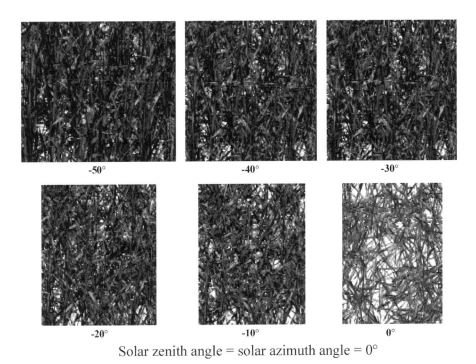

-50°   -40°   -30°

-20°   -10°   0°

Solar zenith angle = solar azimuth angle = 0°

Figure 3.24   Simulated barley reflectance at different viewing zenith angles (negative angles correspond to those in the relative azimuth angle of 180°). Solar zenith angle and solar azimuth angle are zero.

After generating the canopy field, the ARARAT ray tracer can be used to simulate a canopy radiation field given the illumination conditions (direct solar lights plus diffuse sky radiance). One example is shown in Fig. 3.24, demonstrating what the canopy looks like at different view zenith angles. It is extremely valuable to understand the characteristics of the canopy radiation field given the canopy configuration and leaf optical properties. It can also be used for validating other simple models, training inversion models for retrieving biophysical parameters from remote sensing imagery.

3.5.1.5   SPRINT Model

To overcome the prohibitive computer memory and computation time requirements of ordinary Monte Carlo ray tracing for kilometer-scale scenes, Goel and Thompson (2000) developed a "universal" BRDF SPRINT (*s*preading of *p*hotons for *r*adiation *int*erception) model. This model generates radiation photon paths just as the traditional Monte Carlo method. The major difference is that at randomly chosen steps, the photon partially spreads out as a continuous wave from each collision point. When it spreads, it contributes to BRDF at all viewing directions. Thus, this model is approximately 1000 times faster than a traditional Monte Carlo ray tracing method.

3.5.2 Radiosity Models

Radiosity methods are widely used in computer graphics for realistic scene rendering, and have also found application in canopy reflectance modeling (Borel et al. 1991, Goel et al. 1991). A major advantage of the method is that once a solution is found for radiative transport, canopy reflectance can be simulated at any view angle. Although various acceleration methods can be applied, a major limitation of the method is the initial computational load in forming the view factor matrix and solving for radiative transport. This is particularly true for very complex scenes involving a large number of scattering primitives. However, radiosity remains a useful technique in scenes characterized by a relatively simple set of primitives.

The conceptual difference between radiative transfer (RT) modeling method and the radiosity method can be well understood by comparing Figs. 3.25 and 3.6. The RT method is based on the average optical properties and configuration of the canopy field (1D) or a cell in the three-dimensional case. This average process eliminates the discrete nature of the leaves and creates a continuous medium description. In contrast, the radiosity method retains the discrete nature of each leaf as a reflecting and transmitting surface. The information about the location, shape, and orientation of all leaves is retained in the radiosity equation. These two methods are compared in Table 3.1.

Radiosity, also termed *radiant existence* (Nicodemus et al. 1977), describes the amount of total energy leaving a surface per unit time per unit area (flux density). For performing radiosity analysis, the canopy field is divided into discrete surface areas (facets). The radiosity equation describes an equilibrium radiation energy balance within an enclosure that contain N discrete surfaces ($I = 1, 2, \ldots, 2N$) (Borel et al. 1991):

$$B_i\, dS_i = E_i\, dS_i + \chi_i \sum_{j=1}^{2N} \int_{S_j} B_j F_{dS_j \to dS_i}\, dS_j \qquad (3.99)$$

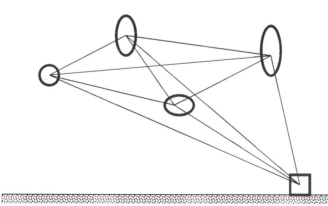

Figure 3.25 Conceptual illustration of the radiosity method. The circle, ellipses, and square represent the individual discrete objects considered in the radiosity model.

**TABLE 3.1   Comparison of Radiative Transfer Modeling and
Radiosity Modeling**

Radiative Transfer	Radiosity
Volume scattering	Surface reflection and transmission
Continuous medium	Discrete and oriented surfaces
Averaged scattering phase function	Explicit scattering characteristics
No physically based spatial correlations of leaves	Spatial correlations retained
No holes or clumps in canopy	Holes and clumps describable
Multiple scattering	Multiple scattering
Integrodifferential equation	System of coupled linear equations

Source: Gerstl and Borel (1992).

where

$$\chi_i = \begin{cases} \rho_i, & \text{if } \left(\vec{n}_i \cdot \vec{n}_j\right) < 0 \\ \tau_i, & \text{if } \left(\vec{n}_i \cdot \vec{n}_j\right) > 0 \end{cases} \tag{3.100}$$

and B_i = radiosity of differential area dS_i: sum of the emitted, re-flected, and transmitted radiative energy flux (W m^{-2})

E_i = emitted radiosity from differential area dS_i

ρ_i = hemispherical reflectance of differential area dS_i

τ_i = hemispherical transmittance of differential area dS_i

$\vec{n}_i$ = the normal vector on a surface i pointing outward

$F_{dS_j \to dS_i}$ = view factor or form factor: fraction of radiative energy leaving the infinitesimal surface dS_j that reaches another infinitesimal surface dS_i

N = number of discrete surfaces, where $2N$ is the number of (single-sided) surface components S_i

Equation (3.99) states that the radiosity leaving a differential area dS_i equals the sum of emitted light plus reflected and/or transmitted fractions of radiosities transferred onto dS_i from all other surface elements dS_j and integrated over all these other surfaces. Figure 3.26 shows the radiosity relations described in Eq. (3.99).

The radiosity algorithm developed by Borel et al. (1991) includes five steps, each of which will be discussed from Sections 3.5.2.1–3.5.2.5, the text of which draws heavily from their paper.

3.5.2.1 *Generating the 3D Scene*
This step is similar to that used in Monte Carlo ray tracing method (see Section 3.5.1.2). We also very briefly mentioned it in Section 1.3.1.

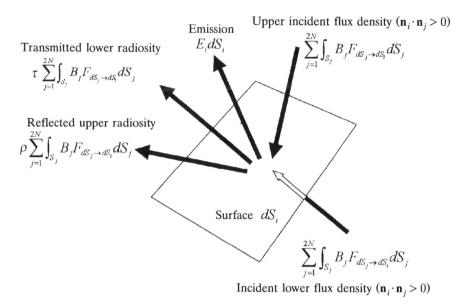

Figure 3.26 Radiosity relations for a surface patch [after Borel et al. (1991)].

3.5.2.2 Calculating the Emission for All Surfaces in the Scene

For vegetation canopies, all surfaces (or patches) that are directly illuminated by solar radiation are considered to be emitting surfaces. These illuminated surface patches then become the emission terms in the radiosity equations. The emission E_i on surface i at local coordinates (x, y) is given by

$$E_i(x, y) = \chi_i \, \text{HID}(x, y) E_0 |\vec{n}_i \vec{s}| \qquad (3.101)$$

where $\text{HID}(x, y) = $ ("hiding") occlusion factor (it is zero if another surface is in the direct line of sight between location (x, y) on the surface i and the Sun, and 1 otherwise)

$E_0 = $ incident solar irradiance

$\vec{s} = $ vector pointing in the solar illumination direction

The average emission for one specific surface can be determined by

$$E_i = \frac{1}{S_i} \int_{S_i} E_i(x, y) \, dx \, dy, \qquad i = 1, \ldots, 2N \qquad (3.102)$$

3.5.2.3 Computing the View Factors

One of the most challenging tasks in the radiosity algorithm is the calculation of view factors. Some view factors for specific shapes and orientations can be

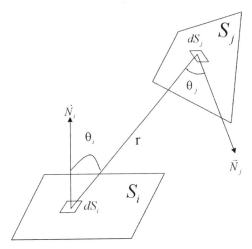

Figure 3.27   Illustration of the viewing factor.

calculated analytically, such as parallel directly opposed rectangles and disks, infinitely long parallel cylinders, or a sphere to a disk. However, under most conditions, numerical approximations are needed. For two infinitesimal surfaces, the viewing factor can be expressed (Sparrow and Cess 1978):

$$F_{dS_i \to dS_j} = \frac{\cos \theta_i \cos \theta_j \, dS_j}{\pi r^2} \tag{3.103}$$

where r is the distance between two differential elements dS_i and dS_j. For any two facets S_i and S_j, an integration over two surfaces is needed (see Fig. 3.27). The final formula can be written

$$F_{S_i \to S_j} = \frac{1}{S_i} \int_{S_i} \int_{S_j} \frac{\cos \theta_i \cos \theta_j \, dS_i \, dS_j}{\pi r^2} \tag{3.104}$$

In Eq. (3.104) the view factor does not account for possible occlusion of parts of one surface by some object or another surface between S_i and S_j. An additional term HID to account for occluding or hiding objects is needed:

$$F_{S_i \to S_j}^{\text{HID}} = \frac{1}{S_i} \int_{S_i} \int_{S_j} \frac{\cos \theta_i \cos \theta_j \, \text{HID} \, dS_i \, dS_j}{\pi r^2} \tag{3.105}$$

It is still very difficult to calculate view factors using Eq. (3.104) or (3.105) for complex geometries. Approximation methods, such as Nusselt's method, known as the *unit sphere method*, or "hemicube" method, have been developed. Details are omitted here but can be found in the literature (e.g., Cohen and Greenberg 1985, Borel et al. 1991, Goel et al. 1991).

3.5.2.4   Solving the Radiosity Equation

After determining the view factors, the next important issue is to solve the radiosity equation. The radiosity equation (3.99) can be written for finite areas or patches if one can assume that the radiosity and emission terms remain constant over all finite surface patches S_i:

$$B_i = E_i + \chi_i \sum_{j=1}^{2N} B_j F_{ij} \tag{3.106}$$

where the simplified notation for the view factor $F_{S_i \to S_j} \equiv F_{ij}$ is used. Equation (3.106) is a system of linear algebraic equation with known emission E_i, view factors F_{ij}, and reflectance/transmittance properties χ_i. The unknown quantities are the radiosities B_i. This equation system can be written in matrix form:

$$
\begin{bmatrix}
(1-\chi_1 F_{11}) & -\chi_1 F_{12} & \cdots & -\chi_1 F_{1(2N)} \\
-\chi_2 F_{21} & (1-\chi_2 F_{22}) & \cdots & -\chi_2 F_{2(2N)} \\
\vdots & \vdots & \ddots & \vdots \\
-\chi_{2N} F_{(2N)1} & -\chi_{2N} F_{(2N)2} & \cdots & (1-\chi_{2N} F_{(2N)(2N)})
\end{bmatrix}
\begin{bmatrix}
B_1 \\ B_2 \\ \vdots \\ B_{2N}
\end{bmatrix}
=
\begin{bmatrix}
E_1 \\ E_2 \\ \vdots \\ E_{2N}
\end{bmatrix}
$$

$$\tag{3.107}$$

If N is small, it is easy to solve this linear equation system. Unfortunately, N is very large for a reasonably complex scene. Therefore, an iteration method is needed. The Gauss–Seidel method has been most often used (Goral et al. 1984).

3.5.2.5   Rendering the Scene for a Given Viewpoint and Calculating BRF

The radiosity results can be displayed with computer graphics based on different algorithms, such as the Z-buffer algorithm, the depth sort algorithm, and the ray tracing algorithm. For land surface characterization, we need to calculate the bidirectional reflectance factor (BRF):

$$\mathrm{BRF}(\vec{n}_v) = \frac{\sum_i (B_i/\pi) |\vec{n}_i \vec{n}_v| A_i^v}{\sum_i (1/\pi) |\vec{n}_i \vec{n}_v| A_i^v} \tag{3.108}$$

where $\vec{n}_v$ is the viewing direction and A_i^v is the area of surface i seen by the viewer. Integration of BRF produces albedo under a specific illumination condition [see Eq. (1.10)].

3.5.2.6 Applications

Borel and Gerstl (1994) applied an analytic solution of the radiosity equation to compute vegetation indices, reflectance spectra, and the spectral bidirectional reflectance distribution function for simple canopy geometries. They showed that nonlinear spectral mixing occurs as a result of multiple reflection and transmission from surfaces. They also developed a simple model to predict the reflectance spectrum of binary and ternary mineral mixtures of faceted surfaces, and validated the two-facet model by measurements of the reflectance.

Goel et al. (1997) investigated the feasibility of estimating leaf size and crown size, for deciduous and coniferous trees, from canopy reflectance in the hotspot region using their radiosity model. For deciduous trees, as represented by aspen trees, it appears that under certain conditions one can estimate leaf size if we specify optimal Sun-view geometry, wavelength, and index, which can minimize the impacts of other structural parameters (e.g., leaf area index, leaf angle distribution, and interplant spacing) for the most accurate estimation of leaf size. However, for coniferous trees an accurate estimation of crown size seems unlikely, except possibly for sparsely spaced canopies.

Qin and Gerstl (2000) used a modified extended L-system method to generate a 3D realistic scene and a radiosity graphics combined method to calculate the radiation regime based on the 3D structures rendered with MELS. The 3D simulation tool is then evaluated using field measurements of both plant structure and spectra collected near Las Cruces, NM. The modeled scene reflectance was compared with measurements from three platforms (ground tower, and satellite) at various scales (from the size of individual shrub component to satellite pixels of kilometers), and an excellent agreement was found with measured reflectances at all sampling scales tested. As an example of the model's application, they used the model output to examine the validity of a linear mixture scheme over the Jornada semidesert scene. The results showed that the larger the sampling size (at least larger than the size of the shrub component), the better the hypothesis is satisfied because of the unique structure of the Jornada scene: dense plant clumps (shrub component) sparsely scattered on a predominantly bare soil background.

Qin et al. (2002) explored the potential of canopy reflectance distributions in the hotspot region for characterizing leaf geometry (leaf size and shape) of grass and crop canopies with their radiosity model. Their results from simulated hotspot reflectance demonstrate that for both canopies, leaf geometry is estimable by using normalized reflectance within $\pm 4-8°$ (or $\pm 2-4°$) around the hotspot direction in the principal cone (or principal plane). However, the center position and angular width of the optimal sampling region are affected by the number of noise factors, such as leaf area index, leaf angle distribution for leaf canopies, plus row structure for row-plant crop canopies, and their variation ranges. In most cases, normalized spectral reflectance in the near-infrared region at a high solar zenith angle in the principal component produces the most reliable results.

3.6 SUMMARY

We have discussed three types of modeling techniques for calculating canopy reflectance: radiative transfer, geometric optical, and computer simulation methods (e.g., Monte Carlo ray tracing and radiosity). The radiative transfer modeling technique is most suitable for dense vegetation canopies. Although three-dimensional radiative transfer models were developed to handle heterogeneity, geometric optical models are more suitable for handling sparse vegetation canopies with regularly shaped crowns. Efforts have been made to integrate these two modeling techniques together to generate the so-called hybrid canopy model. Computer simulation models have been applied primarily for understanding of the radiation regime and validation of some simplified models. More recent studies have also attempted to apply them to the retrieval of biophysical parameters, which are discussed in detail in Chapter 8.

All three modeling techniques were discussed, but the emphasis was on radiative transfer modeling. The radiative transfer formulation and numerical and approximate solutions were introduced. Note that all these solutions to canopy radiative transfer equations may be used for other media, such as atmosphere, soil, and snow. Various modeling techniques for calculating leaf optics were also outlined since leaf optical properties are required by all canopy models.

Although there have been a few attempts to compare models with laboratory measurements (e.g., Liang et al. 1997) and conduct model intercomparisons (e.g., Pinty et al. 2001), further efforts on model calibration and validations are definitely needed.

This chapter provides the theoretical foundation to develop various practical algorithms for estimating land surface biophysical variables discussed in Chapter 8.

REFERENCES

Allen, W. A. (1973), Transmission of isotropic light across a dielectric surface in two and three dimensions, *J. Opt. Soc. Am.* **63**: 664–666.

Allen, W. A., Gausman, H. W., and Richardson, A. J. (1970), Plant canopy irradiance specified by the Duntley equations, *J. Opt. Soc. Am.* **60**: 372–376.

Allen, W. A., Gausman, H. W., and Richardson, A. J. (1973), Willstatter-Stoll theory of leaf reflectance evaluated by ray-tracing, *Appl. Opt.* **12**: 2448–2453.

Allen, W. A., Gausman, H. W., Richardson, A. J., and Thomas, J. R. (1969), Interaction of isotropic light with a compact plant leaf, *J. Opt. Soc. Am.* **59**: 1376–1379.

Andrieu, B., Kiriakos, S., and Jaggard, K. W. (1992), Estimation of chlorophyll content of leaves using reflectance measurements or digitalized photographs of leaves taken in the laboratory, *Agronomie* **12**: 477–485.

Anisimov, O. A. and Menzhulin, G. V. (1981), The problem of modeling the radiation regime in the plant cover, *Sov. Meteorol. Hydrol.* 70–74.

Antyufeev, V. S. and Marshak, A. L. (1990), Inversion of monte carlo model for estimating vegetation canopy parameters, *Remote Sens. Envir.* **33**: 201–209.

Borel, C. C. and Gerstl, S. A. W. (1994), Nonlinear spectral mixing models for vegetative and soil surfaces, *Remote Sens. Envir.* **47**: 403–416.

Borel, C. C., Gerstl, S. A. W., and Powers, B. J. (1991), The radiosity method in optical remote sensing of structured 3-D surfaces, *Remote Sens. Envir.* **36**: 13–44.

Bunnik, N. J. J. (1978), *The Multispectral Reflectance of Shortwave Radiation by Agricultural Crops in Relation with Their Morphological and Optical Properties*, Pudoc, Wageningen, The Netherlands.

Campbell, G. S. (1984), Extinction coefficients for radiation in plant canopies calculated using an ellipsoidal inclination angle distribution, *Agric. Forest Meteorol.* **36**: 317–321.

Card, D. H. (1987), A simplified derivation of leaf normal spherical coordinates, *IEEE Trans. Geosci. Remote Sens.* **25**: 884–887.

Chelle, M. and Andrieu, B. (1999), Radiative models for architectural modeling, *Agronomie* **19**: 225–240.

Chen, J. M. and Leblanc, S. G. (2001), Multiple-scattering scheme useful for geometric optical modeling, *IEEE Trans. Geosci. Remote Sens.* **39**: 1061–1071.

Chen, J., Li, X., Nilson, T., and Strahler, A. (2000), Recent advances in geometrical optical modelling and its applications, *Remote Sens. Rev* **18**: 227–262.

Cohen, M. F. and Greenberg, D. P. (1985), The hemi-cube: A radiosity solution for complex environments, *Comput. Graph.* **19**: 31–40.

Dawson, T., Curran, P., and Plummer, S. (1998), LIBERTY: Modeling the effects of leaf biochemistry on reflectance spectra, *Remote Sens. Envir.* **65**: 50–60.

Dawson, T., Curran, P., North, P., and Plummer, S. (1999), The propagation of foliar biochemical absorption features in forest canopy reflectance: A theoretical analysis, *Remote Sens. Envir.* **67**: 147–159.

De Reffye, P. and Houllier, F. (1997), Modelling plant growth and architecture: Some recent advances and applications to agronomy and forestry, *Current Sci.* **73**: 984–992.

de Wit, C. T. (1965), *Photosynthesis of Leaf Canopies*, Pudoc, Wageningen, The Netherlands, Agric. Research Report 663.

Dickinson, R. E. (1983), Land surface processes and climate-surface albedos and energy balance, *Adv. Geophys.* **25**: 305–353.

Dickinson, R. E., Pinty, B., and Verstraete, M. M. (1990), Relating surface albedos in gcm to remotely sensed data, *Agric. Forest Meteorol.*, **52**: 109–131.

Diner, D. J. and Martonchik, J. V. (1984a), Atmospheric transfer of radiation above an inhomegenous non-lambertian reflective ground, I: Theory, *J. Quant. Spectrosc. Radiat. Transfer* **31**: 97–125.

Diner, D. J. and Martonchik, J. V. (1984b), Atmospheric transfer of radiation above an inhomegenous non-lambertian reflective ground, II: Computational considerations and results, *J. Quant. Spectrosc. Radiat. Transfer* **32**: 279–304.

Disney, M. I. (2002), *Improved Estimation of Surface Biophysical Parameters through Inversion of Linear BRDF Models*, Ph.D. thesis. Univ. London.

Disney, M. I., Lewis, P., and North, P. R. J. (2000), Monte Carlo ray tracing in optical canopy reflectance modeling, *Remote Sens. Rev.*, **18**: 163–196.

Esposito, L. (1979), Extensions to the classical calculation of the effect of mutual shadowing in diffuse reflection, *Icarus* **39**: 69–80.

Fournier, C. and Andrieu, B. (1999), Adel-maize: An L-system based model for the integration of growth processes from the organ to the canopy. Application to regulation of morphogenesis by light availability, *Agronomie* **19**: 313–327.

Franklin, J. and Strahler, A. (1988), Invertible canopy reflectance modeling of vegetation structure in semiarid woodland, *IEEE Trans. Geosci. Remote Sens.* **26**: 809–825.

Franklin, J. and Turner, D. L. (1992), The application of a geometric optical canopy reflectance model to semiarid shrub vegetation, *IEEE Trans. Geosci. Remote Sens.* **30**: 293–301.

Fukshansky, L. (1991), Photon transport in leaf tissue: Applications in plant physiology, in *Photon-Vegetation Interactions*: *Applications in Optical Remote Sensing and Plant Ecology*, R. B. Myneni and J. Ross, eds., Springer-Verlag, pp. 253–302.

Ganapol, B. D., Johnson, L. F., and Bond, B. (1999), LCM2: A coupled leaf/canopy radiation transfer model, *Remote Sens. Envir.* **70**: 153–166.

Gastellu-Etchegorry, J. P., Zagolski, F., and Romier, J. (1996a), A simple anisotropic reflectance model for homogeneous multilayer canopies, *Remote Sens. Envir.* **57**: 22–38.

Gastellu-Etchegorry, J. P., Demarez, V. P., and Zagolski, F. (1996b), Modeling radiative transfer in heterogeneous 3D vegetation canopies, *Remote Sens. Envir.* **58**: 131–156.

Gerstl, S. A. W. and Zardecki, A. A. (1985), Discrete-ordinates finite-element method for atmospheric radiative transfer and remote sensing, *Appl. Opt.* **24**: 81–93.

Gerstl, S. A. W. and Borel, C. C. (1992), Principles of the radiosity method versus radiative transfer for canopy reflectance modeling, *IEEE Trans. Geosci. Remote Sens.* **30**: 271–275.

Glassner, A. (1995), *Principles of Digital Image Synthesis*, Morgan Kaufmann Publishers.

Gobron, N., Pinty, B., Verstraete, M. M., and Govaerts, Y. (1997), A semidiscrete model for the scattering of light by vegetation, *J. Geophys. Res.* **102**: 9431–9446.

Goel, N. S. (1988), Models of vegetation canopy reflectance and their use in estimation of biophysical parameters from reflectance data, *Remote Sens. Rev.* **4**: 1–222.

Goel, N. S. and Strebel, D. E. (1984), Simple beta distribution representation of leaf orientation in vegetation canopies, *Agron. J.* **76**: 800–803.

Goel, N. S. and Grier, T. (1986a), Estimation of canopy parameters for inhomogeneous vegetation canopies from reflectance data. I. Two-dimensional row canopy, *Int. J. Remote Sens.* **7**: 665–681.

Goel, N. S. and Grier, T. (1986b), Estimation of canopy parameters from inhomogeneous vegetation canopies from reflectance data, II: Estimation of leaf area index and percentage of ground cover for row canopies, *Int. J. Remote Sens.* **7**: 1263–1286.

Goel, N. S. and Grier, T. (1988), Estimation of canopy parameters from inhomogeneous vegetation canopies from reflectance data, III: TRIM: A model for radiative transfer in heterogeneous three-dimensional canopies, *Remote Sens. Environ.* **25**: 255–293.

Goel, N. S. and Thompson, R. L. (2000), A snapshot of canopy reflectance models and a universal model for the radiation regime, *Remote Sens. Rev.* **18**: 197–225.

Goel, N. S., Rozehnal, I., and Thompson, R. I. (1991), A computer graphics based model for scattering from objects of arbitrary shapes in the optical region, *Remote Sens. Envir.* **36**: 73–104.

Goel, N. S., Qin, W. H., and Wang, B. Q. (1997), On the estimation of leaf size and crown geometry for tree canopies from hotspot observations, *J. Geophys. Res.* **102**: 29543–29554.

Goral, C., Torrance, K. E., Greenberg, D., and Battaile, B. (1984), Modeling the interaction of light between diffuse surfaces, *Proc. SIGGRAPH* **18**: 213–222.

Goudriaan, J. (1977), *Crop Micrometeorology: A Simulation Study*, Simulation Monographs.

Govaerts, Y. M., Jacquemoud, S., Verstraete, M. M., and Ustin, S. L. (1996), Three-dimensional radiation transfer modeling in a dicotyledon leaf, *Appl. Opt.* **35**: 6585–6598.

Gutschick, V. P. and Weigel, F. W. (1984), Radiative transfer in vegetative canopies and other layered media: Rapidly solvable exact integral equation not requiring Fourier resolution, *J. Quant. Spectrosc. Radiat. Transfer* **31**: 71–82.

Hall, F., Shimabukuro, Y. E., and Huemmich, K. (1997), Remote sensing of forest structure in boreal stands of picea mariana using mixture decomposition and geometric reflectance models, *Ecol. Appl.* **5**: 993–1013.

Hapke, B. W. (1981), Bidirectional reflectance spectroscopy 1. Theory, *J. Geophys. Res.* **86**: 3039–3054.

Herman, B. W. and Browning, S. R. (1965), A numerical solution to the equation of radiative transfer, *J. Atmos. Sci.* **22**: 559–566.

Hoffer, R. M. (1978), Biological and physical considerations in applyingcomputer-aided analysis techniques to remote sensor data, in *Remote Sensing: The Quantitative Approach*, P. H. Swain and S. M. Davis, eds., McGraw-Hill, pp. 227–289.

Idso, S. B. and de Wit, D. T. (1970), Light relations in plant canopies, *Appl. Opt.* **9**: 177–184.

Jackson, R. D., Reginato, R. J., Pinter, P. J. J., and Idso, S. B. (1979), Plant canopy information extraction from composite scene reflectance of row crops *Appl. Opt.* **18**: 3775–3782.

Jacquemoud, S. (1993), Inversion of the PROSPECT + SAIL canopy reflectance models from aviris equivalent spectra: Theoretical study, *Remote Sens. Envir.* **44**: 281–292.

Jacquemoud, S. and Baret, F. (1990), PROSPECT: A model of leaf optical properties spectra, *Remote Sens. Envir.* **34**: 75–91.

Jacquemoud, S., Baret, F., Andrieu, B., Danson, F. M., and Jaggard, K. (1995), Extraction of vegetation biophysical parameters by inversion of the PROSPECT + SAIL models on sugar beet canopy reflectance data. Application to TM and AVIRIS sensors, *Remote Sens. Envir.* **52**: 163–172.

Jupp, D. L. B. and Strahler, A. H. (1991), A hotspot model for leaf canopies, *Remote Sens. Envir.* **38**: 193–210.

Jupp, D. L. B., Walker, J., and Penridge, L. K. (1986), Interpretation of vegetation structure in landsat mss imagery: A case study in distributed semi-arid eucalypt woodland. Part 2. Model based analysis, *J. Envir. Manage.* **23**: 35–57.

Jupp, D. L. B., Walker, J., and Penridge, L. K. (1996), Interpretation of vegetation structure and growth changes in woodlands and forests: The challenge for remote sensing and the role of geometric-optical modelling, in *The Use of Remote Sensing in the Modelling of Forest Productivity*, H. L. Gholz, K. Nakane, and Shimoda, eds., Kluwer Academic Publishers, pp. 75–108.

Kimes, D. S. (1991), Radiative transfer in homogeneous and heterogeneous vegetation canopies, in *Photon-Vegetation Interactions: Applications in Optical Remote Sensing and Plant Physiology*, R. B. Myneni and J. Ross, eds., Springer-Verlag, pp. 339–388.

Kimes, D. S. and Kirchner, J. A. (1982), Radiative transfer model for heterogeneous 3d scenes, *Appl. Opt.* **21**: 4119–4129.

King, M. D. (1987), Determination of the scaled optical thickness of clouds from reflected solar radiation measurement, *J. Atmos. Sci.* **44**: 1734–1751.

Knyazikhin, Y. V., Marshak, A. L., and Myneni, R. B. (1992), Interaction of photons in a canopy of finite-dimensional leaves, *Remote Sens. Envir.* **39**: 61–74.

Kuusk, A. (1985), The hotspot effect of a uniform vegetation cover, *Sov. J. Remote Sens.* **3**: 646–658.

Kuusk, A. (1991a), The hot spot effect in plant canopy reflectance, in *Photon-Vegetation Interactions: Applications in Optical Remote Sensing and Plant Physiology*, R. B. Myneni and J. Ross, eds., Springer-Verlag, pp. 140–159.

Kuusk, A. (1994), A multispectral canopy reflectance model, *Remote Sens. Envir.* **50**: 75–82.

Kuusk, A. (1995a), A fast invertible canopy reflectance model, *Remote Sens. Envir.* **51**: 342–350.

Kuusk, A. (1995b), A Markov chain model of canopy reflectance, *Agric. Forest Meteorol.*, **76**: 221–236.

Kuusk, A. (1998), Monitoring of vegetation parameters on large areas by the inversion of a canopy reflectance model, *Int. J. Remote Sens.* **19**: 2893–2905.

Lacaze, R. and Roujean, J. L. (2001), G-function and hot spot (GHOST) reflectance model—application to multi-scale airborne polder measurements, *Remote Sens. Envir.* **76**: 67–80.

Lewis, P. (1999), Three-dimensional plant modelling for remote sensing simulation studies using the botanical plant modelling system, *Agronomie* **19**: 185–210.

Lewis, P. and Boissard, B. (1997), The use of 3D plant modelling and measurement in remote sensing, *Proc. 7th ISPRS*, Courchevel, France, pp. 319–326.

Li, X. and Strahler, A. (1985), Geometric-optical modeling of a coniferous forest canopy, *IEEE Trans. Geosci. Remote Sens.* **23**: 705–721.

Li, X. and Strahler, A. (1986), Geometric-optical bi-directional reflectance modeling of a coniferous forest canopy, *IEEE Trans. Geosci. Remote Sens.* **24**: 906–919.

Li, X. and Strahler, A. H. (1992), Geometric-optical bidirectional reflectance modeling of the discrete crown vegetation canopy: Effect of crown shape and mutual shadowing, *IEEE Trans. Geosci. Remote Sens.* **30**: 276–292.

Li, X., Woodcock, C., and Davis, R. (1994), A hybrid geometric optical-radiative transfer approach for modeling albedo and directional reflectance of discontinuous canopies, *IEEE Trans. Geosci. Remote Sens.* **33**: 466–480.

Li, X., Strahler, A. H., and Friedl, M. (1999), A conceptual model for effective directional emissivity from nonisothermal surface, *IEEE Trans. Geosci. Remote Sens.* **37**: 2508–2517.

Liang, S. L. and Strahler, A. H. (1993a), The calculation of the radiance distribution of the coupled atmosphere-canopy, *IEEE Trans. Geosci. Remote Sens.* **31**: 491–502.

Liang, S. and Strahler, A. H. (1993b), An analytic BRDF model of canopy radiative transfer and its inversion, *IEEE Trans. Geosci. Remote Sens.* **31**: 1081–1092.

Liang, S. and Strahler, A. H. (1994a), A stochastic radiative transfer model of a discontinuous vegetation canopy, *Proc. IGARSS.* Pasadena, California, 1626–1628.

Liang, S. and Strahler, A. H. (1994b), A four-stream solution for atmospheric radiative transfer over an non-lambertian surface, *Appl. Opt.* **33**: 5745–5753.

Liang, S. and Strahler, A. H. (1995), An analytic radiative transfer model for a coupled atmosphere and leaf canopy, *J. Geophys. Res.* **100**: 5085–5094.

Liang, S. and Lewis, P. (1996), A parametric radiative transfer model for sky radiance distribution, *J. Quant. Spectrosc. Radiat. Transfer* **55**: 181–189.

Liang, S., Strahler, A. H., Jin, X., and Zhu, Q. (1997), Comparisons of radiative transfer models of vegetation canopies and laboratory measurements, *Remote Sens. Envir.* **61**: 129–138.

Lin, Y. C. and Sarabandi, K. (1999), A monte carlo coherent scattering model for forest canopies using fractal-generated trees, *IEEE Trans. Geosci. Remote Sens.* **37**: 440–451.

Liou, K. N. (1974), Analytic two-stream and four-stream solutions for radiative transfer, *J. Atmos. Sci.* **31**: 1473–1475.

Ma, Q. L., Ishimaru, A., Phu, P., and Kuga, Y. (1990), Transmission, reflection, and polarization of an optical-wave for a single leaf, *IEEE Trans. Geosci. Remote Sens.* **28**: 865–872.

Maier, S. W., Ludeker, W., and Gunther, K. P. (1999), SLOP: A revised version of the stochastic model for leaf optical properties, *Remote Sens. Envir.* **68**: 273–280.

Marshak, A. L. (1989), The effect of the hot spot on the transport equation in plant canopies, *J. Quant. Spectrosc. Radiat. Transfer* **42**: 615–630.

Mech, R. and Prusinkiewicz, P. (1996), Visual models of plants interacting with their environment, *Proc. SIGGRAPH'96* **30**: 397–410.

Myneni, R. B. and Ganapol, B. D. (1991), A simplified formulation of photon transport in leaf canopies with finite dimensional scatters, *J. Quant. Spectroscp. Radiat. Transfer* **46**: 135–140.

Myneni, R. B. and Williams, D. L. (1994), On the relationship between FAPAR and NDVI, *Remote Sens. Envir.* **49**: 200–211.

Myneni, R. B., Asrar, G., and Kanemasu, E. T. (1987), Light scattering in plant canopies: The method of successive orders of scattering approximation (SOSA), *Agric. For. Meteorol.* **39**: 1–12.

Myneni, R. B., Gutschick, V. P., Shultis, J. K., Asrar, G., and Kanemasu, E. T. (1988), Photon transport in vegetation canopies with anisotropic scattering, part IV: Discrete-ordinates finite difference exact kernel technique for photon transport in slab geometry for the two-angle problem, *Agric. Forest Meteorol.* **42**: 101–120.

Myneni, R. B., Asrar, G., and Gerstl, S. A. W. (1990a), Radiative transfer in three-dimensional leaf canopies, *Transport Theory Stat. Phys.* **19**: 205–250.

Myneni, R. B., Ross, J., and Asrar, G. (1990b), A review on the theory of photon transport in leaf canopies, *Agric. Forest Meteorol.* **45**: 1–153.

Myneni, R. B. and Ross, J., Eds. (1991), *Photon-Vegetation Interactions: Applications in Optical Remote Sensing and Plant Physiology*, Springer-Verlag.

Myneni, R. B., Marshak, A. L., and Knyazikhin, Y. V. (1991), Transport theory for a leaf canopy of finite-dimensional scattering centers, *J. Quant. Spectrosc. Radiat. Transfer* **46**: 259–280.

Myneni, R. B., Hall, F. G., Sellers, P. J., and Marshak, A. L. (1995), The interpretation of spectral vegetation indices. *IEEE Trans. Geosci. Remote Sens.* **33**: 481–486.

Ni, W., Li, X., Woodcock, C. E., Roujean, J. L., and Davis, R. (1997), Transmission of solar radiation in boreal conifer forests: Measurements and models, *J. Geophys. Res.* **102**: 29555–29566.

Ni, W., Li, X., Woodcock, C. E., Caetano, R., and Strahler, A. H. (1999), An analytical model of bidirectional reflectance over discontinuous plant canopies, *IEEE Trans. Geosci. Remote Sens.* **37**: 1–13.

Nicodemus, F. E., Richamond, J. C., Hsia, J. J., Ginsberg, I. W., and Limperis, T. (1977), *Geometrical Considerations and Nomenclature for Reflectance*, NBS Monograph (U.S.) 160.

Nilson, T. and Kuusk, A. (1989), A reflectance model for the homogeneous plant canopy and its inversion. *Remote Sens. Envir.* **27**: 157–167.

Nilson, T. and Peterson, U. (1991), A forest canopy reflectance model and a test case, *Remote Sens. Envir.* **37**: 131–142.

Ni-Meister, W., Jupp, D. L. B., and Dubayah, R. (2001), Modelling lidar waveforms in heterogeneous and discrete canopies, *IEEE Trans. Geosci. Remote Sens.* **39**: 1943.

Norman, J. M. and Welles, J. M. (1983), Radiative transfer in an array of canopies, *Agron. J.* **75**: 81–488.

North, P. R. J. (1996), Three-dimensional forest light interaction model using a monte carlo method, *IEEE Trans. Geosci. Remote Sens.* **34**: 946–956.

Perttunen, J., R., S., Nikinmaa, E., Salminen, H., Saarenmaa, H., and Vkev, J. (1996), Lignum: A tree model based on simple structural units *Ann. Botany* **77**: 87–98.

Peterson, D. L., Aber, J. D., Matson, P. A., Card, D. H., Swanberg, N., Wessman, C., and Spanner, M. (1988), Remote sensing of forest canopy and leaf biochemical contents, *Remote Sens. Envir.* **24**: 85–108.

Pinty, B., Verstraete, M. M., and Dickinson, R. E. (1990), A physical model for the bidirectional reflectance of vegetation canopies—Part 2: Inversion and validation, *J. Geophys. Res.* **95**: 11767–11775.

Pinty, B., Gobron, N., Widlowski, J. L., Gerstl, S. A. W., Verstraete, M. M., Antunes, M., Bacour, C., Gascon, F., Gastellu, J. P., Goel, N., Jacquemoud, S., North, P., Qin, W., and Thompson, R. (2001), Radiation transfer model intercomparison (RAMI) exercise, *J. Geophys. Res.* **106**: 11,937–11,956.

Prusinkiewicz, P. (1996), Virtual plants: New perspectives for ecologists, pathologists and agricultural scientists, *Trends Plant Sci.* **1**: 33–38.

Prusinkiewicz, P. (1999), A look at the visual modelling of plants using L-systems, *Agronomie* **19**: 211–224.

Prusinkiewicz, P. and Lindenmayer, A. (1990), *The Algorithmic Beauty of Plants*, Springer-Verlag.

Qin, W. and Jupp, D. L. B. (1993), An analytical and computationally efficient reflectance model for leaf canopies, *Agric. Forest Meteorol.* **66**: 31–64.

Qin, W. and Xiang, Y. (1994), On the hotspot effect of leaf canopies: Modeling study and influences of leaf shape, *Remote Sens. Envir.* **50**: 95–106.

Qin, W. and Goel, N. S. (1995), An evaluation of hotspot models for vegetation canopies, *Remote Sens. Rev.*, **13**: 121–159.

Qin, W. and Xiang, Y. (1997), An analytical model for bidirectional reflectance factor of multicomponent vegetation canopies, *Sci. China (Series C)* **40**: 305–315.

Qin, W. and Liang, S. (2000), Plane-parallel canopy radiation transfer modeling: Recent advances and future directions, *Remote Sens. Rev.*, **18**: 281–306.

Qin, W. H. and Gerstl, S. A. W. (2000), 3-D scene modeling of semidesert vegetation cover and its radiation regime, *Remote Sens. Envir.* **74**: 145–162.

Qin, W. H., Gerstl, S. A. W., Deering, D. W., and Goel, N. S. (2002), Characterizing leaf geometry for grass and crop canopies from hotspot observations: A simulation study, *Remote Sens. Envir.* **80**: 100–113.

Richardson, A. J., Wiegand, C. L., Gausman, H. W., Cuellar, J. A., and Gerberman, A. H. (1974), Plant, soil and shadow reflectance components of row crops, *Photogramm. Eng. Remote Sensing* **41**: 1401–1407.

Richter, T. and Fukshansky, L. (1996), Optics of a bifacial leaf.1. A novel combined procedure for deriving the optical parameters, *Photochem. Photobiol.* **63**: 507–516.

Room, P., Hanan, J., and Prusinkiewicz, P. (1996), Virtual plants: New perspectives for ecologists, pathologists and agricultural scientists, *Trends Plant Sci.* (update), **1**: 33–38.

Ross, J. (1981), *The Radiation Regime and Architecture of Plant Stands*, Dr. W. Junk Publishers

Ross, J. and Nilson, T. (1975), Radiation exchange in plant canopies, in *Heat and Mass Transfer in the Biosphere. I. Transfer Processes in Plant Environment*, D. A. de Vries and N. H. Afgan, eds., Halsted Press (Wiley), pp. 327–336.

Ross, J. K. and Marshak, A. L. (1991), Monte carlo methods, in *Photon-Vegetation Interactions: Applications in Optical Remote Sensing and Plant Physiology*, R. Myneni and J. Ross, eds., Springer-Verlag, pp. 441–467.

Roujean, J. L. (2000), A parametric hot spot model for optical remote sensing applications, *Remote Sens. Envir.* **71**: 197–206.

Sellers, P. (1985), Canopy reflectance, photosynthesis and transpiration, *Int. J. Remote Sens.* **6**: 1335–1372.

Shabanov, N. V., Knyazikhin, Y., Baret, F., and Myneni, R. B. (2000), Stochastic modeling of radiation regime in discontinuous vegetation canopies, *Remote Sens. Envir.* **74**: 125–144.

Shultis, J. K. and Myneni, R. B. (1988), Radiative transfer in vegetation canopies with an isotropic scattering, *J. Quant. Spectrosc. Radiat. Transfer* **39**: 115–129.

Slater, P. N. (1980), *Remote Sensing: Optics and Optical Systems*, Addison-Wesley.

Sparrow, E. M. and Cess, R. D. (1978), *Radiation Heat Transfer*. Hemisphere.

Stamnes, K., Tsay, S. C., Wiscombe, W., and Jayaweera, K. (1988), Numerically stable algorithm for discrete-ordinate-method radiative transfer in multiple scattering and emitting layered media, *Appl. Opt.* **27**: 2502–2509.

Stern, F. (1964), Transmission of isotropic radiation across an interface between two dielectrics, *Appl. Opt.* **3**: 111–113.

Strahler, A. H. (1997), Vegetation canopy reflectance modeling - recent developments and remote sensing perspectives, *Remote Sens. Rev.*, **15**: 179–194.

Strahler, A. H. and Jupp, D. L. B. (1990), Modeling bidirectional reflectance of forests and woodlands using Boolean models and geometric optics, *Remote Sens. Envir.* **34**: 153–166.

Strebel, D. E., Goel, N. S., and Ranson, K. J. (1985), Two-dimensional leaf orientation distributions, *IEEE Trans. Geosci. Remote Sens.* **23**: 640–647.

Suits, G. H. (1972), The calculation of the directional reflectance of vegetative canopy, *Remote Sens. Envir.* **2**: 117–125.

Sun, G. and Ranson, K. J. (1995), A three-dimensional radar backscatter model of forest canopies, *IEEE Trans. Geosci. Remote Sens.* **33**: 372–382.

Sun, G. and Ranson, K. J. (2000), Modeling lidar returns from forest canopies, *IEEE Trans. Geosci. Remote Sens.* **38**: 2617–2626.

Tucker, C. J. (1977), Asymptotic nature of grass canopy spectral response, *Appl. Opt.* **16**: 1151–1157.

van de Hulst, H. C. (1968), Asymptotic fitting, a method for solving anisotropic transfer problems in thick layers, *J. Comput. Phys.* **3**: 291–306.

Vanderbilt, V. C. and Grant, L. (1985), Plant canopy specular reflectance model, *IEEE Trans. Geosci. Remote Sens.* **23**: 722–730.

Verhoef, W. (1984), Light scattering by leaf layers with application to canopy reflectance modeling: The SAIL model, *Remote Sens. Envir.* **16**: 125–141.

Verstraete, M. M., Pinty, B., and Dickinson, R. E. (1990), A physical model of the bidirectional reflectance vegetation canopies, I: Theory, *J. Geophys. Res.* **95**: 11755–11765.

Woodcock, C. E., Collins, J., Jakabhazy, V. D., Li, X., Macomber, S. A., and Wu, Y. (1997), Inversion of the Li–Strahler canopy reflectance model for mapping forest structure, *IEEE Trans. Geosci. Remote Sens.* **35**: 405–414.

Woodcock, C. E., Collins, J. B., Gopal, S., Jakabhazy, V. D., Li, X., Macomber, S., Ryherd, S., Harward, V. J., Levitan, J., and Wu, Y. C. (1994), Mapping forest vegetation using Landset TM imagery and a canopy reflectance model, *Remote Sens. Envir.* **50**: 240–254.

Yamada, N. and Fujimara, S. (1991), Nondestructive measurement of chlorophyll pigment content in plant-leaves from 3-color reflectance and transmittance, *Appl. Opt.* **30**: 3946–973.

4

Soil and Snow Reflectance Modeling

Soil and snow are spatially extensive land surface covers, and they often serve as the lower boundary condition of vegetation canopies. Soil and snow play very important interactive roles in the surface radiation budget because of their high albedo values. The physical and chemical properties of soil, such as soil carbon, nutrition, phosphorus, bulk density, and particle size distribution, are required by agricultural, ecological, and hydrological models. The snow geophysical properties of interest include the snow-covered area, water equivalent, and grain size. Modeling soil and snow radiation regimes will help us develop the advanced algorithms for estimating their properties.

There are important similarities between modeling the bidirectional reflectance distribution functions of soils and snow. These similarities include the fact that they are both characterizations of dense particulate media that can be models as spheroids, surface roughness on the order of millimeters to centimeters, and anisotropic diffuse and direct solar illumination irradiance. This chapter introduces their reflectance models together, including mainly radiative transfer and geometric optical models. More details on the recent progress are available from a review paper by Nolin and Liang (2000).

Section 4.1 introduces the basic concepts of the single scattering properties of a single particle. To calculate multiple scattering of radiation by snow and soil in natural environments, one must begin with knowledge of the single particle properties that govern the ensemble properties of the bulk medium. For both snow and soil, the individual particles are often characterized as simple spheres but may also be represented as other geometric shapes (cylinders, hexagonal plates) or as irregularly shaped particles. In this section, individual particle characterizations and single scattering properties for snow and soil particles are examined.

Quantitative Remote Sensing of Land Surfaces. By Shunlin Liang
ISBN 0-471-28166-2 Copyright © 2004 John Wiley & Sons, Inc.

Section 4.2 discusses commonly used methods for calculating multiple scattering from snow and soil. These methods fall into two categories as they do in atmospheric and canopy radiative transfer: approximate and numerical solutions to the equation of radiative transfer. Section 4.3 briefly introduces the modeling method based on geometric optical principles primarily for soils. Section 4.4 presents some examples of estimating snow properties (mainly grain size) from optical remote sensing data. Section 4.5 discusses various research topics in bidirectional reflectance modeling of snow and soils that warrant additional research in the respective areas.

4.1 SINGLE SCATTERING PROPERTIES OF SNOW AND SOIL

As introduced in Section 2.3, the single particle scattering and absorbing quantities that are required for calculation of multiple scattering are (1) the single scattering albedo, ω, which represents the probability that light incident on a particle will be scattered; and (2) the particle scattering phase function $P(\theta)$, where θ is the scattering angle.

If snow and soil particles can be assumed to behave as spheres and the size distribution and refractive indices of components comprising the particulate media are known, then Mie theory (van de Hulst 1980, Bohren and Huffman 1983) can provide the optical parameters of individual particles. Section 2.3.2 has discussed this topic and the Mie codes have also be included in the accompanying CD-ROM. For axially symmetric nonspherical particles with size parameters less than about 130, Mishchenko (1991, 1993) has developed the T-matrix approach to compute the optical parameters. For nonspherical particles that are large relative to the wavelength, the geometric optics method can be used to compute the single scattering albedo and asymmetry parameter. This topic has been discussed in Section 2.3.

4.1.1 Optical Properties of Snow

Snow may be modeled as a layered particulate medium composed of ice spheres in air. The presence of liquid water may also be accounted for but is typically neglected because the refractive indices of liquid and frozen water are nearly identical in the reflected solar spectrum (Dozier 1989b) and, unless standing water is present, the volume of liquid water in the snowpack rarely exceeds 6%. Using the refractive indices of ice (Warren, 1984, Kou et al. 1994) and an optically equivalent ice sphere radius, Mie theory may be used to calculate single particle scattering and absorption.

In the context of remote sensing, an appropriate means of characterizing grain size is to use the "optically equivalent sphere" having the same volume-to-surface ratio as ice grains measured in snow samples (Dobbins and

Jizmagian 1966, Dozier et al. 1987). With Mie theory it is assumed that snow grains behave as spheres, and although snow grains are usually not spherical using the optically equivalent sphere can approximate the ensemble properties (see Section 2.3.3). Mugnai and Wiscombe (1980) showed that a collection of nonoriented spheroids produce the same scattering results as spheres. Grenfell and Warren (1999) have shown that using the volume:surface area ratio to define the optically equivalent sphere provides an accurate representation of a collection of nonspherical ice particles. Effective snow grain radii typically range in size from about 50 μm for fresh, cold snow to 1000 μm representing wet snow, or grain clusters.

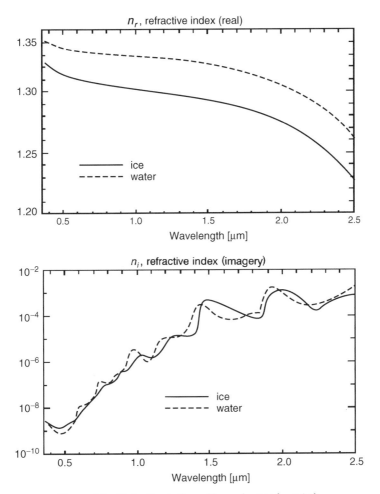

Figure 4.1   Refractive indices of ice and water ($n_r + in_i$).

The most important optical property of ice, which causes spectral variation in the reflectance of snow in the visible and near-infrared wavelengths, is that the absorption coefficient (i.e., the imaginary part of the refractive index) varies by seven orders of magnitude at wavelengths of 0.4–2.5 μm. Normally, the index of refraction is expressed as a complex number, $n_r + in_i$. Figure 4.1 shows the real and imaginary parts of the refractive index for ice and water. The digital files of the refractive indices of ice and water are included in the CD-ROM. The important properties (Dozier 1989b) include (1) the spectral variation in the real part n_r is small and the difference between ice and water may not be significant; (2) the absorption coefficient n_i of ice and water are very similar, except for the region between 1.35 and 1.75 μm, where ice is slightly more absorptive; (3) in the visible wavelengths, n_i is very small and ice is transparent, and (4) in the near-infrared wavelengths ice is moderately absorptive, and this absorption increases with wavelength.

4.1.2 Optical Properties of Soils

Determining the optical properties of soils is a complicated process. The mineral and organic components of the soil must be known and their relative fractions estimated. Egan and Hilgeman (1979) have measured the refractive indices of various soil types. One example of the measured soil refractive index is given in Table 4.1. As mentioned above, Mie scattering theory has been successfully used to calculate scattering properties of spheroidal particles such as liquid water and ice; however, it has been less successful in applications to soils. Problems arise in using Mie theory for soils from two sources: particle shape and particle size. A large fraction of soil particles have nonspherical shapes for which Mie theory fails. If particles cannot be approximated as spheres, different algorithms are needed to calculate the optical parameters n_i of the medium. For particles whose radius is comparable to the wavelength, Mishchenkos T-matrix method has been used to compute soil optical parameters (Liang 1997, Liang and Mishchenko 1997).

TABLE 4.1 Soil Refractive Index ($n = n_r + in_i$) of Illite Particles from Fithian, IL

Wavelength (μm)	n_r	n_i
0.500	1.415	0.05029
0.600	1.411	0.04289
0.700	1.399	0.05393
0.817	1.395	0.05448
0.907	1.391	0.05393
1.000	1.387	0.05448
1.015	1.387	0.05340

Source: Egan and Hilgeman (1979).

A second aspect of this problem is that most soil particles are much larger than the visible and near-IR wavelengths. When the size parameter χ [defined by Eq. (2.33)], exceeds ~ 50, it is computationally expensive to calculate the scattering properties of a single sand particle using Mie theory. More computationally efficient methods are needed. One possibility is the *ray tracing approach* (van de Hulst 1981), which considers diffraction, reflection, and refraction of a large particle. This has been tested for ice crystals and polydisperse spheres (Liou and Hansen 1971, Yang and Liou 1995) and has been used to evaluate the effects of large nonspherical particles on soil-polarized bidirectional reflectance (Takano and Liou, 1989a, 1989b, 1995). Leroux et al. (1998) used geometric optics to compute the polarized bidirectional reflectance for hexagonal ice crystals. They found improved agreement with measurements when hexagonal crystals were used rather than spherical particles.

Anomalous diffraction theory has proved accurate for large particles (Ackerman and Stephens 1987, Zege and Kokhanovskiy 1989). The *method of complex angular momentum* (Nussenzveig and Wiscombe 1991, Nussenzveig 1992) is both fast and accurate in calculating Mie scattering and absorption efficiencies for spherical particles that are large in comparison to the wavelength. The *high-energy approximation* is also a useful method for large spheres (Perrin and Chiappetta 1985; Chen 1987). For large particles, these approximation techniques have been shown to be accurate and more computationally efficient than Mie calculations. However, intercomparison of these approximation techniques has not been performed, nor have these methods been evaluated for soils applications.

Determining the appropriate grain size distribution to use is also difficult. Calculating light-scattering properties for ensembles of soil particles is a necessary yet daunting task for practical remote sensing applications. The soil particle size distribution can be represented by different functions, such as the lognormal (Shirazi and Boersma 1984, Buchan 1989) and Rosin distributions (Dapples 1975). Liang (1997) used the gamma distribution in his study using radiative transfer to calculate the depth of soil contributing to measured reflectance, also termed "sensible" depth. Lognormal and gamma distribution functions for characterizing aerosol size distributions have been given in Section 2.3.3 for aerosol particles. The Rosin distribution follows the Weibull probability law and can be written as

$$f(x) = cb^c x^{c-1} e^{-(x/b)^c} \qquad (4.1)$$

where b and c are two parameters; c determines the shape of the probability functions, namely, skewness and kurtosis; and b is an arbitrary scaling factor of particle size. Figure 4.2 illustrates the impacts of these two parameters on the distribution function. Note that the lognormal distribution function and the Robin distribution appear similar but not identical.

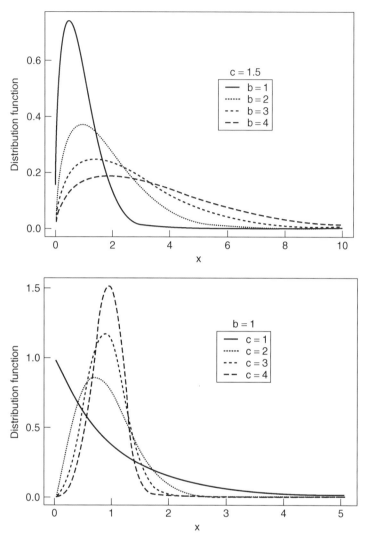

Figure 4.2 Soil particle distribution with different parameters in the Rosin distribution function defined in Eq. (4.1).

4.2   MULTIPLE SCATTERING SOLUTIONS FOR ANGULAR REFLECTANCE FROM SNOW AND SOIL

The use of the different approaches described here for calculating multiple scattering from particulate media are by no means balanced in their use by the snow and soil communities. The three categories of solutions are approximate, numerical, and geometric optical. Geometric optics have been used by both the snow and soil communities, but predominately by the latter. This is because soil roughness effects can be well represented with this method and soil roughness is the one easily measurable physical parameter that may also be obtained by model inversion.

4.2.1 Approximate Solutions

The multiple scattering of light from both soil and snow can be calculated using Chandrasekhar's (1960) radiative transfer equation or variants thereof. The radiative transfer equations used for snow are described by Wiscombe and Warren (Wiscombe and Warren, 1980, Warren and Wiscombe, 1980) and for soils by Liang and Townshend (1996a, 1996b). The basic form is exactly the same as Eq. (2.3). Most analytic radiative transfer models that simplify multiple scattering are based on the two-stream approximation. For flux calculations only, a two-stream approximation to the analytic solution for the equation of radiative transfer allows us to quickly solve for the diffuse upwelling flux from either snow or soil.

4.2.1.1 Snow

For snow, a delta-Eddington approximation has been widely used (Wiscombe and Warren 1980) that characterizes the strong forward scattering of snow throughout the optical region. The physical depth and bulk density of the snowpack affect scattering and absorbing properties via the optical depth τ. Optical depth is a dimensionless quantity and, for snow, is related to the snowpack density, depth, snow grain size, and wavelength of the observation:

$$\tau = \frac{3W_{eq}Q_{ext}}{4r\rho_{ice}} = \frac{3\rho_{snow}zQ_{ext}}{4r\rho_{ice}} \quad (4.2)$$

where W_{eq} is the water equivalent depth of the snowpack, defined as the geometric depth z times the snow density, Q_{ext} is the extinction efficiency factor defined in Eq. (2.49). The density of ice ρ_{ice} is about 917 kg/m³. For the optical depth to be dimensionless, W_{eq} is in kilograms per cubic meter and r, the radius of the equivalent snow grain, is in meters. Dozier (1989b) gives a practical definition of an optically semi-infinite snowpack as one whose directional hemispheric reflectance is within 1% of the reflectance of that for $\tau = \infty$. Typical snow water equivalent depths for an optically thick snowpack at a wavelength of 0.46 μm are 20 mm for a particle radius of 50 μm and 200 mm for a particle radius of 1000 μm. Smaller ice grains increase the number of scattering events, requiring less snow for the layer to appear optically thick.

   In the following, we will first introduce the two-stream radiative transfer model developed by Wiscombe and Warren (1980), and present the calculated surface albedo dependent on snow particle size and water equivalence.

A revised model that considers the rough surface developed by Choudhury and Chang (1981) is also briefly discussed.

Wiscombe and Warren (1980) applied the delta-Eddington approximation, one of the two-stream approximations discussed in Section 2.4.3, to calculate the directional hemispheric reflectance (i.e., planar albedo) $R(\theta_0)$ at the solar zenith angle θ_0. The formula is expressed for total optical depth τ under the direct-beam illumination at the top of the snow surface:

$$R(\theta_0) = \frac{2\gamma_2 \xi \exp\left(-\dfrac{\tau^*}{\mu_0}\right)\left(r_s - \dfrac{\omega^* R^*}{1 - \xi^2 \mu_0^2}\right) + \omega^*(Q^+ P^+ - Q^- P^-)}{Q^+(\gamma_1 + \xi) - Q^-(\gamma_1 - \xi)} \qquad (4.3)$$

where the scaled optical depth τ^*, single scattering albedo ω^*, and asymmetry parameter g^* are related to their original parameters as follows:

$$\tau^* = (1 - \omega g^2)\tau$$

$$\omega^* = \frac{(1 - g^2)\omega}{1 - g^2 \omega} \qquad (4.4)$$

$$g^* = \frac{g}{1 + g}$$

The parameters characterizing the Eddington approximation are copied here from Table 2.9:

$$\gamma_1 = \frac{7 - \omega^*(4 + 3g^*)}{4}$$

$$\gamma_2 = -\frac{7 - \omega^*(4 - 3g^*)}{4} \qquad (4.5)$$

$$\gamma_3 = \frac{2 - 3g^* \mu_0}{4}$$

$$\gamma_4 = 1 - \gamma_3$$

The remaining parameters are given by

$$Q^{\pm} = \exp(\pm \xi \tau^*)[\gamma_2 - r_s(\gamma_1 \pm \xi)]$$

$$P^{\pm} = \frac{\alpha_2 \pm \xi \gamma_3}{1 \pm \xi \mu_0}$$

$$R^* = \mu_0(r_s \alpha_1 - \alpha_2) + r_s \gamma_4 + \gamma_3 \qquad (4.6)$$

$$\alpha_1 = \gamma_1 \gamma_4 + \gamma_2 \gamma_3$$

$$\alpha_2 = \gamma_2 \gamma_4 + \gamma_2 \gamma_3$$

$$\xi = \sqrt{\gamma_1^2 - \gamma_2^2}$$

where r_s is the Lambertian reflectance of the underlying snow.

If snowpacks are very deep, the directional hemispheric reflectance can be calculated from the "semiinfinite" medium:

$$R(\theta_0) = \frac{\omega^*[\gamma_3(\xi + \gamma_1 - \gamma_2) + \gamma_2]}{(\xi + \gamma_1)(1 + \xi\mu_0)} \tag{4.7}$$

For a snow surface under a natural condition, both direct and diffuse radiation will be illuminated. Thus, the total albedo is the weighted average

$$\bar{\alpha} = w\bar{R} + (1 - w)R(\theta_0) \tag{4.8}$$

where w is the fraction of diffuse radiation and

$$\bar{R} = 2\int_0^{90°} R(\theta_0) \cos\theta_0 \sin\theta_0 \, d\theta_0 \tag{4.9}$$

The following figures show the spectral characteristics of the reflectance of snow using the two-stream approximation by Wiscombe and Warren (1980).

Figure 4.3 shows the spectral direct albedo of semi-infinite pure snow at the visible and near-infrared wavelengths, for snow grain radii of 50–1000 μm, representing a range from new snow to spring snow, although depth hoar and grain clusters in coarse spring snow can exceed 0.5 cm in radius. The solar zenith angle $\theta_0 = 60°$. In general, increasing the grain size affects the albedo much more in the near-IR spectrum than in the visible region. The albedo is sensitive to grain size mainly at 1.0–1.3 μm.

Figure 4.4 shows the relationship of snow reflectance to snow–water equivalence (snow depth), for a black substrate: $r_s = 0$. For grain radius $r = 1.0$ mm, the reflectance is perceptibly reduced when the snow amount is reduced to 100 mm snow–water equivalence. The top curve in each case is for semi-infinite depth.

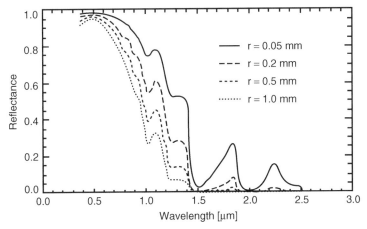

Figure 4.3 Snow albedo dependent on snow grain size [modified from Wiscombe and Warren (1980)].

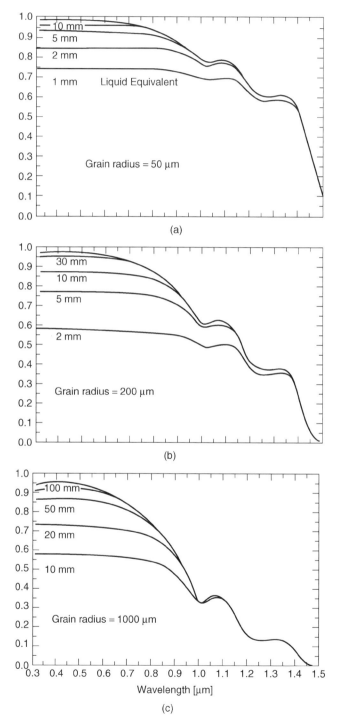

Figure 4.4 Relationship between snow reflectance and snow water equivalence [from Wiscombe and Warren (1980)].

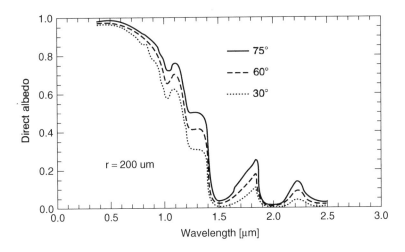

Figure 4.5 Snow reflectance dependent on the solar zenith angle.

Figure 4.5 illustrates the snow direct albedo dependent on the solar zenith angle. The snow depth is assumed to be semi-infinite. It is evident that the increased solar zenith angles lead to larger snow reflectance.

Choudhury and Chang (1981) developed a snow reflectance model that consists of two components: snow volume reflectance and surface specular reflectance, which is illustrated in Fig. 4.6. The volume reflectance is similar

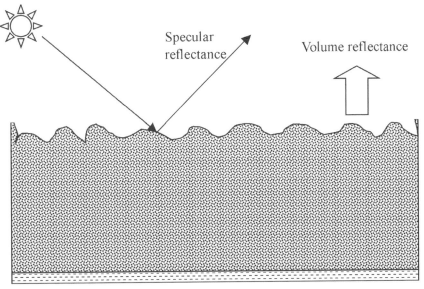

Figure 4.6 A conceptual representation of the snow reflectance model of Choudhury and Chang (1981).

to that by Warren and Wiscombe (1980). The surface specular reflectance term accounts for surface roughness characterized by the isotropic Gaussian facets using rough-surface scattering theory (Sancer 1969). Their formula is given below by using the same nomenclature as employed in the Wiscombe–Warren model:

$$R(\mu_0) = f(\mu_0)(1 - w) + [1 - (1 - w)f(\mu_0)]\bar{R} \qquad (4.10)$$

where the first term is denoted as the surface specular reflectance and the second term is the volume reflectance term, w is the fraction of diffuse radiation, $\bar{R}$ is defined by Eq. (4.9), and

$$f(\mu_0) = \int_0^{2\pi}\int_0^1 f_s(\mu_0, \phi_0, \mu, \phi)\mu\, d\mu\, d\phi \qquad (4.11)$$

and the surface directional reflectance is expressed by

$$f_s(\mu_0, \phi_0, \mu, \phi) = \frac{\sec^4\alpha \, \exp\left(-\dfrac{\tan^2\alpha}{2s^2}\right)S(\theta, \theta_0)\rho(\psi)}{8\pi s^2\mu\mu_0} \qquad (4.12)$$

where s is the variance of surface slope,

$$\tan\alpha = \frac{\sqrt{\sin^2\theta + \sin^2\theta_0 - 2\sin\theta \sin\theta_0 \cos(\phi - \phi_0)}}{\mu + \mu_0} \qquad (4.13)$$

$$\cos\psi = \sqrt{\frac{1 + \mu\mu_0 - \sin\theta \sin\theta_0 \cos(\phi - \phi_0)}{2}} \qquad (4.14)$$

and $\rho(\psi)$ is the Fresnel reflectance for angle ψ:

$$\rho(\psi) = \frac{1}{2}\left[\left(\frac{\sqrt{\varepsilon - \sin^2\psi} - \cos\psi}{\varepsilon - \sin^2\psi + \cos\psi}\right)^2 + \left(\frac{\varepsilon\cos\psi - \sqrt{\varepsilon - \sin^2\psi}}{\varepsilon\cos\psi + \sqrt{\varepsilon - \sin^2\psi}}\right)^2\right] \qquad (4.15)$$

ε is the dielectric constant for the ice grains. The shadowing function for the rough surface is written as

$$S(\theta, \theta_0) = \frac{1}{1 + C_0 + C_1} \qquad (4.16)$$

where

$$C_0 = \frac{1}{2}\left[\sqrt{\frac{2s^2}{\pi}}\tan\theta \exp\left(-\frac{\cot^2\theta}{2s^2}\right) - \operatorname{erfc}\left(\frac{\cot\theta}{s\sqrt{2}}\right)\right] \qquad (4.17)$$

$$C_1 = \frac{1}{2}\left[\sqrt{\frac{2s^2}{\pi}}\tan\theta_0 \exp\left(-\frac{\cot^2\theta_0}{2s^2}\right) - \operatorname{erfc}\left(\frac{\cot\theta_0}{s\sqrt{2}}\right)\right] \qquad (4.18)$$

where erfc is the complementary error function.

It should be pointed out that while the two-stream approach can give reasonably accurate estimates of the change in directional hemispherical reflectance with the solar zenith angle (Wiscombe and Warren 1980, Choudhury and Chang 1981, Dozier 1989a, 1989b), it does not provide the angular reflectance values needed to characterize the snow BRDF.

4.2.1.2 Soil

A widely used model of soil reflectance in this category of approximate solutions to the equation of radiative transfer is the Hapke model of bidirectional reflectance (Hapke 1981, 1984, 1986):

$$R(\theta_i, \theta_v, \phi) = R_1(\theta_i, \theta_v, \phi) + R_M(\theta_i, \theta_v, \phi)$$

$$= \frac{\omega}{4\pi} \frac{\cos \theta_i}{\cos \theta_i + \cos \theta_v} \{P(\xi)[1 + B(\xi)]$$

$$+ [H(\cos \theta_i) H(\cos \theta_v) - 1]\} \qquad (4.19)$$

where ω is the single scattering albedo; θ_i, θ_v, and ϕ are the solar zenith angle, viewing zenith angle, and relative azimuth angle, respectively; ξ is the phase angle defined in Eq. (2.21) with different notation [i.e., $\cos \xi = \cos \theta_i \cos \theta_v + \sin \theta_i \sin \theta_v \cos(\phi)$]; and $H(\mu)$ is an approximation to the Chandrasekhar H function:

$$H(\mu) = \frac{1 + 2\mu}{1 + 2\sqrt{1 - \omega\mu}} \qquad (4.20)$$

In the Hapke model [Eq. (4.19)], the single scattering component (the first term of the right side of the equation) is modified to account for the hotspot effect, and the multiple scattering radiance is assumed to be isotropic and is expressed by the approximate Chandrasekhar H function. The hotspot effect is represented by a semiempirical formula based on the shadow hiding principle

$$B(\xi) = \frac{B_0}{1 + \tan(\xi/2)/h} \qquad (4.21)$$

where B_0 and h are the empirical parameters related to the height and width of the hotspot peak; h can be represented by the half-angular width at half-maximum magnitude (HWHM) of the hotspot peak: $h = \tan(\text{HWHM}/2)$. According to the shadowing theory, particles that are large in comparison to the wavelength and opaque cast shadows on neighboring particles. When the viewing direction matches the illumination direction, all shadows are hidden by the particles that cast them and thus the local brightness reaches its

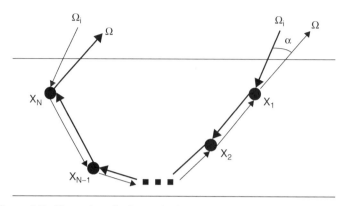

Figure 4.7   Illustration of coherent backscattering for soil reflectance hotspot.

maximum. When the viewing direction moves away from the illumination direction, shadows can be seen and the detected brightness sharply decreases.

In later studies, it was found (Hapke et al. 1996, Liang and Mishchenko 1997) that the hotspot effect of the soil reflectance is also caused by coherent backscattering. The basic principle of coherent backscattering is illustrated in Fig. 4.7. The discrete random medium is illuminated by a scalar plane wave in the direction Ω_i. Two partial waves associated with the same incident wave travel through the same group of N scatters, denoted by $x_1, x_2, \ldots, x_N$, but in opposite directions. If α is relatively large, the average effect of interference will approach zero. However, coherence is completely preserved at the backscattering direction, which results in an enhanced backscattering peak.

In the backscattering direction (zero phase), the magnitudes of the electric fields associated with the two light paths are equal ($A_1 = A_2$) and the reflected intensity is $|A_1 + A_2|^2 = 4A^2$. In contrast, the combined intensity would be $|A_1|^2 + |A_2|^2 = 2A^2$ in the absence of coherence. The enhancement factor obviously is 2. However, if a path involves only one particle, there is no reversed path; in other words, single scattering does not contribute to coherent enhancement. Thus, it is easy to write the enhancement factor ξ, which is defined as the ratio of the total backscattering radiance ($I_d + I_c$) to the incoherent (diffuse) radiance (I_d):

$$\alpha = \frac{I_d + I_c}{I_d} \tag{4.22}$$

where the coherent radiance (I_c) is equal to the total incoherent radiance minus single scattering radiance (I_1). Thus:

$$\alpha = 2 - \frac{I_1}{I_d} \qquad (4.23)$$

Therefore, the enhancement factor is always smaller than 2. Note that this formula is valid only in the exact backscattering direction. Shadow hiding is for single scattering, whereas coherent backscattering is for multiple scattering. Liang and Mishchenko (1997) demonstrated that for a soil composed of fine particles where shadowing theory cannot account for hotspot effects, coherent scattering of a plane wave enhances reflectance in the backward direction and that the width and height of the backscattering peak depend on refractive index, particle shape, and filling factor. The hotspot peaks caused by these three factors may have different magnitudes and angular widths but overlap together.

This Hapke model was originally developed to model angular reflectances from planetary surfaces and has since been used by the terrestrial community to characterize bidirectional reflectance from soils. A number of inversion experiments have been carried out to retrieve soil physical properties using the Hapke model (e.g., Pinty et al. 1989, Jacquemoud et al. 1992). Jacquemound et al. (1992) inverted these parameters of the Hapke model from 26 soil samples with a modified phase function:

$$P(\xi, \xi') = 1 + a\cos\xi + b\frac{3\cos^2\xi - 1}{2} + c\cos\xi' + d\frac{3\cos^2\xi' - 1}{2} \qquad (4.24)$$

where $a, b, c,$ and d are coefficients and $\cos\xi' = \cos\theta_i\cos\theta_v - \sin\theta_i\sin\theta_v\cos(\phi)$, ξ', represents the antiphase angle between the specular and the viewing direction). The retrieved parameters are listed in Table 4.2. Only one parameter (i.e., ω) is wavelength-dependent.

One approximation made in the Hapke model is that the multiple scattering component is isotropic regardless of the actual phase function of the medium. Hapke inferred that backscattering from planetary particulate surfaces indicated a negative value for the single-particle asymmetry parameter. Mischenko (1994) and Mishchenko and Macke (1997) have questioned the accuracy of the Hapke model based on laboratory measurements and Monte Carlo simulations. Their measurements and model output indicate that Hapke's negative asymmetry parameters (backscattering) are the result of numerical inaccuracies that are generated through Hapke's inversion approach. Mishchenko and Macke (1997) hypothesize that the observed backscattering results from surface roughness at the microscopic and macroscopic scales, rather than as an intrinsic property of the mineral particles themselves.

Liang and Townshend (1996b) modified the Hapke model with multiple scattered radiance approximated by the same isotropic H-function expressed

TABLE 4.2 Inverted Parameters of the Hapke Model from 26 Soil Samples at Five Different Wavelengths

No.	Type	ω					h	Phase Function			
		538 nm	631 nm	851 nm	1768 nm	2209 nm		a	b	c	d
1	Clay	0.16	0.22	0.27	0.35	0.28	0.05	1.31	0.48	0.12	0.07
2		0.16	0.20	0.26	0.33	0.26	0.08	1.33	0.38	0.29	−0.04
3		0.73	0.77	0.82	0.85	0.80	0.20	−1.77	0.96	0.20	0.13
4		0.30	0.37	0.43	0.54	0.51	0.08	1.18	0.67	0.08	−0.13
5		0.31	0.37	0.44	0.54	0.52	0.09	1.44	0.64	0.18	−0.05
6		0.66	0.74	0.80	0.86	0.84	0.00	0.06	0.53	0.83	−0.06
7		0.30	0.36	0.41	0.52	0.50	0.13	1.54	0.77	0.28	−0.01
8		0.31	0.37	0.42	0.53	0.52	0.11	1.64	0.66	0.31	−0.02
9		0.70	0.77	0.82	0.88	0.87	0.11	0.06	0.10	0.54	0.02
10	Sand	0.71	0.81	0.87	0.83	0.64	0.27	0.28	0.00	0.23	−0.04
11		0.84	0.89	0.93	0.89	0.76	0.10	−0.36	0.22	0.42	−0.06
12		0.81	0.87	0.91	0.95	0.92	0.01	0.75	0.53	0.54	−0.27
13		0.92	0.95	0.96	0.98	0.96	0.00	−0.03	0.31	0.09	0.09
14		0.81	0.87	0.91	0.95	0.93	0.09	0.74	0.12	0.32	−0.14
15		0.87	0.91	0.94	0.96	0.95	0.18	0.17	0.06	0.50	−0.06
16	Peat	0.06	0.09	0.24	0.40	0.22	0.02	0.77	0.37	0.19	0.01
17		0.06	0.11	0.33	0.55	0.32	0.01	0.28	0.19	−0.14	0.15
18		0.16	0.23	0.51	0.69	0.46	0.00	−0.57	0.32	0.08	0.09
19		0.09	0.14	0.32	0.62	0.53	0.09	1.20	0.51	0.16	−0.02
20		0.10	0.16	0.38	0.70	0.62	0.09	0.91	0.37	0.14	−0.05
21		0.15	0.25	0.56	0.81	0.72	0.04	0.94	0.03	0.55	0.01
22		0.09	0.14	0.31	0.63	0.55	0.06	1.28	0.65	0.17	0.04
23		0.10	0.16	0.36	0.69	0.62	0.16	0.99	0.12	−0.01	0.15
24		0.19	0.28	0.57	0.86	0.82	0.01	0.6	0.42	0.16	0.09
25	Pozzolana	0.55	0.57	0.57	0.63	0.59	0.27	0.79	−0.06	0.14	0.01
26	Pebbles	0.24	0.27	0.28	0.27	0.25	0.09	1.11	0.53	0.33	−0.11

Source: Jacquemoud et al. (1992).

in the original Hapke model, but it is more accurate than the original Hapke model because of the exact inclusion of the double-scattering component. Mathematically, the total reflectance is the sum of three components:

$$R = R_1 + R_2 + R_M \qquad (4.25)$$

where reflectance due to single-scattering (R_1) and double-scattering (R_2) can be exactly calculated, while reflectance with multiple-scattering (R_M, higher than double-scattering) is approximated by the H-function. Most existing parametric soil directional reflectance models have not incorporated sky radiance component in an effective manner. The Hapke model assumes monochromatic collimated illumination, that is, no diffuse irradiance radiance at the surface. However, under normal atmospheric conditions, sky radiance may range from 5–40 percent of the total downward radiance

(Ranson et al. 1984), and the sky radiance distribution is anisotropic (Coulson 1988, Liang and Lewis 1996). Liang and Townshend (1996b) have incorporated arbitrary directional distribution of the sky radiance.

In further improvements to the Hapke model, Liang and Townshend (1996a) incorporated a new multiscattering formula based on a four-stream approximation (Liang and Townshend 1996a). This latter model provides

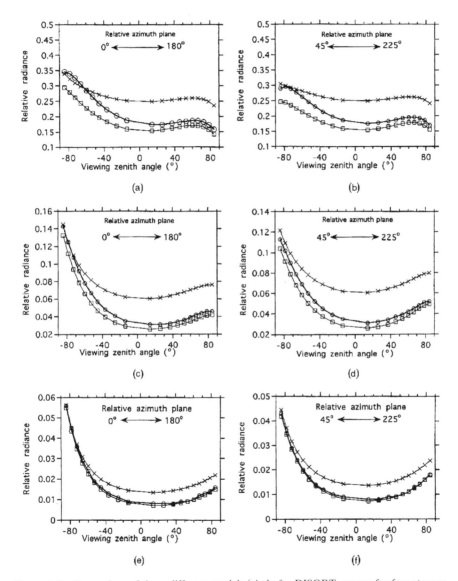

Figure 4.8 Comparison of three different models (circle for DISORT; square for four-stream; cross for the original Hapke model) for calculating soil reflectance with different single scattering albedos: 0.9 for (a), (b); 0.6 for (c), (d); 0.3 for (e), (f). [From Liang and Townshend (1996a), *Int. J. Remote Sens.* Reproduced by permission of Taylor & Francis, Ltd.]

more accurate solutions than the original and earlier improved Hapke models. Results from the *dis*crete *o*rdinates *r*adiative *t*ransfer (DISORT) model (Stamnes et al. 1988) were used as a baseline for these comparisons. Figure 4.8 compares outputs of the three models (DISORT, the original Hapke model, and the four-stream model) with the same phase function but different single-scattering albedo ω. The soil is assumed to be illuminated

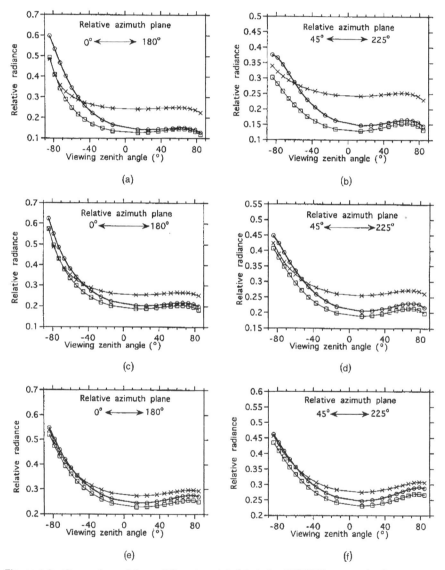

Figure 4.9 Comparison of three different models (circle for DISORT; square for four-stream; cross for the original Hapke model) for calculating soil reflectance with different asymmetry parameters: 0.7 for (a), (b); 0.5 for (c), (d); 0.3 for (e), (f). [From Liang and Townshend (1996a), *Int. J. Remote Sens.* Reproduced by permission of Taylor & Francis, Ltd.]

only by uncollimated solar radiance. The solar zenith angle is 30°, and solar azimuth angle is 0°. The phase function is expressed by one-term Henyey–Greenstein function with the asymmetry parameter $g = 0.65$. The larger single scattering albedo indicates stronger multiple scattering. When the single scattering albedo is 0.9, the relative error of the present model is about 10%, but the Hapke model is much worse. When the single scattering albedo becomes smaller, both parametric models become more accurate. Figure 4.9 presents the same comparisons with different asymmetry parameters. The solar zenith angle and the solar azimuth angle are 50° and 0°, respectively. When the asymmetry parameter is larger, indicating that scattering is more anisotropic, the present model performs much better than does the Hapke model. When scattering by the soil particles is close to isotropic, both parametric models perform almost equally well. These figures show that when multiple scattering dominates, our four-stream model underestimates the upwelling radiance, especially in the large viewing angles in the forward directions. However, it predicts angular patterns very well. The original Hapke model always overestimates the upwelling radiance, particularly with large asymmetry parameter (i.e, scattering is very anisotropic) and strong multiple scattering. The angular pattern predicted by the original Hapke model is far away from DISORT's because of its assumption of isotropic scattering for the multiscattering component. Overall, the new four-stream model has a much better accuracy than does the original Hapke model.

4.2.2 Numerical Solutions

Unlike the radiative transfer approximations, methods such as discrete ordinates and adding–doubling are numerical methods for determining bidirectional reflectance quantities of snow and soil. The adding–doubling method was first introduced by van de Hulst (1980). It assumes superimposed plane parallel layers with solar illumination entering the top layer. The angular reflectance of the combined layers is the sum of the reflectances of each very thin individual layer for which the single scattering properties have been calculated. The discrete ordinates method allows numerous computational streams for calculation of angular radiance quantities. Its basic principles are discussed in Section 2.4.2.2. Originally designed and used for cloud scattering problems, DISORT (Stamnes et al. 1988) has also been effective for modeling angular reflectances from snow and soil. Mishchenko et al. (1999) recently developed a numerical code (refl.f) for calculating bidirectional reflectance of soil and snow. Both DISORT and refl.f are included in the accompanying CD-ROM.

One question that arises when applying Mie theory to snow (and soil) is whether the particles behave as independent scatterers (i.e., as completely incoherent scatterers). If they do act as separate spheres, then classical Mie theory may be used to define the angular scattering pattern. For snow grains to behave as independent scatterers, they need to be randomly organized and sufficiently distant from one another that they do not interfere with each others illumination or scattering. Wiscombe and Warren (1980) found that the single scattering albedo of snow was insensitive to close packing. This can be explained by the fact that light refraction through the ice particles dominates scattering in snow and refraction is not sensitive to dense packing. The authors also demonstrated from some limited measurements that any near-field effects would be relatively minor. However, complete tests of this assumption remain to be performed.

The nature of snow grain metamorphism creates a tendency toward spherical grains. Indeed, for equitemperature metamorphism in which the surface free energy of the snow grain is minimized, the equilibrium shape is a sphere (Colbeck 1982). In wet snow, crystals rapidly become very rounded, often forming clusters. Conditions where snow grains are markedly nonspherical include the transient case of fresh snow (hexagonal crystals, needles, etc.), surface hoar created by condensation of water vapor on the snow surface (faceted, cup-shaped crystals), and depth hoar induced by strong temperature gradients (faceted, cup-shaped crystals). Surface hoar occurs under restricted surface energy balance conditions and is not common, while the more common depth hoar would have virtually no effect on albedo if the hoar layer is buried by more than 10 cm (as is typically the case).

Jin and Simpson (1999) have demonstrated the differences between using the exact Mie phase function for an ice particle and the Henyey–Greenstein phase function. They found that, although both result in the same asymmetry parameter, the exact Mie phase function predicts a smaller backscattering peak and a slightly smaller forward scattering peak than does the Henyey–Greenstein phase function. However, this backscattering peak typically does not appear in measurements, and it remains unclear if its inclusion in a model gives a more accurate result. Mishchenko (1994) demonstrated that for moderately and nonabsorbing particles that are large relative to the wavelength, the asymmetry parameter may be used without modification. He emphasized that for grain radius:wavelength ratios (r/λ) between 0.01 and 10, the asymmetry parameter value is sensitive to the density of the particulate medium. The minimum value of r/λ is about 20 (where $r = 50$ μm and $\lambda = 2.5$ μm), indicating that, for virtually all combinations of grain size and wavelength, snowpack density does not influence the asymmetry parameter. According to Mishchenko (1994), effects of particle nonsphericity are also quite small in that range and may be neglected. Thus, for snow grains in the optical region, problems associated with dense packing of particles and nonsphericity have a negligible effect on the asymmetry parameter.

4.3 GEOMETRIC OPTICAL MODELING

For models involving soils, mathematical expressions are used to describe the geometry of the surface, and the bidirectional reflectance is calculated from parameterizations of the reflectance and transmittance of the three-dimensional objects that make up the scattering medium. For instance, with plant canopies, geometric optics models distribute incident and reflected radiation within a three-dimensional layered canopy composed of cone-shaped trees (Li and Strahler 1985, 1986, 1992). Otterman and Weiss (1984) modeled vegetation as vertical cylinders over a Lambertian soil surface. More details about this topic have been given in Section 3.4.

There have only been a few applications of geometric optical modeling efforts that apply strictly to a soil surface. Norman et al. (1985) developed a geometric model to simulate soil aggregates by cuboids. Cooper and Smith (1985) simulated "row" and "clump" soil radiation using the Monte Carlo method in which the soil surface height varies periodically with the cosine function in one or two directions. Irons et al. (1992) modeled soil particles as opaque spheres in a regularly spaced grid over an isotropically scattering horizontal surface. Individual spheres have a Lambertian reflectance equal to that of the horizontal surface, but anisotropic scattering was simulated by calculating the sunlit and shadowed proportions of the viewing area for particular illumination and viewing geometries. Their simulated bidirectional reflectance factors were in good agreement with field measurements, although larger differences were observed for larger solar zenith angles.

Cierniewski and his collaborators have been responsible for developing many of the commonly used analytic soil models that incorporate geometric optics (Cierniewski 1987, Cierniewski 1989, Cierniewski and Courault 1993, Cierniewski and Verbrugghe 1994, 1997; Cierniewski et al. 1996). In their work, a soil is modeled as a set of regularly-spaced equally-sized opaque spheroids on a horizontally flat surface (Fig. 4.10). The models predict directional reflectances in the principal plane (the plane that includes both the Sun and the target) by calculating the areas of sunlit facets, shaded facets, and their radiant existence. Their models have not accurately predicted soil bidirectional reflectance factors in all Sun-viewing geometries, but appear promising.

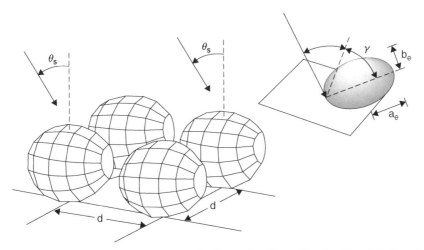

Figure 4.10 Geometric optical modeling of soil particles. [From Cierniewski and Verbrugghe (1997), *Int. J. Remote Sens*. Reproduced by permission of Taylor & Francis, Ltd.]

4.4   INVERSION OF SNOW PARAMETERS

Many snow parameters can be estimated from remotely sensed observations, such as snow albedo, snow grain size, and equivalent water content. Snow albedo is discussed in Chapter 9, and we mainly report several studies on estimation of grain size in the rest of this section.

Snow grain size is the primary parameter controlling broadband albedo (Wiscombe and Warren 1980); hence its estimate is crucial for calculating the snowpack's absorption of solar radiation. Because rate of grain growth is exponentially proportional to snow temperature, changes in grain size are useful indicators of thermodynamic processes in the snowpack. Changes in snow grain size can help identify ice sheet surface features, such as melt areas, snow dunes, and blue ice regions, and often indicate changes in snow energy balance (Nolin and Dozier 2000).

From a series of laboratory experiments, Hyvarinen and Lammasniemi (1987) related changes in a reflectance ratio of near-infrared bands, centered at 1.03 and 1.26 μm, to changes in average ice grain diameter for three size classes of ice particles. Dozier and Marks (1987) used Landsat thematic mapper (TM) data to classify snow-covered regions into relatively fine-grained new snow and older, coarser-grained snow, but they made no ground truth measurements at the times of the overpasses. Bourdelles and Fily (1993) and Fily et al. (1997) also tried to map the snow grain size using TM imagery.

Painter et al. (1998) developed a technique to improve spectral mixture analysis of a snow-covered area in alpine regions through the use of multiple snow endmembers. Snow reflectance in near-infrared wavelengths is sensitive to snow grain size, while in visible wavelengths it is relatively insensitive. Snow-covered alpine regions often exhibit large surface grain size gradients due to changes in aspect and elevation. The sensitivity of snow spectral reflectance to grain size translates these grain size gradients into the spectral nature of snow, which must be accounted for by use of multiple snow endmembers of varying grain size. Results from AVIRIS (airborne visible–infrared imaging spectrometry) imagery were verified with a high-spatial-resolution aerial photograph demonstrating equivalent accuracy. Analysis of fraction under/overflow and residuals confirmed mixture analysis sensitivity to grain size gradients.

Nolin and Dozier (2000) developed an inversion technique for estimating the grain size in a snowpack's surface layer from the hyperspectral AVIRIS data. Using a radiative transfer model, the method relates an ice absorption feature, centered at 1.03 μm, to the optically equivalent snow grain size. Figure 4.11 shows the snow spectral reflectance and continuum reflectance for the absorption features centered at 1.03 μm, derived from AVIRIS measurements. The scaled area associated with the absorption region is calculated by

$$A = \int_\lambda \frac{R_c - R_b}{R_c} \tag{4.26}$$

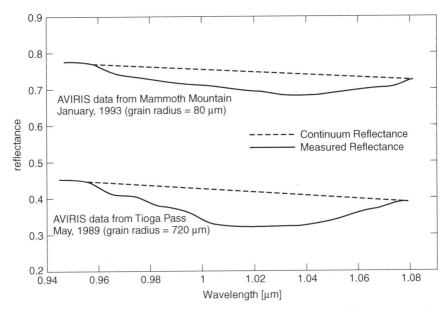

Figure 4.11   The 1.03-μm absorption feature from two AVIRIS snow reflectance spectra for Mammoth Mountain in 1993 and Tioga Pass in 1989. The continuum reflectance is plotted as the solid line across the top of each feature. Derived grain sizes are shown for each location. [From Nolin and Dozier (2000), *Remote Sens. Environ.* Copyright © 2000 with permission from Elsevier.]

where R_c is the continuum reflectance and R_b is the measured reflectance at the deepest part of the absorption band. This scaled area A can be well linked with the snow grain size through radiative transfer modeling (see Fig. 4.12). Ground measurements have been used to verify that this approach can provide very accurate results. The advantage of scaling the band depth by the continuum reflectance is that the grain size estimates become independent of the absolute magnitude of reflected radiance and therefore are not sensitive to topography. The solar illumination angle changes the magnitude of the reflectance spectrum, but changes neither the shape nor the relative depth of the absorption feature.

Nakamura et al. (2001) measured the spectral reflectance of snow with a known grain size distribution experimentally in a cold room under successive metamorphism in the wavelength region 280–2500 nm. Two one-time sequential experiments showed that the spectral reflectance of snow decreased when the snow metamorphosed or aged from new to granular snow by way of compacted snow under the influence of intermittent radiation and freezing. This experimental result was due to the growth of snow grains, as was revealed by measuring the grain size on microscopic photographs taken using the aniline method. It was experimentally confirmed that smaller grain sizes have larger reflectances. It was also found that metamorphism was observed

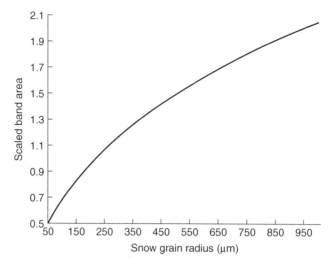

Figure 4.12   The nonlinear relationship between snow grain radius and scaled band area. Scaled dimensionless band areas were computed with DISORT. [From Nolin and Dozier (2000), *Remote Sens. Environ.* Copyright © 2000 with permission from Elsevier.]

only in the uppermost snow layer, which was about 3 cm thick. This led to the discovery that the snow reflectance was strongly affected by the texture of the uppermost layer of the snow surface.

Li et al. (2001) used a comprehensive forward radiative transfer model to construct a snow grain size retrieval algorithm that relies on the use of NIR radiances. Data collected by the airborne visible–infrared imaging spectrometer (AVIRIS) at the wavelengths 0.86, 1.05, 1.24, and 1.73 μm are used to retrieve snow grain size. On the basis of a single-layer (homogeneous) snow model, the retrieved snow grain size appears to depend on wavelength. It reveals that this apparent wavelength dependence occurs because (1) the snow grain size generally increases with depth and (2) the photon penetration depth decreases with increasing wavelength. The results show that the wavelength dependence of the photon penetration depth can be used to retrieve the depth dependence of the snow grain size.

4.5   PRACTICAL ISSUES

In this section, we list and discuss various topics in bidirectional reflectance modeling of snow and soils that warrant additional research in the respective areas. The research topics in each field are prioritized as an incentive for new and continued investigations (Nolin and Liang 2000).

4.5.1 Snow and Soil Surface Roughness

Snow surface roughness is a concern for characterizing snow BRDF, especially for regions such as Antarctica, where the effects of wind erosion on surface features (termed "sastrugi") can be substantial. Warren et al. (1998) measured the effects of oriented snow surface roughness on measured HDRF at the South Pole station in Antarctica. There, sastrugi cause roughness on the order of 10–25 cm in height for a feature that is several meters long. They found that bidirectional reflectance measurements best represented the albedo when viewed at nadir or within 20° of nadir. When sastrugi were perpendicular to the solar beam irradiance, forward scattering was significantly decreased. In addition to the BRDF patterns being affected by the presence of sastrugi, slight albedo reductions were also measured. This latter effect results from the change in effective illumination angle when the Sun illuminates the tilted surfaces of the sastrugi. Warren et al. also postulate that "light trapping" of photons emerging from a sastrugi feature and being intercepted by another may also be responsible for small albedo decreases. The authors offered an albedo parameterization for such snow surfaces and have shown that it is valid for viewing zenith angles less than 50°. Satellite measurements that are close to nadir have the best chance of producing accurate albedo measurements over snow with oriented surface roughness.

Leroux et al. (1998) have also modeled snow surface roughness effects on albedo and found that sastrugi have a darkening effect that may reduce albedo by as much as 10%. They used a photometric roughness model based on a model by Roujean et al. (1992) that uses vertical protrusions arranged on a flat scattering surface. Leroux and Fily modified the model by orienting the protrusions in a single direction, rendering the shadowed areas not completely dark and by providing the flat underlying surface with appropriate bidirectional reflectance characteristics. They found that although there was good qualitative agreement between measurements and model output, the quantitative agreement was lacking because of errors in model representation of the surface and measurement accuracy. Clearly, more measurements and model refinements are needed to derive accurate means of correcting satellite-derived albedos over rough snow surfaces.

Soil surface roughness is a great concern since the sensible depth is quite small in the visible and near-infrared spectra. Hapke (1984) developed a simple formula to account for this surface roughness. Liang et al. (1997) compared different shadowing functions for rough surfaces, and more details are presented in Section 7.4. Shoshany (1992) developed a simple method to calculate the directional reflectance in the principal plane for surfaces with roughness described by a repetition of microstructural features. Shoshany later (1993) described the effects of soil surface roughness on field measurements of surface HDRF. Shoshanys method appears to be valid for ideal characterizations of soil roughness. Despan et al. (1999) developed a BRDF model that treats soil surfaces as random Gaussian rough surfaces. They calculated statistical expressions for shadowing factor and effective slope for rough soil surfaces and were able to validate the model for a range of surface roughness conditions. At the present time, there is no consensus on an algorithm that accounts for all types of surface roughness and further research in this area is needed.

4.5.2   Mixed Snow Pixels

Another priority task for snow BRDF research is to improve our interpretation of image data containing mixtures of land cover types. Coarse and medium spatial resolution sensors such as AVHRR and MISR produce off-nadir/multiangle views of inhomogeneous surfaces. For hydrologic modeling, we need to isolate the albedo of the snow cover from the mixture, while for climate modeling, it is the albedo of the mixed pixel that is required. Thus, an estimate of the albedo of single components in a mixture as well as the albedo of complex mixtures of materials is needed. Although we are able to estimate the fraction of snow cover in an image pixel (Nolin et al. 1993, Rosenthal and Dozier 1996), there have been no validated efforts to estimate surface albedo from mixtures of image components such as snow, vegetation, and soil. Hu et al. (1997) describe a kernel-driven semi-empirical approach using the algorithm for MODIS bidirectional reflectance anisotropy of the land surface (AMBRALS) BRDF model. In that approach, each land surface cover type has a unique kernel that approximates the BRDF derived from radiative transfer or geometric optical models. Weighting factors that are applied to the individual kernels are derived from measurements of combinations of the different elements. At the present time, snow is not well represented in this approach. The combined effects of forward and backward scattering surface covers may introduce nonlinearities into the calculation of composite BRDFs and mixed-pixel albedos. While the kernel approach is appealing in its simplicity, a physically based model that can account for nonlinear effects of spatially inhomogeneous pixels is the desired goal.

4.5.3   Thin Snow versus Dirty Snow

Other research issues include better characterization of the effects of light absorbing particulates in the snow and effects of thin snow on BRDF. Both of these effects are known to reduce the forward scattering peak. DISORT model runs have been performed to test snow BRDF sensitivity to light absorbing particulates (such as dust and soot) and thin snow. Minute amounts of absorbing impurities have been shown to reduce snow reflectance in the visible wavelengths, where ice is highly transparent. Figure 4.13 shows that soot concentrations as low as 0.1 ppmw (parts per million by weight) are enough to reduce reflectance. The effect of the absorbing impurities is apparently enhanced when they are inside the snow grains because refraction

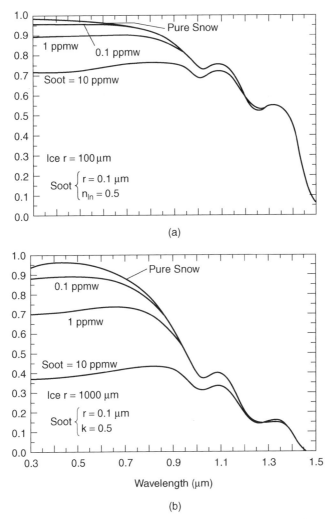

Figure 4.13 Snow direct albedo versus soot. The snow depth is semi-infinite, and the solar zenith angle is 60° [from Warren and Wiscombe (1980)].

focuses the light on the absorbers. Soot in the snow was found to reduce the forward scattering peak in the visible wavelengths, but has very little effect in the near infrared. Additional measurements are needed using an instrument such as PARABOLA III, which has high angular resolution in both the azimuth and zenith axes. Questions remain about the types, concentrations, and depth distributions of absorbing particulates and their effect on snow BRDF. It is unlikely that all such cases can be measured, so the use of a model such as DISORT should be able to provide insight into a wide range of possible ice/soot configurations.

4.5.4 Soil Inversion with Ancillary Information

As reviewed by Irons et al. (1989) and Moran et al. (1997), many studies have reported linkages between surface reflectance and soil properties, such as organic matter, calcium carbonate content, nutrient status, iron oxide content, and texture class. Lack of constraint makes it impossible to retrieve all these soil properties from a single set of directional reflectance measurements. However, a tractable strategy might be to use ancillary data to fix the more temporally or spatially stable soil parameters and use the directional reflectance data to estimate the more variable components. For example, county soil surveys at $1:12,000 - 1:24,000$ scales or greater might be very useful for mapping the variability of soil classes. Information on the associations between soil properties and specific vegetation types can also help control the problem (Korolyuk and Shcherbenko 1994). Muller and James (1994) suggested that the uncertainty in mapping soil particle size caused by differences in soil roughness, moisture, and vegetation cover can be minimized using multitemporal imagery for soil classification.

4.5.5 Soil Sensible Depth

A challenging issue for remote sensing of soils is how deep are significant light interactions are with the particulate medium. Secondary aspects of this issue include spectral variability in light transmittance, effects of particle size, and bulk density. Liang (1997) investigated the sensible depth using a coupled atmosphere–soil radiative transfer model. The sensible depth was determined by examining the downward hemispheric transmittance profile, the hemispherical reflectance, and the bidirectional reflectance. This modeling effort used a range of solar zenith angles and wavelengths and evaluated the effects of different particle size distributions and particle shapes. Under ordinary conditions the sensible optical depth is ≈ 3, corresponding to a geometric depth of $\sim 4-5$ times the particle's effective radius.

4.5.6 Soil Moisture Conditions

Surface moisture condition for soils is another issue of continuing interest. Evaluations of soil moisture effects have a long history. A number of measurements have been made to discern the dependence of visible and near-infrared soil reflectance on moisture content (Irons et al. 1989, Bedidi et al. 1992). Figure 4.14 shows the impacts of soil moisture on reflectance based on the measurements of a silt loam soil (Bowers and Hanks 1965). The number above each curve represents the percent moisture content. The general theoretical approach to account for the moisture effect was developed by Angstrom (1925):

$$\rho = \frac{\rho_0}{n^2(1 - \rho_0) + \rho_0} \tag{4.27}$$

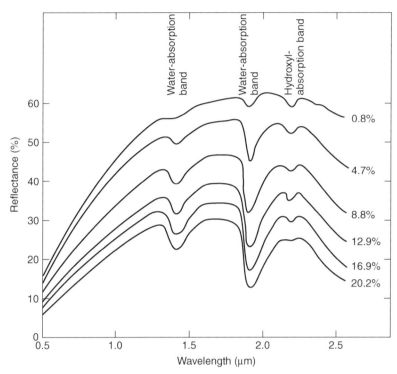

Figure 4.14 Measured dependence of soil reflectance on soil moisture content [from Bowers and Hanks (1965)].

where ρ is the reflectance of a wet soil surface, ρ_0 is the reflectance of a dry soil surface, and n is the refractive index of liquid water. Further extensions and applications of this method have been developed (Palmer and Williams 1974, Lekner and Dorf 1988). Other studies have reported that soil surface reflectance in the near infrared is highly sensitive to soil moisture. As yet, however, soil moisture effects cannot be explicitly treated in a radiative transfer model. What is required is to include soil moisture content in the calculation of optical properties of the soil medium. Without an idea of surface soil moisture conditions, it is impossible to correctly convert soil reflectance measurements into soil albedo information. Microwave remote sensing can provide a direct measurement of the surface soil moisture. For example, spatially distributed and multitemporal observations of surface soil moisture can be estimated from the advanced microwave scanning radiometer (AMSR-E) on the ADEOS-II satellite that was launched on December 14, 2001.

Moreover, a comprehensive model is needed that can account for vegetation–soil moisture interactions (Jackson et al. 1996). Only limited modeling efforts that focus on particular types of vegetation (Lin et al. 1994) have attempted to develop a soil moisture algorithm. Experimental studies (Genda and Okayama 1978, Curran 1979) have linked the polarized directional reflectance with soil moisture conditions, and such an approach shows promise. Spaceborne POLDER data offer an opportunity to explore this possibility further with the additional polarization measurements. Use of both a vegetation index and surface temperature measurements may also be an effective approach, which will be discussed extensively in Section 13.2.3.

4.6 SUMMARY

Both snow and soil are particulate media, similar to the atmosphere. Therefore, atmospheric radiative transfer theory can be applied to snow and soil. However, snow and soil particulars are densely packed, and their resulting reflectance patterns are quite different. In this chapter, most of the discussion has been devoted to introducing the single scattering properties of snow and soil, which are the basis for radiative transfer calculations. Most approximate and numerical solutions are similar to those in atmosphere- and canopy-radiative transfer, so no details are provided. It is interesting to note that geometric optical modeling was used mainly for soils, although it is also suitable for snow.

The inversion of snow properties was discussed in Section 4.4, with emphasis on estimating grain size. In section 4.5 we discussed various practical issues associated with snow/soil modeling, areas that need to be explored further. Since both dry soil and snow have high albedo values, they provide large feedback to the atmosphere. In particular, we need to develop tractable analytic formulas of broadband albedo. This topic will be further explored in Chapter 9.

REFERENCES

Ackerman, S. A. and Stephens, G. L. (1987), The absorption of solar radiation by cloud droplets: An application of anomalous diffraction theory. *J. Atmos. Sci.* **44**: 1574–1588.

Angstrom, A. (1925), The albedo of various surfaces of ground, *Geogr. Ann.* **7**: 323–342.

Bedidi, A., Cervelle, B., Madeira, J., and Pouget, M. (1992), Moisture effects on visible spectral characteristics of lateritic soils, *Soil Sci.* **153**: 129–141.

Bohren, C. F. and Huffman, D. R. (1983), *Absorption and Scattering of Light by Small Particles*, Wiley.

Bourdelles, B. and Fily, M. (1993), Snow grain-size determination from Landsat imagery over Adelie, Antarctica, *Ann. Glaciol.* **17**: 86–92.

Bowers, S. A. and Hanks, R. J. (1965), Reflection of radiant energy from soils, *Soil Sci.* **100**: 130–138.

Buchan, G. D. (1989), Applicability of the simple lognormal model to particle-size distribution in soils, *Soil Sciences* **147**: 155–161.

Chandrasekhar, S. (1960), *Radiative Transfer*, Dover Publications.

Chen, T. W. (1987), Scattering of light by a stratified sphere in high energy approximation, *Appl. Opt.* **26**: 4155–4158.

Choudhury, B. and Chang, A. (1981), On the angular variation of solar reflectance of snow, *J. Geophys. Res.* **86**: 465–472.

Cierniewski, J. (1987), A model for soil surface roughness influence on the spectral response of bare soils in the visible and near-infrared range, *Remote Sens. Envir.* **123**: 97–115.

Cierniewski, J. (1989), The influence of the viewing geometry of bare rough soil surfaces on their spectral response in the visible and near-infrared range, *Remote Sens. Environ.* **127**: 135–142.

Cierniewski, J. and Courault, D. (1993), Bidirectional reflectance of bare soil surfaces in the visible and near-infrared range, *Remote Sens. Rev.* **7**: 321–339.

Cierniewski, J. and M. Verbrugghe, M. (1994), A geometrical model of soil bidirectional reflectance in the visible and near-infrared range, *Proc. 6th Int. Symp. Physical Measurements and Signatures in Remote Sensing*, France, International Society for Photogrammetry and Remote Sensing (Commission VII, Working Group 1), pp. 635–642.

Cierniewski, J. and Verbrugghe, M. (1997), Influence of soil surface roughness on soil bidirectional reflectance, *Int. J. Remote Sens.* **18**: 1277–1288.

Cierniewski, J., Baret, F., Verbrugghe, M., Hanocq, J. F., and Jacquemoud, S. (1996), Geometrical modeling of soil bidirectional reflectance incorporating specular effects, *Int. J. Remote Sens.* **17**: 3691–3704.

Colbeck, S. C. (1982), An overview of seasonal snow metamorphism, *Rev. Geophys. Space Phys.* **20**: 45–61.

Cooper, K. D. and Smith, J. A. (1985), A Monte Carlo reflectance model for soil surfaces with three-dimensional structure, *IEEE Trans. Geosci. Remote Sens.* **23**: 668–673.

Coulson, K. L. (1988), *Polarization and Intensity of Light in the Atmosphere*, A. Deepak Publishing.

Curran, P. J. (1979), The use of polarized panchromatic and false-color infrared film for the monitoring of soil surface moisture, *Remote Sens. Envir.* **8**: 249–266.

Dapples, E. C. (1975), Laws of distribution applied to sand sizes, *Geolog. Soc. Am. Mem.* **142**: 37–61.

Despan, D., Bedidi, A., and Cervelle, B. (1999), Bidirectional reflectance of rough bare soil surfaces, *Geophys. Res. Lett.* **26**: 2777–2780.

Dobbins, R. A. and Jizmagian, G. S. (1966), Optical scattering cross sections for polydispersions of dielectric spheres, *J. Opt. Soc. Am.* **56**: 1345–1350.

Dozier, J. (1989a), Spectral signature of alpine snow cover from the Landsat Thematic Mapper, *Remote Sens. Envir.* **28**: 9–22.

Dozier, J. (1989b), Remote sensing of snow in visible and near-infrared wavelengths, in *Theory and Applications of Optical Remote Sensing*, G. Asrar, ed., Wiley, pp. 527–547.

Dozier, J. and Marks, D. (1987), Snow mapping and classification from Landsat Thematic Mapper data, *Ann. Glaciol.* **9**: 97–103.

Dozier, J., Davis, R. E., and Perla, R. (1987), On the objective analysis of snow microstructure, in *Avalanche Formation, Movement and Effects*, B. Salm and H. Gubler, eds., International Association of Hydrological Sciences, pp. 49–59.

Egan, W. G. and Hilgeman, T. W. (1979), *Optical Properties of Inhomogeneous Materials: Applications to Geology, Astronomy, Chemistry, and Engineering*, Academic Press.

Fily, M., Bourdelles, B., Dedieu, J. P., and Sergent, C. (1997), Comparison of in situ and Landsat Thematic Mapper derived snow grain characteristics in the Alps, *Remote Sens. Envir.* **59**: 452–460.

Genda, H. and Okayama, H. (1978), Estimation of soil moisture and components by measuring the degree of spectral polarization with a remote sensing simulator, *Appl. Opt.* **17**: 3439–3443.

Grenfell, T. C. and Warren, S. G. (1999), Representation of a nonspherical ice particle by a collection of independent spheres for scattering and absorption of radiation, *J. Geophys. Res.* **104**: 31697–31709.

Hapke, B. (1984), Bidirectional reflectance spectroscopy. 3. Correction for macroscopic roughness, *Icarus* **59**: 41–59.

Hapke, B. (1986), Bidirectional reflectance spectroscopy. 4: The extinction coefficient and the opposition effect, *Icarus* **67**: 264–280.

Hapke, B., Dominick, D., Nelson, R., and Smythe, W. (1996), The cause of the hot spot in vegetation canopies and soils: Shadow-hiding versus coherent backscattering, *Remote Sens. Envir.* **58**: 63–68.

Hapke, B. W. (1981), Bidirectional reflectance spectroscopy 1. Theory, *J. Geophys. Res.* **86**: 3039–3054.

Hu, B., Lucht, W., Li, X., and Strahler, A. (1997), Validation of kernel-driven semiempirical models for the surface bidirectional reflectance distribution function of land surfaces, *Remote Sens. Envir.* **62**: 201–214.

Hyvarinen, T. and Lammasniemi, J. (1987), Infrared measurement of free-water content and grain size of snow, *Opt. Eng.* **26**: 342–348.

Irons, J., Campbell, G., Norman, J., Graham, D., and Kovalick, W. (1992), Prediction and measurement of soil bidirectional reflectance, *IEEE Trans. Geosci. Remote Sens.* **30**: 249–260.

Irons, J. R., Weismiller, R. A., and Petersen, G. W. (1989), Soil reflectance, in *Theory and Applications of Optical Remote Sensing*, G. Asrar, ed., Wiley, pp. 66–106.

Jackson, J., Schmugge, T., and Engman, E. (1996), Remote sensing applications to hydrology: Soil moisture, *Hydrolog. Sci. J.* **41**: 517–530.

Jacquemoud, S., Baret, F., and Hanocq, J. F. (1992), Modeling spectral and bidirectional soil reflectance, *Remote Sens. Envir.* **41**: 123–132.

Jin, Y. and Wang, Y. (1999), A genetic algorithm to simultaneously retrieve land surface roughness and soil moisture, *Proc. 25th Annual Conf Exhibition of the Remote Sensing Society Earth Observation: From Data to Information*, Univ. Wales at Cardiff and Swansea.

Jin, Z. and Simpson, J. (1999), Bidirectional anisotropic reflectance of snow and sea ice in AVHRR channel 1 and 2 spectral regions. Part I: Theoretical analysis, *IEEE Trans. Geosci. Remote Sens.* **37**: 543–554.

Korolyuk, T. V. and Shcherbenko, H. (1994), Compiling soil maps on the basis of remotely-sensed data digital processing: Soil interpretation, *Int. J. Remote Sens.* **15**: 1379–1400.

Kou, L., Labrie, D., and Chylek, P. (1994), Refractive indices of water and ice in the 0.65- to 2.5-μm spectral range, *Appl. Opt.* **32**: 3531–3540.

Lekner, J. and Dorf, M. C. (1988), Why some things are darker when wet, *Appl. Opt.* **27**: 1278–1280.

Leroux, C., Deuze, J. L., Goloub, P., C., S., and Fily, M. (1998), Ground measurements of the polarized bidirectional reflectance of snow in the near-infrared spectral domain: Comparisons with model results, *J. Geophys. Res.* **103**: 19721–19731.

Li, W., Stamnes, K., Chen, B. Q., and Xiong, X. Z. (2001), Snow grain size retrieved from near-infrared radiances at multiple wavelengths, *Geophys. Res. Lett.* **28**: 1699–1702.

Li, X. and Strahler, A. (1985), Geometric-optical modeling of a coniferous forest canopy, *IEEE Trans. Geosci. Remote Sens.* **23**: 705–721.

Li, X. and Strahler, A. (1986), Geometric-optical bi-directional reflectance modeling of a coniferous forest canopy, *IEEE Trans. Geosci. Remote Sens.* **24**: 906–919.

Li, X. and Strahler, A. H. (1992), Geometric-optical bidirectional reflectance modeling of the discrete crown vegetation canopy: Effect of crown shape and mutual shadowing, *IEEE Trans. Geosci. Remote Sens.* **30**: 276–292.

Liang, S. (1997), An investigation of remotely sensed soil depth in the optical region, *Int. J. Remote Sens.* **18**: 3395–3408.

Liang, S. and Lewis, P. (1996), A parametric radiative transfer model for sky radiance distribution, *J Quant. Spectrosc. Radiat. Transfer* **55**: 181–189.

Liang, S. and Townshend, J. R. G. (1996a), A parametric soil BRDF model: A four-stream approximation, *Int. J. Remote Sens.* **17**: 1303–1315.

Liang, S. and Townshend, J. R. G. (1996b), A modified hapke model for soil bidirectional reflectance, *Remote Sens. Envir.* **55**: 1–10.

Liang, S. and Mishchenko, M. I. (1997), Calculation of soil hot spot effects using coherent backscattering theory, *Remote Sens. Envir.* **60**: 163–173.

Liang, S., Lewis, P., Dubayah, R., Qin, W., and Shirey, D. (1997), Topographic effects on surface bidirectional reflectance scaling, *J. Remote Sens.* **1**: 82–93.

Lin, D., Wood, E., Saatchi, S., and Beven, K. (1994), Soil moisture estimation over grass covered areas using airSAR, *Int. J. Remote Sens.* **15**: 2323–2343.

Liou, K. N. and Hansen, J. E. (1971), Intensity and polarization for single scattering by polydisperse spheres: A comparison of ray optics and Mie theory, *J. Atmos. Sci.* **28**: 995–1004.

Mishchenko, M. I. (1991), Light scattering by randomly oriented axially symmetric particles, *J. Opt. Soc. Am.* **A8**: 871–882.

Mishchenko, M. I. (1993), Light scattering by size-shape distributions of randomly oriented axially symmetric particles of a size comparable to a wavelength, *Appl. Opt.* **32**: 4652–4666.

Mishchenko, M. I. (1994), Asymmetry parameters of the phase function for densely packed scattering grains, *J. Quant. Spectrosc. Radiat. Trans.* **52**: 95–110.

Mishchenko, M. I. and Macke, A. (1997), Asymmetry parameters of the phase function for isolated and densely packed spherical particles with multiple internal inclusions in the geometric optics limit, *J. Quant. Spectrosc. Radiat. Transfer* **57**: 767–794.

Mishchenko, M. I., Dlugach, J. M., Yanovitskijb, E. G., and Zakharovac, N. T. (1999), Bidirectional reflectance of flat, optically thick particulate layers: An efficient radiative transfer solution and applications to snow and soil surfaces, *J. Quant. Spectrosc. Radiat. Transfer* **63**: 409–432.

Moran, M. S., Inoue, Y., and Barnes, E. (1997), Opportunities and limitations for image-based remote sensing in precision crop management, *Remote Sens. Envir.* **61**: 319–346.

Mugnai, A. and Wiscombe, W. J. (1980), Scattering of radiation by moderately nonspherical particles, *J. Atmos. Sci.* **37**: 1291–1307.

Muller, E. and James, M. (1994), Seasonal variation and stability of soil spectral patterns in a fluvial landscape, *Int. J. Remote Sens.* **15**: 1885–1900.

Nakamura, T., Abe, O., Hasegawa, T., Tamura, R., and Ohta, T. (2001), Spectral reflectance of snow with a known grain-size distribution in successive metamorphism, *Cold Regions Sci. Technol.* **32**: 13–26.

Nolin, A. and Liang, S. (2000), Progress in directional reflectance modeling and applications for surface particulate media: Snow and soils, *Remote Sens. Rev.* **18**: 307–342.

Nolin, A. W. and Dozier, J. (2000), A hyperspectral method for remotely sensing the grain size of snow, *Remote Sens. Envir.* **74**: 207–216.

Nolin, A. W., Dozier, J., and Mertes, L. A. K. (1993), Mapping alpine snow using a spectral mixture modeling technique, *Ann. Glaciol.* **17**: 121–124.

Norman, J. M., Welles, J. M., and Walter, E. A. (1985), Contrast among bidirectional reflectance of leaves, canopies and soils, *IEEE Trans. Geosci. Remote Sens.* **23**: 659–667.

Nussenzveig, H. M. (1992), *Differential Effects in Semiclassical Scattering*, Cambridge Univ. Press.

Nussenzveig, H. M. and Wiscombe, W. J. (1991), Complex angular momentum approximation to hard-core scattering, *Phys. Rev.* **43A**: 2093–2112.

Otterman, J. and Weiss, G. H. (1984), Reflections from a field of randomly located vertical protrusions, *Appl. Opt.* **23**: 1931–1936.

Painter, T. H., Roberts, D. A., Green, R. O., and Dozier, J. (1998), The effect of grain size on spectral mixture analysis of snow-covered area from AVIRIS data, *Remote Sens. Envir.* **65**: 320–332.

Palmer, K. F. and Williams, D. (1974), Optical properties of water in the near infrared, *J. Opt. Soc. Am.* **64**: 1107–1110.

Perrin, J. M. and Chiappetta, P. (1985), Light scattering by large particles I: A new theoretical description in the eikonal picture, *Optica Acta*, **32**: 907–921.

Pinty, B., Verstraete, M. M., and Dickinson, R. E. (1989), A physical model for predicting bidirectional reflectances over bare soil, *Remote Sens. Envir.* **27**: 273–288.

Ranson, K. J., Biehl, L. L., and Daughtry, C. S. T. (1984), *Soybean Canopy Reflectance Modeling Data Sets*. Lab. Appl. Remote Sens., West Lafayette, IN, LARS Tech. Rep. 671584.

Rosenthal, W. and Dozier, J. (1996), Automated mapping of montane snow cover at subpixel resolution from the landsat Thematic Mapper, *Water Resour. Res.* **32**: 115–130.

Roujean, J. L., Leory, M., Podaire, A., and Deschamps, P. Y. (1992), Evidence of surface reflectance bidirectional effects from a NOAA/AVHRR multi-temporal data set, *Int. J. Remote Sen.* **13**: 685–698.

Sancer, M. I. (1969), Shadow-corrected electromagnetic scattering from a randomly rough surface, *IEEE Trans. Antennas Propag.* **17**: 577–585.

Shirazi, M. A. and Boersma, L. (1984), A unifying quantitative analysis of soil texture, *Soil Sci. Soc. Am. J.* **48**: 142–147.

Shoshany, M. (1992), A simulation of directional reflectance distributions for various surface microstructures, *Int. J. Remote Sens.* **13**: 2355–2361.

Shoshany, M. (1993), Roughness-reflectance relation of bare desert terrain: An empirical study, *Remote Sens. Envir.* **45**: 15–27.

Stamnes, K., Tsay, S. C., Wiscombe, W., and Jayaweera, K. (1988), Numerically stable algorithm for discrete-ordinate-method radiative transfer in multiple scattering and emitting layered media, *Appl. Opt.* **7**: 2502–2509.

Takano, Y. and Liou, K. N. (1989a), Solar radiative transfer in cirrus clouds. Part I: Single-scattering and optical properties of hexagonal ice crystals, *J. Atmos. Sci.* **46**: 3–19.

Takano, Y. and Liou, K. N. (1989b), Solar radiative transfer in cirrus clouds. Part II: Theory and computation of multiple scattering in an anisotropic medium, *J. Atmos. Sci.* **46**: 20–36.

Takano, Y. and Liou, K. N. (1995), Solar radiative transfer in cirrus clouds. Part III: Light scattering by irregular ice crystals, *J. Atmos. Sci.* **52**: 818–837.

van de Hulst, H. C. (1980), *Multiple Light Scattering, Tables, Formulas and Applications*, Vols. 1, 2, Academic Press.

Warren, S. and Wiscombe, W. (1980), A model for the spectral albedo of snow. II. Snow containing atmospheric aerosols, *J. Atmos. Sci.* **37**: 2734–2745.

Warren, S. G. (1984), Optical constants of ice from the ultraviolet to the microwave, *Appl. Opt.* **23**: 1206–1225.

Warren, S. G., Brandt, R. E., and Hinton, P. O. (1998), Effect of surface roughness on bidirectional reflectance of antarctic snow, *J. Geophys. Res.* **103**: 25789–25807.

Wiscombe, W. and Warren, S. (1980), A model for the spectral albedo of snow. I. Pure snow, *J. Atmos. Sci.* **37**: 2712–2733.

Yang, P. and Liou, K. (1995), Light scattering by hexagonal ice crystals: Comparison of finite-difference time domain and geometrical optics models, *J. Opt Soc. Am.* **A12**: 162–176.

Zege, E. P. and Kokhanovskiy, A. A. (1989), Approximation of the anomalous diffraction of coated spheres, *Atmos. Oceanic Phys.* **25**: 883–887.

5

Satellite Sensor Radiometric Calibration

Calibration and instrument characterization are essential for any satellite or airborne remote sensing device. Prelaunch instrument characterization, onboard calibration, vicarious calibration, and interinstrument cross-calibration are all critical components of a calibration system. In this chapter, we will mainly introduce the basic concepts and present calibration results for both TM and AVHRR.

Section 5.1 introduces some background information, including the importance of sensor calibration, basic concepts, and principles. Section 5.2 presents various postlaunch calibration methods that are currently being used. Section 5.3 shows the calibration results for both Landsat 4/5 TM and NOAA AVHRR, which are essential for use of these datasets in quantitative computations.

5.1 BACKGROUND

The Working Group on Calibration and Validation (WGCV) of the international Committee on Earth Observation Satellites (CEOS) defines remote sensing calibration as the process of quantitatively defining the system response to known controlled signal inputs (Belward 1999). The main fundamental aspects that need to be calibrated are the sensor system's response to electromagnetic radiation as a function of

- Wavelength and/or spectral band (spectral response)
- The intensity of the input signals (radiometric response)
- Different locations across the instantaneous field of view and/or the overall scene (spatial response or uniformity)

Quantitative Remote Sensing of Land Surfaces. By Shunlin Liang
ISBN 0-471-28166-2 Copyright © 2004 John Wiley & Sons, Inc.

- Different integration times and lens or aperture settings
- Unwanted signals such as stray light and leakage from other spectral bands

Estimating land surface bio/geophysical variables accurately from remotely sensed data relies largely on the quality of the data and, in particular, on the accuracy of radiometric calibration. *Radiometric calibration* is a process that converts recorded sensor voltages or digitized counts to an absolute scale of radiance that is independent of the image forming characteristics of the sensor. This process can be a relative or an absolute calibration. *Absolute calibration*, for a linear sensor, is performed by ratioing the digital numbers (DNs) from the sensor, with the value of an accurately known, uniform-radiance field at its entrance pupil. *Relative calibration* is determined by normalizing the outputs of the detectors to a given, often average, output from all the detectors in the band.

Those who build remote sensing devices must have accurate measurements of a sensor's radiometric properties before that sensor is sent into space—this is usually called *preflight calibration*. This calibration may change in space in response to variations in the environment surrounding the sensor in a spaceborne environment. Examples include outgassing, the bombardment by energetic particles from space, variation in the filter transmittance and spectral response, and slow deterioration of the electronic system. *In-flight absolute calibration* is usually performed on a routine basis for the thermal infrared channels to allow for precise temperature information, but the solar channels used for imaging on most operational satellites do not have onboard calibration capabilities mainly because of the limitations in satellite power, weight, and space. Some orbiting satellites even have simple onboard calibration systems, but they change in sensitivity with time. *Post-launch calibration* data have to be obtained from vicarious calibration techniques. *Vicarious calibration* usually refers to techniques that make use of natural or artificial sites on the surface of the Earth for the postlaunch calibration of sensors.

The usual approach to sensor calibration starts with the formulation of a sensor calibration model. The simplest calibration model is the linear formula that links the sensor output (DN) to the radiance L at the entrance pupil of the sensor

$$Y = AL \tag{5.1}$$

where Y represents the DN values and A is the matrix of the absolute calibration coefficients. Matrix A can be determined from the accurate preflight measurements, then monitored on orbit by onboard calibration devices using secondary or tertiary standard light sources (lamps or the Sun), and finally vicarious methods using images of specific well-known ground targets or the Moon (Dinguirard and Slater 1998). Since preflight calibration coefficients are provided by the sensor builders and thermal infrared sensors can be accurately calibrated using onboard devices, we focus on presenting the vicarious calibration techniques and results for solar channels (visible and near-infrared spectra) in the following sections. More detailed discussions on this subject can be found in other books (e.g., Chen 1997) or many other journal papers cited in the following sections.

None of the techniques we discuss appear to have the potential for satisfying the requirement of knowing the characteristics of the instruments to better than a few percent over several years, especially when measurement continuity, combined with expected satellite lifetimes, requires that the instruments be on different platforms. A possible solution to this problem has been discussed for some time within the EOS terra and aqua platform communities, namely, performing spacecraft calibration attitude maneuvers to allow the instruments to view the Moon and/or cold space. The Moon provides an on-orbit radiance target whose properties are essentially invariant over time, while cold space provides an on-orbit zero level reference for detectors and electronics, particularly in the thermal infrared.

5.2 POSTLAUNCH CALIBRATION METHODS

There are roughly two typical postlaunch calibration techniques. The first is to fly an aircraft with a calibrated radiometer that measures the spectral radiance of the target observed by the satellite in the same illumination and observing directions (Smith et al. 1988, Kriebel and Amann 1993, Abel et al. 1993, Green and Shimada 1997). It is often involved in simultaneous radiometric measurements of spatially and spectrally homogeneous Earth targets. This method is usually called the *radiance-based calibration method*. The *reflectance-based method* requires an accurate measurement of the spectral reflectance of the ground target and measurement of spectral extinction depths and other meteorological variables. Figure 5.1 shows a number of methods used to completely characterize the reflectance of a homogeneous ground site and retrieve sufficient atmospheric optical parameters to derive accurate TOA radiances over the wavelength channel(s) of the instrument, as it images the scene. Radiance-based and reflectance-based calibration methods are discussed by Slater et al. (1987).

Although these may be the most direct methods, they are relatively expensive and complex and cannot be used to calibrate historical data (Kaufman and Holben 1993). Another technique is to compare the observed radiance with radiative transfer calculations using well-known physical characteristics of the atmosphere and surface targets. Most of the methods discussed below largely belong to this category. Atmospheric radiative transfer is detailed in Chapter 2.

The surface targets used for postlaunch calibration include ocean, desert, cloud, snow, dry·lake, ice sheet, and the Moon.

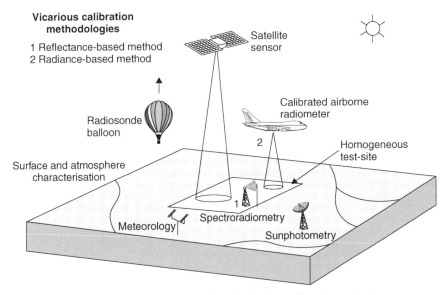

Figure 5.1   Illustration of postlaunch vicarious calibration methods.

5.2.1   Ocean Calibration

Ocean calibration relies on either molecular scattering in the atmosphere over the ocean or the Sun glint. For a cloudless air mass with a small amount of haze that is far away from the ocean glint, the major contribution (~ 70–80%) to upward radiance over deep oceans in the visible part of the spectrum is from molecular scattering in the atmosphere (Fraser and Kaufman 1986). To reduce the influence of other variables (e.g., aerosol scattering, foam and glint from the ocean water), particular viewing conditions need to be chosen. These conditions include deep oceans to get clear water, large viewing and solar zenith angles to increase the photon travel pathlength, and viewing the western direction to avoid specular reflection. The non-Rayleigh component of the scattering is deduced from the signals in the near-IR spectrum where Rayleigh scattering is negligible. Note that Rayleigh scattering has been discussed in Section 2.3.1. This method has been used to calibrate Meteosat (Fraser and Kaufman 1986), SPOT (Dilligeard et al. 1996), POLDER (Hagolle et al. 1997), and AVHRR (Kaufman and Holben 1993).

Approximately 87% of glint radiance is due to specular reflectance (Kaufman and Holben 1993). This reflectance cannot be theoretically established with the same accuracy as molecular scattering because of its dependence on wind speed and wave structure, but it is independent of the radiation wavelength and therefore can be used to determine the relative calibration of the near-IR bands to the visible bands. Good conditions usually correspond to wind speeds between 2 and 5 m/s. This approach has to be performed over as many glitter images as possible.

5.2.2 Desert Calibration

Desert sites have been widely used for sensor calibration since they have a stable spectral response over time. Because of their high reflectances, the atmospheric effect on the upward radiance is relatively minimal. They are also spatially uniform. Their temporal instability without atmospheric correction has been determined to be less than 1–2% over a year. Several sites have been used for sensor calibration, including the Libyan desert (Staylor 1990, Kaufman and Holben 1993) for calibrating AVHRR data, the North Africa desert for calibrating SPOT imagery (Henry et al. 1993), and the Egyptian desert, which was identified by the international remote sensing community for sensor intercalibration (Dinguirard and Slater 1998).

For calibration of high-resolution imagery, the White Sands Missile Range test site in New Mexico has been extensively used since the 1980s. It is located in the desert southwest of the United States in a region of low aerosol loading and an elevation of 1.2 km.

5.2.3 Clouds

Very-high-altitude (10-km) bright clouds are good validation targets in the visible and near-IR spectra because of their high spectrally consistent reflectance (Vermote and Kaufman 1995). If the clouds are very high, we do not need to correct aerosol scattering and water vapor absorption as both aerosol and water vapor are distributed near the surface. Only Rayleigh scattering and ozone absorption need to be considered. This method has been found (Hagolle et al. 1997, Lafrance et al. 2002) to give a 4% uncertainty for the intercalibration of the POLDER spectral bands.

5.2.4 Others

Dry lakes and other large homogeneous areas are also used for calibration, for example, Railroad Valley Playa, a dry lakebed in Nevada (USA) with a composition dominated by clay and Rogers dry lake at Edwards Air Force Base in California.

Permanent ice sheets are another target that have been used for calibration. Loeb (1997) developed calibration curves for NOAA9 AVHRR channel 1 and 2 reflectance using the permanent ice sheets of Greenland and Antarctica. Tahnk and Coakley (2001) determined the sensor degradation coefficients for NOAA14 AVHRRs first two channels using the permanent ice sheets of the central Antarctica.

Cloud shadows over water have been used for calibrating high-resolution sensors (Reinersman et al. 1998). This cloud shadow method uses the difference between the total radiance values observed at the sensor for these two regions of sunlit and shadowed, thus removing the nearly identical atmospheric radiance contributions to the two signals (e.g., path radiance and Fresnel reflected skylight). What remains is due largely to solar photons backscattered from beneath the sea to dominate the residual signal. Normalization by the direct solar irradiance reaching the sea surface and correction for some second-order effects provides the remote sensing reflectance of the ocean at the location of the neighbor region, providing a known ground target spectrum for use in testing the calibration of the sensor. A similar approach may be useful for land targets if horizontal homogeneity of scene reflectance exists about the shadow.

The Moon is another calibration target. The stability of its reflectance is extremely high, but its radiance is not. The Moon can be used to (1) check the in-flight stability of a solar diffuser and (2) provide a direct calibration of the sensor (Dinguirard and Slater 1998). From 1995 to 1998, the USGS and the Northern Arizona University Department of Physics and Astronomy constructed an observatory in Flagstaff, Arizona dedicated to making long-term radiometric measurements of the Moon. The purpose of this ongoing program with respect to EOS calibration is to utilize the radiometric stability of the lunar surface to provide long-term, on-orbit calibration and cross-calibration of EOS and non-EOS sensors flown on similar and different platforms. Currently, accurate measurements of the radiance and irradiance of the Moon are made at a number of wavelengths in the 348–2385 nm wavelength region using two telescopic imaging systems. The observatory measurements are used to produce exoatmospheric radiance images of the Moon that can be compared with orbiting spacecraft lunar observations. The lunar radiometric data are being archived in the NASA Goddard Space Flight Center (GSFC) Distributed Active Archive Center (DAAC).

The uncertainties and constraints of different vicarious calibration methods are compared in Table 5.1.

5.3 CALIBRATION COEFFICIENTS FOR LANDSAT TM AND AVHRR REFLECTIVE BANDS

When satellite data are ordered from the data center, radiometric calibration coefficients are usually provided in the data header files for most of sensors. For Landsat 4/5 TM and NOAA AVHRR, sensor degradation has been a serious problem. Their calibration coefficients are documented below for easy reference.

5.3.1 Landsat TM

5.3.1.1 Absolute Calibration of the Reflective Bands

The preflight calibration coefficients for converting DN of both Landsat 4 and 5 to radiance L (W m^{-2} sr^{-1} μm^{-1}) in the reflective bands are given in

TABLE 5.1 Comparisons of Different Vicarious Calibration Methods

Calibration Methods: Type	Uncertainties	Constraints
Test sites:		
Absolute	3.5% reflectance-based	Expensive
	2.8% radiance-based	Needs ground instrumentation
		Needs good atmospheric conditions
		Specific sensor program in most cases
Rayleigh scattering:		
Absolute	5%for SPOT XSL and	Specific geometric condition
	2–3.5% for POLDER	Need very good atmospheric conditions
	blue bands	Not applicable to longer wavelengths
		Easier with larger FOV (greater occurrence)
Stable desert:		
Multitemporal and multisensor	About 3%	Specific programs
		Needs clear-sky imagery
Clouds:		
Interband	4% on POLDER	Specific images of high clouds
		Needs suitable geometric conditions
Glitter:		
Interband	1–2% on POLDER	Specific geometric conditions
		Wind speeds between 2–5 m/s, no clouds
The Moon:		
Multitemporal	Expected 2%	Cannot provide calibration near the top of the dynamic range for land observing sensors
		Specific programming and viewing conditions
Absolute	Expected 2%	As above
		More radiometric verification needed
		Require low uncertainty calibration of the Moon

Source: Modified from Dinguirard and Slater (1998).

TABLE 5.2   Preflight Calibration Coefficients for Landsat 4 and 5 TM Sensors

Spectral Bands	α	β
1	0.602	-1.5
2	1.17	-2.8
3	0.806	-1.2
4	0.815	-1.5
5	0.108	-0.37
7	0.057	-0.15

Source: Price (1989).

Table 5.2 for the following linear equation

$$L = \alpha \cdot DN + \beta \qquad (5.2)$$

For Landsat 4, which was launched on July 16, 1982, Markham and Barker (1986) proposed post-calibration for converting DN to radiance using the same formula as (5.2), but the coefficients are calculated using the radiance dynamic ranges as $\alpha = (L_{max} - L_{min})/255$ and $\beta = L_{min}$, where the dynamic ranges are as given in Table 5.3 Note that since L_{min} is negative under most conditions, it is physically meaningless as the low end of the dynamic range and we have to use them in the statistical sense.

Landsat 5 was launched on March 1, 1984 and operated until 2001. During the latter portion of its lifetime, there was no official validation effort. Use of the coefficients provided on tape to convert DN to radiance (and then to reflectance) can lead to problems in subsequent analyses. Investigators have found that the top-of-atmosphere (TOA) reflectances over time got too low to allow for proper atmospheric correction—even a purely Rayleigh atmosphere over a zero reflectance surface yields larger values of TOA reflectance than one gets from the calibrated image values.

TABLE 5.3   TM Postcalibration Dynamic Ranges for Landsat 4 Data ($W\ m^{-2}\ sr^{-1}\ \mu m^{-1}$)

Band	Before 8/1/1983		8/1/1983–1/1/1984		1/15/1984–10/1/1991	
	L_{min}	L_{max}	L_{min}	L_{max}	L_{min}	L_{max}
1	-1.52	158.42	0.00	142.86	-1.50	152.1
2	-2.84	308.17	0.00	291.25	-2.80	296.8
3	-1.17	234.63	0.00	225.00	-1.20	204.3
4	-1.51	224.32	0.00	214.29	-1.50	206.2
5	-0.3	32.42	0.00	30.00	-0.37	271.9
6	2.0	15.64	4.84	12.40	1.238	15.60
7	-0.15	17.00	0.00	15.93	-0.15	14.38

TABLE 5.4 DNs per Unit Radiances (W m^{-2} sr^{-1} μm^{-1}) for Landsat 5 Reflective Bands on Level 0 Data or Equivalent

Dates	Band					
	1	2	3	4	5	7
Preflight	1.555	0.786	1.020	1.082	7.875	14.77
July 08, 1984		0.734	0.955	1.055		
Oct. 28, 1984	1.389	0.732	0.927	1.087	7.024	14.99
May 24, 1985		0.749	0.942	1.045		
Aug. 28, 1995		0.715	0.914	1.121	7.227	15.24
Nov. 16, 1985	1.367	0.716	0.922	1.094	7.506	16.11
March 27, 1987	1.307	0.702	0.891	1.048	7.441	16.18
Feb. 10, 1988	1.304	0.721	0.918	1.059	7.351	16.29
Aug. 15, 1992		0.651	0.883	1.048	7.416	15.45
Oct. 21, 1993	1.281	0.683	0.924	1.094	7.477	15.24
Oct. 8, 1994	1.222	0.653	0.887	1.054	6.811	13.166

The calibration coefficients derived from the vicarious results using data from White Sands Missile Range (Thome et al. 1997) are given in Table 5.4. These calibration results were based on data acquired as late as 1994. In the calibration exercises associated with Landsat 7 ETM + around 2000, several studies (Teillet et al. 2001, Vogelmann et al. 2001) concluded that Landsat 5 TM did not degradate much from 1994 to 2000.

On the basis of the dataset in Table 5.4, the Canadian Center for Remote Sensing (CCRS) developed simple linear equations of calibration gain coefficient, which are given in Table 5.5,

$$L = \frac{\text{DN} - \text{offset}}{G} \tag{5.3}$$

where the offset coefficients provided with the TM imagery should be used.

TABLE 5.5 Calibration Gain Coefficients (G) in Eq. (5.3) for Landsat 5 TM

TM Spectral Band	Calibration Gain Coefficient (counts W m^{-2} sr^{-1} μm^{-1})	Characteristic Wavelength (μm)
1	$G = (-3.58 \times 10^{-5}) * d + 1.376$	0.4863
2	$G = (-2.10 \times 10^{-5}) * d + 0.737$	0.5706
3	$G = (-1.04 \times 10^{-5}) * d + 0.932$	0.6607
4	$G = (-3.20 \times 10^{-6}) * d + 1.075$	0.8382
5	$G = (-2.64 \times 10^{-5}) * d + 7.329$	1.677
7	$G = (-3.81 \times 10^{-4}) * d + 16.02$	2.223

d = days since launch (March 1, 1984)

As noted from CCRS (`http://www.ccrs.nrcan.gc.ca/ccrs/rd/ana/calval/landst5_e.html`), these time-dependent equations can be applied to the DN values provided on the tape. Although this is not entirely appropriate because the DN values are not raw data; a better approach has yet to be devised. Nevertheless, using these new characterizations should improve things enough to make it worthwhile.

Note that Landsat 5 TM has an internal calibrator for the reflective bands. Vicarious calibration performed independently by several teams suggested that the internal calibrator was tracking the instrument gain well until about 1988. After that time, the performance of the internal calibrator tends to diverge from the vicarious calibrations. Vogelmann et al. (2001) suggest using the internal calibration as an interpolator for those dates when vicarious calibration information is not available.

5.3.1.2 *Relative Calibration of the Thermal IR Band*

Landsat TM band 6 is a thermal IR band detecting surface thermal emission in the wavelength from 10.4 to 12.5 μm. For absolute calibration, there has been an internal calibration system in all three Landsat satellites (4, 5 and 7). It seems to be working fine in Landsat 7, but did not work appropriately to correct the individual responses of the four detectors in band 6 (Chander et al. 2002). As a result, the images have strips. Chander et al. (2002) developed regression formulas to normalize the responses of these four detectors and remove the image strips. These formulae are used to predict the response of each detector (y) from the band average (x):

For Landsat 4:

$$\begin{cases} \text{Detector 1} & y = 0.9871x + 0.8800 \\ \text{Detector 2} & y = 0.9778x + 0.9929 \\ \text{Detector 3} & y = 1.0212x - 0.3577 \\ \text{Detector 4} & y = 1.0139x - 1.5152 \end{cases} \tag{5.4}$$

For Landsat 5:

$$\begin{cases} \text{Detector 1} & y = 0.9785x + 3.1251 \\ \text{Detector 2} & y = 1.0074x - 1.3159 \\ \text{Detector 3} & y = 0.9751x + 3.3115 \\ \text{Detector 4} & y = 1.039x - 5.1207 \end{cases} \tag{5.5}$$

One destriping example is shown in Fig. 5.2 (Chander et al., 2002). On the left is the original band 6 image of Landsat 5 on which the strips are very obvious. After applying Eq. (5.5), the image quality has been significantly improved (see on right). Note that the relative calibration Eqs. (5.4) and (5.5) are suitable for the raw data (level 0). There is destriping processing in the calibrated level 1 data product.

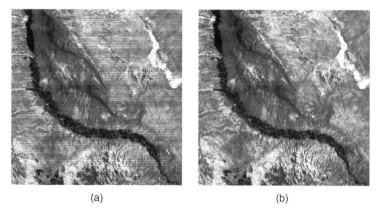

(a)   (b)

Figure 5.2 An example of Landsat band 6 thermal imagery before (striped image) (a) and after (destriped image) (b) calibration. [From Chander et al. (2002), *IEEE Trans. Geosci. Remote Sens.* Copyright © 2002 with permission from IEEE.]

5.3.2 NOAA AVHRR

The NOAA series of satellites (the current sequence is known as the Advanced TIROS-N series) has been in continuous operation since October 1978. In its full configuration it consists of two satellites in complementary near-polar orbits, with one crossing the equator at local solar times of approximately 0730 (morning), and the other at 1430 (afternoon). By convention, the even-numbered satellites normally cover the "morning pass" and the odd-numbered satellites normally cover the "afternoon pass." However, NOAA13 failed on launch, and after years of service NOAA11 failed in orbit, NOAA14 was launched on December 30, 1994 to replace NOAA11 and NOAA13. NOAA15, and NOAA16 were also launched as of 2002.

Archived AVHRR data span the operational lifetime of several satellites; however, sensor degradation has been identified as a major factor affecting the stability of the data quality. Although AVHRR's two reflective sensors are calibrated prior to launch, there is no proper onboard capability for assessing postlaunch sensor degradation of the visible and near-infrared spectral channels. Thus, vicarious calibration is an important process in monitoring the sensor performance. There is evidence that the first two channels degrade in orbit, initially because of outgassing (e.g., water vapor from filter interstices) and launch associated contamination (e.g., rocket exhaust and outgassing), and subsequently because of the continued exposure to the harsh space environment. Use of the prelaunch calibration coefficients would lead to erroneous values of the upwelling radiances and hence of the geophysical products derived from these radiances such as the vegetation indices, Earth's shortwave radiation budget, and the columnar aerosol burden over the global oceans. The degradations of two AVHRR channels of NOAA7, -9, and -11, are shown in Fig. 5.3.

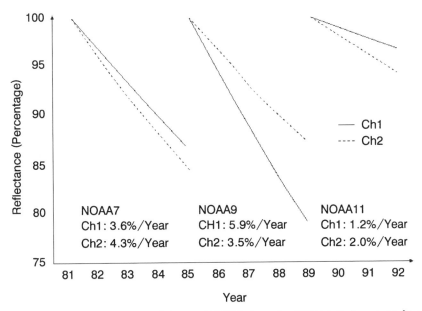

Figure 5.3 Relative degradation of the two AVHRR channels of NOAA-7, -9, and -11. [From Rao and Chen (1995), *Int. J. Remote Sens.* Reproduced by permission of Taylor & Francis, Ltd.]

There have been many efforts to calibrate the two AVHRR solar bands (e.g., Staylor 1990, Kaufman and Holben 1993, Vermote and Kaufman 1995, Rao and Chen 1996). Here we mainly report results from Rao and Chen in their series of studies (Rao and Chen 1996). The satellite lifetimes are listed in Table 5.6. In the following sections we discuss the calibration coefficients for each NOAA satellite (7, 9, 11, 14, and 15), d is the day since satellite launch, and C_{10} represents a DN value with quantization level 10 (see Section 1.2.1).

TABLE 5.6 NOAA Satellite Lifetimes

Spacecraft	Launch Date	Operational Dates
NOAA7	June, 23, 1981	Aug. 24, 1981–Feb. 1, 1985
NOAA9	Dec. 12, 1984	Feb. 25, 1985–Nov. 7, 1988
NOAA11	Sept. 24, 1988	Nov. 8, 1988–April 11, 1995
NOAA12	May 14, 1991	Sept. 16, 1991–Dec. 14, 1998
NOAA14	Dec. 30, 1994	Current (2003)
NOAA15	May 13, 1998	Current
NOAA16	Sept. 21, 2000	Current

5.3.2.1 *NOAA7*

The following formulas involve converting DN values to radiance (W m^{-2} sr^{-1} μm^{-1}).

$$L_1 = 0.5753e^{0.000101\,d}(C_{10} - 36) \tag{5.6}$$

$$L_2 = 0.3914e^{0.00012\,d}(C_{10} - 37) \tag{5.7}$$

5.3.2.2 *NOAA9*

Set 1:

$$L_1 = 0.5415e^{0.000166(\,d-65)}(C_{10} - 37) \tag{5.8}$$

$$L_2 = 0.3832e^{0.000098(\,d-65)}(C_{10} - 39.6) \tag{5.9}$$

Set 2:

$$L_1 = 0.5406e^{0.000166\,d}(C_{10} - 37) \tag{5.10}$$

$$L_2 = 0.3808e^{0.000098\,d}(C_{10} - 39.6) \tag{5.11}$$

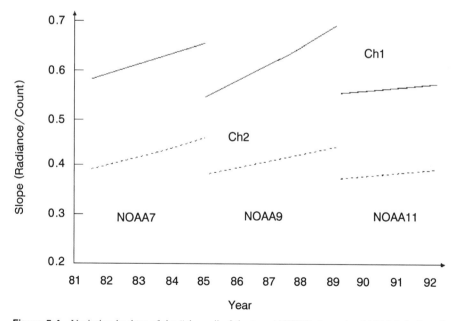

Figure 5.4   Variation in time of the "slopes" of the two AVHRR channels of NOAA-7, -9, and -11. [From Rao and Chen (1995), *Int. J. Remote Sens.* Reproduced by permission of Taylor & Francis, Ltd.]

5.3.2.3 *NOAA11*

$$L_1 = 0.5496 e^{0.000033\,d}(C_{10} - 40) \tag{5.12}$$

$$L_2 = 0.3680 e^{0.000055\,d}(C_{10} - 40) \tag{5.13}$$

Figure 5.4 shows the temporal variations of the slopes of two AVHRR channels of NOAA-7, -9, and -11. Note that not all vicarious studies produced the same results.

5.3.2.4 *NOAA14*

The NOAA14 spacecraft was launched into a nominal Sun-synchronous orbit on December 30, 1994. Because the two solar AVHRR bands do not have the onboard calibration devices, Rao and Chen (1996) developed calibration coefficients using the vicarious calibration method based the data of about one year from the Libyan desert.

The calibration coefficients are given for converting DN to TOA reflectance:

$$\rho_i = S_i(C_{10} - C_0)D^2 \tag{5.14}$$

where $i = 1, 2$ for two bands, D^2 characterizes the distance between the Sun and the Earth that is given in Eqs. (1.12) and (1.13), and

$$S_1 = 0.0000135d + 0.111 \tag{5.15}$$

$$S_2 = 0.0000133d + 0.134 \tag{5.16}$$

C_0 is set equal to 41 counts in both channels.

CCRS recommendations (available at `http://www.ccrs.nrcan.gc.ca/ccrs/rd/ana/calval/noaag14_e.html`) are given below.

The conversion is for radiance L (W m^{-2} sr^{-1} μm^{-1}) at the sensor (top of the atmosphere):

$$L = \frac{DN - 41}{\text{gain}} \tag{5.17}$$

where gain $= A \times d + B$ where d is the days since launch ($d = 0$ for Dec. 30, 1994); A and B are coefficients given in Table 5.7.

Tahnk and Coakley (2001) suggested a set of different calibration coefficients for TOA reflectance:

$$\rho_1 = (-5.35829 \cdot 10^{-9}d^2 + 1.70469 \cdot 10^{-5}d + 0.11414) \cdot (C_{10} - 41) \tag{5.18}$$

TABLE 5.7   Coefficients for Calculating Gain Value for NOAA14 AVHRR Sensor

Time	Band 1		Band 2	
	A	B	A	B
1995	-3.532×10^{-4}	1.796	-6.161×10^{-4}	2.364
1996	-3.055×10^{-4}	1.778	-5.090×10^{-4}	2.325
1997	-2.671×10^{-4}	1.750	-4.275×10^{-4}	2.265
1998	-2.356×10^{-4}	1.715	-3.644×10^{-4}	2.196
1999	-1.209×10^{-4}	1.587	-3.714×10^{-5}	1.883
2000/2001	-1.249×10^{-4}	1.639	-3.837×10^{-5}	1.946

From launch to January 1, 2000

$$\rho_2 = (-1.46883 \cdot 10^{-9}d^2 + 5.59073 \cdot 10^{-6}d + 0.14302) \cdot (C_{10} - 41) \quad (5.19)$$

After January 2, 2000

$$\rho_2 = (4.38569 \cdot 10^{-5}d + 0.06829) \cdot (C_{10} - 41) \quad (5.20)$$

5.3.2.5   NOAA15

Some piecewise linear (PWL) calibration coefficients for NOAA15 AVHRR spectral data in channels 1, 2, and 3A acquired from 1998 and 2002 are presented here.

The NOAA15 AVHRR/3 radiometer has a dual-gain response; thus, there are dual calibration gains for each channel. The recommended PWL coefficients for channels 1 and 2 are based on time-dependent calibration equations provided by Tahnk and Coakley (2001). Their calibration dataset was derived from the analysis of ice sheets in the Antarctic and Greenland.

Unfortunately, the recommended PWL coefficients for channel 3A are still based on prelaunch calibration coefficients provided in Appendix D of the NOAA KLM *User's Guide* on the webpage from NOAA/NESDIS:

For the low-radiance range:

$$L = \frac{DN - 38.9}{47.420} \quad (5.21)$$

For the high-radiance range:

$$L = \frac{DN - 423.7}{7.064} \quad (5.22)$$

where L is TOA radiance (W m^{-2} sr^{-1} μm^{-1}) at the sensor.

TABLE 5.8   Coefficient *A* for Eq. (5.23)

Year	Band 1		Band 2	
	Low	High	Low	High
1998	1.389×10^{-6}	4.832×10^{-7}	3.564×10^{-5}	1.304×10^{-5}
1999	1.390×10^{-6}	4.834×10^{-7}	3.580×10^{-5}	1.310×10^{-5}
2000	1.390×10^{-6}	4.835×10^{-7}	3.600×10^{-5}	1.317×10^{-5}
2001	1.391×10^{-6}	4.837×10^{-7}	3.621×10^{-5}	1.325×10^{-5}
2002	1.391×10^{-6}	4.838×10^{-7}	3.641×10^{-5}	1.332×10^{-5}

For bands 1 and 2:

$$L = \frac{DN - DN_0}{A \times d + B} \tag{5.23}$$

where d = days since launch ($d = 0$ for 5/13/1998). Since the DN_0 and B are consistent for a given radiance range (for low radiance range, $DN_0 = 38.5$ and $B = 3.269$ for band 1, $DN_0 = 40.4$ and $B = 3.564$ for band 2; for the high radiance range, $DN_0 = 336.9$ and $B = 1.137$ for band 1, $DN_0 = 338.8$ and $B = 1.696$ for band 2), Table 5.8 gives only the coefficient A.

Table 5.8 indicates that the AVHRR sensors on NOAA15 are quite stable since the calibration coefficients do not change much with time.

5.3.2.6   *NOAA16*

There are no postflight calibration coefficients available at the time of writing this book. The preflight calibrations are provided below. The NOAA16 AVHRR/3 radiometer also has a dual-gain response; thus, there are dual calibration coefficients for each channel:

Low radiance range:

$$L_1 = \frac{DN - 38.5}{3.653} \tag{5.24}$$

$$L_2 = \frac{DN - 37.9}{5.920} \tag{5.25}$$

$$L_{3A} = \frac{DN - 71.25}{44.944} \tag{5.26}$$

High radiance range:

$$L_1 = \frac{DN - 339.7}{1.250} \tag{5.27}$$

$$L_2 = \frac{DN - 342.8}{2.011} \tag{5.28}$$

$$L_{3A} = \frac{DN - 432.1}{7.142} \tag{5.29}$$

5.4 SUMMARY

Sensor radiometric calibration is a very important process in quantitative remote sensing that converts the digital numbers to TOA radiance. There are three stages of calibration activities: prelaunch, in-flight, and postlaunch. This chapter focuses on the postlaunch vicarious calibration methods based on various targets, such as ocean, desert, cloud, and moon.

In Section 5.3, we provided the calibration coefficients for Landsat TM and NOAA AVHRR sensors. These coefficients indicate that sensor degradation has been a serious problem.

REFERENCES

Abel, P., Guenther, B., Galimore, R. N., and Cooper, J. W. (1993), Calibration results for NOAA-11 AVHRR channel-1 and channel-2 from congruent path aircraft observations, *J. Atmos. Oceanic Technol.* **10**: 493–508.

Belward, A. S. (1999), International co-operation in satellite sensor calibration; the role of the CEOS working group on calibration and validation, *Adv. Space Res.* **23**: 1443–1448.

Chander, G., Helder, D. L., and Boncyk, W. C. (2002), Landsat-4/5 band 6 relative radiometry, *IEEE Trans. Geosci. Remote Sens.* **40**: 206–210.

Chen, H. S. (1997), *Remote Sensing Calibration Systems*: *An Introduction*, A. Deepak Publishing.

Dilligeard, E., Briottet, X., Deuze, J. L., and Saanter, R. (1996), SPOT calibration of XS1 and XS2 channels using Rayleigh scattering over clear oceans, *Proc. SPIE*, **2957**: 373–379.

Dinguirard, M. and Slater, P. N. (1998), Calibration of space-multispectral imaging sensors: A review, *Remote Sens. Envir.* **68**: 194–205.

Fraser, R. S. and Kaufman, Y. J. (1986), Calibration of satellite sensors after launch, *Appl. Opt.* **25**: 1177–1185.

Green, R. O. and Shimada, M. (1997), On-orbit calibration of a multispectral satellite sensor using a high altitude airborne imaging spectrometer, *Adv. Space Res.* **19**: 1387–1398.

Hagolle, O., Goloub, P., Deschamps, P. Y. et al. (1997), Results of polder in-flight absolute calibration, *Proc. SPIE* **3221**: 122–131.

Henry, P., Dinguirard, M., and Bodilis, M. (1993), SPOT multitemporal calibration over stable desert areas, *Proc. SPIE* **1938**: 67–76.

Kaufman, Y. J. and Holben, B. N. (1993), Calibration of the AVHRR visible and near-IR bands by atmospheric scattering, ocean glint and desert reflection, *Int. J. Remote Sens.* **14**: 21–52.

Kriebel, K. T. and Amann, V. (1993), Vicarious calibration of the meteosat visible channel, *J. Atmos. Oceanic Technol.* **10**: 225–232.

Lafrance, B., Hagolle, O., Bonnel, B., Fouquart, Y., and Brogneiz, G. (2002), Interband calibration over clouds for POLDER space sensor, *IEEE Trans. Geosci. Remote Sens.* **40**: 131–142.

Loeb, N. G. (1997), In-flight calibration of NOAA AVHRR visible and near-IR bands over Greenland and Antarctica, *Int. J. Remote Sens.* **18**: 477–490.

Markham, B. L. and Barker, J. L. (1986), Landsat MSS and TM post-calibration ranges, exoatmospheric reflectances and at-satellite temperatures, *EOSAT Landsat Tech. Notes* **1**: 3–8.

Price, J. C. (1989), Calibration comparison for the Landsat 4 and 5 multispectral scanners and thematic mappers, *Appl. Opt.* **28**: 465–471.

Rao, C. R. N. and Chen, J. (1995), Inter-satellite calibration linkages for the visible and near-infrared channels of the advanced very high resolution radiometer on the NOAA-7, -9, and -11 spacecraft, *Int. J. Remote Sens.* **16**: 1931–1942.

Rao, C. R. N. and Chen, J. (1996), Post-launch calibration of the visible and near-infrared channels of the advanced very high resolution radiometer on the NOAA-14 spacecraft, *Int. J. Remote Sens.* **17**: 2743–2747.

Reinersman, P. N., Carder, K. L., and Chen, F. (1998), Satellite-sensor calibration verification with the cloud-shadow method, *Appl. Opt.* **37**: 5541–5549.

Slater, P. N., Biggar, S. F., Holm, R. G., Jackson, R. D., Mao, Y., Moran, M. S., Palmer, J. M., and Yuan, B. (1987), Reflectance- and radiance-based methods for the in-flight absolute calibration of multispectral sensors, *Remote Sens. Envir.* **22**: 11–37.

Smith, G. R., Levin, R. H., Abel, P., and Jacobowitz, H. (1988), Calibration of the solar bands of the NOAA-9 AVHRR using high altitude aircraft measurements, *J. Atmos. Ocean* **5**: 631–639.

Staylor, W. F. (1990), Degradation rates of the AVHRR visible channel for the NOAA-6, -7 and -9 spacecraft, *J. Atmos. Ocean* **7**: 411–423.

Tahnk, W. R. and Coakley, J. A. J. (2001), Update calibration coefficients for NOAA-14 AVHRR channels 1 and 2, *Int. J. Remote Sen.* **22**: 3053–3057.

Teillet, P. M., Barker, J. L., Markham, B. L., Irish, R. R., Fedosejevs, G., and Storey, J. C. (2001), Radiometric cross-calibration of the Landsat-7 ETM + and Landsat-5 TM sensors based on tandem data sets, *Remote Sens. Envir.* **78**: 39–54.

Thome, K., Markham, B., Barker, J., Slater, P., and Biggar, S. (1997), Radiometric calibration of Landsat, *Photogramm. Eng. Remote Sens.* **63**: 853–858.

Vermote, E. and Kaufman, Y. J. (1995), Absolute calibration of AVHRR visible and near-infrared channels using ocean and cloud views, *Int. J. Remote Sens.* **16**: 2317–2340.

Vogelmann, J. E., Helder, D., Morfitt, R., Choate, M. J., Merchant, J. W., and Bulley, H. (2001), Effects of Landsat 5 Thematic Mapper and Landsat 7 Enhanced Thematic Mapper plus radiometric and geometric calibrations and corrections on landscape characterization, *Remote Sens. Envir.* **78**: 55–70.

6

Atmospheric Correction of Optical Imagery

This chapter discusses various practical algorithms for removing atmospheric effects from remotely sensed data for the purpose of recovering surface reflectance. This topic is generally listed as one of the preprocessing steps in many introductory remote sensing books, but it must be considered a critical step in quantitative remote sensing since most inversion algorithms are based on surface reflectance that are retrieved from atmospheric correction. The theoretical foundations for this topic have been discussed in Chapter 2.

Section 6.1 introduces some basic concepts and background information regarding atmospheric correction. Section 6.2 presents representative algorithms for correcting the single-viewing-angle imagery, particularly the nadir viewing imagery (e.g., Landsat TM/ETM +). Section 6.3 presents the algorithm for correcting multiangle imagery, primarily for MISR (multiangle imaging spectroradiometer) imagery. Section 6.4 discusses various algorithms for estimating the total column water vapor content of the atmosphere. The water vapor content has a significant impact on near-IR signals. This section starts with a general overview of different methods, ending with a discussion of the differential absorption technique. A brief description of split-window methods using thermal IR bands is also given.

6.1 INTRODUCTION

A very large portion of optical remotely sensed imagery is severely contaminated by aerosols, clouds, and their shadows. It will be greatly beneficial for land surface characterization if we can remove these atmospheric effects from imagery. As the utility of these data becomes more quantitative, the accurate retrieval of surface reflectance becomes increasingly important. For

Quantitative Remote Sensing of Land Surfaces. By Shunlin Liang
ISBN 0-471-28166-2   Copyright © 2004 John Wiley & Sons, Inc.

example, almost all of the canopy models that can be used for inverting canopy biophysical parameters are based on surface reflectance (see Chapter 3). The procedure for retrieving surface reflectance from remotely sensed imagery is usually called *atmospheric correction* in optical remote sensing.

Atmospheric correction consists of two major steps: atmospheric parameter estimation and surface reflectance retrieval. Assuming a Lambertian surface, and one that is working with near-nadir-viewing optical remotely sensed data, retrieval of surface reflectance is relatively straightforward as long as all atmospheric parameters are known. The typical approach is the *look-up table method* (Kaufman 1989), in which radiative transfer codes are used offline (i.e., tables are created before we run atmospheric correction code) to compute tables for online corrections. Estimation of atmospheric parameters from imagery itself is a more difficult and challenging step.

Atmospheric effects include molecular and aerosol scattering and absorption by gases, such as water vapor, ozone, oxygen, and aerosols (see Section 2.3). Molecular scattering and absorption by ozone, oxygen, and other gases are relatively easy to correct because of the stable concentrations of these elements over both time and space. The most difficult task is to estimate the spatial distributions of aerosols and water vapor directly from imagery.

There is a relatively long history to atmospherically correcting remotely sensed imagery quantitatively. The representative methods for correcting nadir viewing (plus/minus a few degrees from the zenith) multispectral imagery are briefly discussed in the next section. The atmospheric correction algorithms for imagery viewed from multiple angles are discussed in Section 6.3.

6.2 METHODS FOR CORRECTING SINGLE-VIEWING-ANGLE IMAGERY

6.2.1 Invariant-Object Methods

Assume that there are some pixels in a scene whose reflectances are quite stable through time. A linear relationship based on the reflectance of these "invariant objects" can be used to normalize imagery acquired at different times. This method was successfully used in FIFE [the First ISLSCP (International Satellite Land Surface Climatology Project) Field Experiment] TM imagery processing (Hall et al. 1991). It is a relative normalization. If there are simultaneous ground reflectance measurements available or some assumptions about surfaces can be made (Moran et al. 1992, Chavez, 1996), it can be an absolute correction procedure.

Also assume that N "invariant" pixels are identified from all M imagery acquired at different times. If we can select a clear image, say, J, as the reference, all other images can be normalized to image J using a linear regression based on these N pixels. The procedure is illustrated in Fig. 6.1

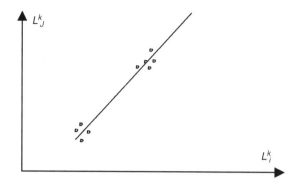

Figure 6.1 Linear regression analysis in the "invariant" object atmospheric correction algorithm.

where L_j represents the radiance of the reference imagery J and L_i denotes radiance of any other images. The resulting equation is

$$L_j^k = a_i^k + b_i^k L_i^k \qquad (6.1)$$

The linear regression analysis for each band k will produce two coefficients a and b that are used for normalizing all other pixels of band k for each image i.

Ideally, we need to identify "invariant" pixels with variable brightness from dark to bright in each band. If we have only either very dark pixels or bright pixels (one of the clusters in the figure), a linear transformation will introduce very large errors since the clustered pixels will not produce statistically significant linear relationships.

On the other hand, normalizing top-of-atmosphere (TOA) radiance is not equivalent to the normalization of surface reflectance because of the nonlinear relationship between the TOA radiance and surface reflectance. This can be easily understood from Eq. (2.115), which is duplicated here for easy reference.

$$L(\mu_v, -\mu_0, \phi) = L_p(\mu_v, -\mu_0, \phi) + \frac{r}{1 - rs} \mu_0 E_0 \gamma(-\mu_0) \gamma(\mu_v) \qquad (6.2)$$

where L_p is called path radiance, r is the surface reflectance, E_0 is the extraterrestrial irradiance, s is the spherical albedo of the atmosphere, μ_0 is the cosine of the solar zenith angle, $\gamma(\mu)$ is the total transmittance from the surface to the sensor, and $\gamma(-\mu_0)$ is the total transmittance from the Sun to the surface.

A better approach is to incorporate the lookup table method. The idea is quite simple and consists of several steps:

- The first step is to create tables of these variables $[L_p, s, E_0 \gamma(\mu_v) \gamma(-\mu_0)]$ for each image using a radiative transfer package (e.g., MODTRAN or 6S) with two free variables (aerosol optical depth and water vapor content). Since aerosol scattering effect is very small in the near-IR bands, and water vapor absorption is weak in the visible bands, it might be possible to simplify the tables using one variable for each band.

• The second step is to determine the spectral reflectance of these "invariant" pixels of the reference image J by assuming the values of aerosol optical depth and water vapor content. The formula can be easily derived from Eq. (6.2):

$$r = \frac{L(\mu_v, -\mu_0, \phi) - L_p(\mu_v, -\mu_0, \phi)}{[L(\mu_v, -\mu_0, \phi) - L_p(\mu_v, -\mu_0, \phi)]s + \mu_0 E_0 \gamma(-\mu_0)\gamma(\mu_v)} \quad (6.3)$$

• The third step is to determine aerosol optical depth and/or water vapor content of other images by searching these tables created from the first step and matching the TOA radiance of these "invariant" pixels. In this step, the surface reflectances of these "invariant" pixels are the same as those in image J. Because of the discrete nature of these tables, a linear interpolation is then needed. The last step is to retrieve surface reflectance of all pixels of all other images as soon as the atmospheric variables $[L_p, s, E_0 \gamma(\mu_r)\gamma(-\mu_0)]$ are known.

This approach sounds very complicated, but does not have any strict requirements of the traditional invariant-objects method; therefore it is a simple, physically-based atmospheric correction method.

6.2.2 Histogram Matching Methods

The histogram matching method is based on the assumption that the surface reflectance histograms of clear and hazy regions are the same. After identifying clear and hazy regions in an image, the histograms of the hazy regions are shifted to match the histogram of their reflectance of the clear regions (Richter 1996a, 1996b).

More specifically, this method is composed of several steps:

• Partitioning of the image into $N_x \times N_y$ sectors.
• Selection of the type of reference target with known reference (i.e., dark water or dense vegetation), identifying these target pixels with interactive thresholding, and determining the optical depths over these pixels.
• Identifying hazy and cloudy pixels interactively using the "tasseled cap" transformation (TCT) (see Section 8.1.1.8).
• Sector-dependent histogram matching of the hazy regions to clear regions for determining the atmospheric visibility in each sector.
• Retrieval of surface reflectance by using the sector averaged atmospheric visibility. This method has been incorporated into the ERDAS image processing software package.

The major limitation of this method is its assumption that the reflectance histogram of a hazy region is the same as the histogram of a clear region, which implies that there are same portions of various land covers in both hazy and clear regions. This implicit assumption is not valid under most conditions, even if both clear and hazy regions are the same landscape types. Additionally, the TCT method for identifying hazy pixels does not always work.

6.2.3 Dark-Object Methods

The classic dark-objects method—one of the oldest and simplest methods of atmospheric correction—assumes that an image has pixels whose surface reflectance is negligible (e.g., in a complete shadow), and the image pixel values of each band are subtracted by its minimum value. Although this method has been extended to assume a minimal surface reflectance, say 1% (Chavez 1988, 1996; Moran et al. 1992), an image could not be corrected on a pixel basis, and also it is statistical in nature.

This method has been significantly improved by incorporating a physically based procedure to correct imagery on a pixel basis. If a scene contains dense vegetation, the middle-IR band around 2.1 μm can be used to identify these dense vegetation pixels, and their reflectances are highly correlated with the reflectances of both blue and red bands. Since dense vegetation canopies have very low reflectance in the visible spectrum, they are also referred to as "dark objects." This method has a long history (Kaufman and Sendra 1988; Popp 1995; Teillet and Fedosejevs 1995; Kaufman et al. 1997a, 1997b, 2000; Liang et al. 1997) and is probably the most popular atmospheric correction method currently in use. It has been used for operationally estimating aerosol optical depth from MODIS (moderate-resolution imaging spectroradiometer) land data (Kaufman et al. 1997) and will also be used for correcting MERIS (medium-resolution imaging spectrometer) data (Santer et al. 1999). It consists of several steps in any "dark object" algorithms that are more or less similar. The procedure used by Liang et al. (1997) for TM imagery is as follows:

- Identify dark pixels by using a low-threshold reflectance value (e.g., 0.05) of the middle-IR band around 2.1 μm (band 7). If we assume that scattering in this band is negligible, its TOA reflectance can be easily converted to surface reflectance by considering the transmittance due to absorption. Those identified pixels could correspond to water or wet soil besides dense vegetation. A vegetation index or other method is needed to extract dense vegetation pixels as the required "dark objects."
- Calculate the surface reflectances of the blue (band 1) and red (band 3) bands using the simple statistical relations:

$$\rho_{\text{red}} = 0.5\rho_{2.1}$$
$$\rho_{\text{blue}} = 0.25\rho_{2.1}$$

(6.4)

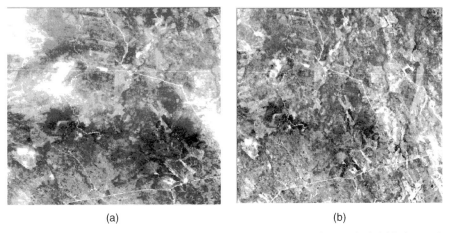

(a)   (b)

Figure 6.2 An example of atmospheric correction using the dark-object method: (a) before and (b) after correction. [From Liang et al. (1997), *J. Geophys. Res.* (*Atmos.*). Reproduced by permission of American Geophysical Union.]

- Determine optical depths of bands 1 and 3 by searching the lookup tables from their TOA radiance based on the predicted surface reflectance of both bands (red and blue) from (6.4).
- Determine the optical depths of other bands by assuming a functional relation of the optical depth

$$\tau_i = a\lambda_i^{-b} \tag{6.5}$$

where τ_i and λ_i are the aerosol optical depth and the central wavelength (μm) for channel i, $1 \leq i \leq 5$. The parameters a and b are estimated from the optical depths of bands 1 and 3; b is often called the *Angstrom index*.

- Estimate spatial distribution of the optical depth by using a moving-window interpolation technique (Fallah-Adl et al. 1997).

One example is given in Fig. 6.2, which compares an original true-color composite image (channels 1, 2 and 3) with a corrected true-color composite image. This TM image, acquired on February 26, 1990, is over Bolivia in South America with the central longitude 63°45′19″W and latitude 17°20′39″S. This image consists of 800 × 800 pixels. It is clear that most of the haze has been removed, and the surface features blocked by haze have been successfully recovered after the atmospheric correction. The overall image contrast is also improved even though a few clouds still remain.

6.2.4 Contrast Reduction Methods

For regions where surface reflectance is very stable, variations of satellite signal acquired at different times may be attributed to variations of atmospheric optical properties. Aerosol scattering reduces variance of the local reflectance. The larger the aerosol loading, the smaller the local variance. Mathematically, the difference (contrast) of the apparent TOA reflectance Δ_{ij}^* at two neighboring pixels is approximately related to the actual surface reflectance difference Δ_{ij}

$$\Delta_{ij}^*(\mu_s, \mu_v, \phi) = \Delta_{ij}(\mu_s) \exp\left(-\frac{\tau}{\mu_v}\right) \tag{6.6}$$

where, $\mu_s = \cos(\theta_s)$; $\mu_v = \cos(\theta_v)$, and θ_s, θ_v, and ϕ are the solar zenith angle, viewing zenith angle, and relative azimuth angle, respectively. $T(\mu_s)$ is the total atmospheric transmittance (direct plus diffuse) in the solar illumination path (Sun to ground); $\exp(-\tau/\mu_v)$ is the direct transmittance in the viewing path (ground to satellite). If the difference is characterized by local radiance, then the apparent TOA variance σ^{*2} is linearly proportional to the surface actual variance σ^2:

$$\sigma^{*2} = \sigma^2 T^2(\mu_s) \exp\left(-\frac{2\tau}{\mu_v}\right) \tag{6.7}$$

Supposing that the actual surface variance is known, the TOA radiance from satellite observations will allow us to estimate the transmittance terms of Eq. (6.7), which depend mainly on aerosol optical depth τ. In this case, we usually can use a mean aerosol model. The actual variance of surface reflectance can be estimated from the clearest imagery.

This method has been successfully applied to desert dust monitoring (Tanré et al. 1988, Tanré, and Legrand 1991). The correlation between the estimated aerosol optical depth using this method, based on land–sea contrast and ground-based measurements, varies from 0.968 to 0.984, and the uncertainty $\Delta\tau$ ranges from 0.08 to 0.12. The assumption of invariant surface reflectance within neighborhoods and high contrast between them limits this approach to be globally applicable. The large uncertainty of this method implies its applicability to limited cases, such as heavy dust and smog.

Similar concepts have been further extended for correcting multiangular imagery, which will be discussed in Section 6.3.

6.2.5 Cluster Matching Method

To overcome a series of problems associated with existing methods, we (Liang et al. 2001) developed a new atmospheric correction algorithm for TM/ETM + imagery in which the key component is to estimate the spatial distribution of aerosol loadings under general conditions. The correction of adjacency effects was also explored. This method has also been extended for correcting other imagery such as MODIS and SeaWiFS (sea-viewing wide-field-of-view sensor) (Liang et al. 2002).

To accurately calculate aerosol effects, we need not only aerosol optical depth, but also single scattering albedo and phase function. The last two variables are determined based on aerosol climatologic data (Hess et al. 1998) (see also Section 2.3.5). This may be updated by EOS Terra data (MODIS and MISR) in the future. In the following section, we describe the procedures for estimating aerosol optical depth, correcting adjacency effects, and surface reflectance retrieval. Interested readers should refer to original publications for additional information and details.

6.2.5.1 *Estimation of Aerosol Optical Depth*
This algorithm is similar to the histogram matching algorithms (Richter 1996a, 1996b); however, instead of matching the histograms of two regions (clear and hazy), it is assumed that the average reflectance of each cover type is the same under different atmospheric conditions (from clear to hazy). Because near-IR (bands 4 and 5) and mid-IR (band 7) bands are much less contaminated by most aerosols, these three bands are used to classify all pixels into specific cover types. Mean reflectance matching is performed in the first three visible bands, separately. A smoothing process is followed for each band to determine its final aerosol optical depth. The procedure is illustrated in Fig. 6.3. The major steps are discussed below.

If dense aerosols and thin clouds exist, scattering effects may contaminate bands 4, 5, and 7. A histogram match processing is first performed to adjust reflectance of these three bands. A clustering analysis using these three bands is also conducted. In our experiments, a *K*-mean clustering analysis with 50 clusters is often used. Other clustering algorithms can be easily employed, and specifying the number of clusters depends on the complexity of landscape. Ideally, the more complex the landscape, the larger the number of clusters are needed. However, our experiments indicated that 20–50 clusters produce similar results, probably because only three bands are available for clustering analysis. More bands will improve the results.

It is assumed that clear regions exist in a scene and these clear regions have small aerosol loadings. There are several options for separating clear regions from hazy regions. The first is to make use of the fourth component of TCT (Crist and Cicone 1984; also see Chapter 8), as used by other algorithms (e.g., Richter 1996a). However, this method fails when surfaces are very bright. Another option is to determine the highest reflectance of each cluster, and the segmentation of high-reflectance produces hazy regions. When hazy regions are not widely distributed in the scene, this method does not work well. The easiest and most reliable option for determining hazy regions is determined by hand drawing (note that there is no need for accurate determination of the boundary between clear and hazy regions). Almost all image processing software packages have a graphics interface that allow users to specify hazy regions conveniently. Of course, this is too time-consuming for massive operational atmospheric correction. Research is still needed to develop such a fully reliable automatic algorithm.

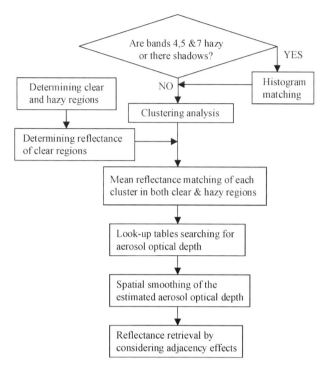

Figure 6.3 Flowchart of an ETM + atmospheric correction algorithm developed by Liang et al. (2001).

Surface reflectances of the visible bands in clear regions are determined by the knowledge of minimum surface reflectance. This step may introduce uncertainty whose magnitude depends on the surface brightness. If low-reflectance surfaces exist in a scene, such as vegetation, water, wet soil, the atmospheric correction error is very small.

The next step is to match average reflectance of each cluster between clear and hazy regions. Given surface reflectance, aerosol optical depth of hazy regions can be determined by searching lookup tables.

Because aerosol optical depth is usually distributed more smoothly than that of surface reflectance, and clusters of hazy atmospheric regions may be present over the surface, a spatial lowpass filter is warranted. In our experiments, an average smoothing process with a 5×5 window works quite well.

Given the distribution of aerosol optical depth, it is straightforward to determine surface reflectance by searching lookup tables.

6.2.5.2 Surface Adjacency Effects

The adjacency effect is caused by complicated multiple scattering in the atmosphere−land surface system. The pixel values of high-resolution imagery over a heterogeneous landscape are affected by their neighboring pixels. As a result, dark pixels appear brighter and bright pixels, darker. The practical implication to remotely sensed data is that the imagery seems hazy and lacks contrast. The pixel value depends largely on its contrasts with neighboring pixel values when spatial resolution increases, especially if the atmosphere is not very clear. As spatial resolution decreases, the mixture problem represents the within-pixel effects. As spatial resolution increases, the adjacency problem represents the between-pixel effects. This mixture problem has been extensively investigated (Maselli 1998, Hu et al. 1999, Chang and Ren 2000). The adjacency effect deserves more attention as we are being faced with a growing amount of high-resolution satellite imagery.

Earlier studies on the adjacency effect are summarized by an excellent review (Kaufman 1989), and only a few studies have reported effectively removing adjacency effects since then (e.g., Kozoderv 1995, Reinersman and Carder 1995, Miesch et al. 1999). These studies can be grouped into two broad categories: (1) using the atmospheric point spread function (PSF) and (2) developing empirical formulas. Different methods have been explored to calculate the atmospheric PSF, including the Monte Carlo simulation (Pearce 1977, Reinersman and Carder 1995, Miesch et al. 1999) and radiosity simulation (Borel and Gerstl 1992).

Given the atmospheric PSF, adjacency effects can be corrected by using the Fourier transform approach (Mekler and Kaufman 1980, 1982; Kaufman 1984; Kozoderv 1995). Although efforts have been made to develop an empirical function for the atmospheric modulation transfer function (MTF) that is the Fourier transform of the atmospheric PSF, most methods for calculating the atmospheric PSF/MTF are computationally expensive. The validity of the atmospheric PSF method for correcting the adjacency effect in high-resolution satellite imagery is also still questionable. When the atmosphere is turbid and where multiple scattering dominates or the surface reflects strongly, there are multiple interactions between the atmosphere and the surface. Thus, upwelling radiance is not simply a convolution of the atmospheric PSF with the surface reflectance. As will be demonstrated later, the adjacency effect is significant only when the multiple interaction between the atmosphere and the surface is dominant.

This is actually a typical three-dimensional (3D) radiative transfer problem (Evans 1998). Theoretically, we are able to calculate the adjacency effects by solving the 3D radiative transfer equation. Unfortunately, solving a 3D radiative transfer equation is very time-consuming. For operational applications, empirical solutions are more appealing. For TM imagery, Kaufman (1984) developed a simple formula of the normalized atmospheric modulation transfer function (MTF) with a relative accuracy of 25% compared with a Monte Carlo simulation result:

$$M^N(k) = 1 - 0.5\tau_R\left[1 - \exp(-2.5kH_R)\right] - 0.7\lambda^{-2}\tau_A\left[1 - \exp(-1.3kH_A)\right]$$

$$(6.8)$$

where λ is the wavelength in micrometers, τ_R and τ_A are the optical depths of Rayleigh and aerosol, and H_R and H_A are the scaling height of Rayleigh and aerosol, respectively. To apply this equation to correct the adjacency effects, a Fourier transformation is needed.

Tanré et al. (1981) derived a TOA reflectance formula for a Lambertian surface that has been incorporated into the 6S code (Vermote et al. 1997):

$$\rho_{\text{TOA}} = \rho_p + \gamma(\theta_0) \frac{\rho \exp\left(-\frac{\tau}{\mu_v}\right) + \rho_e t_d(\theta_v)}{1 - \rho_e S} \tag{6.9}$$

where
ρ_{TOA} = TOA reflectance
ρ_p = pure atmospheric reflectance
θ_0 = solar zenith angle
θ_v = viewing zenith angle [$\mu_v = \cos(\theta_v)$]
$\gamma(\theta_0)$ = total atmospheric transmittance from Sun to surface
τ = optical depth
$t_d(\theta_v)$ = diffuse transmittance from surface to sensor
S = spherical albedo of atmosphere
ρ = target surface reflectance
ρ_e = average surface reflectance around pixel (x, y).

However, the interpretation of the effective reflectance ρ_e in this formulation is not straightforward.

Takashima and Masuda (1996) developed a formula with nine terms that has been incorporated into the ASTER atmospheric correction algorithm. However, when the formula is very complicated, the inversion of surface reflectance becomes more difficult.

In a more recent study, Liang et al. (2001) addressed the surface adjacency effects for the nadir viewing sensors located at the top of the atmosphere above a Lambertian surface. We tried to define an "effective" surface reflectance so that the classic plane-parallel formulas in Chapter 2 can be exactly applied. If a surface is flat and Lambertian underneath a horizontally homogeneous atmosphere, the TOA radiance can be expressed by the classic formula defined in Eq. (6.2). If the surface is not homogeneous, this formula is not valid. This is another typical 3D radiative transfer problem. Our purpose is to develop an empirical formula for calculating the "effective" reflectance of a heterogeneous Lambertian surface so that the exact formula (6.2) is valid for a heterogeneous surface except that the reflectance r is replaced by the "effective reflectance" r_e. The basic approach employed is to run the 3D radiative transfer code (SHDOM) (Evans 1998) over a step function surface and fit an empirical formula of "effective" reflectance.

The SHDOM code has been discussed by Evans (1998) in detail. In the course of solving the 3D radiative transfer equation, SHDOM transforms between the discrete ordinate and spherical harmonic representations. This numerical code can handle surface and atmosphere heterogeneity. It has been widely used in various applications (e.g., Evans et al. 1999, Liang 2000).

On the basis of our exploratory simulations with different aerosol scaling heights, solar zenith angles, and aerosol optical depth, we found that aerosol optical depth is the dominating factor. We also found that Rayleigh scattering causes the secondary order of the adjacency effects, which is consistent with the earlier studies (e.g., Reinersman and Carder 1995, Miesch et al. 1999). Therefore, the numerical simulations were designed with different aerosol optical depths over a step function surface (2D radiative transfer). The surface reflectances were set 0.05 (dense vegetation or water) and 0.8 (snow or bright sand), respectively. The atmosphere was stratified with aerosol scattering coefficient b_a decreasing exponentially with altitude z with a scale height H_a (Gordon and Castano 1989)

$$b_a = b_0 \exp\left(-\frac{z}{H_a}\right) \tag{6.10}$$

where b_0 is the aerosol scattering coefficient at surface. SHDOM code was run over an H_0 (7-km) slab of the atmosphere. Assuming no absorption, the aerosol optical depth is given by

$$\tau = b_0 H_a \left[1 - \exp\left(-\frac{H_0}{H_a}\right)\right] \tag{6.11}$$

Six aerosol optical depths were used: 0.05, 0.3, 0.6, 0.9, 1.2, and 1.5. SHDOM code was run over three solar zenith angles: 10°, 30°, and 50°. For each case, atmospheric parameters, such as L_p, $\gamma(\mu_i)$, $\gamma(\mu_v)$, and $\bar{r}$ in Eq. (5.4), were derived by running the plane-parallel mode of the SHDOM code with two surface reflectances of 0.0 and 0.5. Liang et al. defined an empirical weighting function $g(s)$ whose convolution with the step-function produces the "effective" reflectance

$$r_e = \int g(s) r(s) \, ds \tag{6.12}$$

where $r(s)$ is the true step function surface reflectance and s is the distance from the central location. The fitted empirical function is

$$g(s) = f_1(\tau) \exp(-1.42352s) + f_2(\tau) \exp(-12916.05s) \tag{6.13}$$

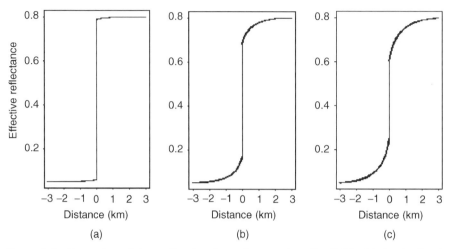

Figure 6.4 Comparison of the predicted surface reflectance from Eq. (6.13) with the simulated surface reflectance (lines) using three-dimensional atmospheric radiative transfer model (points). [From Liang et al. (2001), *IEEE Trans. Geosci. Remote Sens.* Copyright © 2001 with permission of IEEE.]

where s is the distance from the central pixel, τ is the aerosol optical depth, and

$$f_1(\tau) = 0.00289\tau \qquad (6.14)$$

$$f_2(\tau) = 0.0714189\tau^3 - 0.0610574\tau^2 - 0.439108\tau + 0.995683 \quad (6.15)$$

It is evident that the first term on the right side of Eq. (6.13) represents the contribution from the background pixels; the second term represents the contribution from the current pixel and its nearest-neighbor pixels. The background contribution depends largely on the aerosol optical depth. If the aerosol optical depth is small (i.e., the atmosphere is very clear), the major contribution to the pixel value is from the pixel itself. As the optical depth increases, the background contribution (i.e., adjacency effects) becomes larger. The simulated effective reflectance and the fitted value using Eq. (6.13) for three optical depths are shown in Fig. 6.4. It is clear that Eq. (6.13) captures the surface adjacency effects very well.

It is believed that the fitted empirical weighting function [Eqs. (6.14) and (6.15)] from 2D step function surfaces is also suitable for the 3D domain (e.g., an image). The "effective" reflectance of the satellite imagery can be calculated in the discrete form

$$r_e = \frac{\displaystyle\sum_i^N \sum_j^N r_{ij} g_{ij}}{\displaystyle\sum_i^N \sum_j^N g_{ij}} \qquad (6.16)$$

where g_{ij} is the relative contribution from the pixel (i, j) and can be integrated from Eq. (6.13) over pixel (i, j):

$$g_{ij} = \int_{x_i} \int_{y_j} g\left(\sqrt{x^2 + y^2}\right) dx \, dy \tag{6.17}$$

In Eq. (6.16) N is the window size in which all pixels contribute to the apparent value of the central pixel and can be determined by a threshold value α:

$$r_e = \frac{\sum\limits_{i}^{N} \sum\limits_{j}^{N} r_{ij} g_{ij}}{\sum\limits_{i}^{N} \sum\limits_{j}^{N} g_{ij}} \tag{6.18}$$

To examine whether the formula derived from 2D step function surface is valid for a 3D image, we extracted several 30×30-pixel windows from ETM + band 1 imagery acquired on July 28, 1999 at Beltsville, MD. After running SHDOM for three optical depths $(0.3, 0.6, 0.9)$, we compared them with the plane-parallel version with the effective reflectance. In many cases where reflectance in the window is quite homogeneous, the three results are quite similar. When the reflectances in the window vary dramatically, the differences are very large. One example is shown in Fig. 6.5, where all pixels

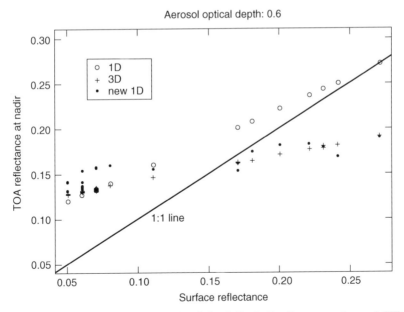

Figure 6.5 Testing the empirical formula of the "effective" reflectance using real ETM + band 1 imagery. [From Liang et al. (2001), *IEEE Trans. Geosci. Remote Sens.* Copyright © 2001 with permission of IEEE.]

were extracted from a window in which there are concrete, houses, and vegetation canopies. The weighing window is 12×12. A total of 36 pixels in the center of the window are compared among three methods. The 1D method in the figure represents the plane-parallel formula with an average (mean) reflectance of the window. The new 1D method is the one we proposed using a weighted average. In this case, the plane-parallel method overestimates high reflectance significantly. Because of the adjacency effects, the bright pixels (houses) tend to be darker. The new approximate formula produces points very close to the 3D simulation results.

6.2.5.3 *Correction Examples*

Figure 6.6 compares three band 1 (blue) images before and after atmospheric correction. These are three 600×600 windows from the same ETM + imagery acquired on November 17, 1999, but they have different surface reflectance and aerosol distribution patterns. The solar zenith angle is 63.51° and the azimuth angle is 162.83°. In these examples, the band 1 : band 4 image ratios were segmented to generate clear and hazy regions. From these figures, we can see that atmospheric correction produces significantly different visual effects. Most of the hazy regions have been cleaned up. Note that all pixels seem brighter after atmospheric correction. The reason is that the dynamic range of pixel values becomes smaller after atmospheric correction, but the display brightness range remains the same.

It is important to point out that the dark-object method fails to correct these three images since no dense vegetation canopies are widely distributed over the agricultural region in the winter season. Use of the histogram-matching algorithm is also inappropriate since landscape of the hazy and clear areas are not exactly the same and the spatial distribution of aerosol optical depth changes dramatically.

6.3   METHODS FOR CORRECTING MULTIANGULAR OBSERVATIONS

There are several airborne and satellite sensors that observe the Earth's environment from different angles (directions) simultaneously, such as ASAS (advanced solid-state array spectroradiometer), MISR (multiangle imaging spectroradiometer) Air-MISR, and POLDER (polarization and directionality of Earth's reflectances). In many cases, images acquired at each direction are corrected individually by assuming a Lambertian surface. This assumption is contradictory to the purpose of the multiangular observations.

MISR atmospheric correction utilizes multiangular imagery simultaneously. Discussed here are mainly the algorithms for correcting the MISR imagery. The following materials are generally quoted from the MISR algorithm theoretical basis document (ATBD) (Diner et al. 2001), readers are referred to the original publication for specific details.

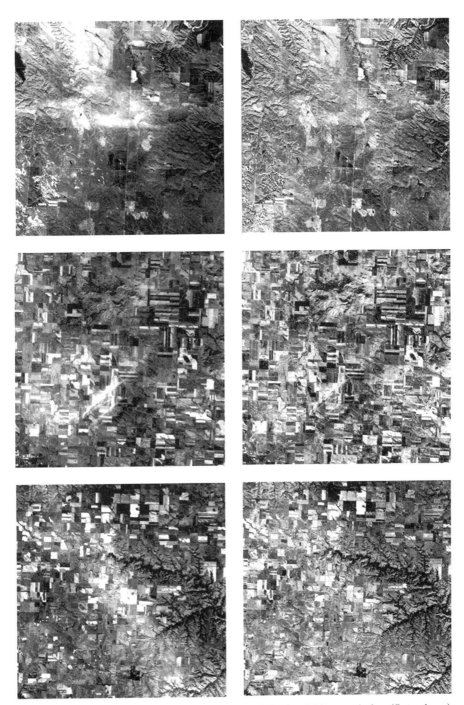

Figure 6.6 Atmospheric correction examples of 3 ETM + band 1 imagery before (first column) and after (second column) atmospheric correction. [From Liang et al. (2001), *IEEE Trans. Geosci. Remote Sens.* Copyright © 2001 with permission of IEEE.]

6.3.1 Estimating Aerosol Optical Depth from MISR Data

When dealing with multiangle observations, there are two algorithms for estimating aerosol optical depth over land: (1) that based on dense vegetation and is similar to the dark-object method discussed in Section 6.2.3 and (2) that for heterogeneous land. The latter does not use the observed radiance directly, but instead uses the presence of spatial contrast to derive an empirical orthogonal function (EOF) representation of the angular variation of the scene reflectance. The idea is similar to the contrast reduction algorithms discussed in Section 6.2.4, but the procedural details are quite different. We will discuss mainly this algorithm. The recent study indicates that the EOF algorithm also works over densely vegetated surfaces (Martonchik et al. 2002)

Let us start with the computation of each reduced sample equivalent reflectance at location x, y for each viewing direction (μ, ϕ) and the solar illumination direction (μ_0, ϕ_0):

$$J_{x,y}(\mu, \mu_0, \phi - \phi_0) = \rho_{x,y}(\mu, \mu_0, \phi - \phi_0) - \rho_{bias}(\mu, \mu_0, \phi - \phi_0) \quad (6.19)$$

where $\rho_{bias}(\mu, \mu_0, \phi - \phi_0)$ denotes an offset equivalent reflectance, numerically the minimum value over all the samples in the region; $\rho_{x,y}(\mu, \mu_0, \phi - \phi_0)$ is the surface reflectance at location (x, y). The scatter matrix is defined as

$$C_{i,j} = \frac{1}{N_{sub}} \sum_{x,y} J_{x,y,i} \cdot J_{x,y,j} \quad (6.20)$$

where the indices i and j are used to denote the viewing geometry and N_{sub} is the number of summed-up subregions. The scatter matrix is a $N_{cam} \cdot N_{cam}$ array, where N_{cam} is the number of available cameras (nominally, 9) in each band with μ_i, ϕ_i defining the viewing geometry of MISR camera i. The eigenvectors of C are solutions to the eigenvector equation given by

$$\sum_{j=1}^{N_{cam}} C_{i,j} \cdot f_{j,n} = \lambda_n \cdot f_{i,n} \quad (6.21)$$

where λ_n is the eigenvalue of $f_{i,n}$. Thus, every N_{cam}-element vector $J_{x,y}$ can be expanded in the following way

$$J_{x,y,i} = \sum_{n=1}^{N_{cam}} A_n^{x,y} \cdot f_{i,n} \quad (6.22)$$

where $A_n^{x,y}$ are the principal components, determined from the orthonormality condition:

$$A_n^{x,y} = \sum_{i=1}^{N_{\text{cam}}} J_{x,y,i} \cdot f_{i,n} \qquad (6.23)$$

In fact, only the first largest eigenvalues (N_{max}) are used in determining aerosol optical depth by the condition

$$\sum_{n=N_{\text{max}}+1}^{N_{\text{cam}}} \lambda_n \leq 0.05\lambda_1 \qquad (6.24)$$

After determining the eigenvalues and eigenvectors, we can estimate the optical depth in the following two steps:

- *The first step* in the retrieval procedure is to compare the following index:

$$\chi_N^2(\tau) = \frac{1}{N_{\text{band}} N_{\text{cam}}} \sum_{l=1}^{N_{\text{band}}} \sum_{j=1}^{N_{\text{cam}}} \frac{\left[\langle \rho_{\text{MISR}}(l,j) \rangle - \rho^b(l,j) + \sum_{n=1}^{N} A_n f_{j,n}(l) \right]^2}{\sigma_{\text{abs}}^2(l,j)}$$

$$(6.25)$$

  where $\langle \rho_{\text{MISR}} \rangle$ is the sample-averaged TOA equivalent reflectance, ρ^b is the TOA equivalent reflectance of path radiance with a black surface (zero reflectance), and σ_{abs} is the absolute radiometric uncertainty in ρ_{MISR}.
- *The second step* is to define an estimate of the aerosol optical depth from the minimum χ^2 for each value of N. For the aerosol model being evaluated, the reported best-fitting optical depth is then computed from a weighted average of all N_{max} optical depths.

After estimation of aerosol optical depth, the following step is to retrieve surface directional reflectance, which will be the topic of the next section.

6.3.2 Retrieval of Surface Directional Reflectance from MISR Data

The MISR team produces four surface spectral reflectance products:

- Hemispherically directional reflectance factor (HDRF, $r_{x,y}$)
- Bihemispherical reflectance (BHR, $A_{x,y}^{\text{hem}}$)

- Bidirectional reflectance factor (BRF, $R_{x,y}$)
- Directional-hemispherical reflectance (DHR, $A_{x,y}^{\text{dir}}$).

Below we discuss the inversion procedures in detail.

6.3.2.1 HDRF and BHR

The *hemispherically directed reflectance factor* (HDRF) ($r_{x,y}$) is defined as the ratio of the radiance reflected from the surface to that from an ideal Lambertian target reflected into the same beam geometry and illuminated under identical atmospheric conditions

$$r_{x,y}(-\mu, \mu_0', \phi - \phi_0')$$

$$= \frac{\frac{1}{\pi} \int_0^1 \int_0^{2\pi} R_{x,y}(-\mu, \mu', \phi - \phi') L_{x,y}^{\text{inc}}(\mu', \mu_0, \phi' - \phi_0) \mu' \, d\mu' \, d\phi'}{\frac{1}{\pi} \int_0^1 \int_0^{2\pi} L_{x,y}^{\text{inc}}(\mu', \mu_0, \phi' - \phi_0) \mu' \, d\mu' \, d\phi'} \quad (6.26)$$

where L^{inc} is the incoming solar radiance (direct plus diffuse) on the surface; $R_{x,y}(\cdot)$ is the surface BRF that will be discussed in the next section in detail. BHR is integration of HDRF over the viewing directions:

$$A_{x,y}^{\text{hem}}(\mu_0) = \frac{1}{\pi} \int_0^1 \int_0^{2\pi} r_{x,y}(-\mu, \mu_0, \phi - \phi_0) \mu \, d\mu \, d\phi \quad (6.27)$$

The retrieval procedure is to first determine the surface-reflected upwelling radiance $L_{x,y}^{\text{refl}(n)}$ iteratively

$$L_{x,y}^{\text{refl}(n+1)}(-\mu, \mu_0, \phi - \phi_0)$$

$$= \left[L_{x,y}(-\mu, \mu_0, \phi - \phi_0) - L^{\text{atm}}(-\mu, \mu_0, \phi - \phi_0) \right] e^{\tau/\mu}$$

$$- 2\pi \cdot e^{\tau/\mu} \int_0^1 T_0(-\mu, -\mu') L_{0,x,y}^{\text{refl}(n)}(-\mu', \mu_0) \, d\mu'$$

$$- \pi \cos(\phi - \phi_0)$$

$$\cdot e^{\tau/\mu} \int_0^1 T_1(-\mu, -\mu') L_{1,x,y}^{\text{refl}(n)}(-\mu', \mu_0) \, d\mu' \quad (6.28)$$

where $L_{0, x, y}^{\text{refl}(n)}$ and $L_{1, x, y}^{\text{refl}(n)}$ are computed from $L_{x, y}^{\text{refl}(n)}$

$$L_{0, x, y}^{\text{refl}(n)}(-\mu, \mu_0)$$

$$= \frac{L_{x, y}^{\text{refl}(n)}(-\mu, \mu_0, \phi_a - \phi_0) \cos(\phi_f - \phi_0) - L_{x, y}^{\text{refl}(n)}(-\mu, \mu_0, \phi_f - \phi_0) \cos(\phi_a - \phi_0)}{\cos(\phi_f - \phi_0) - \cos(\phi_a - \phi_0)}$$

$$(6.29)$$

$$L_{1, x, y}^{\text{refl}(n)}(-\mu, \mu_0) = \frac{L_{x, y}^{\text{refl}(n)}(-\mu, \mu_0, \phi_f - \phi_0) - L_{x, y}^{\text{refl}(n)}(-\mu, \mu_0, \phi_a - \phi_0)}{\cos(\phi_f - \phi_0) - \cos(\phi_a - \phi_0)} \quad (6.30)$$

where ϕ_a and ϕ_f are the two azimuth angles for each fore−aft camera pair; L^{atm} is the atmospheric path radiance, and T_0 and T_1 are computed from the atmospheric transmittance T:

$$T_0(-\mu, -\mu') = \frac{1}{2\pi} \int_0^{2\pi} T(-\mu, -\mu', \phi - \phi') \, d\phi' \quad (6.31)$$

$$T_1(-\mu, -\mu') \cos(\phi - \phi_0) = \frac{1}{\pi} \int_0^{2\pi} T(-\mu, -\mu', \phi - \phi') \cos(\phi' - \phi_0) \, d\phi' \quad (6.32)$$

Both L^{atm} and T are independent of the surface properties. BHR is updated from the following expression:

$$A_{x, y}^{\text{hem}(n)}(\mu_0) = \frac{M_{x, y}^{(n)}(\mu_0)}{E_b(\mu_0) + s \cdot M_{x, y}^{(n)}(\mu_0)} \quad (6.33)$$

where

$$M_{x, y}^{(n)}(\mu_0) = 2\pi \int_0^1 L_{0, x, y}^{\text{refl}(n)}(-\mu, \mu_0) \mu \, d\mu \quad (6.34)$$

In Eq. (6.33), s is the atmosphere spherical albedo, and $E_b(\mu_0)$ is the incident solar irradiance when surface reflectance is zero. Both s and $E_b(\mu_0)$ are independent of the surface properties. The iteration process continues until $A_{x, y}^{\text{hem}}$ convergences. Once $L_{1, x, y}^{\text{refl}}$ and $A_{x, y}^{\text{hem}}$ are determined, HDRF then can be evaluated from the expression

$$r_{x, y}(-\mu, \mu_0, \phi - \phi_0) = \frac{\pi L_{x, y}^{\text{refl}}(-\mu, \mu_0, \phi - \phi_0) \left[1 - A_{x, y}^{\text{hem}}(\mu_0) s \right]}{E_b(\mu_0)} \quad (6.35)$$

6.3.2.2 Retrieving BRF and DHR

The BRF, $R_{x,y}(\cdot)$, is directly proportional to BRDF, $f_{x,y}$, in the following format, which is duplicated from Eq. (1.9) (Nicodemus et al. 1977):

$$R_{x,y}(-\mu, \mu_0, \phi - \phi_0) = \pi \cdot f_{x,y}(-\mu, \mu_0, \phi - \phi_0) \qquad (6.36)$$

Aside from the factor of π, the BRF and BRDF are essentially identical descriptions of the reflectance properties of a surface, and the two terms can be used interchangeably. BRF is determined in the following iterative process:

$$
\begin{aligned}
R_{x,y}^{(n+1)}&(-\mu, \mu_0, \phi - \phi_0) \\
&= \frac{E_{x,y}(\mu_0)}{E_b^{\text{dir}}(\mu_0)} r_{x,y}(-\mu, \mu_0, \phi - \phi_0) \\
&\quad - \frac{2\pi}{e^{-\tau/\mu_0}} \int_0^1 R_{0,\text{model}}^{(n)}(-\mu, \mu') T_0(-\mu_0, -\mu') \, d\mu' \\
&\quad - \frac{\pi \cdot \cos(\phi - \phi_0)}{e^{-\tau/\mu_0}} \int_0^1 R_{1,\text{model}}^{(n)}(-\mu, \mu') T_1(-\mu_0, -\mu') \, d\mu' \\
&\quad - \frac{2 A_{x,y}^{\text{hem}}(\mu_0) E_{x,y}(\mu_0)}{E_b^{\text{dir}}(\mu_0)} \int_0^1 R_{0,\text{model}}^{(n)}(-\mu, \mu') \mu' \, d\mu' \qquad (6.37)
\end{aligned}
$$

where the direct irradiance with the black surface (zero reflectance) is given by

$$E_b^{\text{dir}}(\mu_0) = \mu_0 E_0 \cdot e^{-\tau/\mu_0} \qquad (6.38)$$

E_0 was discussed in Section 1.2. The surface total irradiance is calculated by

$$E_{x,y}(\mu_0) = \int_0^1 \int_0^{2\pi} L_{x,y}^{\text{inc}}(\mu', \mu_0, \phi' - \phi_0) \mu' \, d\mu' \, d\phi' \qquad (6.39)$$

$R_{0,\text{model}}^{(n)}$ and $R_{1,\text{model}}^{(n)}$ are calculated from a BRF model:

$$R_{0,\text{model}}^{(n)}(-\mu, \mu') = \frac{1}{2\pi} \int_0^{2\pi} R_{\text{model}}^{(n)}(-\mu, \mu', \phi - \phi') \, d\phi' \qquad (6.40)$$

and

$$R_{1,\text{model}}^{(n)}(-\mu,\mu')\cos(\phi-\phi_0)$$

$$= \frac{1}{\pi}\int_0^{2\pi}R_{\text{model}}^{(n)}(-\mu,\mu',\phi-\phi')\cos(\phi-\phi_0)\,d\phi' \qquad (6.41)$$

The BRF model is modified from a semiempirical model (Rahman et al. 1993) as introduced in Section 2.2.5

$$R_{\text{model}}(-\mu,\mu_0,\phi-\phi_0) = r_0\left[\mu_0\mu(\mu+\mu_0)\right]^{k-1}\exp[b\cdot p(\Omega)]\left(\frac{2+G-r_0}{1+G}\right) \qquad (6.42)$$

where

$$p(\Omega) = \cos\Omega = -\mu\mu_0 + \sqrt{(1-\mu^2)(1-\mu_0^2)}\;\cos(\phi-\phi_0) \qquad (6.43)$$

and

$$G = \left\{\left(\frac{1}{\mu^2}-1\right)+\left(\frac{1}{\mu_0^2}-1\right)+2\sqrt{\left(\frac{1}{\mu^2}-1\right)\left(\frac{1}{\mu_0^2}-1\right)}\;\cos(\phi-\phi_0)\right\}^{1/2} \qquad (6.44)$$

The iteration continues by changing three parameters in the BRF model (r_0, k, and b) until convergence is achieved. DHR is then computed as follows:

$$A_{x,y}^{\text{dir}}(\mu_0) = \frac{1}{\pi}\int\int R_{x,y}(-\mu,\mu_0,\phi-\phi_0)\mu\,d\mu\,d\phi \qquad (6.45)$$

These algorithms have been used to generate the land surface products operationally.

6.4   METHODS FOR ESTIMATING TOTAL COLUMN WATER VAPOR CONTENT

In order to correct near-IR and middle-IR imagery, water vapor absorption in the atmosphere has to be accounted for. Water vapor mass mixing ratios vary from over 20 g/kg in the tropical boundary layer to less than 1 g/kg in the Arctic winter. The relative humidity may vary between ~ 10% and 100% of the saturation value at any location. The water absorption coefficients are quite stable and accurately known; the critical parameter is the water vapor content, which varies quite dramatically in both space and time. For atmospheric correction of near-IR and middle-IR imagery, only the total column water vapor content of the atmosphere is required in most cases. For thermal IR remote sensing, the vertical profile of water vapor (discussed in Chapter 10) must be known.

There are many different techniques for measuring the vertical profile or total integrated amount of water vapor. A brief description of different methods is given in the next section. Differential absorption techniques using near-IR bands and split-window algorithms using two thermal bands for estimating the total column water vapor amount are discussed in detail.

6.4.1 Overview of Various Techniques

Various ground-based, airborne, and spaceborne techniques for estimating water vapor content are briefly outlined below:

- *Surface Meteorological Observations*. These provide surface measurements with good temporal but poor spatial frequencies.

- *Ground-based LiDAR* (*light detection and ranging*). This provides high-quality data with good vertical and daytime temporal resolution. A pulse of laser light is emitted into the sky, and the amount of return due to backscatter from the atmosphere is measured versus time. With knowledge of the speed of light, one can convert the time into altitude. The number of photons counted for each altitude bin is proportional to the atmospheric density.

- *Ground-Based Global Positioning System* (*GPS*). GPS detects water vapor due to the delay of the signal from the satellite to the receiver. Atmospheric water vapor degrades GPS positioning accuracy and is a large source of error. However, if the position of a receiver is known, the receiver can then detect the errors in the GPS signal and determine with great precision the water vapor content in the atmosphere.

- *Radiosonde* (*Weather Balloons*). Various instrument packages yield differing levels of quality with good vertical resolution data, poor horizontal resolution, and varying temporal resolution. The radiosonde is a balloonborne instrument platform with radio transmitting capabilities. The radiosonde contains instruments capable of making direct in situ measurements of air temperature, humidity, and pressure with height, typically to altitudes of approximately 30 km. To measure humidity, radiosondes carry a *hygristor*—a glass or plastic strip that is coated with lithium chloride (LiCl). This chemical changes electrical resistance when the atmosphere it is exposed to changes. Metal on the edges of the strip enable the electrical resistance of the chemical to be measured.

- *Airplanes*. Programs utilizing research aircraft are expensive; those using commercial aircraft have not been implemented yet. These aircraft could potentially provide global information scattered among airports with varying temporal frequency.

- *Infrared Sounding Sensors*. Sounding is the identification of atmospheric constituents based on detecting the presence of spectral lines associated with a specific molecule. Determining abundance requires additional measurements. Pressure and temperature can be used to determine the line strength. From the line strength, abundance can be derived. The vertical profiles can be derived from channels between 6700 and 8200 nm and around the 13,300 and 18,600 nm windows, which are sensitive to the water vapor continuum. The TOVS [TIROS (television infrared observation/operational satellite) operational vertical sounder] sensor consists of three independent instruments: the high resolution infrared radiation sounder number 2 (HIRS/2), the microwave sounding unit (MSU), and the Stratospheric Sounding Unit (SSU). The HIRS/2 has 20 channels; the MSU has 4 channels. Together the TOVS instruments determine a three-dimensional temperature–moisture picture of the atmosphere. Coverage with TOVS is limited to sensing percipitable water vapor in cloud-free regions. The total column water vapor content of the atmosphere is estimated operationally on the basis of the thermal infrared at 5.7–7.1 μm [geostationary operational environment satellite (GOES) and Meteosat]. But the spatial (horizontal) resolution of this type of products is quite coarse from 5 km (Meteosat) to 8 km (GOES). These sensor systems also have a limited penetration into the lower atmosphere [e.g., 700 hPa (hectopascals) for Meteosat]. Sounders [e.g., along-track scanning radiometer (ATSR), high-resolution infrared radiation sounder (HIRS), or in Defense Meteorological Satellite Program (DMSP)] have vertical resolutions in the kilometer range but do not have imaging capabilities.

- *Microwave Sensors* (*SMMR, SSMI*). These provide total column water vapor data over large ice-free oceans. Vertical resolution is poor. The special sensor microwave imager (SSMI) is a seven-channel, four-frequency, linearly polarized, passive microwave radiometric system. The SSMI has been flown by the DMSP on several satellites. The scanning multichannel microwave radiometer (SMMR) operated on NASA's Nimbus 7 satellite for more than 8 years, from 1978 to 1987.

- *Solar Occultation Methods*. These include SAGE II (stratospheric aerosol and gas experiment number 2) and provide high accuracy and vertical resolution for data in the stratosphere and above. Solar occultation is a technique that determines the profile of the atmosphere by viewing the atmosphere from the side and passively detecting the refraction of the solar radiation. An atmospheric density profile, pressure, and temperature can be obtained using the hydrostatic equation and the equation of state. SAGE II was the solar occultation sensor used to detect water vapor operating onboard the Earth Radiation Budget Satellite (ERBS). SAGE III is producing the vertical profile of water vapor at 0–50 km.

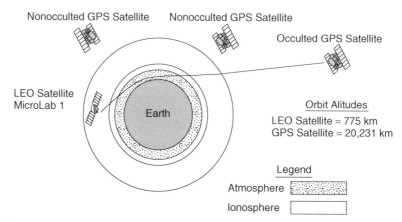

Figure 6.7 Spaceborne GPS technique for estimating atmospheric water vapor content.

- *Spaceborne GPS.* This technique is based on the active limb sounding using radio occultation when the signal path from a GPS satellite to a low Earth orbiting (LEO) satellite intersects the Earth's atmosphere (Fig. 6.7). The principle involved is that the atmosphere acts as a spherical lens, bending and slowing the propagation of signals. Given accurate positions for the two satellites, the atmospheric delay can be determined, which leads to determination of the atmospheric profiles. More details are available in the literature (Elgered et al. 1991, Bevis et al. 1992).

6.4.2 Differential Absorption Technique

The differential absorption technique has been widely used to estimate the total water vapor content of the atmosphere directly from multispectral or hyperspectral imaging systems, such as AVIRIS (Green et al. 1998), POLDER, and MODIS. The general idea is to have one spectral band at the water absorption region (e.g., 0.94 μm) and one or more bands outside the absorption region. The difference among these bands indicates the amount the water vapor in the atmosphere. There are several different algorithms, typically including the narrow/wide algorithm (Frouin et al. 1990), the continuum interpolation band ratio algorithms (Kaufman and Gao 1992, Green et al. 1998), curve fitting algorithm (Gao and Goetz 1990), and the atmospheric precorrected differential absorption algorithm (Schlapfer et al. 1998). In the following, we introduce the continuum interpolation band ratio algorithm (see Fig. 6.8).

The idea of the continuum interpolation band ratio (CIBR) algorithm is to define an index R as the ratio of the radiance received by the sensor at

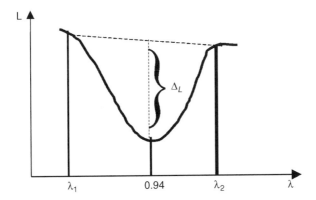

Figure 6.8 Illustration of the CIBR algorithm for estimating water vapor content.

$\lambda_0 = 0.94$ μm (L_{λ_0}) to the linear combination of the radiance at two neighboring bands with the central wavelength at λ_1 and λ_2:

$$R = \frac{L_{\lambda_0}}{C_1 L_{\lambda_1} + C_2 L_{\lambda_2}} \qquad (6.46)$$

where

$$C_1 = \frac{\lambda_0 - \lambda_1}{\lambda_2 - \lambda_1} \qquad (6.47)$$

and

$$C_2 = \frac{\lambda_2 - \lambda_0}{\lambda_2 - \lambda_1} \qquad (6.48)$$

From Fig. 6.8, we can see that when water vapor content increases, $L_{0.94}$ decreases and radiances at the shoulders around 0.94 μm do not change significantly. Thus, the absorption valley becomes deeper, and ΔL increases. Both simulations and experiments reveal that this index R is well correlated to the total column water vapor content of the atmosphere (W) in the following relation:

$$R = \exp(-\alpha W^{\beta}) \qquad (6.49)$$

where α and β are two coefficients.

Tahl and Schonermark (1998) applied this CIBR method to estimate the column water vapor content of the atmosphere from the modular optoelectronic scanner (MOS). MOS was developed and built by the Institute of Space Sensor Technology of the German Aerospace Research Establishment. It is a spaceborne imaging pushbroom spectroradiometer measuring in the visible and near-IR region (400–1010 nm). Bands 11 (867 nm), 12 (940 nm) and 13 (1009 nm) were used to calculate the index R. MODTRAN3 was used to simulate the relationships between the index R and the total path water vapor content (V_p) based on 23,868 cases of different atmospheric and surface conditions:

$$V_p = \begin{cases} \left(-\dfrac{\ln R}{0.592} \right)^{1/0.568} & \text{nonvegetation} \\[2ex] \left(-\dfrac{\ln R}{0.599} \right)^{1/0.575} & \text{vegetation} \end{cases} \tag{6.50}$$

where the total path water vapor content is related to the true column water vapor content (W):

$$W = V_p \left(\frac{1}{\cos \theta_s} + \frac{1}{\cos \theta_v} \right)^{-1} \tag{6.51}$$

Although they separated vegetated surfaces from nonvegetated surfaces in establishing the empirical formulae for estimating water vapor content, Tahl and Schonermark (1998) found from further investigations of the error patterns that their formulas underestimated the water vapor content over dark surfaces. A further correction is required to remove the estimation biases by V_p/f

$$f = \begin{cases} 0.464 + 0.13 \ln \left(\dfrac{C_1 L_{0.867} + C_2 L_{1.009}}{\cos \theta_s} \right) & \text{nonvegetation} \\[2ex] 0.587 + 0.092 \ln \left(\dfrac{C_1 L_{0.867} + C_2 L_{1.009}}{\cos \theta_s} \right) & \text{vegetation} \end{cases} \tag{6.52}$$

where radiances $L_{0.867}$ and $L_{1.009}$ are in units of W m^{-2} μm^{-1} sr^{-1} and C_1 and C_2 are defined in Eqs. (6.47) and (6.48). The validation results using radiosonde measurements indicate that this method produces relative errors smaller than 20% (absolute error 0.27 cm).

The POLDER science team estimated the column water vapor content using the differential absorption technique (Bouffies et al. 1997, Vesperini et al. 1999). Their algorithm was based on the ratio of two bands centered at 910 and 865 nm. A large number of H$_2$O absorption lines exist around 910 nm, and the 865-nm band is not significantly affected by water vapor absorption. On the basis of extensive simulations, their empirical formula is

$$W = \frac{a_0 + a_1 \log(X) + a_2 [\log(X)]^2}{[\cos^{-1} \theta_s + \cos^{-1} \theta_v] P_s} \tag{6.53}$$

where the a_i are the coefficients, X is the reflectance ratio of these two bands ($X = \rho_{910}/\rho_{865}$), P_s is the surface pressure, and θ_s and θ_v are the solar zenith and viewing zenith angles. The validation results (Vesperini et al. 1999) indicate that POLDER estimates over land are compared well with radiosonde measurements with a 3.1-kg m^2 RMS error.

A similar algorithm (Bennartz and Fischer 2001) has been proposed to estimate the column atmospheric water vapor content from the medium-resolution imaging spectrometer (MERIS) in the European Space Agency (ESA)'s *environmental satellite* (ENVISAT). The general form of the empirical equation is

$$W = a_0 + a_1 \log(X) + a_2 [\log(X)]^2 \qquad (6.54)$$

where the a_i are the coefficients and X is the radiance ratio of two bands ($X = L_{900}/L_{885}$). They showed that the algorithm's theoretical accuracy is in the order of 1.7 kg/m^2.

The MODIS algorithms have been discussed by Kaufman and Gao (1992) in detail.

6.4.3 Split-Window Algorithms

For sensors that do not have spectral bands suitable for differential absorption algorithms, alternative techniques are needed. Assuming that there are two thermal-IR bands with different water vapor absorptions (typically one spectrum is in 10–11 μm and the other, 11–12 μm) covariance–variance of these two brightness temperature images may be used to estimate the total water vapor content of the atmosphere. It was first suggested by Kleespies and McMillin (1990) and Jedlovec (1990), and then refined by Harris and Mason (1992), among others, over ocean. Sobrino et al. (Sobrino et al. 1994, 1996, 1999) and Ottle et al. (1997) extended this technique over land. The limitations of this technique have been discussed in the literature (Barton and Prata 1999, Ottle and Francois 1999).

This technique assumes that the cloud-free atmosphere is horizontally homogeneous and that surface temperature and emissivity vary within a neighborhood. Kleespies and McMillin (1990) showed that the ratio of the brightness temperature (T) difference between two neighboring pixels i and j at two bands is equal to the ratio of the atmospheric transmittance (t) and surface emissivity (ε):

$$R_{12,11} = \frac{(\Delta T_{ij})_{12}}{(\Delta T_{ij})_{11}} = \frac{(T_i - T_j)_{12}}{(T_i - T_j)_{11}} = \frac{\varepsilon_{12}}{\varepsilon_{11}} \frac{t_{12}}{t_{11}} \qquad (6.55)$$

where the subscripts 11 and 12 refer to the split-window bands around 11 and 12 μm, respectively. For ocean water, surface emissivity is very close to 1. Thus, $R_{12,11}$ is equal to the ratio of the atmospheric transmittance, which is largely dependent on the total water vapor content since the band transmittance can be expressed by

$$t_\lambda = \exp(-k_\lambda U) \tag{6.56}$$

where k_λ is the absorption coefficient and U is the total content of the absorbing gases. If we assume that the total transmittance due to other atmospheric gases is constant and the exponential function is approximated by the linear function as the first-order approximation, $R_{12,11}$ can be linearly related to the total water vapor content.

Jedlovec (1990) extended this technique by using the ratio of the spatial variance (σ^2) of the brightness temperature

$$\frac{t_{12}^2}{t_{11}^2} = \frac{\sigma_{12}^2 - \sigma_{12,\varepsilon}^2}{\sigma_{11}^2 - \sigma_{12,\varepsilon}^2} \tag{6.57}$$

where σ_ε^2 is the estimated error variance for each band, which can be estimated from structure function analysis or an appropriate variance method. Sobrino et al. (1994) further extended this technique by using the ratio of the covariance to the variance:

$$R_{12,11} = \frac{\sigma_{11,12}}{\sigma_{11}^2} = \frac{\sum(T_{11,i} - \bar{T}_{11})(T_{12,i} - \bar{T}_{12})}{\sum(T_{11,i} - \bar{T}_{11})^2} \tag{6.58}$$

They related this ratio to the water vapor content W based on the similar exponential function with an explicit viewing angle dependence:

$$W = 0.154 - 15.359(\cos\theta \ln R_{12,11}) - 12.85(\cos\theta \ln R_{12,11})^2 \tag{6.59}$$

Using the same ratio index, Ottle et al. (1997) obtained a linear relation

$$W = 12.178 - 11.566 R_{12,11} \tag{6.60}$$

Both formulas have been applied to real AVHRR imagery (Ottle et al. 1997, Sobrino et al. 1999).

When Barton and Prata (1999) applied this variance-based technique to several sets of data from AVHRR, however, they found that no statistically significant correlation exists between the estimated and the measured total water vapor content. Since this technique relies on variance and covariance, it might be subject to different error sources, such as instrument noise, data quantization, and subpixel clouds. This indicates that this technique is not mature enough for operational applications.

Note that there are also several studies in the literature reporting linear relationships between the total water vapor content and the brightness temperature difference (e.g., Dalu 1987, Eck and Holben 1994, Choudhury and Digirolamo 1995), but it appears that the coefficients in their relations are quite variable from one location to another.

6.4.4 Unit Conversions

Because of the peculiarity of the various units for the atmospheric absorber concentration, we provide a small section on the unit conversion. It is under the section of water vapor, but can be also used for other gases as well:

$$1 \text{ preci.cm } H_2O = 1 \text{ g/cm}^2 \text{ liquid } H_2O \qquad (6.61)$$

where 1 g/cm^2 is equal to 1 cm of a column of liquid water and

$$1 \text{ preci.cm } H_2O = 3.34 \times 10^{22} \text{ molecules/cm}^2 \qquad (6.62)$$

The unit used in MODTRAN is cm·atm at standard tremperature and pressure (STP). The conversion into the more common unit of grams per square centimeters is accomplished using

$$1 \text{ g/cm}^2 = 1.244 \times 10^3 \text{ cm·atm STP} \qquad (6.63)$$

For noncondensable gases, for which the units of atm.cm are usually used, the conversion is made

$$1 \text{ atm·cm (gas)} = M/22.4 \times 10^3 \text{ g/cm}^2 \qquad (6.64)$$

at STP (where M is the molecular weight in grams). For M grams of gases, there are 6.02×10^{23} molecules; thus

$$1 \text{ atm·cm (gas)} = 2.69 \times 10^{19} \text{ molecules/cm}^2 \qquad (6.65)$$

6.5 SUMMARY

Most of practical algorithms for estimating land surface variables are based on surface reflectance. Atmospheric correction is a process that converts the top-of-atmosphere (TOA) radiance to surface reflectance. Two major parameters in the atmosphere cause difficulties in atmospheric correction because of their dramatic variations of their concentrations in both space and time: aerosol and water vapor. Aerosol scattering dominates the shortwave band signals, while water vapor mainly affects the near-IR signals. Section 6.2 outlined different methods for correcting aerosol effects from multispectral remotely sensed data, some of which are quite empirical. The most difficult aspect is how to estimate aerosol properties from imagery itself. This is the area that needs further development.

Multiangle remote sensing is a new area in the remote sensing community (Diner et al. 1999, Liang and Strahler 2000). One advantage with multiangular observations is the ability to estimate aerosol properties much better than the traditional single-view multispectral remote sensing. Section 6.3 introduced the MISR atmospheric correction algorithm.

Water vapor estimation methods were evaluated in Section 6.4. Following an overview of various methods for estimating water vapor profiles using sounders or the total water vapor content of the atmosphere, which is more or less ancillary information for water vapor correction, This chapter presented two major techniques based on the differential absorption principles in the near-IR spectrum and the split-window algorithms in the thermal IR spectrum. The differential absorption method requires two or more bands around the water vapor absorption region (~ 0.94 μm) and can be easily incorporated into any new remote sensing system.

REFERENCES

Barton, I. J. and Prata, A. J. (1999), Difficulties associated with the application of covariance-variance techniques to retrieval of atmospheric water vapor from satellite imagery, *Remote Sens. Envir.* **69**: 76–83.

Bennartz, R. and Fischer, J. (2001), Retrieval of columnar water vapour over land from backscattered solar radiation using the Medium Resolution Imaging Spectrometer, *Remote Sens. Envir.* **78**: 274–283.

Bevis, M., Businger, S., Herring, T. A., Rocken, C., Anthes, R. A., and Ware, R. H. (1992), GPS meteorology: Remote sensing of atmospheric water vapor using the global positioning system, *J. Geophys. Res.* **97**: 15787–15801.

Borel, C. C. and Gerstl, S. A. W. (1992), Adjacency-blurring-effect of scenes modeled by the radiosity method, *Proc. SPIE* 620–624.

Bouffies, S., Breon, F. M., Tanre, D., and Dubuisson, P. (1997), Atmospheric water vapor estimate by a differential absorption technique with the POLDER instrument, *J. Geophys. Res.* **102**: 3831–3841.

Chang, C. I. and Ren, H. (2000), An experiment-based quantitative and comparative analysis of target detection and image classification algorithms for hyperspectral imagery, *IEEE Trans. Geosci. Remote Sens.* **38**: 1044–1063.

Chavez, J. P. (1996), Image-based atmospheric corrections—revisited and improved, *Photogramm. Eng. Remote Sens.* **62**: 1025–1036.

Chavez, J. P. S. (1988), An improved dark-object subtraction technique for atmospheric scattering correction of multispectral data, *Remote Sens. Envir.* **24**: 459–479.

Choudhury, B. J. and DiGirolamo, N. E. (1995), Quantifying the effect of emissivity on the relation between AVHRR split window temperature difference and atmospheric precipitable water over land surfaces, *Remote Sens. Envir.* **54**: 313–323.

Crist, E. P. and Cicone, R. C. (1984), A physically-based transformation of thematic mapper data—the TM tasseled cap, *IEEE Trans. Geosci. Remote Sens.* **22**: 256–263.

Dalu, G. (1987), Satellite remote sensing of atmospheric water vapor, *Int. J. Remote Sens.* **7**: 1089–1097.

Diner, D., Asner, G., Davies, R., Knyazikhin, Y., Muller, J. P., Nolin, A., Pinty, B., Schaaf, C., and Stroeve, J. (1999), New directions in earth observing: Scientific applications of multiangle remote sensing, *Bull. Am. Meteorol. Soc.* **80**: 2209–2228.

Diner, D. J., Martonchik, J. V., Borel, C., Gerstl, S. A. W., Gordon, H. R., Myneni, R., Pinty, B., Verstraete, M. M., and Knyazikhin, Y. (1999), *Level 2 Surface Retrieval Algorithm Theoretical Basis Document*, NASA/JPL, JPL D-11401, Rev. D.

Eck, T. F. and Holben, B. N. (1994), AVHRR split window temperature differences and total precipitable water over land surfaces, *Int. J. Remote Sens.* **15**: 567–582.

Elgered, G., Davis, J. L., Herring, T. A., and Shapiro, I. I. (1991), Geodesy by radio interferometry: Water vapor radiometry for estimation of the wet delay, *J. Geophys. Res.* **96**: 6541–6555.

Evans, K. F. (1998), The spherical harmonics discrete ordinate method for three-dimensional atmospheric radiative transfer, *J. Atmos. Sci.* **55**: 429–464.

Evans, K. F., Evans, A. H., Nolt, I. G., and Marshall, B. T. (1999), The prospect for remote sensing of cirrus clouds with a submillimeter-wave spectrometer, *J. Appl. Metorol.* **38**: 514–525.

Fallah-Adl, H., JaJa, J., and Liang, S. (1997), Fast algorithms for estimating aerosol optical depth of Thematic Mapper (TM) imagery, *J. Supercomput.* **10**: 315–330.

Frouin, R., Deschamps, P. Y., and Lecomte, P. (1990), Determination from space of atmospheric total water vapor amounts by differential absorption near 940 nm: Theory and airborne verification, *J. Appl. Meteorol.* **29**: 448–459.

Gao, B. C. and Goetz, A. (1990), Determination of total column water vapor in the atmosphere at high spatial resolution from AVIRIS data using spectral curve fitting and band ratioing techniques, *Proc. SPIE* **1298**: 138–149.

Gordon, H. R. and Castano, D. J. (1989), Aerosol analysis with the coastal zone color scanner: A simple method for including multiple scattering effects, *Appl. Opt.* **28**: 1320–1326.

Green, R. O., Eastwood, M. L., Sarture, C. M., Chrien, T. G., Aronsson, M., Chippendale, B. J., Faust, J. A., Pavri, B. E., Chovit, C. J., Solis, M., Olah, M. R., and Williams, O. (1998), Imaging spectroscopy and the airborne visible/infrared imaging spectrometer (AVIRIS), *Remote Sens. Envir.* **65**: 227–248.

Hall, F. G., Strebel, D. E., Nickeson, J. E., and Goetz, S. J. (1991), Radiometric rectification: Toward a common radiometric response among multidate, multisensor images, *Remote Sens. Envir.* **35**: 11–27.

Harris, A. R. and Mason, I. M. (1992), An extension to the split-window technique giving improved atmospheric correction and total water vapor, *Int. J. Remote Sens.* **13**: 881–892.

Hess, M., Koepke, P., and Schult, I. (1998), Optical properties of aerosols and clouds: The software package OPAC, *Bull. Am. Meteorol. Soc.* **79**: 831–844.

Hu, Y. H., Lee, H. B., and Scarpace, F. L. (1999), Optimal linear spectral unmixing, *IEEE Trans. Geosci. Remote Sens.* **37**: 639–644.

Jedlovec, G. J. (1990), Precipitable water estimation from high-resolution split window radiance measurements, *J. Appl. Meteorol.* **29**: 863–877.

Kaufman, Y. and Gao, B. C. (1992), Remote sensing of water vapor in the near IR from EOS/MODIS, *IEEE Trans. Geosci. Remote Sens.* **30**: 871–884.

Kaufman, Y. J. (1984), Atmospheric effect on spatial resolution of surface imagery: Errata, *Appl. Opt.* **23**: 4164–4172.

Kaufman, Y. J. (1989), The atmospheric effect on remote sensing and its correction, in *Theory and Applications of Optical Remote Sensing*, G. Asrar, ed., Wiley, pp. 336–428.

Kaufman, Y. J. and Sendra, C. (1988), Automatic atmospheric correction, *Int. J. Remote Sens.* **9**: 1357–1381.

Kaufman, Y. J., Karnieli, A., and Tanre, D. (2000), Detection of dust over deserts using satellite data in the solar wavelengths, *IEEE Trans. Geosci. Remote Sens.* **38**: 525–531.

Kaufman, Y. J., Tanré, D., Remer, L., Vermote, E. F., Chu, A., and Holben, B. N. (1997a), Operational remote sensing of tropospheric aerosol over the land from EOS-MODIS, *J. Geophys. Res.* **102**: 17051–17068.

Kaufman, Y. J., Wald, A., Lorraine, L. A., Gao, B. C., Li, R. R., and Flynn, L. (1997b), Remote sensing of aerosol over the continents with the aid of a 2.2 μm channel, *IEEE Trans. Geosci. Remote Sens.* **35**: 1286–1298.

Kleespies, T. J. and McMillin, L. M. (1990), Retrieval of precipitable water from observations in the split window over varying surface temperatures, *J. Appl. Meteorol.* **29**: 851–862.

Kozoderv, V. V. (1995), Correction of space image for atmospheric effects, *Sov. Int. J. Remote Sens.* **3**: 255–271.

Liang, S. (2000), Numerical experiments on spatial scaling of land surface albedo and leaf area index, *Remote Sens. Rev.* **19**: 225–242.

Liang, S. and Strahler, A., eds. (2000), Land surface bidirectional reflectance distribution function (BRDF): Recent advances and future prospects, *Remote Sens. Rev.* **18**: 83–551.

Liang, S., Fang, H., and Chen, M. (2001), Atmospheric correction of landsat ETM + land surface imagery: I. Methods, *IEEE Trans. Geosci. Remote Sens.* **39**: 2490–2498.

Liang, S., Fallah-Adl, H., Kalluri, S., JaJa, J., Kaufman, Y. J., and Townshend, J. R. G. (1997), An operational atmospheric correction algorithm for landsat Thematic Mapper imagery over the land, *J. Geophys. Res.* (*Atmos.*) **102**: 17173–17186.

Liang, S., Fang, H., Chen, M., Shuey, C., Walthall, C., and Daughtry, C. (2002), Atmospheric correction of landsat ETM + land surface imagery: II. Validation and applications, *IEEE Trans. Geosci. Remote Sens.* **40**: 2736–2746.

Liou, K. N. (1980), *An Introduction to Atmospheric Radiation*, Academic Press.

Martonchik, J. V., Diner, D. J., Crean, K. A., and Bull, M. A. (2002), Regional aerosol retrieval results from MISR, *IEEE Trans. Geosci. Remote Sens.* **40**: 1520–1531.

Maselli, F. (1998), Multiclass spectral decomposition of remotely sensed scenes by selective pixel unmixing, *IEEE Trans. Geosci. Remote Sens.* **36**: 1809–1820.

Mekler, Y. and Kaufman, Y. J. (1980), The effect of earth's atmosphere on contrast reduction for a uniform surface albedo and 'two-halves' field, *J. Geophys. Res.* **85**: 4067–4083.

Mekler, Y. and Kaufman, Y. J. (1982), Contrast reduction by the atmosphere and retrieval of nonuniform surface reflectance, *Appl. Opt.* **21**: 310–316.

Miesch, C., Briottet, X., Kerr, Y. H., and Cabot, F. (1999), Monte Carlo approach for solving the radiative transfer equation over mountainous and heterogeneous areas, *Appl. Opt.* **38**: 7419–7430.

Moran, M. S., Jackson, R. D., Slater, P. N., and Teillet, P. M. (1992), Evaluation of simplified procedures for retrieval of land surface reflectance factors from satellite sensor output, *Remote Sens. Envir.* **41**: 169–184.

Nicodemus, F. E., Richmond, J. C., Hsia, J. J., Ginsberg, I. W., and Limperis, T. (1977), *Geometrical Considerations and Nomenclature for Reflectance*, NBS Monograph (U.S.), 160.

Ottle, C. and Francois, C. (1999), Further insights into the use of the split-window covariance technique for precipitable water retrieval, *Remote Sens. Envir.* **69**: 84–86.

Ottle, C., Outalha, S., Francois, C., and Le Maguer, S. (1997), Estimation of total water vapor content from split-window radiance measurements, *Remote Sens. Envir.* **61**: 410–418.

Pearce, W. A. (1977), *A Study of the Effects of the Atmosphere on Thematic Mapper Observations*, Report 004–77, EG@E.

Popp, T. (1995), Correcting atmospheric masking to retrieve the spectral albedo of land surface from satellite measurements, *Int. J. Remote Sens.* **16**: 3483–3508.

Rahman, H., Pinty, B., and Verstraete, M. M. (1993), Coupled surface-atmosphere reflectance (CSAR) model, 2, semiempirical surface model usable with NOAA Advanced Very High Resolution Radiometer data, *J. Geophys. Res.* **98**: 20791–20801.

Reinersman, P. N. and Carder, K. L. (1995), Monte carlo simulation of the atmospheric point-spread function with an application to correction for the adjacency effect, *Appl. Opt.* **34**: 4453–4471.

Richter, R. (1996a), A spatially adaptive fast atmospheric correction algorithm, *Int. J. Remote Sens* **17**: 1202–1211.

Richter, R. (1996b), Atmospheric correction of satellite data with haze removal including a haze/clear transition region, *Comput. Geosci.* **22**: 675–681.

Santer, R., Carrere, V., Dubuisson, P., and Roger, J. C. (1999), Atmospheric correction over land for MERIS, *Int. J. Remote Sens.* 20: 1819–1840.

Schlapfer, D., Borel, C. C., Keller, J., and Itten, K. I. (1998), Atmospheric precorrected differential absorption technique to retrieve columnar water vapor, *Remote Sens. Envir.* **65**: 353–366.

Sobrino, J. A., Li, Z., Stoll, M., and Becker, F. (1994), Improvements in the split window technique for land surface temperature determination, *IEEE Trans. Geosci. Remote Sens.* **32**: 243–253.

Sobrino, J. A., Li, Z. L., Stoll, M. P., and Becker, F. (1996), Multi-channel and multi-angle algorithms for estimating sea and land surface temperature with ATSR data, *Int. J. Remote Sens.* **17**: 2089–2114.

Sobrino, J. A., Raissouni, N., Simarro, J., Nerry, F., and Petitcolin, F. (1999), Atmospheric water vapor content over land surfaces derived from the AVHRR data: Application to the Iberian peninsula, *IEEE Trans. Geosci. Remote Sens.* **37**: 1425–1434.

Tahl, S. and Schonermark, M. V. (1998), Determination of the column water vapor of the atmosphere using backscattered solar radiation measured by the modular optoelectronic scanner (MOS), *Int. J. Remote Sens.* **19**: 3223–3236.

Takashima, T. and Masuda, K. (1996), Operational procedure of atmospheric correction on satellite visible data allowing for the adjacency effect, *Proc. SPIE* 70–81.

Tanré, D. and Legrand, M. (1991), On the satellite retrieval of Saharan dust optical thickness over land: Two different approaches, *J. Geophys. Res.* **96**: 5221–5227.

Tanré, D., Herman, M., and Deschamps, P. Y. (1981), Influence of the background contribution upon space measurements of ground reflectance, *Appl. Opt.* **20**: 3673–3684.

Tanré, D., Deschamps, P. Y., Devaux, C., and Herman, M. (1988), Estimation of saharan aerosol optical thickness from blurring effects in Thematic Mapper data, *J. Geophys. Res.* **93**: 15955–15964.

Teillet, P. M. and Fedosejevs, G. (1995), On the dark target approach to atmospheric correction of remotely sensed data, *Can. J. Remote Sens.* **21**: 374–387.

Vermote, E., Tanre, D., Deuze, J. L., Herman, M., and Morcrette, J. J. (1997), Second simulation of the satellite signal in the solar spectrum: An overview, *IEEE Trans. Geosci. Remote Sens.* **35**: 675–686.

Vesperini, M., Breon, F. M., and Tanré, D. (1999), Atmospheric water vapor content from spaceborne POLDER measurements, *IEEE Trans. Geosci. Remote Sens.* **37**: 1613–1619.

7

Topographic Correction Methods

Topographic correction of remotely sensed imagery over mountain regions is at least as important as atmospheric correction. Topographic shading and shadowing modulates any remote sensing signals and affects the inversion of land surface parameters. Topographic correction requires high-resolution and accurate digital elevation model (DEM) data, which are not available globally. This chapter introduces various methods that have been presented in the literature, from simple methods based on empirical statistical relationships to very sophisticated physical methods. Sections 7.1–7.3 present methods for correcting high-resolution imagery; Section 7.4 presents methods for correcting coarse-resolution imagery. Section 7.6 is devoted mainly to DEM data and its generation.

Section 7.1 briefly introduces the development of topographic correction of remote sensing imagery. Section 7.2 presents some simple, more or less semiempirical, methods. Section 7.3 introduces a physical correction method that has been incorporated into a digital image processing package (IPW). Section 7.4 focuses mainly on correction methods for coarse-resolution imagery that characterizes the slope distributions within one pixel. Section 7.5 provides a general overview of different methods for generating DEMs and current products available to the scientific community.

7.1 INTRODUCTION

Mountain regions represent about one fourth of the Earth's terrestrial surface. They provide goods and services to more than half of humanity and are in the proximity of approximately one-fourth of the global population. The United Nation declared year 2002 as the "International Year of

Quantitative Remote Sensing of Land Surfaces. By Shunlin Liang
ISBN 0-471-28166-2 Copyright © 2004 John Wiley & Sons, Inc.

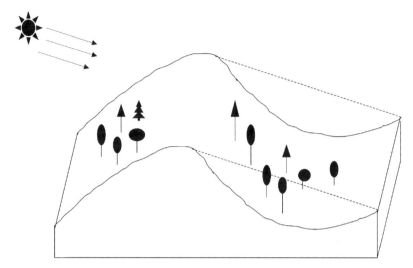

Figure 7.1   Illustration of the shadowing effects in a mountainous area.

Mountains" to increase the awareness of the global importance of mountain ecosystems.

Remote sensing plays a unique role in mountain studies. However, remotely sensed data over mountainous regions are contaminated with topographic shading and shadowing, which are not desirable for land surface characterization. Figure 7.1 illustrates the shadowing effects where surface cover types are the same but their brightness in the sunlit area and the shaded area are quite different. It is important to develop effective algorithms for removing these topographic effects.

Studies on removing the effects of topography on remotely sensed reflectance started in 1980 or before (e.g., Holben and Justice 1980, Proy et al. 1989, Dozier and Frew 1990, Sandmeier and Itten 1997), and some practical algorithms have been developed to correct topographic effects.

The simplest and earliest approaches are ratio algorithms (e.g., Colby 1991). If it is assumed that illumination effects caused by the topography are proportional at different bands, the ratio of two bands can eliminate topographic effects. However, these ratio algorithms generally cannot produce satisfactory results since the radiometric variations caused by topography are wavelength-dependent and their differences at different bands are not simply increased or decreased by a constant. In particular, the ratio is the linear transformation of the original two bands, which may not be desirable for a specific analysis. A further development is generation of the cosine correction algorithms, which are simple but take advantage of DEM data.

The most sophisticated algorithm is probably the one developed by Dozier. It has been incorporated into the Image Processing Workbench (IPW), a digital image processing software system, used for many different applica-

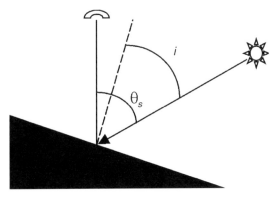

Figure 7.2   Illustration of the sun viewing geometry over a slope.

tions. It is discussed in Section 7.3. Cosine correction algorithms are examined first.

7.2   COSINE CORRECTION METHODS

If we ignore atmospheric influence and adjacency effects, we can correct topographic effects by considering the difference of the direct solar illumination. Cosine correction methods convert the radiance observed over sloped terrain (L_T) to the equivalent value of the horizontal surface (L_H) using the simple relationship:

$$L_H = L_T \left[\frac{\cos(\theta_s)}{\cos(i)} \right]^k \tag{7.1}$$

where θ_s is the solar zenith angle, i the solar illumination angle in relation to the normal on a pixel, and k a measure of the extent to which a surface is a Lambertian; it varies between 0 and 1. For a perfectly Lambertian surface, $k = 1$. k is often called the *Minnaert constant*.

Angle i can be calculated from the following formula (Fig. 7.2):

$$\mu_i = \cos(i) = \cos\theta_s \cos S + \sin S \sin \theta_s \cos(\phi_s - A) \tag{7.2}$$

where A is the azimuth angle of the slope, S is the slope of the pixel, and ϕ_s is the solar azimuth angle. A and S are defined as

$$\tan S \equiv |\Delta_z| = \sqrt{\left(\frac{\partial z}{\partial x}\right)^2 + \left(\frac{\partial z}{\partial y}\right)^2} \tag{7.3}$$

$$\tan A = \frac{\partial z / \partial y}{\partial z / \partial x} \tag{7.4}$$

where $\partial z / \partial x$ and $\partial z / \partial y$ at each point (i, j) are calculated by finite differences

$$\frac{\partial z}{\partial x} = \frac{z_{i+1,j} - z_{i-1,j}}{2\Delta h} \tag{7.5}$$

$$\frac{\partial z}{\partial y} = \frac{z_{i,j+1} - z_{i,j-1}}{2\Delta h} \tag{7.6}$$

where Δh is the horizontal distance of the DEM data.

Teillet et al. (1982) proposed a variant of this formula (7.1):

$$L_H = L_T \frac{\cos(\theta_s) + b/a}{\cos(i) + b/a} \tag{7.7}$$

where a and b are determined from a regression analysis:

$$L_T = a \cos(i) + b \tag{7.8}$$

By examining Eqs. (7.1) and (7.7), we can see that if the k value in (7.1) or (a,b) values in (7.7) are wavelength-independent, the ratio of two bands will eliminate topographic effects. However, they are generally variable for different bands.

Some other statistical correction algorithms using contextual information have also been explored (Gu et al. 1999, Faraklioti and Petrou 2001), but no details are given here. The cosine correction method is very empirical, but very easy to implement and has been widely used.

7.3 IPW METHOD

The IPW method is based mainly on Dozier's work (Dozier and Frew 1981, 1990; Dozier 1989). It formulates the total incoming shortwave irradiance to be the sum of three components: direct solar radiation (F_I), diffuse solar radiation (F_D), and radiation reflected from the neighboring pixels (F_T) (see Fig. 7.3).

$$F = F_I + F_D + F_T \tag{7.9}$$

The direct solar irradiance on a horizontal surface is simply expressed as

$$\mu_s E_0 \exp\left(\frac{-\tau_0}{\mu_s}\right) \tag{7.10}$$

where τ_0 is the total optical depth of the atmosphere and μ_s the cosine of

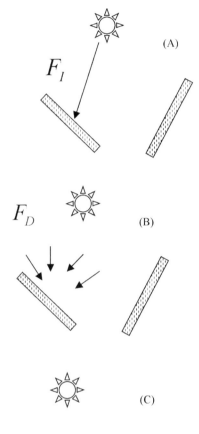

(A)

F_I

(B)

F_D

(C)

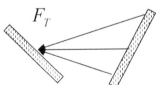

F_T

Figure 7.3 The total incoming flux at each pixel is the sum of three components: direct solar flux, diffuse sky radiation, and reflected radiation from the neighboring pixels.

the solar zenith angle. E_0 has been discussed in Section 1.2. For a sloped pixel

$$F_I = \mu_i E_0 \exp\left(\frac{-\tau_0}{\mu_s}\right) \tag{7.11}$$

where μ_i is defined by Eq. (7.2).

To model diffuse irradance on varying terrain given the diffuse irradiance on a horizontal surface $[F^\downarrow(\tau_0)]$, a sky-viewing factor (V_d) is determined that gives the ratio of diffuse sky irradiance at a point relative to that on an

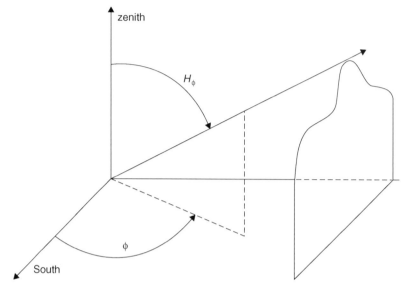

Figure 7.4 Horizon angle H_ϕ for direction ϕ. [From Dozier and Frew (1990), *IEEE Trans. Geosci. Remote Sens.* Copyright © 1990 with permission of IEEE.]

unobstructed horizontal surface

$$F_D = V_d F^\downarrow(\tau_0) \tag{7.12}$$

where accounts for the slope and the orientation of the point, the portion of the overlying hemisphere visible to the point, and the anisotropy of the diffuse irradiance:

$$V_d \approx \frac{1}{2\pi}\int_0^{2\pi}\left[\cos S \sin^2 H_\phi + \sin S \cos(\phi - A)(H_\phi - \sin H_\phi \cos H_\phi)\right]d\phi$$

where H_ϕ is the horizontal angle (see Fig. 7.4). For an unobstructed horizontal surface

$$H_\phi = \frac{\phi}{12} \tag{7.13}$$

Note there are also other approximations in the literature, including those by Kondratyev (1977)

$$V_d = \frac{1 + \cos S}{2} \tag{7.14}$$

and Temps and Coulson (1977):

$$V_d = \frac{1 + \cos S}{2} \left(1 + \sin^3 \theta_s \, \mu_s^2\right) \left[1 + \sin^3 \left(\frac{S}{2}\right)\right] \qquad (7.15)$$

For each pixel, the radiation reflected from surrounding terrain can be accounted for by calculating an average reflected radiation term and adjusting this through a terrain configuration factor (C_t). C_t should include both the anisotropy of the radiation and the geometric effects between a particular location and each of the other terrain locations that are mutually visible. By assuming that the radiation reflected off the terrain is isotropic and given that V_d for an infinitely long slope is $(1 + \cos S)/2$, C_t is then approximated by

$$C_t \approx \frac{1 + \cos S}{2} - V_d \qquad (7.16)$$

The reflected radiation from the surrounding terrain is then

$$F_T = C_t F^{\downarrow}(\tau_0) \bar{R} \qquad (7.17)$$

where $\bar{R}$ is the equivalent reflectance of the surrounding terrain.

The idea of topographic correction is to convert the observed radiance (L) of each pixel to reflectance. Thus, the reflectance for each pixel is

$$R = \frac{\pi L}{F} = \frac{\pi L}{\mu_i E_0 \exp(-\tau_0/\mu_s) + \left(V_d + C_t \bar{R}\right) F^{\downarrow}(\tau_o)} \qquad (7.18)$$

This approach has been widely used for a variety of applications (e.g., Dubayah 1992, Dubayah 1995, Dubayah and Loechel 1997).

However, we also need to note the limitations of this method. The IPW topographic correction algorithm assumes an isotropic sky radiance distribution, which is usually quite different from the actual one (Liang and Lewis 1996). Pixels that have the same slope may receive different incoming solar radiation when they have different orientations (azimuth angles). Thus, this assumption can result in serious errors in the retrieval of surface reflectance from remote sensing imagery. Moreover, the IPW topographic correction algorithm does not provide a solution to estimate the incoming solar radiation that cannot be accurately computed from the climatologic conditions of the atmosphere.

7.4 SHADOWING FUNCTION ALGORITHM

It is not practical to directly correct [e.g., see the article by Dubayah (1992)] topographic effects from coarse-resolution imagery such as AVHRR, MISR, and MODIS using fine-resolution DEM data. This is mainly because of the required computational effort and the poor quality of existing DEM data, and, in particular, the fact that each pixel may contain many different slopes. The slope distribution must also be considered. Some of these problems may be avoided if we parameterize the elevation probability distribution within a pixel and use this as an analytic correction model.

For coarse spatial resolution imagery, such as AVHRR, the effect of interactions among pixels becomes smaller. However, the outgoing radiance from an extended topographic surface will be different from that of a flat surface at the same elevation because the amount of incoming radiation will be modified by shadowing, shading, and view factor effects.

Dubayah (1992) has related effective or apparent reflectance to true reflectance using topographic factors, and Dubayah et al. (1990) calculated aggregate topographic factors for various size regions. There is the potential for using such aggregate topographic factors to correct coarse-scale imagery, especially if the surface BRDF is known.

The reflectance from the topographic surface can then be expressed in a form solvable for true surface reflectance as

$$r_R(\Omega_0, \Omega) = r(\Omega_{0e}, \Omega_e) S(\Omega_0, \Omega) \qquad (7.19)$$

where $r_R(\Omega_0, \Omega)$ is the effective or apparent reflectance and $r(\Omega_{0e}, \Omega)$ the true reflectance. Ω_0 and Ω are the solid angles denoting the illumination direction and viewing direction, respectively. Ω_{0e} and Ω_e are equivalent angles of Ω_0 and Ω by considering the roughness effects. $S(\Omega_0, \Omega)$ is the shadowing function.

If the surface is assumed to be Lambertian with the constant reflectance r_0, then

$$r_R(\Omega_0, \Omega) = r_0 S(\Omega_0, \Omega) \qquad (7.20)$$

There is a considerable body of literature that discusses shadowing on a randomly rough surface, but most papers deal with specularly reflecting facets, usually in connection with either analyses of sea glitter or the special case of radar backscatter. One example of these models was developed by Sancer (1969) for snow reflectance and has been discussed in Section 4.2.1.1. Hapke (1984) developed a shadowing function based on the following assumptions:

- All relevant objects are large compared to wavelength of light so that geometric optics is applicable.
- The mean slope of the rough surface is assumed to be reasonably small.
- Multiple scattering of light from one macroscopic surface facet to another is neglected.

- The surface is made of facets titled at a variety of angles that have no preferred direction in the azimuth but can be described by a Gaussian distribution in the zenith angle.

Given a Gaussian surface with an average slope $\bar{\theta}$, the Hapke shadowing function (Hapke 1984) can be expressed in two separate cases: (1) $\theta_0 < \theta$ and (2) $\theta_0 \geq \theta$, where θ_0 and θ are solar zenith angle and viewing zenith angle, respectively.

Case 1: When $\theta_0 < \theta$, we obtain

$$S(\mu_0, \mu) = \frac{\mu_e(\xi)}{\mu_e(0)} \frac{\mu_0}{\mu_{0e}(0)} \frac{\chi(\bar{\theta})}{1 - f(\xi) + f(\xi)\chi(\bar{\theta})[\mu_0/\mu_{0e}(0)]} \quad (7.21)$$

$$\mu_{0e}(\xi) = \chi(\bar{\theta})\left[\mu_0 + \sqrt{1 - \mu_0^2}\,\tan(\bar{\theta})\right.$$

$$\left. \times \frac{\cos(\xi)E_2(\theta) + \sin^2(\xi/2)E_2(\theta_0)}{2 - E_1(\theta) - \xi/\pi E_1(\theta_0)}\right] \quad (7.22)$$

$$\mu_e(\xi) = \chi(\bar{\theta})\left[\mu + \sqrt{1 - \mu^2}\,\tan(\bar{\theta})\frac{E_2(\theta) - \sin^2(\xi/2)E_2(\theta_0)}{2 - E_1(\theta) - \xi/\pi E_1(\theta_0)}\right] \quad (7.23)$$

Case 2: When $\theta_0 \geq \theta$, we obtain

$$S(\mu_0, \mu) = \frac{\mu_e(\xi)}{\mu_e(0)} \frac{\omega_0}{\mu_{0e}(0)} \frac{\chi(\bar{\theta})}{1 - f(\xi) + f(\xi)\chi(\bar{\theta})[\mu_0/\mu_e(0)]} \quad (7.24)$$

$$\mu_e(\xi) = \chi(\bar{\theta})\left[\mu + \sqrt{1 - \mu^2}\,\tan(\bar{\theta})\right.$$

$$\left. \times \frac{\cos(\xi)E_2(\theta_0) + \sin^2(\xi/2)E_2(\theta)}{2 - E_1(\theta_0) - \xi/\pi E_1(\theta)}\right] \quad (7.25)$$

$$\mu_{0e}(\xi) = \chi(\bar{\theta})\left[\mu_0 + \sqrt{1 - \mu_0^2}\,\tan(\bar{\theta})\frac{E_2(\theta_0) - \sin^2(\xi/2)E_2(\theta)}{2 - E_1(\theta_0) - \xi/\pi E_1(\theta)}\right] \quad (7.26)$$

where

$$\chi(\bar{\theta}) = \frac{1}{\sqrt{1 + \pi \tan^2(\bar{\theta})}} \tag{7.27}$$

$$E_1(x) = \exp\left(-\frac{2}{\pi}\cot(\bar{\theta})\cot(x)\right) \tag{7.28}$$

$$E_2(x) = \exp\left(-\frac{1}{\pi}\cot^2(\bar{\theta})\cot^2(x)\right) \tag{7.29}$$

$$f(\xi) = \exp(-2\tan(\xi/2)) \tag{7.30}$$

and ξ is the phase angle.

The mean slope angle $\bar{\theta}$ is defined by

$$\tan(\bar{\theta}) = \frac{2}{\pi}\int_0^{\pi/2}\alpha(\zeta)\tan(\zeta)\,d\zeta \tag{7.31}$$

where $a(\zeta)$ is the normalized slope distribution function.

Note that the cosine values of the angles have been defined as positive in the expressions above. This shadowing model has been incorporated into the Hapke soil reflectance model described in Section 4.2.1.2 for various applications.

Other shadowing functions (e.g., Smith 1967, Despan et al. 1999) for rough random surfaces are based on similar principles.

Extensive comparisons of different models are not available in the literature. To evaluate the analytic functions, Liang et al. (1997) compared both Hapke and Smith functions (Smith 1967) and the IPW approach (Section 7.3) with two computer simulation models: the ray tracing Monte Carlo simulation model (Lewis and Muller 1992, Burgess and Pairman 1997, Lewis 1999), and the radiosity simulation model (Goel et al. 1991). Comparisons were made on both simulated Gaussian topographic fields and a real DEM (see Fig. 7.5).

The Monte Carlo ray tracer (ARARAT) (Lewis and Muller 1992, Lewis 1999) was used to simulate the directional reflectance of the height fields. The reflectance of the central quarter of each DEM was calculated with an orthographic projection, giving directional reflectance. This allows for any "edge" effects in the spatial filtering applied to the DEMs, and also allows for scattering outside the field of view of the sensor to be taken into account. The area of the DEM that was viewed was kept constant by increasing the aspect ratio of the orthographic camera by $1/\cos(\theta_v)$, where θ_v is the viewing zenith angle.

The numerical ray tracing simulations were performed with a directional component and an isotropic component of illumination. This provides infor-

Figure 7.5 Shaded DEM data.

mation on the bidirectional reflectance of the surface (from the former) and the directional hemispherical reflectance (from the latter). In addition, the reflectance field can be broken down into components as a function of scattering order. This allows for an analysis of the single scattered and multiple scattered components separately. Information is also available on the proportion of sunlit and shadowed surfaces viewed under a particular illumination and viewing geometry, which can also aid in understanding the results. The simulations were performed for each of these DEMs, with 50,000 primary rays per pixel.

In radiosity simulation, the central point of each pixel is connected to generate a network of triangle facets. The viewing factors are calculated based on the slope and azimuth angle of each facet. The radiosity equations are solved in an iterative way. Different rendering techniques are used to determine the reflectance at specific viewing directions. The basic principles and applications to vegetation canopy have been discussed in Section 3.5.2.

From a series of data analyses and from other comparisons, we can conclude from that study that

1. All models indicate that topography has a significant effect on scaled bidirectional reflectance at the large solar zenith angles over rough surfaces.
2. Different models produce different predictions. Further developments are highly desirable for accurate radiation scaling.

3. Two analytic models predict the trend very well even for non-Gaussian surfaces. The prediction accuracies are very close. The Smith model fails for very rough surfaces at the large solar zenith angles.

7.5 DEM DATA AND GENERATION

The effectiveness of topographic correction depends largely on the quality of the *digital elevation model* (DEM), a term that is used generically here, and is also used to refer to digital terrain models or digital surface models. A DEM is a digital file consisting of terrain elevations for ground positions at regularly spaced horizontal intervals. The U.S. Geological Survey (USGS) produces five different digital elevation products. Although all are identical with respect to the manner in which the data are structured, each varies in sampling interval, geographic reference system, areas of coverage, and accuracy; the primary differing characteristic is the spacing, or sampling interval, of the data. These five DEM products are listed in Table 7.1. Note that 1 arc second is approximately 30 meters (m).

Over the years, the USGS has collected digital elevation data using a number of production strategies including manual profiling from photogrammetric stereomodels; stereomodel digitizing of contours, digitizing topographic map contour plates, converting hypsographic and hydrographic tagged vector files, and performing autocorrelation via automated photogrammetric systems (Osborn et al. 2001). Of these techniques, the derivation of DEMs from vector hypsographic and hydrographic data produces the most accurate data and is the preferred method.

The highest-resolution global DEM dataset available to the civil community is GTOPO30, with a horizontal resolution of 30 arc seconds (approximately 1 km) of latitude and longitude and variable accuracy of as poor as 2 km horizontal and ± 650 m vertical. The U.S. NIMA (National Imagery and Mapping Agency) dataset with coverage outside the United States is available only to the U.S. military community, and provides only partial coverage of the land surface.

Stereographic processing of visible imagery from several satellites (e.g., SPOT and ASTER) can produce DEMs with 20 m horizontal (x, y) resolution. Absolute root mean square (RMS) vertical (z) accuracies for SPOT

TABLE 7.1   USGS DEM Data Products

DEM Products	Data Spacing
7.5—minute (minute of arc)	30×30 meters
1—degree	3×3 arc seconds
2—arc second	2×2 arc seconds
15—minute (Alaska)	2×3 arc seconds
7.5—minute (Alaska)	1×2 arc seconds

DEMs are typically in the order of 8–15 m (Davis et al. 2001). DEMs available directly from SPOT Image, Inc. have a 20-m (x, y) resolution and a 7–11-m relative RMS z accuracy. The ASTER team (Welch et al. 1998) is producing DEMs, each covering 60×60 km on the ground, at 15-m resolution, accurate to within ± 7 to ± 10 m [root mean square error (RMSE)]. ASTER DEMs are expected to meet map accuracy standard for scales from $1:50,000$ to $1:250,000$. A lengthy review on generating DEM from various spaceborne visible and near-IR data is given by Toutin (2001).

It is anticipated that the commercial satellite programs incorporating solid-state sensor systems designed for mapping applications will produce image data with approximately 1 m resolution suitable for at large scale mapping (Fritz 1996).

An application of synthetic aperture radar (SAR) interferometry is to generate Z coordinates of the DEM with high accuracy ($< \pm 10$ m) and to detect small changes in elevation (Gens and Van Genderen 1996). Once an interferogram is unwrapped, the altitude of each point on the interferogram is calculated, owing to the two satellite positions. Digital elevation maps can be generated only after calibration procedures that involve precise estimation of the baseline and least squares fitting to ground control points. Even with this procedure, areas can remain within the coverage of an image that has unknown height, as a result of terrain distortion effects produced by steep slopes (in mountainous areas). LightSAR was a proposed lightweight synthetic aperture imaging radar satellite that will use advanced technologies for research, land management, and emergency response applications. LightSAR can provide all-weather, day–night, multiband, dual-polarization images of most of Earth. The proposed interferometric configuration would allow development of high-resolution digital elevation maps. Unfortunately, this program was later cancelled.

ICESat (ice, cloud, and land elevation satellite) is the benchmark EOS mission to achieve the EOS requirements for measuring the ice sheet mass balance, cloud and aerosol heights, optical densities, vegetation, and land topography. Some DEM over ice-covered surfaces will be generated. The progress of the ICESat project can be obtained from their Website at `http://icesat.gsfc.nasa.gov/index.html`.

Since the mid-1990s, airborne scanning or topographic LiDAR has provided additional capabilities of DEM generation. LiDAR compares favorably with several competing and complementary technologies primarily because of its accuracy, active sensor capability, and ability to penetrate through foliage (Fowler 2001).

7.6 SUMMARY

Topographic correction is an area that has not been well explored in land surface remote sensing. One reason may be associated with inaccurate digital elevation models (DEMs). Most DEMs were generated from airborne photogrammetry or digitized from existing topographic maps, from which it is very hard to distinguish tree height and topography beneath the trees. With the airborne or spaceborne LiDAR missions, high-quality topography data might be available in the near future. Clearly, this area deserves more research.

In Sections 7.2 and 7.3, simple correction methods were presented to correct high-resolution remote sensing imagery. These methods have been widely used in practical applications since they are simple and easy to implement over high-resolution imagery, such as TM. In Section 7.4, a typical shadowing function was presented for correcting coarse resolution imagery. These formulas can be used to account for the effects of soil surface roughness on directional reflectance, which was discussed in Chapter 4. This topic still requires extensive research. The global DEM data were reviewed briefly in the last section.

REFERENCES

Burgess, D. W. and Pairman, D. (1997), Bidirectional reflectance effects in NOAA AVHRR data, *Int. J. Remote Sens* 18: 2815–2825.

Colby, J. D. (1991), Topographic normalization in rugged terrain, *Photogramm. Eng. and Remote Sens.*, **57**: 531–537.

Despan, D., Bedidi, A., and Cervelle, B. (1999), Bidirectional reflectance of rough bare soil surfaces, *Geophys. Res. Lett.* **26**: 2777–2780.

Dozier, J. (1989), Spectral signature of alpine snow cover from the Landsat Thematic Mapper, *Remote Sens. Envir.* **28**: 9–22.

Dozier, J. and Frew, J. (1981), Atmospheric corrections to satellite radiometric data over rugged terrain, *Remote Sens. Envir.* **11**: 191–205.

Dozier, J. and Frew, J. (1990), Rapid calculation of terrain parameters for radiation modeling from digital elevation data, *IEEE Trans. Geosci. Remote Sens.* **28**: 963–969.

Dubayah, R. (1992), Estimating net solar radiation using Landast Thematic Mapper and digital elevation data, *Water Resour. Res.* **28**: 2469–2484.

Dubayah, R. (1995), Topographic solar radiation models for GIS, *Int. J. Geogr. Inform. Syst.* **9**: 405–419.

Dubayah, R. and Loechel, S. (1997), Modeling topographic solar radiation using GOES data, *J. Appl. Meteor.* **36**: 141–154.

Faraklioti, M. and Petrou, M. (2001), Illumination invariant unmixing of sets of mixed pixels, *IEEE Trans. Geosci. Remote Sens.* **39**: 2227–2234.

Fowler, R. (2001), Topographic lidar, in *Digital Elevation Model Techniques and Applications: The DEM Users Manual*, D. F. Maune, ed., The American Society for Photogrammetry and Remote Sensing, pp. 207–236.

Fritz, L. W. (1996), The era of commercial earth observation satellites, *Photogramm. Eng. Remote Sens.* **62**: 39–45.

Gens, R. and Van Genderen, J. L. (1996), Review article: SAR interferometry issues, techniques, applications, *Int. J. Remote Sens.* **17**: 1803–1835.

Goel, N. S., Rozehnal, I., and Thompson, R. I. (1991), A computer graphics based model for scattering from objects of arbitrary shapes in the optical region, *Remote Sens. Envir.* **36**: 73–104.

Gu, D. G., Gillespie, A. R., Adams, J. B., and Weeks, R. (1999), A statistical approach for topographic correction of satellite images by using spatial context information, *IEEE Trans. Geosci. Remote Sens.* **37**: 236–246.

Hapke, B. (1984), Bidirectional reflectance spectroscopy. 3. Correction for macro-scopic roughness, *Icarus* **59**: 41–59.

Holben, B. N. and Justice, C. O. (1980), The topographic effect on spectral response from nadir-pointing sensors, *Photogramm. Eng. Remote Sens.* **46**: 1191.

Kondratyev, K. Y. (1977), *Radiation Regime on Inclined Surfaces*, World Meteorologi-cal Organization, Note Technique 152, MF 79 N11613.

Lewis, P. (1999), Three-dimensional plant modelling for remote sensing simulation studies using the botanical plant modelling system, *Agronomie* **19**: 185–210.

Lewis, P. and Muller, J. P. (1992), The advanced radiometric ray tracer: APARAT for plant canopy reflectance simulation, *Proc. XVII Congress ISPRS*, pp. 26–34.

Liang, S. and Lewis, P. (1996), A parametric radiative transfer model for sky radiance distribution, *J. Quant. Spectrosc. Radiat. Transfer* **55**: 181–189.

Liang, S., Lewis, P., Dubayah, R., Qin, W., and Shirey, D. (1997), Topographic effects on surface bidirectional reflectance scaling, *J. Remote Sens.* **1**: 82–93.

Osborn, K., List, J., Gesch, D., Crowe, J., Merrill, G., Constance, E., Mauck, J., Lund, C., Caruso, V., and Kosovich, J. (2001), National Digital Elevation Program (NDEP), *Digital Elevation Model Techniques and Applications: The DEM Users Manual*, D. F. Maune, ed., The American Society for Photogrammetry and Remote Sensing, pp. 83–120.

Proy, C., Tanre, D., and Deschamps, P. Y. (1989), Evaluation of topographic effects in remotely sensed data, *Remote Sens. Envir.* **30**: 21–32.

Sancer, M. I., (1969), Shadow-corrected electromagnetic scattering from a randomly rough surface. *IEEE Transactions on Antennas and Propagation*, **17**: 577–585.

Sandmeier, S. and Itten, K. I. (1997), A physically-based model to correct atmospheric and illumination effects in optical satellite data of rugged terrain, *IEEE Trans. Geosci. Remote Sens.* **35**: 708–717.

Smith, B. G. (1967), Lunar surface roughness: Shadowing and thermal emission, *J. Geophys. Res.* **72**: 4059–4067.

Teillet, P. M., Guindon, B., and Goodenough, D. G. (1982), On the slope-aspect correction of multispectral scanner data, *Can. J. Remote Sens.* **8**: 84–106.

Temps, R. C. and Coulson, K. L. (1977), Solar radiation incident upon slopes of different orientations, *Solar Energy* **19**: 179–184.

Toutin, T. (2001), Elevation modeling from satellite visible and infrared (VIR) data, *Int. J. Remote Sens.* **22**: 1097–1125.

Welch, R., Jordan, T., Lang, H., and Murakami, H. (1998), ASTER as a source for topographic data in the late 1990's, *IEEE Trans. Geosci. Remote Sens.* **36**: 1282–1289.

8

Estimation of Land Surface Biophysical Variables

Many biophysical variables such as leaf area index (LAI), canopy roughness, fractional photosynthetically active radiation (FPAR), and phenology (plant life cycle) are critical inputs to many climate and ecological models (Sellers et al. 1986, Running and Coughlan 1988, Bonan 1993, Prince and Goward 1995). The stability, repeat measurement capability, and global coverage of remote sensing techniques has led to widespread use of its measurements to obtain these variables in studies of land surface and atmospheric processes. Hyperspectral remote sensing provides us with a powerful means to monitor biochemical variables of vegetation canopies. All algorithms for estimating these biophysical and biochemical parameters can be classified into three groups: statistical methods, physical methods, and hybrid methods. *Statistical methods* (Section 8.1) are based mainly on a variety of vegetation indices. *Physical algorithms* (Sections 8.2–8.4) rely on inverting canopy reflectance models. A new trend is to combine statistical and physical methods, which is referred to as a *hybrid algorithm* (Section 8.5) in this book.

In Section 8.1, I present various vegetation indices and their functional relationships with biophysical/biochemical variables. Although most indices are based on two or three bands, they are grouped into two subsections: multispectral (Section 8.1.1) and hyperspectral (Section 8.1.2). Hyperspectral vegetation indices use much narrower bands and are used primarily for detecting biochemical properties of vegetation. The spatial signatures are explored in Section 8.1.3 in terms of a variogram. An operational procedure for estimating LAI and FPAR from AVHRR data is introduced in Section 8.1.4.

Sections 8.2–8.4 concern physical inversion techniques. Section 8.2 is devoted to traditional optimization methods in conjunction with a canopy radiative transfer model. Optimization algorithms consist of derivative and

Quantitative Remote Sensing of Land Surfaces. By Shunlin Liang
ISBN 0-471-28166-2 Copyright © 2004 John Wiley & Sons, Inc.

nonderivative algorithms. This section introduces the basic principles of the nonderivative algorithms, and derivative algorithms are introduced in Section 11.3. Section 8.3 briefly discusses the basic concepts and applications of the generic algorithm. It is an alternative approach for determining global minimization. Section 8.4 presents the lookup table method, which has been used for retrieving biophysical variables from remote sensing data operationally. Although it has limitations, it is simple and easy to implement.

Section 8.5 introduces hybrid algorithms that combine statistical and physical methods. The physical part is associated with extensive simulations using canopy reflectance models, and the statistical part relies on nonparametric regression techniques for linking vegetation variables with remote sensing data. Three nonparametric regression methods are briefly introduced: artificial neural network, projection pursuit, and regression tree. Section 8.6 provides some general comments on comparing different methods. Their advantages and limitations are briefly evaluated.

8.1 STATISTICAL METHODS

In this section, we explore both spectral and spatial signatures for estimating land surface variables statistically. For easy reference, we discuss multispectral and hyperspectral methods separately, in Sections 8.1.1 and 8.1.2, respectively. From a mathematical perspective, these methods are not significantly different. Section 8.1.3 discusses the spatial methods.

8.1.1 Multispectral Vegetation Indices

Different vegetation indices (VIs) from multispectral remotely sensed data have been proposed and widely used in various applications. The most basic assumption is that some algebraic combination of remotely sensed spectral bands can tell us something useful about vegetation structure and the state of the vegetative cover, such as leaf density and distribution, leaf water content, age, mineral deficiencies, and parasitic attacks. A good vegetation index should be very sensitive to these factors. On the other hand, other factors also affect the spectral reflectance, such as soil properties, solar illumination and other atmospheric conditions, and sensor viewing geometry. A good vegetation index should be very insensitive to these factors. Therefore, to find the ideal vegetation index, one must determine those factors that affect vegetation indices on the basis of a physical understanding of the interactions between electromagnetic radiation, the atmosphere, the vegetative cover, and the soil background (Bannari et al. 1995). Physical modeling of the atmosphere and surface radiation regimes has been discussed in Chapters 2–4 in detail.

Vegetation indices have been used to predict various land surface biophysical variables (y). The typical functions are of the following formats:

$$y = \sum_{i=0}^{n} a_i VI^i \tag{8.1}$$

or

$$y = a + bVI^c \tag{8.2}$$

or

$$y = a \ln(b - VI) + c \tag{8.3}$$

Following is a list of representative vegetation indices. The design of a good vegetation index can be improved by examining the geometric relations in the feature space. Details of this concept can be found in the literature (e.g., Liang et al. 1989, Verstraete and Pinty 1998). Note that vegetation indices can be calculated based on digital numbers (DN) values originally from

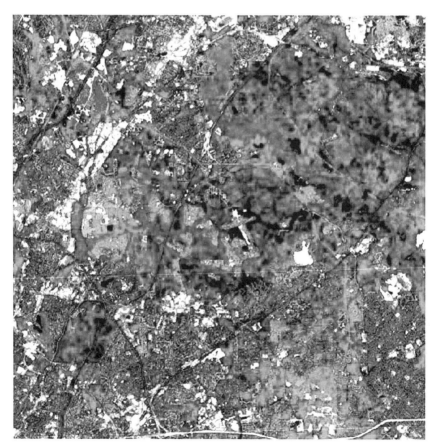

Figure 8.1 ETM + color composite imagery acquired on August 2, 2001 over Beltsville, MD (USA).

Reflectance, Aug 2, 2001

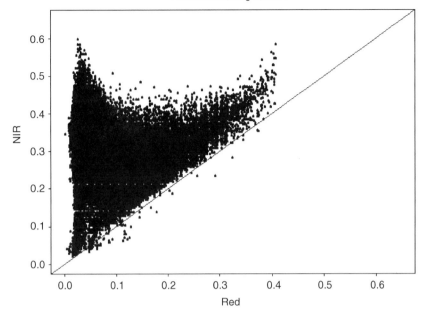

Figure 8.2 Scatterplot of the red/near-IR spectral space from Fig. 8.1.

remote sensors, TOA radiance/reflectance, or surface radiance/reflectance. Ideally, surface reflectance is the best variable for calculating vegetation indices if atmospheric correction can be performed over the imagery. However, in reality, most people calculate VI based on TOA radiance and/or reflectance. In the following text, ρ_n, ρ_r, ρ_b represent the reflectance of near-IR, red, and blue bands, respectively.

8.1.1.1 Soil Line Concept

Soil reflectance depends on many different factors, as discussed in Chapter 4. If we place all soil pixels in the red and near-IR spectral space, however, they usually are distributed alone a line. This implies that soil reflectance in red and that in near-IR are highly correlated with a positive correlation coefficient. Over a vegetated surface, this line, called *soil-line*, often constitutes the base of a triangle shape. Fig. 8.1 is a Landsat ETM + color composite image over Beltsville, MD (USA). It was taken during the growing season, and the scene consists of several different cover types, including green vegetation canopies and soils. Fig. 8.2 is its scatterplot in the red and near-IR spectral space. Those pixels with higher near-IR reflectance and lower red reflectance are dense vegetation canopies; the base of the "triangle" shape is the soil line.

This soil line is characterized by a linear equation, $\rho_n = b + \gamma \cdot \rho_r$, where b and γ are the intercept and slope of this soil line. In Fig. 8.2, we have

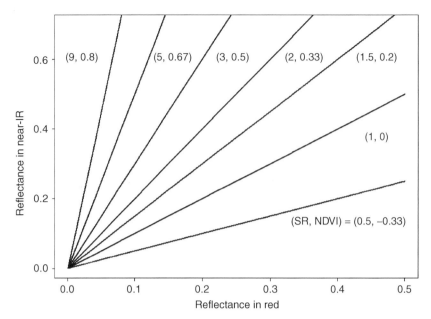

Figure 8.3 NDVI and SR isolines. The first bracketed number represents a SR value; the second, a NDVI value.

$b = 0.044$ and $\gamma = 1.05$. Baret et al. (1993) derived similar coefficients ($b = 0.037$ and $\gamma = 1.176$). Many vegetation indices have been designed on the basis of this concept of the soil line, as discussed below.

8.1.1.2   Normalized Difference Vegetation Index (NDVI)

$$\text{NDVI} = \frac{\rho_n - \rho_r}{\rho_n + \rho_r} \qquad (8.4)$$

This is one of the earliest vegetation indices that have been widely used in various applications. NDVI responds to changes in amount of green biomass, chlorophyll content, and canopy water stress. The isoline of NDVI is illustrated in Fig. 8.3. All points along each isoline that might be caused due to variable soil background or atmospheric conditions having the same NDVI value.

It is simple and easy to implement, and can be effective in predicting surface properties when vegetation canopy is not too dense or too sparse. If a canopy is too sparse, background signal (e.g., soil) can change NDVI significantly. If the canopy is too dense, NDVI saturates because red reflectance does not change much but near-IR reflectance still increases when the canopy becomes denser.

In the following, we briefly describe how NDVI can be linked with some of the surface variables. Assuming that the relationship between NDVI and LAI is linear and the maximum NDVI value in a season corresponds to the maximum LAI of vegetation cover (Justice 1986), LAI can be inferred from the normalized NDVI values:

$$LAI_i = LAI_{max} \frac{NDVI_i - NDVI_{min}}{NDVI_{max} + NDVI_{min}} \qquad (8.5)$$

where max, min and i are the maximum, minimum, and period values observed, respectively. The relationship is not always linear since the vegetation indices approach a saturation level asymptotically for LAI ranging from 2 to 6, depending on the type of vegetation cover, and environmental conditions (e.g., Carlson and Ripley 1998). However, by assuming a nonlinear relationship, the LAI estimates from NDVI are then highly dependent on certain factors such as canopy geometry, leaf and soil optical properties, Sun position, and cloud coverage. The variation of NDVI as a function of LAI can be expressed by a modified Beer's law (Baret and Guyot 1991):

$$NDVI = NDVI_\infty + (NDVI_s - NDVI_\infty) \exp(-K_{NDVI} \cdot LAI) \qquad (8.6)$$

where $NDVI_s$ is the vegetation index corresponding to that of the bare soil, which is the asymptotic value when LAI approaches zero, $NDVI_\infty$ is the asymptotic value of NDVI when LAI tends toward infinity, and K_{ndvi} is the coefficient that controls the slope of the relationship (extinction coefficient). By determining $NDVI_s$, $NDVI_\infty$, and K_{ndvi}, we can estimate LAI from NDVI via Eq. (8.6) easily.

In addition, the green vegetation fraction F_g has been derived from AVHRR NDVI using a simple linear relationship with an assumption of dense vegetation (high leaf area index) (Gutman and Ignatov 1998)

$$F_g = \frac{NDVI_i - NDVI_{min}}{NDVI_{max} - NDVI_{min}} \qquad (8.7)$$

where $NDVI_{min} = 0.04$ and $NDVI_{max} = 0.52$ were prescribed as the global constants (Gutman and Ignatov 1998) as a first approximation. The values of F_g should be restricted to range between 0 and 1. Note that if NDVI is calculated using surface reflectance, the maximum NDVI value could be much larger.

Absorbed photosynthetically active radiation (APAR) is related to LAI using a nonlinear function (Asrar et al. 1984):

$$APAR = 93.5[1 - \exp(-0.9LAI)] \qquad (8.8)$$

If LAI is estimated using either VI or physically based inversion method, we can estimate APAR using Eq. (8.8). We can also estimate LAI given that the APAR can be determined through other approaches. Note that these two constants in Eq. (8.8) were derived from a specific condition.

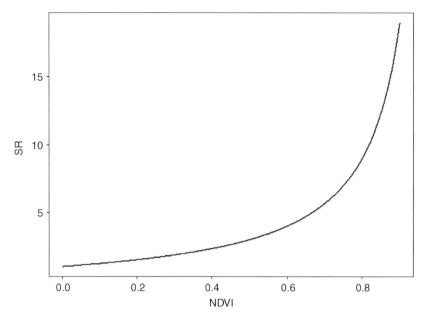

Figure 8.4 Relationship between NDVI and SR.

8.1.1.3 Simple Ratio (SR)

Simple ratio (SR) is defined as

$$\text{SR} = \frac{\rho_n}{\rho_r} \tag{8.9}$$

The isoline of SR is also shown in Fig. 8.3 with NDVI. In fact, they are related functionally:

$$\text{NDVI} = \frac{\text{SR} - 1}{\text{SR} + 1} = 1 - \frac{2}{\text{SR} + 1} \tag{8.10}$$

Their relationship is visually displayed in Fig. 8.4. As we mentioned, NDVI may approach saturation level when LAI is very large. From this figure, we can see that when NDVI is large, a small variation of NDVI results in large change of SR value, which implies that SR is much slower to saturate than NDVI when a canopy is very dense. SR can enhance the contrast between soil and vegetation while minimizing the effects of the illumination conditions (Baret and Guyot 1991), but their effectiveness is reduced by the variable soil reflectance underneath the canopy. Some new indices were then proposed to eliminate the soil effects. For example, Chen (1996) proposed a new index that is related to SR in a similar way:

$$\text{MSR} = \frac{\rho_n/\rho_r - 1}{\sqrt{\rho_n/\rho_r} + 1} = \frac{\text{SR} - 1}{\sqrt{\text{SR}} + 1} \tag{8.11}$$

where MSR = modified simple ratio. The purpose was to derive an index that can be linearly related to LAI and FPAR. SR has been also used for estimating crop LAI (Gardner and Blad 1986, Wiegand et al. 1992).

8.1.1.4 Soil-Adjusted Vegetation Index (SAVI)

Huete (1988) suggests soil-adjusted vegetation index (SAVI) based on NDVI and various observed data to eliminate the effects from background soils:

$$\text{SAVI} = \frac{(\rho_n - \rho_r)(1+L)}{\rho_n + \rho_r + L} \tag{8.12}$$

where L is the coefficient that should vary with vegetation density, ranging from 0 for very high vegetation cover to 1 for very low vegetation cover. It is obvious that if $L = 0$, then SAVI is equivalent to NDVI. Since we seldom know the vegetation density, it is difficult for us to optimize this index. In the following, we would like to introduce some variants of SAVI.

Qi et al. (1994) proposed an empirical way to determine the coefficient L using both NDVI and WDVI:

$$L = 1 - 2a\text{NDVI} \times \text{WDVI} \tag{8.13}$$

where a is a coefficient [$a = 1.6$ was used by Qi et al.] and the weighted difference vegetation index (WDVI) is defined as (Clevers 1989):

$$\text{WDVI} = \rho_n - \gamma\rho_r \tag{8.14}$$

where γ is the slope of the soil line. Thus, we do not need to know the vegetation density a priori. However, they found that SAVI with the empirically determined parameter L still cannot remove the soil effects. They further suggested a self-adjusting L and resulted in a new index, called the modified SAVI (MSAVI):

$$\text{MSAVI} = \rho_n + 0.5 - \sqrt{(\rho_n + 0.5)^2 - 2(\rho_n - \rho_r)} \tag{8.15}$$

The signal-to-noise ratio (SNR) was higher for the MSAVI than that of other vegetation indices (e.g., NDVI and SAVI). It not only increases the vegetation dynamic responses, but also further reduces the soil background variations.

The isoline of WDVI is shown in Fig. 8.5. WDVI as an index has been linked with other land surface variables. Clevers (1989) linked WDVI with LAI and FPAR in the following expressions:

$$\text{LAI} = -\frac{1}{a}\ln\left(1 - \frac{\text{WDVI}}{\text{WDVI}_\infty}\right) \tag{8.16}$$

$$\text{FPAR} = \left[1 - \exp(-K_{\text{par}}\text{LAI})\right]\text{FPAR}_\infty \tag{8.17}$$

where a is a parameter, K_{par} is the extinction coefficient for PAR, and

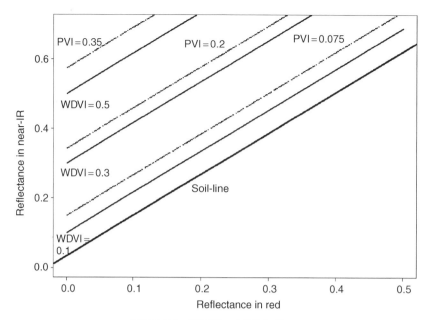

Figure 8.5   WDVI and PVI isolines.

$WDVI_\infty$ and $FPAR_\infty$ are the asymptotically limiting values [$FPAR_\infty = 0.94$ has been suggested by Baret and Guyot (1991)].

Major et al. (1990) used a simple canopy reflectance to show that canopy near-IR reflectance can be expressed as a linear function of canopy red reflectance. They suggested a second version of the SAVI (SAVI2) that models the vegetation isoline behavior by using the ratio b/γ as the soil adjustment factor:

$$SAVI2 = \frac{\rho_n}{\rho_r + b/\gamma} \tag{8.18}$$

where γ and b respectively are the slope and intercept of the soil line. Its isoline is given in Fig. 8.6. If the soil line penetrates through the origin in the red/near-IR spectral space (i.e., $b = 0$), SAVI2 becomes SR. In a comparison study, Broge and Leblanc (2000) found that this index is least affected by background reflectance for estimating both LAI and canopy chlorophyll density, and is the best predictor of LAI among a dozen vegetation indices.

Baret and Guyot (1991) developed a transformed SAVI (TSAVI) also using the soil line concept:

$$TSAVI = \frac{\gamma(\rho_n - \gamma\rho_r - b)}{\gamma\rho_n + \rho_r + \gamma b + X(1 + \gamma^2)} \tag{8.19}$$

where X is the adjustment factor to minimize soil noise (they used 0.08). It is a modification of the SAVI to compensate for soil variability due to changes in solar zenith angle, LAI, and leaf angle distribution (LAD).

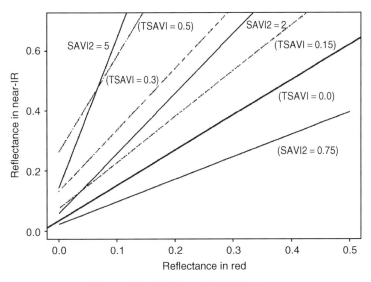

Figure 8.6   SAVI2 and TSAVI isolines.

In fact, TSAVI was built on an earlier index, called the *perpendicular vegetation index* (PVI).

$$PVI = \frac{\rho_n - \gamma\rho_r - b}{\sqrt{\gamma^2 + 1}} \tag{8.20}$$

It is more effective to eliminate differences in soil background when LAI is low (in arid and semiarid environments). PVI has a behavior similar to that of WDVI defined earlier since $PVI = (WDVI - b)/\sqrt{\gamma^2 + 1}$. If the soil line coincides with the 1:1 line in the red/near-IR spectral space (i.e., $b = 0$ and $\gamma = 1$), PVI and WDVI are identical.

It is interesting to interpret some of the indices graphically (see Fig. 8.7). PVI measures the distance of one pixel to the soil line, but several other indices (SR, NDVI, and TSAVI) measure the angle of the vegetation pixel in the red/near-IR space in reference to the soil line.

8.1.1.5   Global Environment Monitoring Index (GEMI)

Pinty and Verstraete (1992) proposed an index, called the *global environment monitoring index* (GEMI):

$$GEMI = \frac{\eta(1 - 0.25\eta) - (\rho_r - 0.125)}{1 - \rho_r} \tag{8.21}$$

where

$$\eta = \frac{2\left(\rho_n^2 - \rho_r^2\right) + 1.5\rho_n + 0.5\rho_r}{\rho_n + \rho_r + 0.5} \tag{8.22}$$

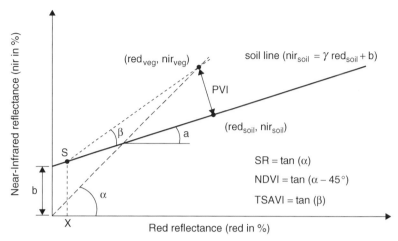

Figure 8.7   Graphic interpretation of several vegetation indices.

The purpose is to eliminate soil and, in particular, atmospheric effects. However, Qi et al. (1994) demonstrated that soil noise caused GEMI to break down at low vegetation covers, and that all the vegetation indices designed to minimize the effect of the atmosphere have increased sensitivity to the soil, which make these indices completely unsuitable for arid regions.

8.1.1.6 *Soil and Atmospherically Resistant Vegetation Index (SARVI)*

Kaufman and Tanré (1992) developed an index called the *soil and atmospherically resistant vegetation index* (SARVI) to minimize both soil- and atmosphere-induced variations in the VI:

$$\text{SARVI} = \frac{(\rho_n - \rho_{rb})(1 + L)}{\rho_n + \rho_{rb} + L} \qquad (8.23)$$

where

$$\rho_{rb} = \rho_r - \beta(\rho_b - \rho_r) \qquad (8.24)$$

The formula looks similar to SAVI, but red reflectance is replaced by the linear combination of both red and blue reflectance. The purpose is to minimize both atmosphere and canopy background. β is used to stabilize the index to variations in aerosol content. After extensive simulations, they found $\beta = 1$ to be the optimal value under most conditions except the aerosol type in Sahel dust, where $\beta = 0.5$. Note that all reflectances are assumed to be corrected for Rayleigh scattering and ozone absorption.

Huete et al. (1997) further developed a new index to eliminate both atmosphere and canopy background variations:.

$$\text{SARVI2} = \frac{2.5(\rho_n - \rho_r)}{1 + \rho_n + 6\rho_r - 7.5/\rho_b} \qquad (8.25)$$

They found that NDVI is more sensitive to FPAR and SARVI is more sensitive to structural canopy parameters such as LAI and leaf morphology.

SARVI2, which is now called the enhanced vegetation index (EVI), has been used for producing the MODIS product operationally (Huete et al. 2002).

8.1.1.7 *Aerosol-Free Vegetation Index (AFVI)*

Karnieli et al. (2001) proposed a new index similar to NDVI but the red band is replaced by the mid-infrared band centered at either 2.1 μm or 1.6 μm:

$$\text{AFVI}_{2.1} = \frac{\rho_{\text{NIR}} - 0.5\rho_{2.1}}{\rho_{\text{NIR}} + 0.5\rho_{2.1}} \tag{8.26}$$

$$\text{AFVI}_{1.6} = \frac{\rho_{\text{NIR}} - 0.66\rho_{1.6}}{\rho_{\text{NIR}} + 0.66\rho_{1.6}} \tag{8.27}$$

where ρ_i are the mid-IR TOA reflectance. The rationale is that mid-IR bands are transparent to most aerosols except dust since the aerosol particle sizes are much smaller than the mid-IR wavelength and the aerosol scattering is negligible. On the other hand, surface reflectance of mid-IR bands is highly correlated to visible bands. For example, they found that the following relations of surface reflectances are statistically significant: $\rho_{0.469} = 0.25\rho_{2.1}$, $\rho_{0.555} = 0.33\rho_{2.1}$, $\rho_{0.645} = 0.5\rho_{2.1}$, $\rho_{0.645} = 0.66\rho_{1.6}$. Thus AFVI is almost identical to NDVI when the atmosphere is very clear. However, AFVI is not sensitive to smog, anthropogenic pollution, or volcanic plumes, but NDVI is still affected by these aerosol types. Note that this index was originally called AFRI by its authors.

One example is given here to illustrate that AFVI is insensitive to the effect of smoke. Figure 8.8 shows an ETM + band 1 image over New Mexico

Figure 8.8 ETM + band 1 acquired on May 9, 2000 over New Mexico (USA), indicating a forest fire around the Los Alamos National Laboratory.

NDVI

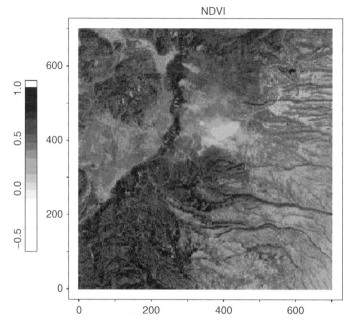

Figure 8.9 NDVI image from the ETM + imagery in Fig. 8.8.

AFVI1

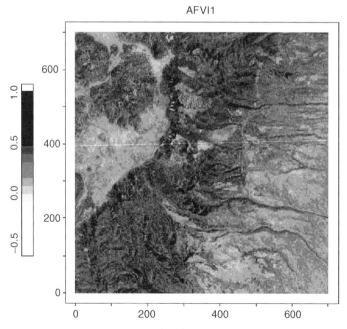

Figure 8.10 AFVI image using Eq. (8.27) from the ETM + imagery in Fig. 8.8.

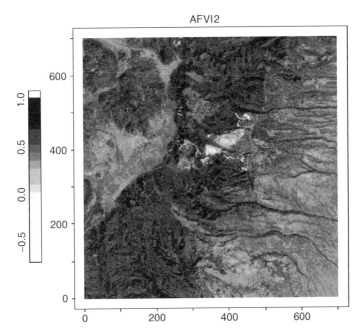

Figure 8.11 AFVI image using Eq. (8.26) from the ETM + imagery in Fig. 8.8.

(USA), where a large forest fire occurred around the Los Alamos National Laboratory. From NDVI imagery (Fig. 8.9), we still can see the impacts of smoke. However, AFVI imagery using both Eqs. (8.26) and (8.27) (Figs. 8.10 and 8.11, respectively) are not sensitive to smoke from forest fires.

8.1.1.8 "Tasseled Cap" Transformation

Most of the indices described above are based on the ratios of two or more bands, which eventually are nonlinear. In fact, many indices based on the linear transformations have been widely used in various applications.

Similar to the principal-component transformation, the "tasseled cap" transformation (Crist and Cicone 1984) converts the multispectral bands into a new feature space that corresponds to physical characterization of ground covers. For TM data of Landsat 4 and 5, the first four components are called brightness, greenness, wetness, and haze. The coefficients of these linear formulas for TOA reflectance are given in Tables 8.1 and 8.2 (Crist et al. 1986).

The first three indices in Tables 8.1 and 8.2 reveal information on vegetation type, stage of development, and condition, and soil type and moisture status. The fourth component is related to the amount of aerosols in the imagery. The readers interested in the derivation details are referred to the original publication (Crist and Cicone 1984). Eventually, these coefficients can be used for Landsat ETM + imagery. Huang et al. (2002) proposed a new set of coefficients for ETM + TOA reflectance, listed in Table 8.3.

TABLE 8.1   Landsat 4 TM "Tasseled Cap" Transformation Coefficients

Feature	Coefficients					
	TM1	TM2	TM3	TM4	TM5	TM7
Brightness	0.3037	0.2793	0.4743	0.5585	0.5082	0.1863
Greenness	−0.2848	−0.2435	−0.5436	0.7243	0.0840	−0.1800
Wetness	0.1509	0.1973	0.3279	0.3406	−0.7112	−0.4572
Haze	0.8832	−0.0819	−0.4580	−0.0032	−0.0563	0.0130

TABLE 8.2   Landsat 5 TM "Tasseled Cap" Transformation Coefficients

Feature	Coefficients						
	TM1	TM2	TM3	TM4	TM5	TM7	Intercept
Brightness	0.2909	0.2493	0.4806	0.5568	0.4438	0.1706	10.3695
Greenness	−0.2728	−0.2174	−0.5508	0.7221	0.0733	−0.1648	−0.7310
Wetness	0.1446	0.1761	0.3322	0.3396	−0.6210	−0.4186	−3.3828
Haze	0.8461	−0.0731	−0.4640	−0.0032	−0.0492	0.0119	0.7879

TABLE 8.3   Landsat 7 ETM + "Tasseled Cap" Transformation Coefficients

Feature	Coefficients					
	ETM + 1	ETM + 2	ETM + 3	ETM + 4	ETM + 5	ETM + 7
Brightness	0.3561	0.3972	0.3904	0.6966	0.2286	0.1596
Greenness	−0.3344	−0.3544	−0.4556	0.6966	−0.0242	−0.2630
Wetness	0.2626	0.2141	0.0926	0.0656	−0.7629	−0.5388
Fourth	0.0805	−0.0498	−0.1950	−0.1327	−0.5752	−0.7775
Fifth	−0.7252	−0.0202	0.6683	0.0631	−0.1494	−0.0274
Sixth	0.4000	−0.8172	0.3832	0.0602	−0.1095	0.0985

"Tasseled Cap" transformation (TCT) has been used extensively. Collins and Woodcock (1996) found that change in the TCT wetness index is the most reliable single indicator of forest change. Dymond et al. (2002) found that phenological change of forest is most accurately captured by combining image differencing and TCT indices. Todd et al. (1998) used the first three TCT indices to estimate aboveground biomass on the shortgrass steppe of eastern Colorado from Landsat TM imagery for two grazing treatments (moderately grazed or ungrazed). Macomber and Woodcock (1994) employed TCT to invert several forest stand parameters, including tree size and canopy cover for each conifer stand, from reflectance values in Landsat TM imagery. The difference in cover estimates between the dates forms the basis for stratifying stands into mortality classes, which were used as both themes in a map and the basis of the field sampling design.

TABLE 8.4 Parameter Values of Eqs. (8.28) and (8.29)

Sensor	a_1	a_2	a_3	a_4	a_5	a_6
MERIS	-37.013	0.25709	4.9387	1.6011	2.5824	-0.17412
GLI	0.28787	0.33326	-0.00739	-0.13661	0.32259	0.098367
VEGETATION	0.37598	0.50132	-0.01091	-0.17150	0.29464	0.11009

Sensor	a_7	a_8	a_9	a_{10}	a_{11}	a_{12}
MERIS	-25.702	1.7533	11.851	9.6473	0.65856	0.29154
GLI	—	—	—	—	—	—
VEGETATION	—	—	—	—	—	—

8.1.1.9 *FPAR Index*

In the previous sections, vegetation indices have been designed to represent the green vegetation amount by eliminating the effects of a variety of factors (e.g., soil, atmosphere, Sun sensor geometry). The same index is usually related to various land surface variables. Gobron et al. (2000) developed an index specifically for FPAR for three different sensors (MERIS, GLI, and VEGETATION) based on radiative transfer simulations. Here we simply call it the FPAR index. The FPAR index is represented by the top of canopy reflectance of both red (ρ_R) and near-IR (ρ_{IR}) bands:

For MERIS:

$$\frac{a_1 \rho_R^2 + a_2 \rho_{IR}^2 + a_3 \rho_R \rho_{IR} + a_4 \rho_R + a_5 \rho_{IR} + a_6}{a_7 \rho_R^2 + a_8 \rho_{IR}^2 + a_9 \rho_R \rho_{IR} + a_{10} \rho_R + a_{11} \rho_{IR} + a_{12}} \qquad (8.28)$$

For VEGETATION/GLI:

$$\frac{a_1 \rho_{IR} - a_2 \rho_R - a_3}{(a_4 - \rho_R)^2 + (a_5 - \rho_{IR})^2 + a_6} \qquad (8.29)$$

where a_i are the coefficients given in Table 8.4, and ρ_R and ρ_{IR} are the canopy reflectance of band red and band near-IR, respectively. Instead of determining the canopy reflectance from atmospheric correction, Gobron et al. (2000) developed simple indices to transform TOA reflectance R_i to canopy reflectance ρ_i:

For MERIS:

$$\rho_{red} = b_1 R_B^2 + b_2 R_R^2 + b_3 R_B R_R + b_4 R_B + b_5 R_R + b_6 \qquad (8.30)$$

$$\rho_{IR} = c_1 R_B^2 + c_2 R_{IR}^2 + c_3 R_B R_{IR} + c_4 R_B + c_5 R_{IR} + c_6 \qquad (8.31)$$

TABLE 8.5   Parameter Values of Eqs. (8.30) and (8.32)

Sensor	b_1	b_2	b_3	b_4	b_5	b_6
MERIS	-9.32524	-3.71573	10.76393	0.35599	0.95988	-0.00333
GLI	-11.120	-0.028923	1.7721	0.098161	11.058	—
VEGETATION	-12.877	-0.019822	1.0180	0.13832	14.593	—

TABLE 8.6   Parameter Values of Eqs. (8.31) and (8.33)

Sensor	c_1	c_2	c_3	c_4	c_5	c_6	c_7	c_8	c_9	c_{10}
MERIS	10.86	0.0524	4.296	-2.219	0.8837	0.0448	—	—	—	
GLI	-1.85	1.107	7.062	-0.6697	23.02	0.4077	2.607	-0.0312	14.90	36.32
VEGETATION	-1.12	2.169	0.2929	4.2614	65.13	-204.3	-0.132	0.0109	23.81	5.593

For VEGETATION/GLI:

$$\rho_{\text{red}} = b_1(R_B + b_2)^2 + b_3(R_R + b_4)^2 + b_5 R_B R_R \qquad (8.32)$$

$$\rho_{\text{IR}} = \frac{c_1(R_B + c_2)^2 + c_3(R_{\text{IR}} + c_4)^2 + c_5 R_B R_{\text{IR}}}{c_6(R_B + c_2)^2 + c_8(R_{\text{IR}} + c_9)^2 + c_{10} R_B R_{\text{IR}}} \qquad (8.33)$$

where b_i and c_i respectively are the coefficients listed in Tables 8.5 and 8.6, and R_B, R_R, and R_{IR} are respectively TOA reflectance of bands blue, red, and near-IR with the angular correction

$$R_i = \frac{r_i}{f(\theta_i, \theta_v, \phi, \Psi_i)} \qquad (8.34)$$

where r_i are TOA BRF, $f(\cdot)$ is the BRDF function defined by (2.17), and the parameter sets $\Psi_i \in (\rho_{0i}, k_i, b_i)$ for these three bands are given in Table 8.7.

TABLE 8.7   Parameter Values of the BRDF Function [Eq. (2.18)]

Sensor	Band	ρ_{0i}	k_i	b_i
MERIS	Blue	0.20	0.69650	-0.05818
	Red	0.20	0.95946	-0.02910
	Near-IR	0.40	0.79069	0.01576
GLI	Blue	-0.13515	0.45696	0.01813
	Red	-0.65625	0.78673	0.12335
	Near-IR	0.63484	0.87758	-0.00264
VEGETATION	Blue	-0.25910	0.46011	0.02979
	Red	-0.53764	0.77449	0.12335
	Near-IR	0.62335	0.88468	-0.00219

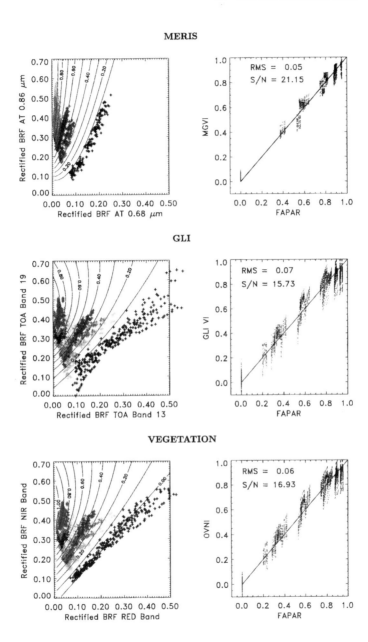

Figure 8.12 Result of the optimization process to derive the vegetation indices for the MERIS (top panels), the GLI (medium panels), and the VEGETATION (bottom panels) sensors. The left panels represent the isolines of vegetation indices in their corresponding spectral spaces. The right panels display the relationships between the index values and the true FAPAR values used in the training dataset. [From Gobron et al. (2000), *IEEE Trans. Geosci. Remote Sens.* Copyright © 2000 with permission of IEEE.]

The overall performance of the proposed method to the design of optimal vegetation indices is illustrated in Fig. 8.12. It shows the clustering of the rectified BRF values for a given biome into a more restricted spectral domain (left panels). This permits us to derive the optimal index formulas so that their isolines can provide a better fit of the same FAPAR values in this rectified space. As a result, the relationships between the three indices and the FAPAR exhibit higher signal-to-noise ratio and lower RMS values than does the NDVI, by a factor of ~ 3.

8.1.1.10 Comparisons and Applications

More comparisons should be made to evaluate which vegetation indices work better under certain conditions, although there have been limited reports in the literature (e.g., Broge and Leblanc 2000). On the basis of the ETM + imagery shown in Fig. 8.1, we calculated the following indices, shown in Fig. 8.13. There are some visual differences. The soil line is represented by this equation ($\rho_n = 0.044 + 1.05\rho_r$). Numerical comparisons are shown in Fig. 8.14. There are large scattering patterns, indicating that these indices are sensitive to different environmental conditions to different degrees.

The linkage of these indices with land surface variables were briefly discussed with respect to the individual index above. We now discuss other applications using these vegetation indices. For corn canopies, Gardner and Blad (1986) developed empirical models based on ground measurements and data collected by airborne sensors characterized by the wavebands similar to those of the thematic mapper (TM):

$$LAI = 0.416 + 0.2553 \, SR \tag{8.35}$$

$$LAI = -1.248 + 5.839 \, NDVI \tag{8.36}$$

$$LAI = -0.0305 + 1.9645 \ln (SR) - 0.1577 \, SR \tag{8.37}$$

R^2 values vary from 0.76 to 0.85 and the standard error, from 0.6 to 0.79.

For wheat canopies, Wiegand et al. (1992) fitted ground biophysical measurements using six vegetation indices with different forms of equations. They found that LAI can be best estimated from SR by linear equations, from NDVI and TSAVI by exponential equations, and from PVI equally well by power and quadratic equation forms. The R-squared values ranged from 0.72 to 0.86, and the root mean square errors (RMSE), from 0.63 to 0.90. Samples obtained from several studies and revealed a good relationship between NDVI and FPAR as shown in Fig. 8.15.

FPAR is sometimes denoted as FAPAR (fraction of absorbed photosynthetically active radiation by green vegetation). Plant growth is directly related to the absorbed PAR that can be calculated by FPAR and incident PAR (see Section 13.2.4.4). Incident PAR can be estimated from remote sensing (Pinker et al. 1995). FPAR can be estimated by vegetation indices or other approaches (see Section 8.4). Physical model simulations have indicated that FPAR–VI relationships are quite linear (Choudhury 1987, Sellers 1987, Roujean and Bréon 1995). Field measurements showed both

NDVI

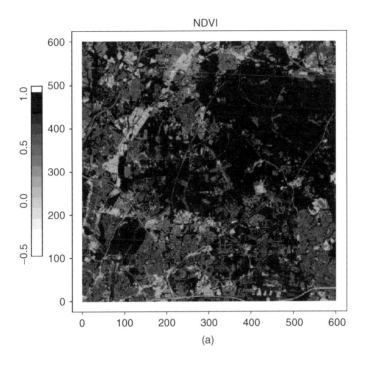

(a)

SAVI2

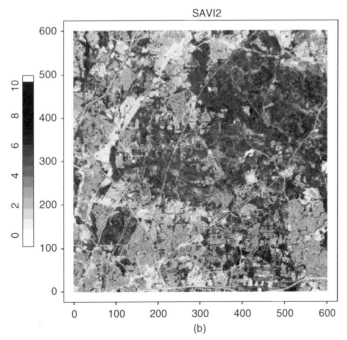

(b)

Figure 8.13 Calculated VI images from the ETM + imagery shown in Fig. 8.1: (a) NDVI; (b) SAVI2; (c) TSAVI; (d) WDVI.

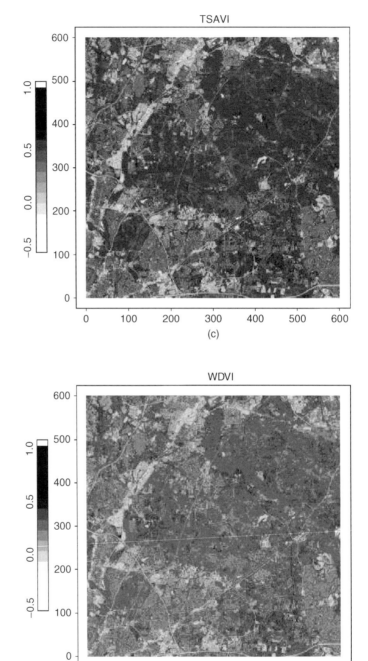

Figure 8.13 (*Continued*) Calculated VI images from the ETM + imagery shown in Fig. 8.1: (c) TSAVI; (d) WDVI.

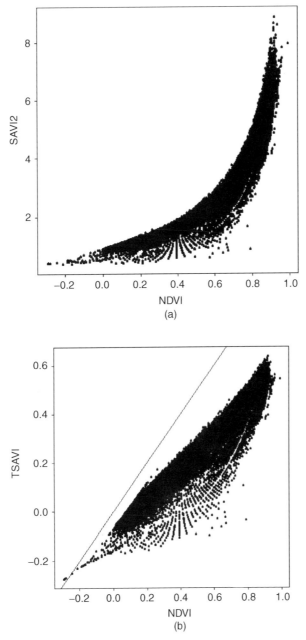

Figure 8.14   Comparisons of NDVI with (a) SAVI2, (b) TSAVI shown in Fig. 8.13.

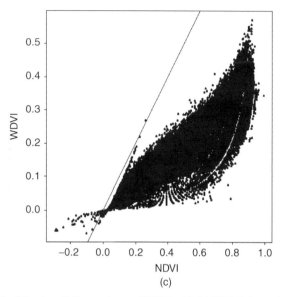

Figure 8.14 (*Continued*) Comparisons of NDVI with (c) WDVI shown in Fig. 8.13.

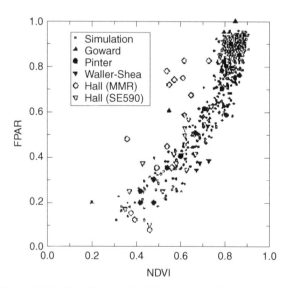

Figure 8.15 Complied relationship between NDVI and FPAR.

linear (Wiegand et al. 1991, Daughtry et al. 1992) and nonlinear relations (Wiegand et al. 1991, 1992, Ridao et al. 1998). Some examples are

$$\text{FPAR} = -0.45 + 1.449 \text{ NDVI}, \qquad R^2 = 0.72, \quad \text{RMSE} = 0.13 \quad (8.38)$$

$$\text{FPAR} = 0.173 \text{ SR}^{0.573}, \qquad R^2 = 0.77, \quad \text{RMSE} = 0.12. \quad (8.39)$$

FPAR can be related to LAI and estimated by different vegetation indices. Wiegand et al. (1992) related FPAR to LAI in the exponential form:

$$\text{FPAR} = 1 - e^{-\text{LAI}}, \qquad R^2 = 0.952, \quad \text{RMSE} = 0.054 \quad (8.40)$$

In mapping regional LAI from AVHRR data for calculating net primary production (NPP) and carbon cycle, Liu et al. (1997) used different equations in conjunction with land cover maps:

For deciduous forest:

$$\text{LAI} = 0.475 \, (\text{SR} - 2.781) \qquad\qquad (8.41)$$

For coniferous forest:

$$\text{LAI} = 1.188 \, (\text{SR} - B_c) \qquad\qquad (8.42)$$

For mixed forest:

$$\text{LAI} = 0.592 \, (\text{SR} - B_m) \qquad\qquad (8.43)$$

For cropland, grassland and others:

$$\text{LAI} = 0.325 \, (\text{SR} - 1.5) \qquad\qquad (8.44)$$

where SR is the simple ratio; B_c and B_m are background SR trajectories for coniferous forest and mixed forest. They are calculated from $B_c = 0.1(1.2 \times 10^{-10} \, D^5 - 1.1 \times 10^{-7} \, D^4 + 4.1 \times 10^{-5} \, D^3 - 6.8 \times 10^{-3} \, D^2 + 5.4 \times 10^{-1} \, D - 15)$, and $B_m = (B_c + 2.781)/2$, where D is the day of year. The error in a single LAI value is estimated to be $\pm 25\%$.

Over Beltsville, MD, we have collected ground LAI data during several Landsat 7 overpasses under clear-sky conditions (Fang and Liang in press). The forest green LAI are calculated according to measurements of both growing season and winter. Four vegetation indices are compared with ground-measured LAI data: NDVI, WDVI, SAVI2, and TSAVI. Three quantities are used to calculate these indices: TOA reflectance, TOA radiance, and surface reflectance. Surface reflectance values are retrieved from ETM + imagery using a physically based atmospheric correction algorithm (Liang et al. 2001) discussed in Chapter 6. All three of these figures are shown in Figs. 8.16–8.18.

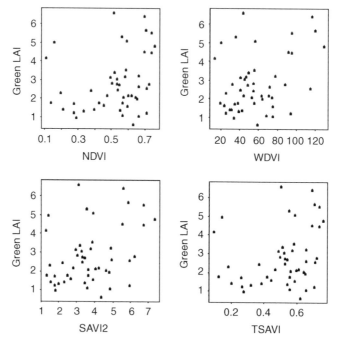

Figure 8.16 Ground measured LAI versus VI calculated from TOA radiance of ETM + imagery over Beltsville, MD (USA).

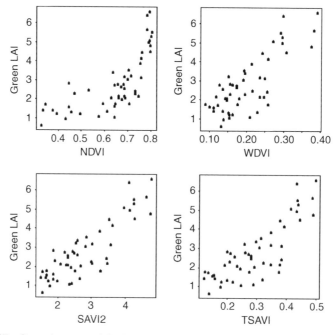

Figure 8.17 Ground measured LAI versus VI calculated from TOA reflectance converted from the TOA radiance in Fig. 8.16.

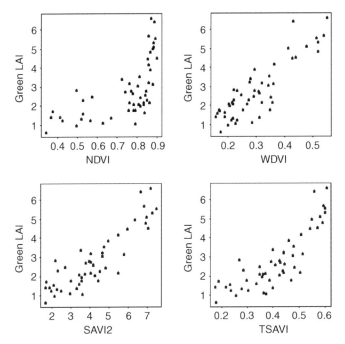

Figure 8.18 Ground measured LAI versus VI calculated from surface reflectance that is determined from atmospheric correction of TOA radiance in Fig. 8.16.

Since the atmospheric conditions were very clear, the VI from TOA reflectance (Fig. 8.17) and surface reflectance (Fig. 8.18) appear very similar, and the relationships between VI and LAI are reasonably good. However, the relationships between LAI and VI calculated from TOA radiance look much worse (Fig. 8.16), and the patterns are quite different from those in the other two figures. It demonstrates the need for converting DN to reflectance for calculating various VIs.

From Figs. 8.17 or 8.18, we can see that the NDVI–LAI relationship is nonlinear, and NDVI is certainly saturated with large LAI values. However, WDVI and SAVI2 reveal much stronger linear relationships with LAI. Because LAI data were collected over several cover types at different dates with the different Sun-sensor geometries and atmosphere and surface conditions, all points are widely spread. There will be large uncertainties if these VI-related statistical relationships are used for predicting LAI. More accurate methods are needed.

8.1.2 Hyperspectral Vegetation Indices

Hyperspectral remote sensing, also known as *imaging spectroscopy*, is a relatively new technology for the Earth environment. Imaging spectroscopy has been used in the laboratory by physicists and chemists for over 100 years for

TABLE 8.8   Selected Application Areas of Hyperspectral Remote Sensing

Areas	Applications
Atmosphere	Water vapor, cloud properties, aerosols
Ecology	Chlorophyll, leaf water, cellulose, pigments, lignin
Geology	Mineral and soil types
Coastal waters	Chlorophyll, phytoplankton, dissolved organic materials, suspended sediments
Snow/Ice	Snow-cover fraction, grain size, melting
Biomass burning	Subpixel temperatures, smoke
Commercial	Mineral exploration, agriculture and forest production

identification of materials and their composition. The concept of hyperspectral remote sensing began in the mid-1980s. Multispectral datasets are usually composed of about 5–10 bands of relatively large bandwidths (70–400 nm), whereas hyperspectral datasets are generally composed of about 100–200 spectral bands of relatively narrow bandwidths (5–10 nm). The typical airborne hyperspectral remote sensing systems include AVIRIS, compact airborne spectrographic imager (CASI), airborne imaging spectroradiometer for applications (AISA), and airborne imaging spectrometer (AIS). The first civil spaceborne hyperspectral sensor, Hyperion, on a EO-1 satellite, has collected a vast amount of data around the world.

Many applications can take advantage of hyperspectral remote sensing (see Table 8.8). Many biophysical and geophysical variables are estimated on the basis of various vegetation indices, some of which are introduced in the following paragraphs.

Before discussing the details, we should provide a cautionary note on estimating biochemical concentrations at the leaf and canopy levels using hyperspectral remote sensing. This has been a very difficult subject. Most of the indices described in the following paragraphs are quite empirical and verified by a limited amount of observations. More studies are definitely needed. As found in a simulation study using a three-dimensional radiative transfer model (Gastellu-Etchegorry and Bruniquel-Pinel 2001), the empirical indices are significantly influenced by canopy structure and viewing direction. There is a long way to go before such indices can be successfully incorporated into operational applications.

8.1.2.1 *Chlorophyll Absorption Ratio Index (CARI)*

Chlorophyll content is of particular significance to precision agriculture since it is an indicator of photosynthesis activity, which is related to the nitrogen concentration in green vegetation and serves as a measure of the crop response to nitrogen application. As is well known, green band reflectance is highly sensitive to leaf chlorophyll content (see Fig. 3.11). Kim et al. (1994) found that the ratio of 550:700 nm reflectance is constant at the leaf level regardless of the difference in chlorophyll concentration and then proposed the CARI index to calculate chlorophyll concentration:

$$CARI = \frac{R_{700}}{R_{670}} CAR = \frac{R_{700}}{R_{670}} \frac{670a + R_{670} + b}{\sqrt{a^2 + 1}} \qquad (8.45)$$

where

$$a = \frac{R_{700} - R_{550}}{150} \qquad (8.46)$$

and

$$b = R_{550} - 550a \qquad (8.47)$$

CAR represents the distance from the baseline spanned by the green reflectance peak at 550 nm and the reflectance at 700 nm.

8.1.2.2 Triangular Vegetation Index (TVI)

Green vegetation canopies have a very unique feature of reflectance spectra around the green, red, and near-IR bands. Green reflectance peak due to chlorophyll causes peak reflectance in the green band and absorption valleys in the red band, while leaf tissue causes a large increase in reflectance in the near-IR band. The region in the spectral space looks like a triangle (see Fig. 8.19). Both chlorophyll absorption, causing a decrease of red reflectance, and leaf tissue abundance, causing increased NIR reflectance, will increase the total area of the triangle. Broge and Leblanc (2000) suggested the triangular vegetation index (TVI) by using reflectance at three bands to represent the area of the triangle defined by the green peak, the chlorophyll absorption

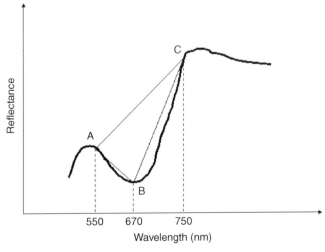

Figure 8.19 Triangular shape of the vegetation reflectance spectra in green-red/NIR region.

minimum, and the NIR shoulder in spectral space:

$$\text{TVI} = \frac{120(R_n - R_g) - 200(R_r - R_g)}{2} \tag{8.48}$$

where R_r, R_g, and R_n, are the red, green, and near-IR band reflectances.

8.1.2.3 BNC and BNA

Near-IR spectrometry has been used to estimate biochemical concentrations for a long time. One technique is to examine the absorption features of the canopy reflectance and then link them to foliage biochemical concentrations. Kokaly and Clark (1999) explored two methods for estimating foliar biochemical concentration of leaves based on the depth of the absorption feature: band depth normalized to (band depth at) the center of the absorption feature (BNC) and the band depth normalized to area of absorption feature (BNA).

BNC measures the depth of the waveband of interest from the continuum line at the center of the absorption feature:

$$\text{BNC} = \frac{1 - \dfrac{R}{R_i}}{1 - \dfrac{R_c}{R_{ic}}} \tag{8.49}$$

where R = reflectance of sample at waveband of interest
R_i = reflectance of continuum line at waveband of interest
R_c = reflectance of sample at absorption feature center
R_{ic} = reflectance of continuum line at absorption feature center:

BNA measures the depth of the waveband of interest from the continuum line, relative to the area (A) of the absorption feature:

$$\text{BNA} = \frac{1 - \dfrac{R}{R_i}}{A} \tag{8.50}$$

Figure 8.20 illustrates the calculation of both indices (Curran et al. 2001), where $R = 0.489$, $R_i = 0.524$, $R_c = 0.415$, and $R_{ic} = 0.495$. Therefore, band depth is 0.07 relative reflectance at wavelength λ and 0.16 relative reflectance at the center of the absorption feature. Given that the area of the absorption feature $A = 19.13$ relative reflectance in reciprocal nanometers (nm^{-1}), then, for waveband centered at λ, BNC = 0.41 and BNA = 0.004.

Curran et al. (2001) further compared these two indices with the standard first derivative reflectance spectra using the spectra/biochemical datasets from early and late in the growing season and found that all three indices could be used to estimate total cholorophyll, nitrogen, cellulose, and sugar. In addition, the BNC methodology could be used to estimate cholorophyll a + b and lignin and water.

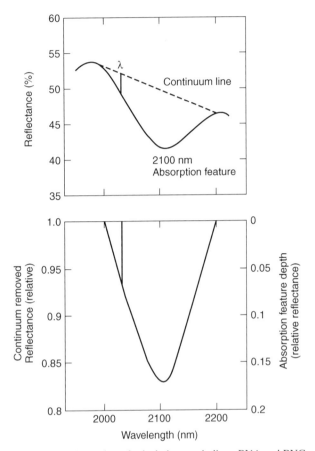

Figure 8.20   Illustration of calculating two indices: BNA and BNC.

8.1.2.4 *Indices Based on the Red Edge Reflectance Characteristics*

The red edge reflectance characteristics of vegetation have been the subject of many studies (see references below), all of which have shown that the observed blueshift and redshift of the red edge inflection point (REIP) is related to plant growth conditions. REIP can be defined as the wavelength around 720 nm at which the first derivative of the spectral reflectance curve reaches its maximum value. REIP is sometimes referred to as *red edge position* (REP). REIP shifts toward shorter wavelength (blueshift) indicate a decrease in green vegetation density, and REIP shifts toward longer wavelengths (redshift) are likely associated with in increase in green vegetation density. This phenomenon is caused by polymer forms of chlorophyll adding closely spaced absorption bands to the far red shoulder of the main chlorophyll a band. At the onset of senescence, the mesophyll structures in the plant tissue (effective near-infrared detectors) begin to collapse. Meanwhile, leaf chlorophyll decreases, causing red reflectance to increase. These combined effects cause a blueshift of REIP. There are different techniques to parameterize this spectral shift.

All of these vegetation indices are dependent on the presence of the red edge feature one way or another. The red edge is due to the distinctive spectral properties of green leaves containing pigments that strongly absorb at visible wavelengths while remaining highly transparent in the near infrared. It is also due to strong scattering from leaf cell walls as photons try to travel from the water-rich cells into the air-filled intercellular spaces. A high red edge value indicates healthy vegetation; a low value indicates senescence, disease, or damaged foliage. The bands in the red and near infrared (i.e., 670 and 865 nm) provide vegetated surface identification owing to their positions on either side of the red edge, marking the transition between chlorophyll absorption and cellulose reflectance. Clearly, the ideal target plant for this index is one with large leaves with high water content and high pigmentation, and little soil or stem structure visible in the view. Six measurement techniques and indices based on the red edge are presented below.

Gaussian Model. The spectral reflectance curve for vegetation canopies exhibits a consistent shape in the red edge region characterized by a relatively low reflectance in the red spectrum, followed by a sharp reflectance increase and then reaches an asymptotic reflectance plateau. It has been suggested that the spectral shape of the red edge reflectance can be approximated by one half of an inverted Gaussian function.

An inverted Gaussian model can be used to describe the variation of reflectance as a function of wavelength $R(\lambda)$:

$$R(\lambda) = R_s - (R_s - R_0)e^{(\lambda - \lambda_0)^2 / 2\sigma^2} \tag{8.51}$$

where R_s is the shoulder reflectance at the near-IR plateau, usually at 780–800 nm; R_0 is the minimum reflectance in the chlorophyll trough at approximately 670 nm; λ_0 is the wavelength of this minimum; and σ is the Gaussian shape parameter such that $\text{REIP} = \lambda_0 + \sigma$. This is illustrated in Fig. 8.21.

In practice, we can set $\lambda_0 = 670$ and use a nonlinear fitting procedure to determine R_s, R_0, and σ. This model has been used for estimating canopy biochemical parameters (e.g., Bonham-Carter 1988, Miller et al. 1990).

Polynomial Function Model. A high-order polynomial will capture potential asymmetry of the red edge. The general formula of a N-order polynomial is

$$R(\lambda) = \sum_{i=0}^{N} c_i \lambda^i \tag{8.52}$$

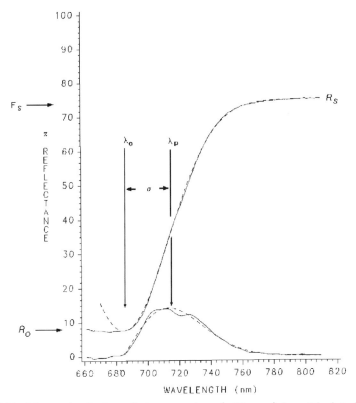

Figure 8.21 Measured red edge reflectance spectrum (solid curve) for a 7-leaf stack and its fitted inverted Gaussian model representation (dashed curve) illustrating the model parameters; the lower curves represent the measured and modeled first derivative curves. [From Miller et al. (1990), *Int. J. Remote Sens.* Reproduced by permission of Taylor & Francis, Ltd.]

where c_i are the coefficients. REIP can be determined by identifying one of the roots of the second derivative of the polynomial $[\partial^2 R(\lambda)/\partial\lambda = 0]$, which is closer to 720 nm.

Interpolation Techniques. Dawson and Curran (1998) presented a technique for rapid location of the position of the red edge based on Lagrangian interpolation. The technique, using three wavebands centered around the maximum first derivative of a vegetation reflectance red edge, fits a parabola through the first-derivative values. A second derivative is then performed on the Lagrangian equation to determine the REIP at the local maximum. Mathematically this is expressed as

$$\text{REIP} = \frac{A(\lambda_i + \lambda_{i+1}) + B(\lambda_{i-1} + \lambda_{i+1}) + C(\lambda_{i-1} + \lambda_i)}{2(A + B + C)} \qquad (8.53)$$

where

$$A = \frac{D_{\lambda_{i-1}}}{(\lambda_{i-1} - \lambda_i)(\lambda_{i-1} - \lambda_{i+1})} \tag{8.54}$$

$$B = \frac{D_{\lambda_i}}{(\lambda_i - \lambda_{i-1})(\lambda_i - \lambda_{i+1})} \tag{8.55}$$

$$C = \frac{D_{\lambda_{i+1}}}{(\lambda_{i+1} - \lambda_{i-1})(\lambda_{i+1} - \lambda_i)} \tag{8.56}$$

and the first derivative is approximated by the band reflectance difference

$$D_{\lambda_i} = (R_{\lambda_{i+1}} - R_{\lambda_i})/(\lambda_{i+1} - \lambda_i)$$

Derivative Techniques. Because of the unique features of vegetation canopy reflectance spectra around the red edge, derivative techniques can be used to distinguish canopy from soil and then further relate the derivative values to biophysical variables. Li et al. (1993) found that the second derivatives of the soil reflectance spectra are almost constant around zero, but the second derivative values of the canopy reflectance around the red edge are very large. This is illustrated in Fig. 8.22.

Second derivatives for the discrete wavebands can be approximated by

$$R''(\lambda_i) = \frac{2}{\lambda_{i+1} - \lambda_{i-1}} \left[\frac{R(\lambda_{i+1}) - R(\lambda_i)}{\lambda_{i+1} - \lambda_i} - \frac{R(\lambda_i) - R(\lambda_{i-1})}{\lambda_i - \lambda_{i-1}} \right] \tag{8.57}$$

Obviously, at least three bands are needed.

Li et al. found that the second derivatives of tall grass prairie canopies centered on the red and NIR wavebands are highly related to LAI regardless of which of two totally different backgrounds (burned and unburned) they were on. They demonstrated that the second derivative was much better than the index of the simple ratio for mapping LAI values (Section 8.2.1.2).

"Red Edge" LAI Estimation Models. Reflectance of vegetation canopy significantly increases at the edges of red and near-infrared spectra, which is usually called "red edge." Laboratory measurements revealed that the REIP wavelength is positively related to leaf chlorophyll content. Further studies have related REIP to LAI and other biophysical variables. An empirical model is given below (Danson and Plummer 1995)

$$R_i = R_{780} + 1.5 R_{673} \tag{8.58}$$

where R_i is the reflectance at wavelength i

$$\text{REIP} = 700 + \frac{R_i - R_{700}}{R_{740} - R_{700}} (740 - 700) \tag{8.59}$$

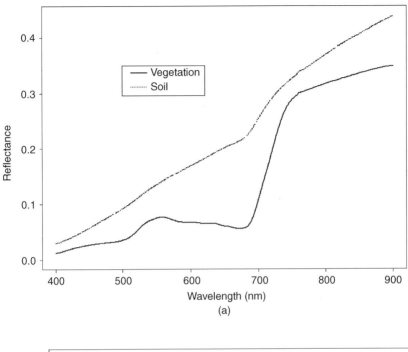

(a)

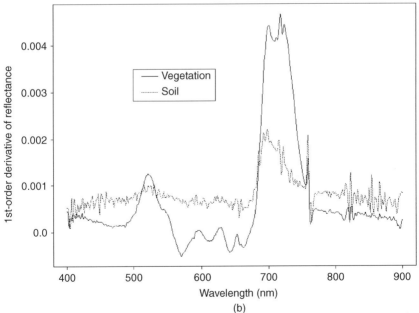

(b)

Figure 8.22 Derivative technique (a) shows the original reflectance spectra of soil and vegetation measured over Beltsville, MD; (b) and (c) are their first-order and second-order derivative spectra.

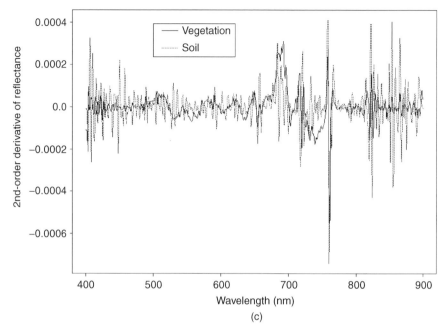

Figure 8.22 (*Continued*).

The empirical equation is given by

$$REIP = 710.1LAI^{0.0084} \qquad (8.60)$$

"Red Edge" Chlorophyll Concentration Estimation. The relationship between REP (in nm) and chlorophyll concentration was used to develop a predictive regression for the estimation of canopy chlorophyll concentration ([Cl] in mg/g) using CASI (Curran and Hay 1986):

$$[Cl] = -32.13 + 0.05REP \qquad (8.61)$$

Jago et al. (1999) further explored this type of relationship using both ground and airborne spectra and found that

- There were strong and statistically significant relationships between canopy chlorophyll concentration and both field and airborne sensor measures of REP. These correlation coefficients were respectively 0.84 and 0.80 for field measurements at grassland and winter wheat sites and 0.73 and 0.86 for airborne sensor measurements at these sites.
- The airborne sensor data were used to estimate canopy chlorophyll concentration with an RMS error of 0.42 mg/g ($\pm 12.69\%$ of mean) and 2.09 mg/g (± 16.4 of mean) for grassland and winter wheat field sites, respectively.

8.1.2.5 Indices Based on Spectral Continuum Measures

The spectral continuum has been calculated according to the shape and area of the troughs spanned by the spectral continuum. This technique has been used to detect narrow absorption features in the spectra, specifically, physical fingerprints of minerals (Ben-dor and Kruse 1995). It can also be used to identify and quantify any materials that exhibit a discrete absorption feature such as chlorophyll in live vegetation as demonstrated in the previous section. Broge and Leblanc (2000) suggested the chlorophyll absorption continuum index (CACI):

$$\text{CACI} = \sum_{\lambda_i}^{\lambda_n} \left(R_1 + i\frac{dR}{d\lambda}\Delta\lambda_i - R_i \right) \Delta\lambda_i \tag{8.62}$$

This utilizes all reflectance of the chlorophyll absorption continuum (~ 550 nm to ~ 730 nm) to the spanned area. CACI is similar to the TVI in the sense that both indices represent the area spanned by the spectral reflectance between the green peak and the NIR plateau.

8.1.2.6 Photochemical Reflectance Index (PRI)

Gamon et al. (1992) introduced a photochemical reflectance index (PRI) using two blue narrowband reflectances:

$$\text{PRI} = \frac{R_{570} - R_{531}}{R_{570} + R_{531}} \tag{8.63}$$

which is associated with radiation use efficiency (RUE). PRI decreases with increasing photosynthetic efficiency. It provides a potential means to remotely detect RUE in leaves and (more controversially) in canopies (Peñuelas et al. 1994, 1995a).

8.1.2.7 Structure-Independent Pigment Index (SIPI)

The structure-independent pigment index (SIPI) is defined as (Peñuelas et al. 1995b)

$$\text{SIPI} = \frac{R_{800} - R_{445}}{R_{800} - R_{680}} \tag{8.64}$$

SIPI uses both blue and red bands to assess the proportion of total photosynthetic pigments to chlorophyll (carotenoid:chlorophyll ratio) and the NIR band to account for the structural changes that may coincide with pigment changes.

8.1.2.8 Indices for Estimating Leaf Moisture Content

At wavelengths greater than 1000 nm, the normally strong water absorption features disappear and reflectance increases when leaves dry out. The reflectance peak between 1530 and 1720 nm provides an especially accurate indication of leaf water content (Fourty and Baret 1997).

Peñuelas et al. (1993) describe a method to measure plant water status, using the *water index* (WI), which is a ratio of two near-infrared bands: 970 and 900 nm. If the plant water status of the plant is known, the WI could become an indicator of cell wall elasticity of leaves (Peñuelas et al. 1996). Bands near 900 and 970 nm are not found on regular broadband radiometers but can be found on spectrometers.

Ceccato et al. (2001) use the reflectance ratio of 1600:820 nm to predict the leaf water content per leaf area X:

$$\frac{R_{1600}}{R_{820}} = 0.666 + \frac{1.0052}{1 + 1159\,X} - 6.976\,X \qquad (8.65)$$

where X is the unit g/cm^2. Although this index is not very sensitive to the high water content, the fitted R^2 value is as high as 0.92. Laboratory measurements performed on five different leaf species (Hunt and Rock 1989) have also shown a good relationship between the EWT and a moisture stress index calculated as the ratio between reflectances measured at 1600 and 820 nm.

8.1.3   Spatial Signatures and Applications

In the previous two sections, spectral signatures of multipectral and hyperspectral remote sensing were used to estimate land surface variables. We now discuss how to use spatial characteristics of the remote sensing data for the same purpose.

The spatial structure of a landscape largely reflects the composition of different components within it and their optical properties. The spatial signature of the remote sensing imagery also depends on the spatial resolution of the sensor and its spatial response (see Section 1.3.5.2) besides the surface variations. Generally speaking, the coarser the spatial resolution, the less variation there is between pixels in the image (Woodcock and Strahler 1987, Jupp et al. 1988, 1989).

In remote sensing, it is acknowledged that traditional statistics have severe limitations since they do not take into account the spatial dependencies of pixel values, which are generally spatially autocorrelated, nonstationary, nonnormal, irregularly spaced, and discontinuous. Spatial statistics techniques are needed to characterize the spatial signatures on the imagery and relate them to surface properties. Geostatistics, as part of spatial statistics, have been widely explored in remote sensing. We briefly present some basic concepts and some applications; interested readers should consult books on this subject for further details (e.g., Cressie 1993, Bailey and Gatrell 1995, Goovaerts 1997).

We will discuss three major concepts: variograms, autocovariance, and autocorrelation. Let us start with variograms. A *variogram*, sometimes called *semivariogram*, is defined as the variance of the difference between the variable at two locations. For reflectance imagery $r(x)$, the variogram $\gamma(h)$

can be defined as

$$\gamma(h) = \frac{1}{2N} \sum_{i=1}^{N} [r(x_i) - r(x_i + h)]^2 \tag{8.66}$$

where N is the number of pairs of pixels $\{r(x_i), r(x_i + h)\}$ at locations $\{x_i, x_i + h\}$ separated by the fixed lag h. The variogram generally increases with distance and can be modeled by some simple statistical formulas. Typical ones include

$$\text{Circular: } \gamma(h) = \begin{cases} \frac{2a}{\pi} \left[\frac{h}{b} \sqrt{1 - \left(\frac{h}{b}\right)^2} + \arcsin\frac{h}{b} \right] & 0 \leq h \leq b \\ a & h > b \end{cases} \tag{8.67}$$

where a and b are two parameters to characterize spatial sill and range, and both are positive.

$$\text{Exponential: } \gamma(h) = a\left[1 - \exp\left(-\frac{3h}{b} \right) \right] \tag{8.68}$$

$$\text{Hole: } \gamma(h) = \begin{cases} 0 & h = 0 \\ a\dfrac{1 - \sin(2\pi h/b)}{\sin(2\pi h/b)} & h \neq 0 \end{cases} \tag{8.69}$$

$$\text{Rational quadratic: } \gamma(h) = a\frac{19(h/b)^2}{1 + 19(h/b)^2} \tag{8.70}$$

$$\text{Spherical: } \gamma(h) = \begin{cases} a\left[\frac{3h}{2b} - \frac{1}{2}\left(\frac{h}{b}\right)^3 \right] & 0 \leq h \leq b \\ a & h > b \end{cases} \tag{8.71}$$

Tetraspherical:

$$\gamma(h) = \begin{cases} \frac{2a}{\pi} \left[\frac{h}{b} \sqrt{1 - \left(\frac{h}{b}\right)^2} + \frac{2h}{3b}\left(1 - \left(\frac{h}{b}\right)^2 \right)^{3/2} + \arcsin\frac{h}{b} \right] & 0 \leq h \leq b \\ a & h > b \end{cases} \tag{8.72}$$

$$\text{Pentaspherical: } \gamma(h) = \begin{cases} a\left[\frac{15h}{8b} - \frac{5}{4}\left(\frac{h}{b}\right)^3 + \frac{3}{8}\left(\frac{h}{b}\right)^3 \right] & 0 \leq h \leq b \\ a & h > b \end{cases} \tag{8.73}$$

$$\text{Stable: } \gamma(h) = a\left[1 - \exp\left(-3\left(\frac{h}{b}\right)^c \right) \right] \tag{8.74}$$

where $0 \leq c \leq 2$. If $c = 2$, it is called a *Gaussian model*.

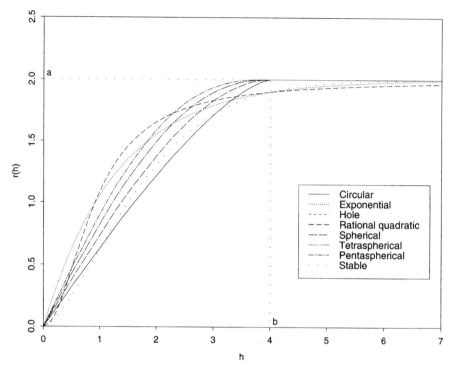

Figure 8.23   Illustration of the variogram functions defined from Eqs. (8.67)–(8.74).

Figure 8.23 illustrates the basic shapes of these variogram functions. On the basis of the experimental variogram figure, we need to select a specific variogram function in the data analysis. The parameters in these functions can be then estimated by fitting a specific function to the calculated variogram from the imagery.

Note that the functions presented above are isotropic and depend only on the lag (h). So they are one-dimensional in nature. Variograms may display directionality, and thus a two-dimensional function is more appropriate. If the 2D function is the multiplication of two functions in each x and y direction, we call it *separable*. A more realistic nonseparable and anisotropic model has also been used (Isaaks and Srivastava 1989):

$$\gamma\left(h_x, h_y\right) = \exp\left(-\sqrt{\alpha h_s^2 + \beta h_y^2}\right)$$

There are several good introductory articles on variograms in the context of remote sensing (e.g., Curran 1988, Woodcock et al. 1988a, 1988b).

Two other functions that are highly related to the variogram: autocovariance and autocorrelation:

Autovariance Function. This is expressed as

$$C(h) = \frac{1}{N} \sum_{i=1}^{N} \left[r(x_i) r(x_i + h) - \bar{r}_{-h} \bar{r}_h \right]^2 \qquad (8.75)$$

where $\bar{r}_{-h}$ and $\bar{r}_{+h}$ are the mean values:

$$\bar{r}_{-h} = \frac{1}{N} \sum_{i=1}^{N} r(x_i) \qquad (8.76)$$

and

$$\bar{r}_h = \frac{1}{N} \sum_{i=1}^{N} r(x_i + h) \qquad (8.77)$$

The autocovariance function is actually related to variogram through the following formula:

$$\gamma(h) = C(0) - C(h) \qquad (8.78)$$

Eventually, autocovariance can also have similar simple functions expressed from Eqs. (8.67)–(8.74).

Autocorrelation. The autocovariance function depends on the pixel values and can be normalized by the standard deviations:

$$\rho(h) = \frac{C(h)}{\sqrt{\sigma_{-h}^2 \sigma_h^2}} \qquad (8.79)$$

where the variance can be calculated by

$$\sigma_{-h}^2 = \frac{1}{N} \sum_{i=1}^{N} \left[r(x_i) - \bar{r}_{-h} \right]^2 \qquad (8.80)$$

$$\sigma_h^2 = \frac{1}{N} \sum_{i=1}^{N} \left[r(x_i + h) - \bar{r}_h \right]^2 \qquad (8.81)$$

Semivariograms have been widely applied in quantitative remote sensing (Curran and Atkinson 1998), such as in studying the structure and understanding the nature of spatial variation in remote sensing images (Woodcock et al. 1988b, Ramstein and Raffy 1989); in forestry to analyze forest stand structure (Cohen et al. 1990, St-Onge and Cavayas 1995), in estimation of structural damage in balsam fir stands (Franklin and Turner 1992), in sampling of ground data to be correlated with remotely sensed data (Atkinson and Emery 1999), and in estimating biomass. Atkinson et al. (1994) estimated biomass (primary variable) using a secondary variable (NDVI). The experimental variograms of biomass and NDVI and their cross-variograms were computed and fitted to models. They concluded that this method had utility as a means to estimate the variable of interest (dry biomass). Phinn et al. (1996) also tried to map the biomass distribution. They used high-resolution airborne digital video image data, biomass field measurements, and variograms in order to map biomass for some semiarid plant communities in southern New Mexico. They determined spatial characteristics of vegetation by analyzing digital images at varying pixel sizes using semivariograms.

8.1.4   An Operational Statistical Method

Sellers et al. (1994) propose a method to estimate LAI and FPAR parameter fields from a global $1° \times 1°$ multitemporal NDVI dataset for climate studies. This dataset has been included in the ISLSCP Initialive I Global Data Sets CD-ROM (`http://daac.gsfc.nasa.gov/CAMPAIGN_DOCS/ISLSCP /islscp_il.html`). The general ideas are outlined below; readers are referred to the original paper for details. It consists of two major parts, each consisting of several steps.

8.1.4.1   NDVI Correction

1. *Fourier Wave Adjustment.* Since there are many errors and outliers in the original dataset, Fourier series analysis was used to adjust the outliers in the NDVI series. A weighted least squares method was used to fit a truncated Fourier series to the observed NDVI profiles of the seasonally variable cover types

$$\text{NDVI}_i = \sum_{j=1}^{m} \left[a_j \cos\left(\frac{(i-1)2\pi}{n} \right) + b_j \sin\left(\frac{(i-1)2\pi}{n} \right) \right] \quad (8.82)$$

where a and b are the Fourier coefficients, n is the number of points in the sequence, and m is the number of harmonics. Here $n = 12$ for 12 monthly NDVI values, and $m = 3$.

2. *Solar Zenith Angle Adjustment.* The NDVI varies with solar zenith angle as a result of increased atmospheric pathlengths and surface BRDF. Lacking a calibrated, physically-based model, a simple empirical procedure was used to account for solar zenith angle effects. NDVI data from any viewing angles were converted into the equivalent value at the nadir viewing

$$\text{NDVI}_0 = \frac{(\text{NDVI}_\theta - \text{NDVI}_{5,\theta})(\text{NDVI}_{98,0} - \text{NDVI}_{5,0})}{\text{NDVI}_{98,\theta} - \text{NDVI}_{5,\theta}} + \text{NDVI}_{5,0} \quad (8.83)$$

where 98 and 5 respectively rerepresent the 98th percentile values of vegetation distributions and the 5th percentile values of bare soil distributions. Details are available from the paper (Sellers et al. 1994).

3. *Interpolation of Missing Data*. In the original dataset, all pixels were excluded whenever the surface radiative temperature falls below 273 K. Thus, missing data were a serious problem at high latitudes during the Northern Hemisphere winter. They were simply replaced by the data in earlier months based on land cover information.

4. *Reconstruction of NDVI Data Classified as Tropical Evergreen Broadleaf*. Because of the consistent cloud coverage and hazy atmospheric conditions, NDVI values were also very low for the tropical evergreen broadleaf. In order to avoid including low values, the maximum NDVI value over the year is selected.

8.1.4.2 *FPAR and LAI Calculations*

After correcting NDVI, we calculate FPAR at time i from SR in a linear relationship

$$FPAR_i = \frac{(SR_i - SR_{min,i})(FPAR_{max,i} - FPAR_{min,i})}{SR_{max,i} - SR_{min,i}} \quad (8.84)$$

where SR denotes for the simple ratio of near-IR and red reflectance.

The relationship between FPAR and green LAI for uniform, homogeneous vegetation is given by a logarithm equation:

$$LAL_i - LAL_{max,i} \frac{\log(1 - FPAR_i)}{\log(1 - FPAR_{max,i})} \quad (8.85)$$

For clustered vegetation, the relationship is linear:

$$LAI_i = \frac{LAL_{max,i}FPAR_i}{FPAR_{max,i}} \quad (8.86)$$

So far we have discussed various statistical methods for estimating land surface variables that rely on two or three bands. With multispectral and hyperspectral remote sensing, we ought to take advantages of many different bands. On the other hand, we need to incorporate our understanding and knowledge of surface radiation regimes presented in Chapters 3–5 into the estimation process. We present several physically-based inversion methods below. Let us start with the traditional optimization technique.

8.2   OPTIMIZATION INVERSION METHOD

Optimization problems are made up of three basic ingredients:

- An objective function that we want to minimize or maximize. For instance, in fitting experimental data to a user-defined canopy model, we might *minimize the total deviation* of observed data from predictions based on the model. Various canopy models have been presented in Chapter 3 and soil/snow reflectance models, in Chapter 4.
- A set of *unknowns or variables* that affect the value of the objective function. In the problem of fitting the data, the unknowns are the *parameters* that define the model, such as LAI, LAD, and leaf optics.
- A set of *constraints* that allow the unknowns to take on certain values but exclude others. For retrieving canopy biophysical parameters, for instance, we can constrain LAI to be nonnegative.

The optimization problem is then to *find values of the variables that minimize or maximize the objective function while satisfying the constraints*.

In the traditional optimization inversion approach, iterative techniques are employed to estimate the biophysical variables. This requires a canopy reflectance model. The measured (retrieved) canopy reflectance data (R_j) are also needed. An optimization method is used to minimize the merit function $F(\Psi)$, which looks like

$$F(\Psi) = \sqrt{\sum_{i=1}^{n} w_i [R_i - f_i(\Psi)]^2} \qquad (8.87)$$

where $f(\Psi)$ is the canopy model with the parameter set Φ and w is the weighting vector. If we have a prior knowledge about the uncertainty of a subset of the measured (retrieved) canopy reflectance data, smaller weights can be assigned to those samples. Otherwise, all weights can be set as 1.

Many different types of optimization methods are available—some require derivative information, and others do not. They are available in standard libraries, such as *Numerical Algorithms Group* (NAG 1990) and *Numerical Recipes* (Press et al., 1989). A multidimensional optimization algorithm adjusts the free parameters until the merit function is minimized. Broadly, these algorithms may be classified according to their reliance on the model's partial derivatives. Generally, merit functions (and hence the models) that are nondifferentiable, or for which finite difference derivatives are computationally expensive, are minimized most efficiently with non-derivative-based optimization algorithms. Merit functions with analytic or computationally inexpensive derivatives may be inverted with more efficient algorithms that require derivative information. The algorithms that require derivative information are discussed in Section 11.3.

Most accurate, physically-based models are nondifferentiable and relatively computationally expensive. Thus, a non-derivative-based optimization algorithm is usually required. All are generally available in standard libraries such as *Numerical Algorithms Group* (NAG 1990) and *Numerical Recipes* (Press et al. 1989). The basic approach of each method is briefly described below.

The simplex method (e.g., AMOEBA) requires initial specification of $p + 1$ simplex vertices in the p-parameter space. Beginning with the vertex producing the largest merit function value (poorest fit), the algorithm attempts to find a lower merit function position by moving the vertex through the opposite face of the simplex (i.e., a "reflection"). If the new merit function value is lower than those of all other vertices, a larger reflection is attempted. If the original reflection does not represent an improvement over the second worst vertex, the simplex contracts by moving the worst vertex closer to the others. This is continued repetitively. When all $p + 1$ vertices produce fits to within a user-defined tolerance, the algorithm terminates. Simplex methods are particularly useful for discontinuous functions or functions subject to numerical inaccuracies or noise (*Numerical Algorithms Group*, 1990).

The conjugate direction set method [e.g., POWELL from Press et al., (1989)] begins from a single initial position and conducts single line minimizations, accurate to within a user-defined tolerance, in each of the current p conjugate directions in order to arrive at the minimum for a given iteration. It then compares the current minimum to the previous iteration's minimum. If the difference between the two estimates is below a second user-defined tolerance, the program terminates. If the difference is greater than the tolerance, the conjugate directions are redefined according to the vector between the two minima, and another iteration begins. This algorithm is often used when there are a large number of free parameters.

This optimization approach has been widely used by the BRDF community. Earlier works have been evaluated in the literature (Goel 1989; Liang and Strahler 1993, 1994; Kimes et al. 2000). Bicheron and Leroy (1999) applied this inversion method to POLDER data for retrieving both LAI and FPAR. The results showed that the LAI is retrieved with a fair degree of accuracy with the RMS difference between model results and observations of 0.70, better than that obtained with a semiempirical LAI–vegetation index relation. The daily FAPAR was also retrieved accurately, with an RMS difference between measured and modeled FAPAR of 0.097. Jacquemoud et al. (2000) compared several canopy reflectance models. Inversions using the optimization procedure were conducted in successive stages where the number of retrieved parameters was reduced. No significant difference could be observed between the three examined models. Overall, the leaf mesophyll structure parameter and leaf dry matter content couldn't be estimated. The chlorophyll content, the leaf area index, and the mean leaf inclination angle yielded better results. Gemmell et al. (2002) inverted boreal forest cover and tree crown transparency [a function of crown leaf area index defined by Baret and Guyot (1991)] from Landsat TM data and a limited amount of ancillary information using a hybrid geometric radiative transfer model. Inversion using this approach performed better than did a number of empirical spectral features. However, the method could not provide accurate information on the stand basis. The results indicated that a lack of stand-specific ancillary information on the needle and background reflectances was a serious limitation in the utility of the approach. Improved methods for obtaining ancillary information seem necessary. In this context, using existing forest inventory information to help estimate some of the key ancillary parameters is considered a possible way forward.

The major limitation associated with this method is that it is very computationally expensive and therefore difficult to apply operationally for regional and global study.

8.3 GENERIC ALGORITHM (GA)

One issue in the traditional optimization method discussed in the previous section is that the success of the inversion often depends on the initial values provided. If the initial values are far away from the "true" values, the iterative algorithm may not converge. One solution is to run the iterative algorithms several times by providing multiple sets of initial values and then find the best solution, which will, of course, increase the computational time significantly. Generic algorithm might be an alternative solution.

The fundamental concept of the generic algorithm (GA) is based on the concept of natural selection in the evolutionary process, which is accomplished by genetic recombination and mutation (Goldberg 1989). Applications of GA to a variety of optimization problems in remote sensing have been demonstrated only since the 1990s. Genetic algorithms have been developed for retrieval of land surface roughness and soil moisture (Jin and Wang 1999, Wang and Jin 2000). Lin and Sarabandi (1999) used GA as a global search routine to characterize the input parameters (such as tree density, tree height, trunk diameter, and soil moisture) of a forest stand. The inversion was tested with measured single polarized SAR data. Zhuang and Xu (2000) tried to retrieve LAI from thermal infrared multiangle data with GA. But the output LAI differed greatly from field data (retrieved 2.9 vs. field 1.4). A genetic algorithm was applied to the numerical optimization of a crop growth model using AVHRR data (de Wit 1999). The "synthetic" model output was compared with the "measured" AVHRR signal and the goodness of fit was used to adjust the crop model parameters in order to find a better set of parameters. Two free parameters, the initial amount of soil moisture and the emergence date of the crop, were selected as independent variables. LAI, as predicted by the crop growth model, was an intermediate parameter used to calculate the weighted difference vegetation index (de Wit 1999). Fang et al. (2003) applied the GA technique to retrieve surface LAI from optical remote sensing data.

8.4 LOOKUP TABLE METHODS

One solution to overcoming the huge demand of computational time of the traditional optimization inversion method is the lookup table method. This method has become popular in various disciplines. This technique is briefly mentioned in Chapter 6 on atmospheric correction.

Lookup Table Based Approach to LAI & FPAR Estimation From Remote Observations

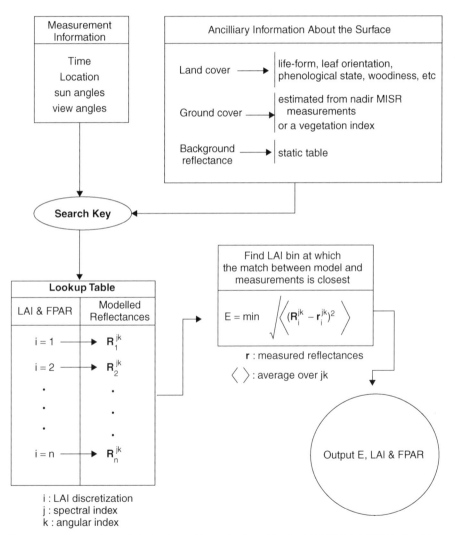

Figure 8.24 Flowchart of the lookup table method used in the MODIS LAI/FPAR algorithm [from Myneni et al. (1999)].

TABLE 8.9 NDVI and the Corresponding Values of LAI and FPAR in the MODIS Backup Algorithm

NDVI	Biome 1 LAI	FPAR	Biome 2 LAI	FPAR	Biome 3 LAI	FPAR	Biome 4 LAI	FPAR	Biome 5 LAI	FPAR	Biome 6 LAI	FPAR
0.025	0	0	0	0	0	0	0	0	0	0	0	0
0.075	0	0	0	0	0	0	0	0	0	0	0	0
0.125	0.3199	0.1552	0.2663	0.1389	0.2452	0.132	0.2246	0.1179	0.1516	0.07028	0.1579	0.08407
0.175	0.431	0.2028	0.3456	0.1741	0.3432	0.1774	0.3035	0.1554	0.1973	0.08922	0.2239	0.1159
0.225	0.5437	0.2457	0.4357	0.2103	0.4451	0.2192	0.4452	0.218	0.2686	0.1187	0.324	0.1618
0.275	0.6574	0.2855	0.5213	0.2453	05463	0.2606	0.574	0.271	0.3732	0.1619	0.4393	0.2121
0.325	0.7827	0.3283	0.6057	0.2795	0.6621	0.3091	0.7378	0.3395	0.5034	0.2141	0.5629	0.2624
0.375	0.931	0.3758	0.6951	0.3166	0.7813	0.3574	0.878	0.393	0.6475	0.2714	0.664	0.3028
0.425	1.084	0.419	0.8028	0.3609	0.8868	0.3977	1.015	0.4425	0.7641	0.32	0.7218	0.333
0.475	1.229	0.4578	0.9313	0.4133	0.9978	0.4357	1.148	0.4839	0.9166	0.3842	0.8812	0.393
0.525	1.43	0.5045	1.102	0.4735	1.124	0.4754	1.338	0.5315	1.091	0.4402	1.086	0.4599
0.575	1.825	0.571	1.31	0.535	1.268	0.5163	1.575	0.5846	1.305	0.4922	1.381	0.5407
0.625	2.692	0.6718	1.598	0.6039	1.474	0.566	1.956	0.6437	1.683	0.568	1.899	0.6458
0.675	4.299	0.8022	1.932	0.666	1.739	0.6157	2.535	0.6991	2.636	0.702	2.575	0.7398
0.725	5.362	0.8601	2.466	0.7388	2.738	0.7197	4.483	0.8336	3.557	0.7852	3.298	0.8107
0.775	5.903	0.8785	3.426	0.822	5.349	0.8852	5.605	0.8913	4.761	0.8431	4.042	0.8566
0.825	6.606	0.9	4.638	0.8722	6.062	0.9081	5.777	0.8972	5.52	0.8697	5.303	0.8964
0.875	6.606	0.9	6.328	0.9074	6.543	0.9196	6.494	0.9169	6.091	0.8853	6.501	0.9195
0.925	6.606	0.9	6.328	0.9074	6.543	0.9196	6.494	0.9169	6.091	0.8853	6.501	0.9195
0.975	6.606	0.9	6.328	0.9074	6.543	0.9196	6.494	0.9169	6.091	0.8853	6.501	0.9195

Source: Myneni et al. (1999).

This approach has been used for inverting LAI from MODIS and MISR data (Knyazikhin et al. 1998a, 1998b). It has also been used for other studies (e.g., Weiss et al. 2000a). Fig. 8.24 shows a flowchart illustrating this method. The table that contains LAI/FPAR and directional reflectance has to be created in advance using a radiative transfer model or other canopy reflectance model. Because the intervals cannot be infinitely small, interpolation (usually linear) has to take place.

In an ordinary lookup table approach, the dimensions of the table have to be large enough to achieve a high degree of accuracy, which leads to a much slower online search. Moreover, many parameters have to be fixed in the lookup table method. For example, in the MODIS/MISR LAI/FPAR algorithm, all canopies are classified into six categories: grasses and cereal crops, shrubs, broadleaf crops, savannas, broadleaf forests, and needle forests. For each category, some representative values are used as constants for some variables, such as soil reflectance, leaf reflectance, and transmittance in the canopy for creating each table. However, some of these variables may vary quite dramatically and affect surface reflectance significantly. This method is simple and easy to implement.

In the MODIS LAI/FPAR algorithm (Myneni et al. 1999), if the lookup table approach fails for some reason, a backup method using NDVI will be used. Table (8.9) is given for reference.

8.5 HYBRID INVERSION METHODS

We discussed various statistical methods in Section 8.1 and physical methods in Sections 8.2–8.4. Their advantages and disadvantages have been discussed in the corresponding sections. The best approach is to combine the statistical method and the physical method in order to take advantage of their strengths.

A hybrid inversion algorithm is a combination of extensive simulations using a canopy radiative transfer model (physical) and a nonparametric statistical inversion model (statistical). The canopy radiative transfer model is used to generate a database by changing the values of key variables systematically. The nonparametric model (e.g., a feedforward neural network, local regression, projection–pursuit regression method) is used to map the relationship between spectral directional reflectance and various land surface variables. This concept is illustrated in Fig. 8.25.

Nonparametric regression approaches are able to deal with a wide range of regression domains by not imposing any predefined global form to the regression surface. These methods assume some functional form only at a *local level,* meaning that they do not try to fit one single model to all given samples. This methodology leads to a regression surface that is globally nonparametric (i.e., an approximation that does not have a fixed global functional form). This is an important feature if one wishes to develop a tool that is applicable to a wide range of regression problems. Several non-

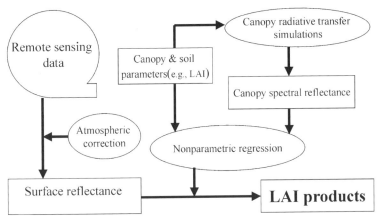

Figure 8.25 Conceptual illustration of the hybrid algorithm for estimating land surface variables (i.e., LAI in this case). [From Liang et al. (2003), *IEEE Trans. Geosci. Remote Sens.* Copyright © 2003 with permission of IEEE.]

parametric regression methods can be used. We briefly introduce three algorithms that have been applied in quantitative remote sensing: neural network, projection–pursuit regression, and regression tree methods.

8.5.1 Neural Network

Artificial neural networks (ANNs) can be seen as highly parallel dynamical systems consisting of multiple simple units that can perform transformations by means of their state response to their input information. A common setup for applying ANNs in regression consists of using a three-layered network with one hidden layer (Fig. 8.26). The ANN approach has been used in different remote sensing applications. The key attribute of ANNs is that they learn by example. In a typical scenario, an ANN is presented iteratively with a set of samples, known as the *training set*, from which the network can learn the values of its internal parameters.

At least four main aspects should be considered in the application of ANNs:

- Preparing the training data from radiative transfer simulations
- Designing the network architecture

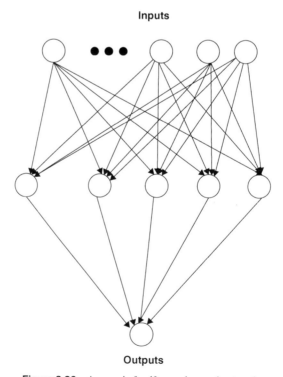

Inputs

Outputs

Figure 8.26 A generic feedforward neural network.

- Estimating the parameters, that is, training a network
- Assessing the performances of the network

The key component is to design ANN architecture that has a critical effect on the results. Various studies have shown that the neural network complexity—mainly the number of free parameters—must be fitted to the problem complexity and the number of available training samples. If the network is too complex, it will learn the training set perfectly (low bias) while generalizing very poorly (high variance). Controlling the complexity is therefore necessary to ensure good generalization. This is a key issue when the training set is small, noisy, and partially inaccurate. However, the training dataset created from the radiative transfer simulation is huge and therefore is not a problem. If ANN is used directly to link ground-measured biophysical variables with spectral reflectance, it might be an issue.

As demonstrated by many studies (e.g., Smith 1993, Baret 1995, Gopal and Woodcock 1996; Baret and Fourty 1997, Jin and Liu 1997, Kimes et al. 1998), ANN provides a very efficient tool that establishes the relationship between the simulated reflectance field and the corresponding biophysical variable of interest. Smith (1993) first trained a backpropagation neural network to invert a simple multiple scattering model to estimate leaf area index from reflectance at three wavelengths and then applied the trained network to satellite observations. He reported estimated errors of less than 30% and indicated that the method appeared to be much less sensitive to initial guesses for the parameters than did other inversion techniques. Baret et al. (1995) compared the use of ANN to the classical VI approach and concluded that this technique was of value, especially when calibrated with a radiative transfer (RT) model simulations rather than experimental observations. Gong et al. (1999) employed an error backpropagation feedforward neural network program to invert LAI and LAD from a canopy reflectance model developed by Liang and Strahler (1993). The test results showed that a relative error between 1% and $\geq 5\%$ was achievable for retrieving one parameter at a time or two parameters simultaneously. Weiss et al. (2000b) validated such techniques over a range of crops by estimating the main canopy biophysical variables, LAI and the fractional coverage.

8.5.2 Projection Pursuit

Projection pursuit regression (Friedman and Stuetzle 1981) applies an additive model to projected variables:

$$\alpha = \sum_{i=1}^{M} f_i\left(w_k^T \mathbf{R}\right) + \varepsilon \tag{8.88}$$

where $\mathbf{R} = (R_1, R_2, \ldots, R_n)$ is the explanatory vector (i.e., spectral reflectance in this particular case). It can be further expressed as

$$\alpha = w_0 + \sum_{i=1}^{M} f_i\left(w_i^T \mathbf{R}\right) \tag{8.89}$$

The "projection" part of the term "projection–pursuit regression" indicates that the carrier vector $\mathbf{R}$ is projected onto the direction vectors $w_1, w_2, \ldots, w_M$, and the "pursuit" part indicates that an optimization technique is used to find "good" direction vectors $w_1, w_2, \ldots, w_M$. For any applications, the important parameter to be determined is M, which depends on the number of variables and the amount of training data. There are some general guidelines to determine its value in Splus (a statistical package), but it is found in our study that the results are not very sensitive when its value is set from 15 to 25.

8.5.3 Regression Tree

A regression tree can be seen as a kind of additive model of the form

$$m(x) = \sum_{i=1}^{l} k_i \times I(x \in D_i) \tag{8.90}$$

where, k_i are constants, $I(\cdot)$ is an indicator function returning 1 if its argument is true and 0 otherwise, and D_i are disjoint partitions of the training data D such that $\bigcup_{i=1}^{l} D_i = D$ and $\bigcap_{i=1}^{l} D_i = \phi$. Models of this type are sometimes called *piecewise constant regression models* as they partition the predictor space x in a set of regions and fit a constant value within each region. An important aspect of tree-based regression models is that they provide a propositional logic representation of these regions in the form of a tree. Each path from the root of the tree to a leaf corresponds to a region. Each inner node of the tree is a logical test on a predictor variable.

Figure 8.27 is a regression tree with a small sample. As there are four distinct paths from the root node to the leaves, this tree divides the input space in four different regions. The conjunction of the tests in each path can

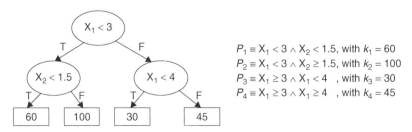

Figure 8.27 A simple regression tree example with a few samples.

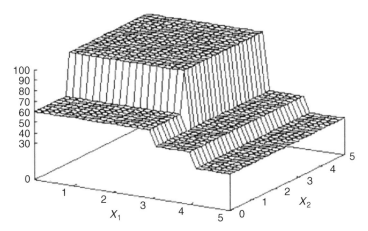

Figure 8.28 Regression surface represented by the regression tree shown in Fig. 8.27.

be regarded as a logical description of such regions, as described above. This tree roughly corresponds to the regression surface (assuming that there were only the predictor variables X_1 and X_2) shown in Fig. 8.28. Using the more concise representation of Eq. (8.90), we obtain

$$m(x) = 60 \times I(X_1 < 3 \wedge X_2 < 1.5) + 100 \times I(X_1 < 3 \wedge X_2 \geq 1.5)$$

$$+ 30 \times I(X_1 \geq 3 \wedge X_2 < 4) + 45 \times I(X_1 \geq 3 \wedge X_2 \geq 4)$$

Regression trees are constructed using a recursive partitioning (RP) algorithm. This algorithm builds a tree by recursively splitting the training sample into smaller subsets. The algorithm has three main components:

- A way to select a split test (the splitting rule)
- A rule to determine when a tree node is terminal (termination criterion)
- A rule for assigning a value to each terminal node

The most common method for building a regression model based on a sample of an unknown regression surface consists in trying to obtain the model parameters that minimize the least squares error criterion; this results in *least squares regression trees*. All input metrics are analyzed across digital number values, and right and left splits are examined. The split that produces the greatest reduction in the residual sum of squares, or *deviance*, is used to divide the data and the process resumes for the two newly created subsets. The regression tree algorithm takes the form

$$D = D_s - D_t - D_u \tag{8.91}$$

where s represents the parent node and t and u are the splits from s. The deviance for nodes is calculated from the equation

$$D_i = \sum_{\text{cases}(j)} \left(y_i - \bar{y}_j\right)^2 \qquad (8.92)$$

for all j cases of y and the mean value of those cases.

The regression tree algorithm is performed as follows. Two samples of training pixels are taken from the training dataset. One is used to grow the regression tree and the other to prune it. Pruning is required because tree algorithms are very robust and delineate even individual pixels isolated in spectral space. By having a set aside of training data, a more generalized tree can be generated. This generalization is achieved by passing the second sample of data down the initial tree. As the data cascade down the tree, the overall sum of squares begins to level out and eventually begins to increase. This indicates an overfitting of the initial tree. For this work, pruning is performed not where the sum of squares begins to increase, but where additional nodes represent a reduction of less than 0.01% of the overall sum of squares for the data. The end result is an easily interpreted hierarchy of splits that allow for a biophysical interpretation of the relationship.

Regression tree methods are known for their simplicity and efficiency when dealing with domains with large numbers of variables and cases. Regression trees are obtained using a fast "divide and conquer" (i.e., greedy) algorithm that recursively partitions the given training data into smaller subsets. The use of this algorithm is the cause of the efficiency of these methods. However, it can also lead to poor decisions at lower levels of the tree due to the unreliability of estimates based on small samples of cases. Methods dealing with this problem turn out to be nearly as important as growing the initial tree. Regression trees are also known for their instability. A small change in the training set may lead to a different choice when building a node, which in turn may represent a dramatic change in the tree, particularly if the change occurs in top-level nodes. Moreover, the function approximation provided by standard regression trees is not smooth, leading to very marked function discontinuities.

Regression trees have been used with remote sensing data (Michaelson et al. 1994, Defries et al. 1997, Prince and Steininger 1999, Hansen et al. 2002). They offer a robust tool for handling nonlinear relationships within remotely sensed datasets. One application is discussed in Section 12.3.2.2.

An increasing number of studies are using this hybrid inversion approach in estimating land surface variables. Weiss and Baret (1999) explored retrieving several biophysical variables from the accumulation of a large swath of satellite data from the VEGETATION/SPOT4 sensor. These variables include leaf area index (LAI), fraction of photosynthetically active radiation (FAPAR), chlorophyll content integrated over the canopy, and gap fraction in any direction for which the gap fraction is theoretically independent of the

LAI. Gong et al. (1999) inverted LAI from the simulated database. Fang and Liang (2003) inverted LAI from ETM + imagery and Liang et al. (2003) estimated LAI and broadband albedo from EO-1 data by incorporating the soil line concept in this procedure. Kimes et al. (2002) inverted a complex 3D DART model for a wide range of simulated forest canopies using POLDER-like data. The model was inverted to recover three forest canopy variables: forest cover, leaf area index, and a soil reflectance parameter. The ranges of these variables were 0.4–1.0, 0.8–9.3, and 0.0–1.0, respectively. Two inversion methods were used: a traditional inversion technique using a modified simplex method, and a hybrid method with a neural network method in combination with an exhaustive variable selection technique. The methods were compared for efficiency, accuracy, and stability. The hybrid method gave relatively accurate solutions to the inversion problem, given a small subset of directional/spectral data using only one to five view angles. An example using this approach for estimating broadband land surface albedo (Liang, 2003) is given in Chapter 9.

8.6 COMPARISONS OF DIFFERENT INVERSION METHODS

There are some good discussions regarding the advantages and disadvantages of these inversion methods by Kimes et al. (2000). However, a rigorous comparison of the accuracy of various inversion methods for physically-based models is not available in the literature. There are several comparisons of the accuracy and computational efficiencies of various traditional inversion methods (e.g., Privette et al. 1994, 1997). However, comparisons between traditional optimization inversion methods, lookup table methods, and hybrid inversion methods are absent in the literature. Consequently, the relative accuracy that can be obtained with each method is speculative until more comparisons are made. A comparison of the accuracy of the various inversion methods, however, is not enough. The stability of the methods must be rigorously compared in terms of the sensitivity of each methods output to uncertainties in measured data and parameters describing the model (e.g., leaf optics, gap probability, solar and viewing geometries).

The traditional optimization inversion methods are computationally intensive and are not appropriate for application on a per pixel basis for regional and global studies. The lookup table and hybrid inversion methods are computationally efficient and can be applied on a per pixel basis. Computational efficiency depends in part on random access memory (RAM). If there is no limitation on RAM, any inversion technique can be used on a per pixel basis given ample computer speed. However, in reality one must find a compromise between memory available and the number of operations needed to process one pixel.

Because of computational constraints, the traditional inversion methods are generally limited to simplified physically based models (e.g., Goel 1989,

Liang and Strahler 1993, Wanner et al. 1995, Braswell et al. 1996). These models have been simplified by reducing the number of variables and/or neglecting or liberally approximating some physical processes (e.g., 1D turbid medium assumption).

The lookup table method and the hybrid inversion method can be applied to the most sophisticated models without any further simplifications. Because the lookup table or the training process (in case of neural networks) is generated prior to the actual application of the method, it does not require the online computation time. Consequently, the hybrid inversion and lookup table methods can be applied to any physically-based model regardless of how the model is formulated. For example, very computationally inefficient models such as the Monte Carlo and ray tracing techniques (see Section 3.5.1) can be used with these methods. In the case of lookup table methods, however, the multidimensional size of the table increases as the number of variables in the model and the discretization of a variable increase.

An advantage of hybrid inversion and lookup table methods is that they do not require any initial guesses to model variables as do the traditional inversion methods. Presently, poor choices in initial guesses can lead to a greater risk of terminating at "false" (nonglobal) minima.

One major advantage of the hybrid inversion method over the lookup table method is that the variations of key variables can be effectively accounted for. The hybrid inversion method can link outputs with a few major input variables, although more variables are used for creating the simulation database. That is because a nonparametric statistical model can nonlinearly project data so that some factors are enhanced and others are compressed. In contrast, the full set of variables has to be tabulated in the ordinary lookup table approach. The hybrid inversion method can also effectively adapt multispectral and multiangle remotely sensed values without slowing down the searching process significantly.

8.7 SUMMARY

One major motivation in quantitative remote sensing is to estimate land surface biophysical variables, such as LAI, FPAR, and various biochemical concentrations. This chapter presents four methods for estimating these biophysical variables. The statistical method described in Section 8.1 is based mainly on different vegetation indices. A comprehensive overview of multispectral and hyperspectral vegetation indices was provided, and their linkages with these biophysical and biochemical variables were also evaluated. Broge and Mortensen (2002) found from extensive comparisons of different indices from measured winter wheat data that most indices have an exponential relationship with green crop area index and canopy chlorophyll density. Further studies are definitely needed to compare different vegetation indices and their ability to predict various biophysical variables.

The traditional optimization inversion method was discussed in Section 8.2. Because of its computational expense, this method has not been used to operationally estimate biophysical variables from satellite observations. One of its key components is the optimization search. In the remote sensing literature, it is interesting to find that all studies employed the optimization method are based on function values, rather than derivatives considering the latter, which is usually much faster. A few function-based optimization methods were briefly presented. More derivative-based methods are presented in Chapter 11.

As an alternative approach, generic algorithms have not been sufficiently explored in remote sensing. The basic principles and some initial applications were provided in Section 8.3.

To reduce the computational requirements, lookup table methods discussed in Section 8.4 have been used in MODIS/MISR LAI estimation. It is simple and fast, but has to predefine many parameters to manage the limited dimension of the table.

Hybrid methods that have the potentials for operational applications have been discussed in Section 8.5. It is anticipated that more and more studies on estimating land surface variables will be based on this approach in the future.

A comprehensive comparison of these methods was presented in Section 8.6. The discussions were mainly conceptual. More studies are needed to compare these methods quantitatively.

Current methods typically involve analyzing single-sensor data, either as a one-time image or as a multitemporal dataset. There have been limited studies on combining optical data with radar to optimize the characterization of land surface variables. The availability of new satellite data will likely place greater emphasis on fusing data using (1) multiple sensors for multiresolution analysis and statistical subsampling approaches or (2) sensor suites, that is, combinations of optical and radar or optical and thermal sensors, that provide collocated and coincident temporal data. A better understanding of the spectral, spatial, and temporal scaling issues and use of high-resolution data as subsampling for moderate-resolution sensors is necessary. A natural extension of these ideas is the so-called data assimilation, discussed in Chapter 11. Within the scheme of data assimilation, it is easy to incorporate various a priori knowledge into the inversion process, which has been explored by Li et al. (2000).

REFERENCES

Asrar, G., Fuchs, M., Kanemasu, E., and Hatfield, J. (1984), Estimating absorbed photosynthetic radiation and leaf area index from spectral reflectance in wheat, *Agron. J.* **76**: 300–306.

Atkinson, P. and Emery, D. (1999), Exploring the relation between spatial structure and wavelength: Implications for sampling reflectance in the field, *Int. J. Remote Sens.* **20**: 2663–2678.

Atkinson, P., Webster, R., and Curran, P. (1994), Cokriging with airborne mss imagery., *Remote Sens. Envir.* **50**: 335–345.

Bailey, T. C. and Gatrell, A. C. (1995), *Interactive Spatial Data Analysis*. Longman.

Bannari, A., Morin, D., Bonn, F., and Huete, A. R. (1995), A review of vegetation indices., *Remote Sens. Rev.* **13**: 95–120.

Baret, F. (1995), Use of spectral reflectance variation to retrieve canopy biophysical characteristics, in *Advances in Environmental Remote Sensing*, M. Darson and S. Plummer, eds., Wiley, pp. 33–51.

Baret, F. and Fourty, T. (1997), Estimation of leaf water content and specific leaf weight from reflectance and transmittance measurements, *Agronomie* **17**: 455–464.

Baret, F. and Guyot, G. (1991), Potential and limits of vegetation indices for LAI and APAR assessment, *Remote Sens. Envir.* **35**: 161–173.

Baret, F., Clevers, J., and Steven, M. (1995), The robustness of canopy gap fraction estimates from red and near-infrared reflectances: A comparison of approaches, *Remote Sens. Envir.* **54**: 141–151.

Baret, F., Jacquemoud, S., and Hanocq, J. F. (1993), The soil line concept in remote sensing, *Remote Sens. Rev.* **7**: 65–82.

Bendor, E. and Kruse, F. A. (1995), Surface mineral mappings of Makhtesh Ramon Negev, Israel using GER 63 channel scanner data, *Int. J. Remote Sens.* **16**: 3529–3553.

Bicheron, P. and Leroy, M. (1999), A method of biophysical parameter retrieval at global scale by inversion of a vegetation reflectance model, *Remote Sens. Envir.* **67**: 251–266.

Bonan, G. B. (1993), Importance of leaf area index and forest type when estimating photosynthesis in boreal forests, *Remote Sens. Envir.* **43**: 303–314.

Bonham-Carter, G. F. (1988), Numerical procedures and computer program for fitting an inverted gaussian model to vegetation reflectance data, *Comput. Geosci.* **14**: 339–356.

Braswell, B. H., Schimel, D. S., Privette, J. L., Moore III, B., Emery, W. J., Sulzman, E., and Hudak, A. T. (1996), Extracting ecological and biophysical information from AVHRR optical measurements: An integrated algorithm based on inverse modeling, *J. Geophys. Res.* **101**: 23335–23348.

Broge, N. H. and Leblanc, E. (2000), Comparing prediction power and stability of broadband and hyperspectral vegetation indices for estimation of green leaf area index and canopy chlorophyll density, *Remote Sens. Envir.* **76**: 156–172.

Broge, N. H. and Mortensen, J. V. (2002), Deriving green crop area index and canopy chlorophyll density of winter wheat from spectral reflectance data, *Remote Sens. Environ.* **81**: 45–57.

Carlson, T. N. and Ripley, D. A. (1998), On the relation between NDVI, fractional vegetation cover, and leaf area index, *Remote Sens. Envir.* **62**: 241–252.

Ceccato, P., Flasse, S., Tarantola, S., Jacquemoud, S., and Gregoire, J. M. (2001), Detecting vegetation leaf water content using reflectance in the optical domain, *Remote Sens. Envir.* **77**: 22–33.

Chen, J. M. (1996), Evaluation of vegetation indices and a modified simple ratio for boreal applications, *Can. J. Remote Sens.* **22**: 229–242.

Choudhury, B. J. (1987), Relationships between vegetation indices, radiation absorption, and net photosynthesis evaluated by sensitivity analysis, *Remote Sens. Envir.* **22**: 209–233.

Clevers, J. (1989), The application of a weighted infra-read vegetation index for estimating leaf area index by correcting for soil moisture, *Remote Sens. Envir.* **29**: 25–37.

Cohen, W., Spies, T., and Bradshaw, G. (1990), Semivariograms of digital imagery for analysis of conifer canopy structure. *Remote Sens. Envir.* **34**: 167–178.

Collins, J. B. and Woodcock, C. E. (1996), An assessment of several linear change detection techniques for mapping forest mortality using multitemporal Landsat TM data, *Remote Sens. Envir.* **56**: 66–77.

Cressie, N. (1993), *Statistics for Spatial Data*, rev. ed. John Wiley.

Crist, E. P. and Cicone, R. C. (1984), A physically-based transformation of thematic mapper data—the TM tasseled cap, *IEEE Trans. Geosci. Remote Sens.* **22**: 256–263.

Crist, E. P., Laurin, R., and Cicone, R. C. (1986), Vegetation and soils information contained in transformed thematic mapper data, *Proc. IGARSS'86*, Zurich, ESA Publications Division, pp. 1465–1470.

Curran, P. and Atkinson, P. (1998), Geostatistics and remote sensing, *Progress Phys. Geogr.* **22**: 61–78.

Curran, P. J. (1988), The semi-variogram in remote sensing: An introduction, *Remote Sens. Envir.* **3**: 493–507.

Curran, P. J. and Hay, A. M. (1986), The importance of measurement error for certain procedures in remote sensing at optical wavelengths, *Photogramm. Eng. Remote Sens.* **52**: 229–241.

Curran, P. J., Dungan, J. L., and Peterson, D. L. (2001), Estimating the foliar biochemical concentration of leaves with reflectance spectrometry: Testing the kokaly and clark methodologies, *Remote Sens. Envir.* **76**: 349–359.

Danson, F. M. and Plummer, S. E. (1995), Red-edge response to forest leaf area index, *Int. J. Remote Sens.* **16**: 183–188.

Daughtry, C. S. T., Gallo, K. P., Goward, S. N., Prince, S. D., and Kustas, W. P. (1992), Spectral estimates of absorbed radiation and phytomass production in corn and soybean canopies, *Remote Sens. Envir.* **39**: 141–152.

Dawson, T. P. and Curran, P. J. (1998), A new technique for interpolating the reflectance red edge position, *Int. J. Remote Sen.* **19**: 2133–2139.

Dawson, T., Curran, P., and Plummer, S. (1998), LIBERTY: Modeling the effects of leaf biochemistry on reflectance spectra, *Remote Sens. Envir.* **65**: 50–60.

DeFries, R. S., Hansen, M., Steininger, M., Dubayah, R., Sohlberg, R., and Townshend, J. R. G. (1997), Subpixel forest cover in central Africa from multisensor, multitemporal data, *Remote Sens. Envir.* **60**: 228–246.

de Wit, A. J. W. (1999), The application of a genetic algorithm for crop model steering using NOAA-AVHRR data (available online from http://cgi.girs.wageningen-ur.nl/cgi/products/publications.htm.).

Dymond, C. C., Mladenoff, D. J., and Radeloff, V. C. (2002), Phenological differences in tasseled cap indices improve deciduous forest classification, *Remote Sens. Envir.* **80**: 460–472.

Fang, H. and Liang, S. (2003), Retrieve LAI from landsat 7 ETM+ data with a neural network method: Simulation and validation study, *IEEE Trans Geosci. Remote Sens.* **41**: 2052–2062.

Fourty, T. and Baret, F. (1997), Vegetation water and dry matter contents estimated from top-of-atmosphere reflectance data: A simulation study, *Remote Sens. Envir.* **61**: 34–45.

Franklin, J. and Turner, D. L. (1992), The application of a geometric optical canopy reflectance model to semiarid shrub vegetation, *IEEE Trans. Geosci. Remote Sens.* **30**: 293–301.

Friedman, J. and Stuetzle, W. (1981), Projection-pursuit regression, *J. Am. Stat. Assoc.* **76**: 183–192.

Gamon, J. A., Penuelas, J., and Field, C. B. (1992), A narrow waveband spectral index that tracks diurnal changes in photosynthetic efficiency, *Remote Sens. Environ.* **41**: 35–44.

Gardner, B. R. and Blad, B. L. (1986), Evaluation of spectral reflectance models to estimate corn leaf area while minimizing the influence of soil background effects, *Remote Sens. Envir.* **20**: 183–193.

Gastellu-Etchegorry, J. P. and Bruniquel-Pinel, V. (2001), A modeling approach to assess the robustness of spectrometric predictive equations for canopy chemistry, *Remote Sens. Envir.* **76**: 1–15.

Gemmell, F., Varjo, J., Strandstrom, M., and Kuusk, A. (2002), Comparison of measured boreal forest characteristics with estimates from tm data and limited ancillary information using reflectance model inversion, *Remote Sens. Envir.* **81**: 365–377.

Gobron, N., Pinty, B., Verstraete, M. M., and Widlowski, J.-L. (2000), Advanced vegetation indices optimized for up-coming sensors: Design, performance, and applications, *IEEE Trans. Geosci. Remote Sens.* **38**: 2489–2504.

Goel, N. (1989), Inversion of canopy reflectance models for estimation of biophysical parameters from reflectance data, in *Theory and Applications of Optical Remote Sensing*, G. Asrar, ed., Wiley, pp. 205–251.

Goldberg, D. E. (1989), *Genetic Algorithms in Search, Optimization and Machine Learning*, Addison-Wesley.

Gong, P., Wang, S. X., and Liang, S. (1999), Inverting a canopy reflectance model using a neural network, *Int. J. Remote Sens.* **20**: 111–122.

Goovaerts, P. (1997), *Geostatistics for Natural Resources Evaluation*, Oxford Univ. Press.

Gopal, S. and Woodcock, C. (1996), Remote sensing of forest change using artificial neural networks, *IEEE Trans. Geosci. Remote Sens.* **34**: 398–404.

Gutman, G. and Ignatov, A. (1998), Derivation of green vegetation fraction from NOAA/AVHRR for use in numerical weather prediction models, *Int. J. Remote Sens.* **19**: 1533.

Hansen, M., DeFries, R. S., Townshend, J. R. G., Sohlberg, R., Dimiceli, C., and Carroll, M. (2002), Towards an operational MODIS continuous field of percent tree cover algorithm: Examples using AVHRR and MODIS data, *Remote Sens. Envir.* **83**: 303–319.

Huang, C., Wylie, B., Yang, L., Homer, C., and Zylstra, G. (2002), Derivation of a tasseled cap transformation based on Landsat-7 at-satellite reflectance, *Int. J. Remote Sens.* **23**: 1741–1748.

Huete, A. R. (1988), A soil-adjusted vegetation index (SAVI), *Remote Sens. Envir.* **25**: 295–309.

Huete, A. R., Liu, H. Q., Batchily, K., and van Leeuwen, W. (1997), A comparison of vegetation indices over a global set of TM images for EOS-MODIS, *Remote Sens. Envir.* **59**: 440–451.

Huete, A., Didan, K., Miura, T., Rodriguez, E. P., Gao, X., and Ferreira, L. G. (2002), Overview of the radiometric and biophysical performance of the MODIS vegetation indices, *Remote Sen. Envir.* **83**: 195–213.

Hunt, E. R. J. and Rock, B. N. (1989), Detection of changes in leaf water content using near- and middle-infrared reflectances, *Remote Sens. Envir.* **30**: 43–54.

Isaaks, E. and Srivastava, R. M. (1989), *An Introduction to Applied Geostatistics*, Oxford Univ. Press.

Jacquemoud, S., Bacour, C., Poilve, H., and Frangi, J. P. (2000), Comparison of four radiative transfer models to simulate plant canopies reflectance: Direct and inverse mode, *Remote Sens. Envir.* **74**: 471–481.

Jago, R. A., Cutler, M. E. J., and Curran, P. J. (1999), Estimating canopy chlorophyll concentration from field and airborne spectra, *Remote Sens. Envir.* **68**: 217–224.

Jin, Y. and Liu, C. (1997), Biomass retrieval from high dimensional active/passive remote data by using artificial neural networks, *Int. J. Remote Sens.* **18**: 971–979.

Jin, Y. and Wang, Y. (1999), A genetic algorithm to simultaneously retrieve land surface roughness and soil moisture, *Proc. 25th Annual Conf. Exhibition of the Remote Sensing Society Earth Observation: From Data to Information*, Univ. Wales at Cardiff and Swansea.

Jupp, D. L. B., Strahler, A. H., and Woodcock, C. E. (1988), Autocorrelation and regularization in digital images I. Basic theory, *IEEE Trans. Geosci. Remote Sens.* **26**: 463–473.

Jupp, D. L. B., Strahler, A. H., and Woodcock, C. E. (1989), Autocorrelation and regularization in digital images II. Simple image models, *IEEE Trans. Geosci. Remote Sens.* **27**: 247–258.

Justice, C. O. (1986), Monitoring east African vegetation using AVHRR data, *Int. J. Remote Sens.* **6**: 1335–1372.

Karnieli, A., Kaufman, Y. J., Remer, L., and Wald, A. (2001), AFRI-aerosol free vegetation index, *Remote Sens. Envir.* **77**: 10–21.

Kaufman, Y. J. and Tanré, D. (1992), Atmospherically resistant vegetation index (ARVI) for EOS-MODIS, *IEEE Trans. Geosci. Remote Sens.* **30**: 261–270.

Kim, M. S., Daughtry, C. S. T., Chappelle, E. W., and McMurtrey, J. E. (1994), The use of high spectral resolution bands for estimating absorbed photonsynthetically active radiation (APAR), *Proc. ISPRS*, Val d'Isere, France, 299–306.

Kimes, D., Gastellu-Etchegorry, J., and Estve, P. (2002), Recovery of forest canopy characteristics through inversion of a complex 3D model. *Remote Sens. Envir.* **79**: 320–328.

Kimes, D., Nelson, R., Manry, M., and A. Fung, A. (1998), Attributes of neural networks for extracting continuous vegetation variables from optical and radar measurements, *Int. J. Remote Sens.* **19**: 2639–2663.

Kimes, D. S., Knyazikhin, Y., Privette, J. L., Abuelgasim, A. A., and Gao, F. (2000), Inversion methods for physically-based models. *Remote Sens. Rev.* **18**: 381–440.

Knyazikhin, Y., Martonchik, J. V., Diner, D. J., Myneni, R. B., Verstraete, M. M., Pinty, B., and Gobron, N. (1998a), Estimation of vegetation canopy leaf area index and fraction of absorbed photosynthetically active radiation from atmosphere-corrected MISR data, *J. Geophys. Res.* **103**: 32239–32256.

Knyazikhin, Y., Martonchik, J. V., Myneni, R. B., Diner, D. J., and Running, S. W. (1998b), Synergistic algorithm for estimating vegetation canopy leaf area index and fraction of absorbed photosynthetically active radiation from MODIS and MISR data, *J. Geophys. Res.* **103**: 32257–32275.

Kokaly, R. F. and Clark, R. N. (1999), Spectroscopic determination of leaf biochemstry using band depth analysis of absorption features and stepwise multiple linear regression, *Remote Sens. Envir.* **67**: 267–287.

Li, Y., Demetriades-Shah, T. H., Kanemasu, E. T., Shultis, J. K., and Kirkham, M. B. (1993), Use of second derivatives of canopy reflectance for monitoring prairie vegetation over different soil background, *Remote Sens. Envir.* **44**: 81–87.

Li, X., Gao, F., Wang, J., and Strahler, A. (2000), A priori knowledge accumulation and its application to linear BRDF model inversion, *J. Geophys. Res.* **106**: 11,925–11,935.

Liang, S. (2003a), Mapping incident photosynthetically active radiation (PAR) from polar-orbiting satellite observations, *J. Geophys. Res.*: submitted.

Liang, S. (2003b), A direct algorithm for estimating land surface broadband albedos from MODIS imagery, *IEEE Trans. Geosci. Remote Sens.* **41**: 136–145.

Liang, S. and Strahler, A. H. (1993), An analytic BRDF model of canopy radiative transfer and its inversion, *IEEE Trans. Geoscie. Remote Sens.* **31**: 1081–1092.

Liang, S. and Strahler, A. H. (1994), Retrieval of surface BRDF from multiangle remotely sensed data, *Remote Sens. Envir.* **50**: 18–30.

Liang, S., Hu, J., and Huang, Q. (1989), Geometry characteristics and mechanisms of band combinations, *Proc. ASPRS/ACSM/Auto-Carto9 Convention.* Vol. 2, pp. 162–171.

Liang, S., Fang, H., and Chen, M. (2001), Atmospheric correction of Landsat ETM + land surface imagery: I. Methods, *IEEE Trans. Geosci. Remote Sens.* **39**: 2490–2498.

Liang, S., Fang, H., Kaul, M., Van Niel, T. G., McVicar, T. R., Pearlman, J., Walthall, C. L., Daughtry, C., and Huemmrich, K. F. (2003), Estimation of land surface broadband albedos and leaf area index from EO-1 ALI data and validation, *IEEE Trans. Geosci. Remote Sens.* **41**: 1260–1268.

Lin, Y. and Sarabandi, K. (1999), Retrieval of forest parameters using a fractal-based coherent scattering model and a genetic algorithm, *IEEE Trans. Geosci. Remote Sens.* **37**: 1415–1424.

Liu, J., Chen, J. M., Cihlar, J., and Park, W. M. (1997), A process-based boreal ecosystem productivity simulator using remote sensing inputs, *Remote Sens. Envir.* **62**: 158–175.

Macomber, S. A. and Woodcock, C. E. (1994), Mapping and monitoring conifer mortality using remote sensing in the Lake Tahoe basin, *Remote Sens. Envir.* **50**: 255–266.

Major, D. J., Baret, F., and Guyot, G. (1990), A ratio vegetation index adjusted for soil brightness, *Int. J. Remote Sens.* **11**: 727–740.

Michaelson, J., Schimel, D. S., Friedl, M. A., Davis, F. W., and Dubayah, R. O. (1994), Regression tree analysis of satellite and terrain data to guide vegetation sampling and surveys, *J. Veg. Sci.* **5**: 673–696.

Miller, J. R., Hare, E. W., and Wu, J. (1990), Quantitative characterization of the vegetation red edge reflectance. 1. An inverted-Gaussian reflectance model, *Int. J. Remote Sens.* **11**: 1755–1773.

Myneni, R. B., Knyazikhin, Y., Zhang, Y., Tian, Y., Wang, Y., Lotsch, A., Privette, J. L., Morisette, J. T., Running, S. W., Nemani, R., Glassy, J., and Votava, P. (1999), *Modis Leaf Area Index (LAI) and Fraction of Photosynthetically Active Radiation Absorbed by Vegetation (fPAR) Product: Algorithm Theoretical Basis Document*, Boston Univ.

NAG (1990), *The NAG Fortran Library*, mark 14, volume 3. NAG, Inc., Downers Grove, IL.

Peñuelas, J., Filella, L., and Gamon, J. A. (1995a), Assessment of photosynthetic radiation use efficiency with spectral reflectance, *New Phytol.* **131**: 291–296.

Peñuelas, J., Baret, F., and Filella, I. (1995b), Semi-empirical indices to assess cartotenoids/chlorophyll a ratio from leaf spectral reflectance, *Photosynthetica* **31**: 221–230.

Peñuelas, J., Fielella, I., Biel, C., Serrano, L., and Save, R. (1993), The reflectance at the 950–970 nm region as an indicator of plant water status, *Int. J. Remote Sens.* **14**: 1887–1905.

Peñuelas, J., Gamon, J. A., Fredeen, A. L., Merino, J., and Field, C. B. (1994), Reflectance indices associated with physiological changes in nitrogen- and water-limited sunflower leaves, *Remote Sens. Environ.* **48**: 135–146.

Phinn, S., Franklin, J., Hope, A., Stow, D., and Huenneke, L. (1996), Biomass distribution mapping using airborne digital video imagery and spatial statistics in a semiarid environment, *J. Envir. Manage.* **47**: 139–164.

Pinty, B. and Verstraete, M. M. (1992), GEMI: A nonlinear index to monitor global vegetation from satellites, *Vegetation* **10**: 15–20.

Press, W. H., Flannery, B. P., Teukolsky, S. A., and Vetterling, W. T. (1989), *Numerical Recipes. The Art of Scientific Computering (Fortran Version)*, Cambridge Univ. Press.

Prince, S. D. and Goward, S. N. (1995), Global primary production: A remote sensing approach, *J. Biogeogr.* **22**: 815–835.

Prince, S. D. and Steininger, M. K. (1999), Biophysical stratification of the Amazon basin, *Global Change Biol.*, **5**: 1–22.

Privette, J. L., Eck, T. F., and Deering, D. W. (1997), Estimating spectral albedo and nadir reflectance through inversion of simple brdf models with AVHRR/MODIS-like data, *J. Geophys. Res.* **102**: 29529–29542.

Privette, J. L., Myneni, R. B., Tucker, C. J., and Emery, W. J. (1994), Invertibility of a 1-D discrete ordinates canopy reflectance model, *Remote Sens. Envir.* **48**: 89–105.

Qi, J., Chehbouni, A., Huete, A. R., Kerr, Y., and Sorooshian, S. (1994), A modified soil adjusted vegetation index (MSAVI), *Remote Sens. Envir.* **48**: 119–126.

Ramstein, G. and Raffy, M. (1989), Analysis of the structure of radiometric remotely sensed images, *Int. J. Remote Sens.* **10**: 1049–1073.

Ridao, E., Conde, J. R., and Minguez, M. I. (1998), Estimating FAPAR from nine vegetation indices for irrigated and nonirrigated faba bean and semileafless pea canopies, *Remote Sens. Envir.* **66**: 87–100.

Roujean, J. L. and Bréon, F. M. (1995), Estimating PAR absorbed by vegetation from bidirectional reflectance measurements, *Remote Sens. Envir.* **51**: 375–384.

Running, S. W. and Coughlan, J. C. (1988), A general model of forest ecosystem processes for regional applications. I. Hydrologic balance, canopy gas exchange and primary production processes, *Ecol. Model.* **42**: 125–154.

Sellers, P., Tucker, C., Collatz, G., Los, S., Justice, C., Dazlich, D., and Randall, D. (1994), A global 1 by 1 NDVI data set for climate studies. Part 2: The generation of global fields of terrestrial biophysical parameters from the NDVI, *Int. J. Remote Sens.* **15**: 3519–3546.

Sellers, P. J. (1987), Canopy reflectance, photosynthesis, and transpiration. II. The role of biophysics in the linearity of their interdependence, *Remote Sens. Environ.* **21**: 143–183.

Sellers, P. J., Mintz, Y., Sud, Y. C., and Dalcher, A. (1986), A simple biosphere model (SIB) for use within general circulation models, *J. Atmos. Sci.* **43**: 505–531.

Smith, J. A. (1993), LAI inversion using a backpropagation neural network trained with a multiple scattering model, *IEEE Trans. Geosci. Remote Sens.* **31**: 1102–1106.

St-Onge, B. A. and Cavayas, F. (1995), Estimating forest stand structure from high resolution imagery using the directional variogram, *Int. J. Remote Sens.* **16**: 1999–2021.

Todd, S. W., Hoffer, R. M., and Milchunas, D. G. (1998), Biomass estimation on grazed and ungrazed rangelands using spectral indices, *Int. J. Remote Sens.* **19**: 427–438.

Verstraete, M. M. and Pinty, B. (1998), Designing optimal spectral indices for remote sensing applications, *IEEE Trans. Geosci. Remote Sens.* **34**: 1254–1265.

Wang, Y. and Jin, Y. (2000), A genetic algorithm to simultaneously retrieve land surface roughness and soil moisture, *J. Remote Sens.* **4**: 90–94.

Wanner, W., Li, X., and Strahler, A. (1995), On the derivation of kernels for kernel-driven models of bidirectional reflectance, *J. Geophys. Res.* **100**: 21077–21090.

Weiss, M. and Baret, F. (1999), Evaluation of canopy biophysical variable retrieval performances from the accumulation of large swath satellite data, *Remote Sens. Envir.* **70**: 293–306.

Weiss, M., Baret, F., Myneni, R. B., Pragnere, A., and Knyazikhin, Y. (2000), Investigation of a model inversion technique to estimate canopy biophysical variables from spectral and directional reflectance data, *Agronomie* **20**: 3–22.

Weiss, M., Baret, F., Leroy, M., Hautecoeur, O., Prevot, L., and Bruguier, N. (2000b), *Validation of Neural Network Techniques for the Estimation of Canopy Biophysical Variables from Vegetation Data, VEGETATION Preparatory Programmer*, final report.

Wiegand, C. L., Richardson, A. J., Escobar, D. E., and Gerbermann, A. H. (1991), Vegetation indices in crop assessment, *Remote Sens. Envir.* **35**: 105–119.

Wiegand, C. L., Maas, S. J., Aase, J. K., Hatfield, J. L., Pinter, P. J. J., Jackson, R. D., Kanemasu, E. T., and Lapitan, R. L. (1992), Multisite analyses of spectral-biophysical data for wheat, *Remote Sens. Envir.* **42**: 1–21.

Woodcock, C. E. and Strahler, A. H. (1987), The factor of scale in remote sensing, *Remote Sens. Envir.* **21**: 311–332.

Woodcock, C. E., Strahler, A. H., and Jupp, D. L. B. (1988a), The use of variograms in remote sensing: Part I. Scene models and simulated images, *Remote Sens. Environ.* **25**: 323–348.

Woodcock, C. E., Strahler, A. H., and Jupp, D. L. B. (1988b), The use of variograms in remote sensing: Part II. Real digital images, *Remote Sens. Environ.* **25**: 349–379.

Zhuang, J. and Xu, X. (2000), Genetic algorithms and its application to the retrieval of component temperature, *Remote Sens. Land Resour.* **1**: 28–33.

9

Estimation of Surface Radiation Budget: I. Broadband Albedo

The Surface Radiation Budget (SRB) determines the amount of radiation absorbed by the land surface. This radiation drives the spatial and temporal variations of various surface phenomena. The net radiation is simply the sum of the shortwave and longwave components. Shortwave radiation is the topic of this chapter, and longwave radiation is discussed in Chapter 10.

On average, the surface absorbs about twice as much solar (e.g., shortwave) radiation as does the atmospheric column. Surface solar absorption is modulated by the surface shortwave albedo, which is the core of this chapter. After presenting some background materials, we examine the basic characteristics of surface broadband albedo in Section 9.2. We then present two approaches of determining surface broadband albedos from remotely sensed observations. Section 9.3 discusses the first one that is based on the detailed physical understanding of different components by performing a series of processing steps (e.g., atmospheric correction, angular modeling, narrow-band–broadband conversion). Section 9.4 describes the second approach that estimates surface broadband albedos directly from TOA radiance without performing all the processing steps in the first approach. The temporal dimensions of surface broadband albedos are discussed in Section 9.5.

9.1 INTRODUCTION

Climate feedback mechanisms are often observed as changes in the Earth Radiation Budget (ERB). This is because Earth's energy balance plays a fundamental role in atmospheric and oceanic circulations, their thermal

Quantitative Remote Sensing of Land Surfaces. By Shunlin Liang
ISBN 0-471-28166-2 Copyright © 2004 John Wiley & Sons, Inc.

conditions, and therefore our climate. Because of the strong influence of ERB on the behavior of other climate variables, it is essential that we monitor the spatial and temporal variations in the radiation energy budget of the Earth.

The radiation energy budget can be monitored at the top of the atmosphere (TOA) and at the surface. Both are equally important to measure in terms of climatic significance. At the TOA, radiation is the only form of energy exchange, and thus TOA radiative fluxes provide a measure of the total energy balance of the Earth–atmosphere system. These fluxes are also important in the study of the radiative impact of clouds on the global climate system. At the TOA, the radiative source of energy for the Earth surface–atmosphere system is a balance between the absorbed solar short-wave radiation and outgoing longwave radiation emitted to space.

A zero-dimensional energy balance model describes the global annual mean equilibrium of the Earth-atmosphere system as

$$0.25(1 - \alpha)S_0 = F = \sigma T_e^4 \tag{9.1}$$

where α is the planetary albedo at TOA, S_0 the solar constant discussed in Section 1.2.6, F the outgoing terrestrial longwave radiation, T_e the radiometric temperature of the Earth system, and σ the Stefan–Boltzmann constant ($\sigma = 5.67 \times 10^{-8}$ W m^{-2} K^{-4}). The factor 0.25 ($= \frac{1}{4}$) arises from the Earth intercepting solar radiation proportionally to its cross section and emitting terrestrial radiation proportionally to its surface. According to the ERBE (Earth Radiation Budget Experiment) satellite measurements (Barkstrom and Smith 1986), $S_0 = 1372$ W/m^2 and $\alpha = 0.295 \pm 0.010$. Thus, $F = 237$ W m^2 and $T_e = 255$ K. The discrepancy between this value and the mean climatological surface temperature ($T_s = 288$ K) is explained by the "greenhouse" effect.

Table 9.1 presents various components of the energy balance at both TOA and surface. Out of the 341 W/m^2 of solar energy available at TOA in the 0.25–4 μm wavelength range, about 30% is reflected back to outer space without a change of wavelength after scattering in the atmosphere and/or reflection at Earth's surface, where the true solar input to the atmosphere is about 239 W/m^2. A large fraction of this reaches the surface and is absorbed by land masses and oceans; only roughly one-quarter of this solar radiation is absorbed within the atmosphere and creates a mean heating of ~ 0.6 K/day. Part of the solar energy input to the surface (~ 184 W/m^2) is returned to the atmosphere by emission of terrestrial longwave radiation in the 4–100 μm wavelength range, but this emission does not fully compensate for the solar flux into the surface. The deficit of ~ 113 W/m^2 is compensated by turbulent transport of latent heat and sensible heat from the surface to the atmosphere. The existence of a radiative balance at TOA and of a radiative imbalance at the surface implies that the atmosphere itself is a net source of

**TABLE 9.1 Radiation Budget at TOA and Surface for
February 1985–January 1989** [a]

Shortwave		Longwave		Net	
Incoming TOA flux	341	—		Net TOA	$+4$[b]
Absorbed TOA flux	239	TOA OLR	235	—	
Surface insolation	184	Surface downward	348	—	
Surface absorbed	160	Surface cooling	47	Net surface	$+113$
Atmosphere absorbed	79	Atmosphere cooling	188	Net atmosphere	-109

[a] In units of W/m^2. TOA values are ERBE estimates, and surface estimates are from surface radiation budget calculations.
[b] Net TOA radiation is a measure of ERBE estimate error, and is probably close to zero in reality.
Source: Hartmann et al. (1999).

terrestrial longwave radiation to compensate for the warming by the solar heating, latent heat release, and sensible-heat flux.

In the preceding discussion, only the values of the radiation budget at TOA are known with some degree of accuracy, owing to some decades of satellite measurements. Estimates of the components of the energy budget at the surface are more difficult to obtain because of the lack of global coverage by conventional observation systems and the large uncertainties in the ongoing tentative determination of these quantities from satellite measurements.

We can see from Table 9.1 that the Earth surface absorbs about twice as much as solar radiation as does the atmosphere. The amount of solar radiation that is absorbed by the surface is modulated by the surface broadband albedo. Land surface albedo is a fundamental component in determining Earth's climate (Cess 1978, Dickinson 1983, Kiehl et al. 1996). It is a parameter needed by both global and regional climate models for computing the surface energy balance. The seasonal and long-term vegetation dynamics that significantly impact the climate are reflected by the dramatic variations of albedo. In semiarid regions, an increase in albedo leads to a loss of radiative energy absorbed at the surface, and convective overturning is reduced. As a result, precipitation decreases. Evaporation may also decrease, further inhibiting precipitation. Similar reductions in precipitation and evapotranspiration have been found for increased albedo in tropical Africa and the Amazon basin (Dirmeyer and Shukla 1994). It has been widely recognized that surface albedo is among the main radiative uncertainties in current climate modeling. Most global circulation models are still using prescribed fields of surface albedo, which are often 5–15% in error (relative error) from place to place and time to time (Dorman and Sellers 1989, Sato et al. 1989). Data from the boreal ecosystem atmosphere study (BOREAS) project revealed that the winter albedos of the forest sites were significantly

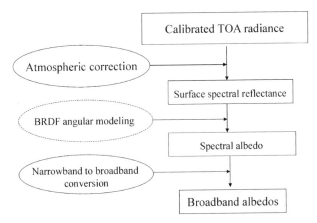

Figure 9.1 Flowchart for estimating land surface broadband albedos from multispectral sensors (e.g., MODIS).

different from those used in the European numerical weather prediction model, which led to a systematic underestimation of the near-surface air temperature (Sellers et al. 1997).

Remote sensing is the only practical means for mapping land surface albedo globally. Broadband albedo is usually estimated from broadband sensors. Measurements obtained with the Nimbus 7 Earth Radiation Budget (ERB) instrument (Jacobowitz et al. 1984), data from the Earth Radiation Budget Experiment (ERBE) (Smith et al. 1986) and the Scanner for the Earth Radiation Budget (ScaRaB) (Kandel et al. 1998) have yielded unique datasets that are a valuable source of knowledge of land surface broadband albedo. Although there are efforts to estimate broadband albedos from single-band sensors, such as GOES or METEOSAT (Pinker 1985; Pinker et al. 1994; Pinty et al. 2000a, 2000b), the accurate determination of land surface broadband albedo from TOA observations requires the knowledge of the atmospheric conditions and surface characteristics, which can be monitored effectively only by multispectral sensors. Narrowband multispectral observations also have much finer spatial resolutions that allow us to characterize both the surface and atmospheric heterogeneity.

The derivation of surface broadband albedos from narrowband observations usually requires several levels of processing (see Fig. 9.1). Important steps include

- Atmospheric correction that converts TOA radiance to surface directional reflectance
- BRDF angular conversion that converts directional reflectance to spectral albedo
- Narrowband–broadband conversion that converts spectral albedos to broadband albedos.

This approach is based on our physical understanding and has been used by the NASA MODIS science team (Schaaf et al. 2002). Atmospheric correction methods have been discussed in Chapter 6. For BRDF modeling, in Chapters 3 and 4 we discussed radiative transfer models of vegetation canopies and soil, which are usually referred to as *physical BRDF models*. The statistical BRDF models are described in Section 2.2. Given BRDF models and their parameters, their integration into albedo is straightforward according to Eqs. (1.10)–(1.11). In Section 9.3, we discuss methods of converting narrowband to broadband albedos.

An alternative solution (Liang 2003) is to estimate surface broadband albedo directly from TOA radiance, discussed in Section 9.4. Because of the strong dependence of broadband albedos of certain land surface covers on the solar zenith angle and their importance in calculating surface energy balance, diurnal cycle modeling is discussed in Section 9.5.

9.2 BROADBAND ALBEDO CHARACTERISTICS

Spectral albedo of land surface has been widely understood, although the terminology can still be quite confusing. It is the angular integration of surface BRDF or BRF over all viewing directions, as defined in Eq. (1.10), and is a measurement of surface inherent reflectivity in the single illustration direction.

Another definition of spectral albedo is ($\alpha(\lambda, \theta_0)$) at wavelength λ, where the solar zenith angle θ_0 is the ratio of upwelling flux (irradiance) $F_u(\lambda, \theta_0)$: to the downward flux (irradiance) $F_d(\lambda, \theta_0)$:

$$\alpha(\lambda, \theta_0) = \frac{F_u(\lambda, \theta_0)}{F_d(\lambda, \theta_0)} \qquad (9.2)$$

In the natural environment, the downward flux is the sum of direct solar flux and the diffuse solar flux that is an integration of sky radiance over all illumination directions. If the sky radiance is negligible, that is, the downward flux can be approximated by the direct solar flux, then Eq. (9.2) is equivalent to Eq. (1.10), and is termed *directional hemispheric reflectance* (DHR). It is also called "black sky" albedo by the MODIS science team since there is no contribution from sky radiance. The numerical difference between Eqs. (1.10) and (9.2) have not been well quantified.

The broadband albedo is defined over a range of the wavelength as the ratio of the surface upwelling flux (F_u) to the downward flux (F_d)

$$\alpha(\theta_i, \Lambda) = \frac{F_u(\theta_i, \Lambda)}{F_d(\theta_i, \Lambda)} = \frac{\int_\Lambda F_d(\theta_i, \Lambda) r(\lambda) \, d\lambda}{\int_\Lambda F_d(\theta_i, \Lambda) \, d\lambda} \qquad (9.3)$$

where $r(\lambda)$ is surface reflectance and Λ denotes the waveband from wavelength λ_1 to wavelength λ_2 $\Lambda \in (\lambda_1, \lambda_2)$. If $\Lambda \in (0.25-5.0 \ \mu m)$, then $\alpha(\theta_i, \Lambda)$ is the total shortwave broadband albedo. The waveranges $\Lambda \in (0.4-0.7 \ \mu m)$ and $\Lambda \in (0.7-5.0 \ \mu m)$ correspond to visible and near-infrared albedos, respectively. Note that the short end ($0.25 \ \mu m$) and the high end ($5.0 \ \mu m$) of the wavelength may appear different in the literature, but it really does not matter much since the incoming solar radiation at two ends approaches zero anyway.

Since broadband albedos have not been widely discussed in the literature, it is helpful to examine the their characteristics first. From Eq. (9.3), we can see that broadband albedo is not the sole measure of surface properties. It depends on the atmospheric conditions through downward fluxes that are the weighting function of the conversion from narrowband to broadband albedos. To illustrate this point quantitatively, let us examine two surface types: deciduous forest and snow. Figure 9.2 shows the dependence of spectral downward flux on solar zenith angle and atmospheric visibility (low visibility representing very turbid atmosphere and high visibility, very clear atmosphere). It is obvious that direct flux increases and diffuse flux decreases when the atmosphere varies from turbid (low visibility) to clear (high visibility). Total spectral downward fluxes decrease under both clear and turbid atmospheric conditions.

Figure 9.3 demonstrates how broadband albedos vary as a function of solar zenith angle and atmospheric visibility where the reflectance spectra of snow and deciduous forest are fixed. It is interesting to see that the total shortwave albedo of deciduous forest increases as solar zenith angle becomes larger, but the total shortwave albedo of snow has the opposite trend. This is because canopy has a much higher reflectance in the near-IR spectrum, but snow has a much higher reflectance in the visible spectrum. When solar zenith angle becomes larger, the downward flux in the visible spectrum is reduced faster than in the near-IR spectrum given a specific atmospheric condition. This indicates that the relative weights are smaller in the visible region and larger in the near-IR region. Total near-IR albedos for both snow and canopy do not change much as solar zenith angle and atmospheric visibility vary.

If the total near-IR albedo is divided into direct and diffuse components, their dependences on solar zenith angle and the atmospheric visibility become much stronger. It is interesting to observe the asymptotic properties of total and diffuse near-IR broadband albedos. When the solar zenith angle is very large or the atmospheric visibility value is very low, total and diffuse near-IR albedos are numerically very close because downward diffuse fluxes actually represent total downward flux under these conditions. For both snow and plant canopies, the total visible albedos (as well as the direct and diffuse components) are almost unchanged as we change the solar zenith angle and atmospheric visibility. Note that in this experiment we assume that both canopy and snow have constant reflectance spectra for different solar zenith angles and atmospheric conditions. In fact, the spectral albedos of both snow

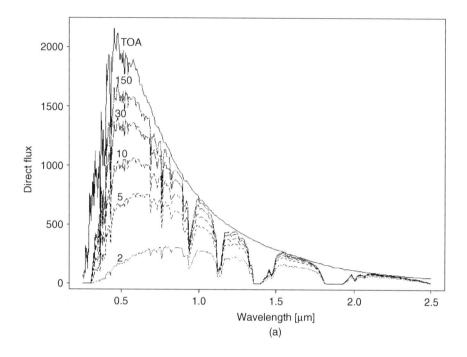

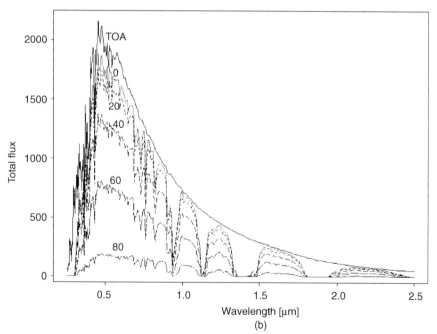

Figure 9.2   Dependence of spectral downward flux (W m^{-2} μm^{-1}) on solar zenith angle [(a), degrees] and atmospheric visibility [(b), in km] [from Liang et al. (1999)].

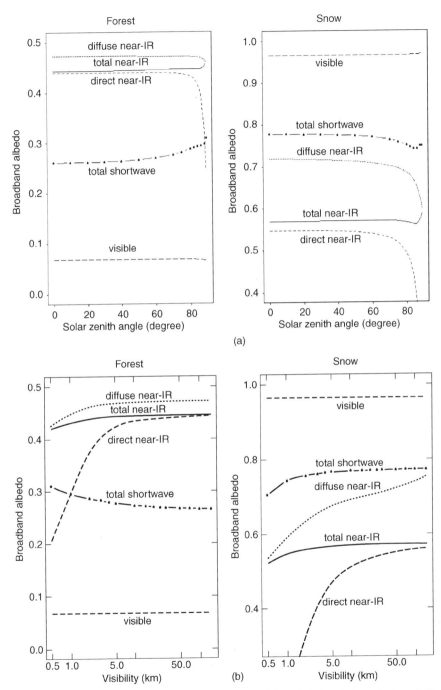

Figure 9.3   Broadband albedo as a function of solar zenith angle (a) and atmospheric visibility (b) [from Liang et al. (1999)].

and canopy depend on solar zenith angle. Thus, canopy and snow visible broadband albedos should change at different solar zenith angles.

Several important features of broadband albedos can be observed in Fig. 9.3. Unless the solar zenith angle and atmospheric visibility value are very large, broadband albedos are relatively stable, which forms the foundation of many studies. We can predict broadband albedos very well even when we do not know the atmospheric conditions. However, the diffuse and direct components are more sensitive to atmospheric conditions. The uncertainties of predicting these separate components are usually much greater than total broadband albedo.

9.3 NARROWBAND–BROADBAND CONVERSION

As just discussed, surface broadband albedos are not the sole measures of surface reflective properties since they also depend on the atmospheric conditions. The downward flux distribution at the bottom of the atmosphere is the weighting function for converting spectral albedos to broadband albedos, and different atmospheric conditions have different downward flux distributions. Thus, surface broadband albedos derived from remotely sensed data under one specific atmospheric condition may not be applicable under other atmospheric conditions.

In an earlier study (Liang et al. 1999), we suggested separating inherent albedo and apparent albedo. Inherent albedo is totally independent of the atmospheric condition, and the apparent albedos are equivalent to those measured by albedometers or pyranometers in the field. If inherent albedos are provided, advanced users may transform them to apparent albedos at any atmospheric conditions that they may need (Lucht et al. 2000).

When the atmospheric downward fluxes are known, integration with inherent albedo will produce much more accurate broadband albedo products. However, most users may not want to implement such a procedure for many practical applications. Given narrowband albedos, the average broadband albedos need to be predicted under general atmospheric conditions. This is what most previous studies have done (e.g., Brest and Goward 1987, Wydick et al. 1987, Saunders 1990, Li and Leighton 1992, Stroeve et al. 1997). The formulas of Brest and Goward (1987), Wydick et al. (1987) and Li and Leighton (1992) are for converting TOA narrowband albedos to surface broadband albedos.

For the two AVHRR reflective bands, Saunders (1990) and Gutman (1994) simply use 0.5 as their conversion coefficients for calculating TOA shortwave albedo. Valiente et al. (1995) derive a conversion formula for calculating surface shortwave broadband albedo based on the earlier version of 6S code and 20 surface reflectance spectra. Russell et al. (1997) have advanced a different formula for converting nadir-view surface reflectance based on a set of ground observations. Stroeve et al. (1997) derive the surface

TABLE 9.2 Coefficients of the Linear Equation (9.4)

a_0	a_1	a_2	References
0.0442	0.441	0.67	Russell et al. (1997)
0.035	0.545	0.32	Valiente et al. (1995)
0.0034	0.34	0.57	Key (2001)
0.0412	0.655	0.216	Stroeve et al. (1997)

broadband albedo over the Greenland ice sheet. Han et al. (1999) calculate surface broadband albedo using a different formula suggested by Key (2001). When we group them as follows, their formulas have the common form:

$$\alpha_{\text{short}} = a_0 + a_1 R_1 + a_2 R_2 \qquad (9.4)$$

where R_1 and R_2 are AVHRR spectral albedos, and the other coefficients are listed in Table 9.2, which demonstrates how different these coefficients are for the same AVHRR sensor. Clearly, further study on this subject is necessary.

All of these conversion studies were based on either field measurements of certain surface types or model simulations. It is impossible to develop a universal formula based only on ground measurements because it is so expensive to collect extensive datasets for different atmospheric and surface conditions. Model simulation is a better approach to develop universal conversion formulas, and ground measurements are certainly valuable for validation.

Most studies mentioned above provide conversion formulas for calculating only the total shortwave broadband albedo. In land surface modeling, the visible and near-infrared (NIR) broadband albedos are needed quite often. Moreover, both total visible and near-IR broadband albedos are further divided into direct and diffuse albedos. For example, in the NASA Goddard Earth observation system–data assimilation system (GEOS-DAS) assimilation surface model (Koster and Suarez 1992), National Center for Atmospheric Research (NCAR) community climate model (Kiehl et al. 1996), and the simple biospheric model (Xue et al. 1991, Sellers et al. 1996), broadband visible and near-IR albedos are further divided into direct and diffuse albedos in both visible and near-IR regions. To calibrate and validate these land surface models, direct and diffuse visible and near-IR broadband albedos should be generated directly from satellite observations.

In a previous study (Liang 2001), we provided simple formulae for calculating average land surface broadband albedos (total shortwave albedo, total, direct, and diffuse visible and near-IR albedos) from a variety of narrowband sensors under various atmospheric and surface conditions. These results are based mainly on extensive radiative transfer simulation using the

Santa Barbara DISORT atmospheric radiative transfer (SBDART) code (Ricchiazzi et al. 1998). The key element of the radiative transfer simulation is the inclusion of representative surface reflectance spectra. In the previous studies, only a few dozens of surface reflectance spectra are usually used in the model simulations (e.g., Li and Leighton 1992, Ustin et al. 1993, Liang et al. 1999) because of computational constraints. In this study, a method was applied that allows us to incorporate as many representative surface reflectance spectra as we want without experiencing the formidable computational burden. In total, we employed 256 surface reflectance spectra in this study, including soil (43), vegetation canopy (115), water (13), wetland and beach sand (4), snow and frost (27), and urban (26), road (15), rock (4), and other cover types (9). Each has different wavelength dependences and magnitudes, from coastal water (low albedos) to snow and frost (high visible albedos). Eleven atmospheric visibility values (2, 5, 10, 15, 20, 25, 30, 50, 70, 100, and 150 km) were used for different aerosol loadings, and five atmospheric profiles of MODTRAN defaults (tropical, midlatitude winter, sub-Arctic summer, sub-Arctic winter, and US62) that also represent different water vapor and other gaseous amounts and profiles were utilized. A range of nine solar zenith angles was simulated from 0° to 80° at 10° increments. SBDART was run at 231 spectral ranges with the increased wavelength increment from 0.0025 μm at the shortest wavelength end to 0.025 μm at the longest wavelength end.

After determining downward fluxes (direct and diffuse), the sensor spectral response functions were integrated with downward flux and surface reflectance spectra to generate narrowband spectral albedos. *Broadband albedo* is simply defined as the ratio of the upwelling flux F_u to the downward flux F_d [see Eq. (9.3)]. A simple regression analysis produces the

TABLE 9.3   Sensor Spectral Bands and Their Spectral Ranges[a]

Sensor	Spectral Bands with Wavelength Ranges						
	1	2	3	4	5	6	7
ALI	(0.43−0.45)/ (0.45−0.52)	0.52−0.61	0.63−0.69	(0.78−0.81)/ (0.85−0.89)	(1.2−1.3)/ (1.55−1.75)	—	2.08−2.35
ASTER [b]	0.52−0.60	0.63−0.69	0.78−0.86	1.6−1.7	2.15−2.18	2.18−2.22	2.23−2.28
AVHRR14	0.57−0.71	0.72−1.01	—	—	—	—	—
GOES8	0.52−0.72	—	—	—	—	—	—
ETM +	0.45−0.51	0.52−0.60	0.63−0.69	0.75−0.90	1.55−1.75	—	2.09−2.35
MISR	0.42−0.45	0.54−0.55	0.66−0.67	0.85−0.87	—	—	—
MODIS	0.62−0.67	0.84−0.87	0.46−0.48	0.54−0.56	1.23−1.25	1.63−1.65	2.11−2.15
POLDER	0.43−0.46	0.66−0.68	0.74−0.79	0.84−0.88	—	—	—
VEGETATION	0.43−0.47	0.61−0.68	0.78−0.89	1.58−1.75	—	—	—

[a] For ALI, there are three additional bands compared to ETM + that are given in the table.
[b] Wavelength ranges for ASTER are 2.29−2.36 and 2.36−2.43 in bands 8 and 9, respectively.

conversion formulas for the following sensors: ALI (advanced land imager), ASTER (advanced spaceborne thermal emission and reflection radiometer), AVHRR (advanced very high-resolution radiometer), GOES (geostationary operational environmental satellite), Landsat7 enhanced thematic mapper plus (ETM +), MISR (multiangle imaging spectroradiometer), MODIS (moderate-resolution imaging spectroradiometer, POLDER (polarization and directionality of Earth's reflectances), and VEGETATION in the SPOT spacecraft. Their spectral wavebands and wavelength ranges are specified in Table 9.3.

This approach is quite straightforward, but the validation results presented in the follow-up study (Liang et al. 2002a) indicate that these formulas can produce the broadband albedos reasonably well with the uncertainty of about 0.02, which meets the required accuracy for land surface modeling.

9.3.1 ALI

The advanced land imager (ALI) is a multispectral sensor aboard on NASA Earth Observer-1. It has similar spatial resolutions to the Landsat 7 ETM + , but has three additional spectral bands. The conversion formulas are

$$\alpha_{\text{short}} = 0.3466r_{1p} - 0.1435r_1 + 0.2278r_2 + 0.0985r_3 - 0.0574r_3$$
$$+ 0.2159r_{4p} + 0.0385r_5 + 0.1139r_{5p} + 0.0620r_7$$

$$\alpha_{\text{vis}} = 0.2812r_{1p} + 0.1248r_1 + 0.3592r_2 + 0.2353r_3$$

$$\alpha_{\text{vis dif}} = 0.3824r_{1p} + 0.1138r_1 + 0.3366r_2 + 0.1681r_3$$

$$\alpha_{\text{vis dir}} = 0.2254r_{1p} + 0.1401r_1 + 0.3707r_2 + 0.2640r_3 \qquad (9.5)$$

$$\alpha_{\text{NIR}} = 0.2917r_4 + 0.2707r_{4p} - 0.0316r_5 + 0.2502r_{5p} + 0.2258r_7$$

$$\alpha_{\text{NIR dif}} = 0.5468r_4 + 0.2012r_{4p} - 0.1531r_5 + 0.1799r_{5p} + 0.2418r_7$$

$$\alpha_{\text{NIR dir}} = 0.2609r_4 + 0.2726r_{4p} - 0.0104r_5 + 0.2615r_{5p} + 0.2179r_7$$

The fitting root mean square error (RMSE) values for the total shortwave, total visible, and total near-IR albedos are 0.0135, 0.0089, and 0.0178, respectively. Validation results using independent ground measurements showed that the RMSE values for these three corresponding albedos are 0.019, 0.019, and 0.021 (Liang et al. 2002a).

9.3.2 ASTER

ASTER (advanced spaceborne thermal emission and reflection radiometer) is a research facility instrument provided by Japan on board the Terra satellite that was launched in December 1999. It has three visible and near-IR bands with 15 m spatial resolution and six short infrared bands with 30 m spatial resolution (see Table 9.3) and five thermal bands. The primary scientific objective of the ASTER mission is to improve understanding of the local- and regional-scale processes occurring on or near the Earth surface and lower atmosphere (Yamaguchi et al. 1998). The conversion formulas are given below:

$$\alpha_{short} = 0.484\alpha_1 + 0.335\alpha_3 - 0.324\alpha_5 + 0.551\alpha_6 + 0.305\alpha_8 - 0.367\alpha_9$$

$$\alpha_{vis} = 0.82\,\alpha_1 + 0.183\alpha_2 - 0.034\alpha_3 - 0.085\alpha_4 - 0.289\alpha_5 + 0.352\,\alpha_6$$
$$+ 0.239\alpha_7 - 0.24\alpha_9$$

$$\alpha_{vis\;dif} = 0.911\,\alpha_1 + 0.089\alpha_2 - 0.04\alpha_3 - 0.109\alpha_4 - 0.388\,\alpha_5 + 0.441\,\alpha_6$$
$$+ 0.316\alpha_7 - 0.303\alpha_9$$

$$\alpha_{vis\;dir} = 0.781\,\alpha_1 + 0.224\alpha_2 - 0.032\,\alpha_3 - 0.07\alpha_4 - 0.257\alpha_5 + 0.308\,\alpha_6 \qquad (9.6)$$
$$+ 0.2\,\alpha_7 - 0.208\,\alpha_9$$

$$\alpha_{NIR} = 0.654\alpha_3 + 0.262\,\alpha_4 - 0.391\,\alpha_5 + 0.5\alpha_6$$

$$\alpha_{NIR\;dif} = 0.835\alpha_3 + 0.033\alpha_4 - 0.191\,\alpha_5 + 0.352\,\alpha_6$$

$$\alpha_{NIR\;dif} = 0.629\alpha_3 + 0.295\alpha_4 - 0.418\,\alpha_5 + 0.517\alpha_6$$

The fitting RMSE values for the total shortwave, total visible, and total near-IR albedos are 0.0135, 0.0089, and 0.0178, respectively (Liang 2001). Validation results using independent ground measurements (Liang et al. 2002a) showed that the RMSE values for these three corresponding albedos are 0.019, 0.019, and 0.021.

9.3.3 AVHRR

Data from AVHRR sensors have been acquired since 1981 and are expected to continue indefinitely. They have been widely used for a variety of applications (Townshend 1994), including the generation of the global albedo products (e.g., Strugnell and Lucht 2001, Csiszar and Gutman 1999). As mentioned in the introduction, many studies have provided AVHRR conversion formulas. Most of these studies were for either very bright surfaces (e.g., snow/ice) or low reflectance surfaces (soil/vegetation), and suggested linear conversions. Our experiments showed that the nonlinear fittings are much better than linear regression since AVHRR has only two bands that do not capture spectral variations of many different cover types very well. One approach is to separate cover types into snow, vegetation, soil, and so on (e.g., Brest and Goward 1987, Song and Gao 1999), which is problematic because the separation of these cover types has errors of its own. Instead, I made use of high-order polynomials in this study:

$$\alpha_{short} = -0.3376\alpha_1^2 - 0.2707\alpha_2^2 + 0.7074\alpha_1\alpha_2 + 0.2915\alpha_1 + 0.5256\alpha_2$$

$$\alpha_{vis} = 0.0074 + 0.5975\alpha_1 + 0.441\alpha_1^2$$

$$\alpha_{vis\ dif} = 0.0093 + 0.519\alpha_1 + 0.5257\alpha_1^2$$

$$\alpha_{vis\ dir} = 0.0051 + 0.6685\alpha_1 + 0.3648\alpha_1^2 \tag{9.7}$$

$$\alpha_{NIR} = -1.4759\alpha_1^2 - 0.6536\alpha_2^2 + 1.8591\alpha_1\alpha_2 + 1.063\alpha_2$$

$$\alpha_{NIR\ dif} = -0.628\alpha_1^2 - 0.3047\alpha_2^2 + 0.8476\alpha_1\alpha_2 + 1.0113\alpha_2$$

$$\alpha_{NIR\ dif} = -1.5696\alpha_1^2 - 0.6961\alpha_2^2 + 1.9679\alpha_1\alpha_2 + 1.0708\alpha_2$$

Derivation of these coefficients was based on NOAA14 AVHRR spectral response functions. We also experimented by using the AVHRR spectral response functions of NOAA7, -9, and -11, and found that the results were not significantly different.

The fitting RMSE values for the total shortwave, total visible, and total near-IR albedos are 0.0156, 0.0216, and 0.030, respectively (Liang 2001). Validation results using independent ground measurements (Liang et al. 2002a) showed that the RMSE values for these three corresponding albedos are 0.021, 0.018, and 0.025, respectively.

There are several studies on converting AVHRR narrowband albedos to the total shortwave broadband albedo (e.g., Valiente et al. 1995, Key 2001, Russell et al. 1997, Stroeve et al., 1997; Song and Gao 1999), most of which were based on experimental data with the limited number of surface types and atmospheric conditions. To verify our approach and results, we tested several formulas using our simulated database. We compared the simulated total shortwave albedo and the predicted shortwave albedo by the formulas of Valiente et al. (1995), Key (2001), and Song and Gao (1999). Key's formula was primarily for snow/ice and does fit our snow/ice samples well, and also some vegetation, soil, wetland, and water samples. The formula of Valiente et al. fits most vegetation/soil samples very well, but overestimates water/wetland and snow samples, probably because they used mainly vegetation/soil reflectance spectra in their simulations. Song and Gao (1999) developed a linear conversion formula whose coefficients are the second-order polynomial functions of NDVI. Their formula overestimates the total short-wave albedo of most vegetation and soil samples, but fits low and very high albedos very well. It is probably because their formula was derived from ground measurements with limited atmospheric and surface conditions.

9.3.4 GOES

The Geostationary Operational Environmental Satellite (GOES) has been commissioned and operated by NOAA. Normally there are two GOES satellites in operation. GOES east is stationed at 75°, and GOES west at 135° west. These provide coverage of most of the Western Hemisphere. GOES has five imagers, but only one band is in the shortwave reflective region. This one-band image has been widely used to provide surface broadband albedo products (e.g., Pinker et al. 1994, 1995). As we will demonstrate below, one band cannot adequately capture the variation of surface reflective properties, and therefore the accuracy of the GOES broadband albedo products is very questionable.

For total shortwave albedo, the conversion formula is

$$\alpha_{\text{short}} = 0.0759 + 0.7712\,\alpha \tag{9.8}$$

For visible albedos, second-order polynomial functions produce better fits:

$$\alpha_{\text{vis}} = -0.0084 + 0.689\alpha + 0.3604\alpha^2$$

$$\alpha_{\text{vis dif}} = -0.006 + 0.6119\alpha + 0.443\alpha^2 \tag{9.9}$$

$$\alpha_{\text{vis dir}} = -0.0111 + 0.7586\alpha + 0.2862\alpha^2$$

The near-IR broadband albedo cannot be predicted by one GOES band albedo simply because the spectral coverage of the band is mainly in the visible spectrum. This also explains why one GOES band cannot predict total shortwave albedo. We used both GOES8 and GOES10 sensor spectral response functions and the results are almost identical.

The fitting RMSE values for the total shortwave and total visible albedos are 0.0557, and 0.0217, respectively (Liang 2001). Validation results using independent ground measurements (Liang et al. 2002a) showed that the RSME values for these two corresponding albedos are 0.024 and 0.016.

9.3.5 LANDSAT TM / ETM +

Landsat thematic mapper (TM) started to acquire imagery in 1982 boarded on Landsat 4 and 5. Landsat 7 was launched in April, 1999 and carries the enhanced thematic mapper plus (ETM +). Both ETM + and TM have the same multispectral reflective wavebands with similar spectral coverages. They have produced long-term, high-resolution multispectral imagery of Earth's land surface globally.

The earlier studies on calculating the total shortwave broadband albedo from Landsat narrowbands (e.g., Brest and Goward 1987) primarily relied on the relations between TOA reflectances and ground measured broadband albedo. Since it is now practical to retrieve surface spectral reflectance by performing atmospheric correction procedures (e.g. Liang et al. 1997, Liang 2001), the equations presented in this study are only suitable for surface spectral albedos:

$$\alpha_{\text{short}} = 0.356\,\alpha_1 + 0.13\,\alpha_3 + 0.373\,\alpha_4 + 0.085\,\alpha_5 + 0.072\,\alpha_7$$

$$\alpha_{\text{vis}} = 0.443\,\alpha_1 + 0.317\alpha_2 + 0.24\alpha_3$$

$$\alpha_{\text{vis dif}} = 0.556\,\alpha_1 + 0.281\,\alpha_2 + 0.163\,\alpha_3$$

$$\alpha_{\text{vis dir}} = 0.39\alpha_1 + 0.337\alpha_2 + 0.274\alpha_3 \qquad (9.10)$$

$$\alpha_{\text{NIR}} = 0.693\,\alpha_4 + 0.212\,\alpha_5 + 0.116\alpha_7$$

$$\alpha_{\text{NIR dif}} = 0.864\alpha_4 + 0.158\,\alpha_7$$

$$\alpha_{\text{NIR dif}} = 0.659\alpha_4 + 0.342\,\alpha_5$$

We also found that the coefficients are not sensitive to the spectral response functions of different Landsat TM sensors. In other words, Eq. (9.10) is equally suitable for Landsat 4 and 5 TM imagery.

The fitting RMSE values for the total shortwave, total visible and total near-IR albedos are 0.0099, 0.0026, and 0.0153, respectively (Liang 2001). Validation results using independent ground measurements (Liang et al. 2002a) showed that the RMSE values for these three corresponding albedos are 0.019, 0.016, and 0.022.

A few studies have been reported to derive surface broadband shortwave albedo from Landsat TM imagery (e.g., Brest and Goward 1987, Duguay and LeDrew 1992, Gratton et al. 1993, Knap et al. 1999). Knap et al. (1999) develop a second-order polynomial formula based on ground measurements of both glacier ice and snow:

$$\alpha_{\text{short}} = 0.726\,\alpha_2 - 0.322\,\alpha_2^2 - 0.051\,\alpha_4 + 0.581\,\alpha_4^2 \qquad (9.11)$$

From our simulated datasets, we found that their formula fits our snow/ice samples quite well, but fits the soil/vegetation samples very poorly. Duguay and LeDrew (1992) develop a linear formula using three TM bands:

$$\alpha_{\text{short}} = 0.526\,\alpha_2 + 0.3139\alpha_4 + 0.112\,\alpha_7 \qquad (9.12)$$

and it fits our data of all cover types rather well.

9.3.6   MISR

The MISR instrument is designed to improve our understanding of the Earth's ecology, environment and climate by providing multiangle observations (Diner et al. 1998). MISR images Earth in nine different view directions (one nadir view, and four off-nadir views pointing to each forward and backward direction) to infer the angular variation of reflected solar radiation within four spectral bands in the visible and near-IR spectra. It takes 7 min for a point on Earth to be observed at all nine angles. This instrument is carried by the Earth Observing System (EOS) spacecraft Terra.

$$\alpha_{\text{short}} = 0.126\,\alpha_2 + 0.343\,\alpha_3 + 0.415\,\alpha_4 + 0.0037$$

$$\alpha_{\text{vis}} = 0.381\,\alpha_1 + 0.334\,\alpha_2 + 0.287\alpha_3$$

$$\alpha_{\text{vis dif}} = 0.478\,\alpha_1 + 0.306\,\alpha_2 + 0.219\alpha_3$$

$$\alpha_{\text{vis dir}} = 0.335\,\alpha_1 + 0.349\alpha_2 + 0.317\alpha_3 \tag{9.13}$$

$$\alpha_{\text{NIR}} = -0.387\alpha_1 - 0.196\,\alpha_2 + 0.504\alpha_3 + 0.83\,\alpha_4 + 0.011$$

$$\alpha_{\text{NIR dif}} = -0.24\alpha_1 + 0.269\alpha_3 + 0.866\,\alpha_4$$

$$\alpha_{\text{NIR dif}} = -0.407\alpha_1 - 0.226\,\alpha_2 + 0.536\alpha_3 + 0.826\,\alpha_4 + 0.012$$

The fitting RMSE values for the total shortwave, total visible, and total near-IR albedos are 0.0157, 0.0026, and 0.0296, respectively (Liang 2001). Validation results using independent ground measurements (Liang et al. 2002a) showed that the RMSE values for these three corresponding albedos are 0.022, 0.015, and 0.027.

9.3.7 MODIS

MODIS is an EOS facility instrument designed to measure biological and physical processes on a global basis every 1–2 days. It will provide long-term observations from which an enhanced knowledge of global dynamics and processes occurring on the surface of Earth and in the lower atmosphere can be derived (King and Greenstone, 1999). It has been carried by Terra since December 1999 and Aqua since May 2002. The MODIS science team is producing surface narrowband spectral albedos as well as three broadband (visible, near-IR and shortwave) albedos (Lucht et al. 2000, Schaaf et al. 2002).

$$\alpha_{\text{short}} = 0.160\,\alpha_1 + 0.291\,\alpha_2 + 0.243\,\alpha_3 + 0.116\,\alpha_4 + 0.112\,\alpha_5 + 0.081\,\alpha_7$$

$$\alpha_{\text{vis}} = 0.331\,\alpha_1 + 0.424\,\alpha_2 + 0.246\,\alpha_4$$

$$\alpha_{\text{vis dif}} = 0.246\,\alpha_1 + 0.528\,\alpha_3 + 0.226\,\alpha_4$$

$$\alpha_{\text{vis dir}} = 0.369\alpha_1 + 0.374\alpha_3 + 0.257\alpha_4$$

$$\alpha_{\text{NIR}} = 0.039\alpha_1 + 0.504\alpha_2 - 0.071\,\alpha_3 + 0.105\alpha_4 + 0.252\,\alpha_5 \tag{9.14}$$
$$+ 0.069\alpha_6 + 0.101\,\alpha_7$$

$$\alpha_{\text{NIR dif}} = 0.085\,\alpha_1 + 0.693\,\alpha_2 - 0.146\,\alpha_3 + 0.176\,\alpha_4 + 0.146\,\alpha_5 + 0.043\,\alpha_7$$

$$\alpha_{\text{NIR dif}} = 0.037\alpha_1 + 0.479\alpha_2 - 0.068\,\alpha_3 + 0.0976\,\alpha_4 + 0.266\,\alpha_5$$
$$+ 0.0757\alpha_6 + 0.107\alpha_7$$

Note that in the previous study (Liang, et al. 1999), we presented a set of coefficients for converting MODIS and MISR narrowband albedos to broadband inherent albedos that need to be converted to apparent albedos for various applications. The MODIS conversion formulas have been adopted for the generation of surface broadband albedos (Lucht et al. 2000). Here we convert them into average broadband apparent albedos under the general atmospheric conditions.

The fitting RMSE values for the total shortwave, total visible and total near-IR albedos are 0.0078, 0.0017, and 0.005, respectively (Liang 2001). Validation results using independent ground measurements (Liang et al. 2002a) showed that the RMSE values for these three corresponding albedos are 0.019, 0.015, and 0.018.

9.3.8 POLDER

The POLDER instrument (Deschamps et al. 1994) aboard the Japanese platform ADEOS (advanced Earth observation satellite) acquired measurements between November 1996 and June 1997. It will be carried again by ADEOS 2, which was launched in November 2002. The multidirectional of the POLDER measurements (12–14 directions), coupled with POLDER information on aerosol and water vapor, offers a unique opportunity to characterize the land surface albedo. Land surface spectral albedos will be provided as the standard products, but we need to convert them into broadband albedos in land surface modeling and applications. Hopefully, the generated conversion coefficients will help extend the applications of the POLDER surface products.

$$\alpha_{\text{short}} = 0.112\,\alpha_1 + 0.388\,\alpha_2 - 0.266\,\alpha_3 + 0.668\,\alpha_4$$

$$\alpha_{\text{vis}} = 0.533\alpha_1 + 0.412\,\alpha_2 + 0.215\alpha_3 - 0.168\,\alpha_4 + 0.0046$$

$$\alpha_{\text{vis dif}} = 0.615\alpha_1 + 0.335\alpha_2 + 0.196\alpha_3 - 0.153\,\alpha_4$$

$$\alpha_{\text{vis dir}} = 0.495\alpha_1 + 0.447\alpha_2 + 0.223\alpha_3 - 0.175\alpha_4 \qquad (9.15)$$

$$\alpha_{\text{NIR}} = -0.397\alpha_1 + 0.451\,\alpha_2 - 0.756\alpha_3 + 1.498\,\alpha_4$$

$$\alpha_{\text{NIR dif}} = -0.209\alpha_1 + 0.279\alpha_2 - 0.21\,\alpha_3 + 1.045\alpha_4$$

$$\alpha_{\text{NIR dif}} = -0.425\alpha_1 + 0.474\alpha_2 - 0.825\alpha_3 + 1.554\alpha_4$$

For the total shortwave albedo, the fit is very good. For visible albedos, we recommend using all four bands. For near-IR bands, the linear fitting is not as good. POLDER has the same number of bands as MISR for land surface albedo determination. Although they are located differently, their fittings appear very similar.

The fitting RMSE values for the total shortwave, total visible, and total near-IR albedos are 0.0149, 0.007, and 0.0254, respectively (Liang 2001). Validation results using independent ground measurements (Liang et al. 2002a) showed that the RMSE values for these three corresponding albedos are 0.022, 0.014, and 0.025.

9.3.9 VEGETATION

The VEGETATION program is maintained jointly by France, the European Commission, Belgium, Italy, and Sweden. It is on board SPOT4, launched on March 24, 1998. It has four spectral bands (see Table 9.3), and the spatial resolution is about 1 km at nadir. There is at least one observation acquired daily at 10:30 local solar time for latitudes higher than 32°. The overall objectives of the VEGETATION system are to provide accurate measurements of basic characteristics of vegetation canopies on an operational basis, either for scientific studies involving both regional and global scales experiments over long time periods, or for systems designed to monitor important vegetation resources, such as crops, pastures, and forests.

$$\alpha_{\text{short}} = 0.3512\,\alpha_1 + 0.1629\alpha_2 + 0.3415\alpha_3 + 0.1651\,\alpha_4$$

$$\alpha_{\text{vis}} = 0.5717\alpha_1 + 0.4277\alpha_2$$

$$\alpha_{\text{vis dif}} = 0.6601\,\alpha_1 + 0.3391\,\alpha_2$$

$$\alpha_{\text{vis dir}} = 0.531\,\alpha_1 + 0.4684\alpha_2 \qquad (9.16)$$

$$\alpha_{\text{NIR}} = 0.6799\alpha_3 + 0.3157\alpha_4$$

$$\alpha_{\text{NIR dif}} = 0.8495\alpha_3 + 0.135\,\alpha_4$$

$$\alpha_{\text{NIR dif}} = 0.6567\alpha_3 + 0.3382\,\alpha_4$$

The fitting RMSE values for the total shortwave, total visible, and total near-IR albedos are 0.0101, 0.0054, and 0.0171, respectively (Liang 2001). Validation results using independent ground measurements (Liang et al. 2002a) showed that the RMSE values for these three corresponding albedos are 0.020, 0.016, and 0.022.

9.4 DIRECT ESTIMATION OF SURFACE BROADBAND ALBEDOS

The MODIS land surface products are derived through a series of steps in the processing chain (Lucht et al. 2000, Schaaf et al. 2002) as illustrated in Fig. 9.1 and 9.4, including atmospheric correction (Vermote et al. 1997b), angular modeling for calculating spectral (narrowband) albedos (Lucht and Roujean 2000) and narrowband–broadband albedo conversions (Liang 2001). The accuracy of the MODIS albedo products depend on the performance of all of these processes. Uncertainties associated with each procedure may be canceled out, but also may accumulate. For example, the MODIS atmospheric correction algorithm (Vermote et al. 1997) requires the aerosol product (Kaufman et al. 1997). Determining the aerosol optical depth over land is based on the dark-object approach that relies on the existence of dense vegetation canopies widely distributed over the scene (Kaufman et al. 1997). The dark-object method does not work well over non-vegetated surfaces (e.g., snow, ice, desert, and winter landscape in the Northern hemisphere) (Liang et al. 1997), but nonvegetated surfaces have larger albedos and thus greater feedback to the atmosphere. "Until EOS determines the surface albedo of all land surfaces to greater accuracy, we cannot adequately quantify the radiative forcing to climate that is associated with changes in land use" (Hartmann et al. 1999).

An alternative scheme is developed to link TOA (top-of-atmosphere) narrowband albedos with three land surface broadband albedos using a feedforward neural network without performing any atmospheric correction (Liang et al. 1999). Earlier studies have explored the similar ideas (Chen and Ohring 1984, Pinker 1985, Koepke and Kriebel 1987, Li and Garand 1994) that linearly related TOA to surface broadband albedos. Surface broadband albedo depends on surface spectral reflectance as well as atmospheric conditions. TOA observations contain information on both surface reflectance and atmospheric optical properties, which implies that it is possible for us to predict the broadband albedos using TOA narrowband albedos without performing any atmospheric corrections. Those earlier studies mainly focus on the linkage between planetary shortwave albedo and surface shortwave albedo. In our earlier study (Liang et al. 1999), we suggested linking several TOA spectral albedos with surface broadband albedos (Fig. 9.4). In the follow-up study, this idea was implemented by using actual MODIS observations (Liang 2003) with significant improvement on several aspects. In Section 9.4.1, I outline this new algorithm. Attention will be paid to the improvements over the previously published algorithm (Liang et al. 1999). Several case studies are reported in Section 9.4.2. The first case is to validate the accuracy of this algorithm. Several MODIS images over a remote sensing

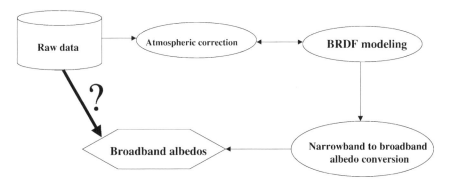

Figure 9.4   Illustration of the need for the hybrid algorithm to estimate broadband albedo directly from TOA observations. [From Liang et al. (2003), *IEEE Trans. Geosci. Remote Sens.* Copyright © 2003 with permission of IEEE.]

test site in Beltsville, MD (USA) were processed and the derived surface broadband albedos were compared with ground measurements that were scaled through Landsat 7 ETM + imagery. The second case demonstrates how this new algorithm can automatically eliminate thick aerosol effects.

9.4.1 Methodology

The basic procedure is based on the hybrid algorithm as described in Section 8.5. It consists of two steps:

- The first step is to conduct extensive radiative transfer simulations using MODTRAN4 (Berk et al. 1998).
- The second step is to link the simulated TOA reflectance with surface broadband albedos using nonparametric regression algorithms.

9.4.1.1 Radiative Transfer Simulations

For a given atmospheric and surface condition, we need to know the directional reflectance at the top of the atmosphere. Many radiative transfer packages are available right now for us to achieve that, such as MODTRAN (Berk et al. 1998), 6S (Vermote et al. 1997) and SBDART (Ricchiazzi et al. 1998). MODTRAN has been extensively used in our studies.

To incorporate as many surface reflectance spectra as possible, a novel approach is required to decouple surface reflectance spectra in the simulations. Specifically, MODTRAN was run 3 times with three surface reflectances (0.0, 0.5, and 0.8) for each atmospheric condition and solar viewing geometry. Assuming that the surface is Lambertian, the downward flux $F(\mu_i)$ and upwelling radiance at the top of atmosphere $L(\mu_i, \mu_v)$ can be expressed by (Liou 1980)

$$F(\mu_i) = F_0(\mu_i) + \frac{r_s}{1 - r_s \bar{r}} \pi \mu_i E_0 \gamma(\mu_i) \bar{r} \qquad (9.17)$$

$$L(\mu_i, \mu_v, \phi) = L_p(\mu_i, \mu_v, \phi) + \frac{r_s}{1 - r_s \bar{r}} \mu_i E_0 \gamma(\mu_i) \gamma(\mu_v) \qquad (9.18)$$

where μ_i is the cosine of solar zenith angle (θ_i), μ_v corresponds to the viewing zenith angle; F_0 and L_p are the upwelling flux and path radiance without the surface contribution (i.e., surface reflectance is zero, $r_s = 0$); $\bar{r}$ is the spherical albedo of the atmosphere, E_0 the TOA downward flux, and $\gamma(\mu_i)$ and $\gamma(\mu_v)$ the total atmospheric transmittance of the solar illumination path and the viewing path, respectively. Results from three MODTRAN runs form six equations that enable us to determine these unknowns [F_0, L_p, $\bar{r}$, $\gamma(\mu_i)$ and $\gamma(\mu_v)$] in the preceding equations. As long as these unknowns are determined, it is straightforward to calculate downward flux at the surface and TOA radiance with any surface reflectance spectra.

In the MODTRAN simulations, 11 atmospheric visibility values (2, 5, 10, 15, 20, 25, 30, 50, 70, 100, and 150 km) were used for different aerosol loadings for each of these four aerosol models (rural, Navy maritime, urban, and troposphere), and five atmospheric profiles (tropical, midlatitude winter, sub-Arctic summer, sub-Arctic winter, and US62). These profiles also represent different water vapor and other gaseous amounts and profiles. The aerosol models and atmospheric profiles used are the defaults in MOD-TRAN. For the operational application of this method, more data should perhaps be included to represent the variable atmospheric conditions.

One major limitation in this simulation study is its assumption of Lambertian surfaces. The major reason for this is that we do not have a good understanding of the directional reflectance properties of various surface types at MODIS resolutions. Both MODIS and MISR teams are producing global land surface BRDF products (Diner et al. 1998, Lucht et al. 2000). Improving this procedure is straightforward as long as we accumulate enough data since many radiative transfer models have the capability to incorporate directional surface reflectance. However, we speculate that the spectral dependence of surface reflectance spectra is the most important aspect.

9.4.1.2 *Statistical Algorithms*
As long as the database is created from the simulations described above, the next step is to link TOA reflectance to land surface broadband albedos. We proposed using a neural network in the previous study (Liang et al. 1999), but have also explored a new method called *projection–pursuit regression* (PPR) in a later study (Liang 2003). Both artificial neural networks (ANN) and PPR methods have been discussed in Sections 8.5.1 and 8.5.2, and no further details are provided here.

9.4.2 Case Studies

Several case studies were conducted (Liang 2003) to evaluate if this direct estimation approach is reasonably accurate in estimating land surface broadband albedos from MODIS imagery. The first case is based on several clear images over the greater Washington DC area where ground measurements were scaled up through Landsat 7 ETM + imagery. The second case demonstrates how this method can retrieve surface broadband albedos by automatically removing a large thick patch of aerosols.

Two case studies were presented (Liang 2003). The first one validated the accuracy of this direct estimation method using ground measurements and high-resolution imagery over our test site at Beltsville, MD. The ground measurements were used to calibrate the high-resolution albedo products from Landsat ETM + imagery. The high-resolution albedo products were then aggregated to register with MODIS imagery. Data from four dates (May 11, Nov. 3, Dec. 5, 2000, and Jan. 22, 2001) demonstrated that this new method is reasonably accurate. Both neural network and PPR consistently produced similar results. The average mean albedo differences from ETM + albedos are smaller than 0.03 (shortwave), 0.01 (visible), and 0.04 (near-IR), and the standard deviations are smaller than 0.04. These numbers are good indications of the accuracy of this new method. The MODIS black-sky albedos products were also used in the comparisons. The results indicate that our new method produces albedo with accuracy comparable to that of the MODIS standard albedo products, and is even better when the surface conditions are highly variable (e.g., snow).

The second case demonstrated that this new method can recover land surface broadband albedos from the heavily contaminated MODIS image of northeast China by aerosol particles without performing any atmospheric correction. In this study, a Lambertian surface has been assumed in the MODTRAN radiative transfer simulation. It is not an inherent limitation of this new algorithm. The reasons for making such an assumption is that it not only significantly simplifies the procedure but also results from the fact that a global surface BRDF database at the one kilometer scale is simply not available at this point. As long as we continue to gain a deeper understanding of the surface non-Lambertian properties and accumulate enough data from MODIS (Lucht et al. 2000, Schaaf et al. 2002) and MISR (Diner et al. 1998, Martonchik et al. 1998) or other sensors, incorporating a non-Lambertian surface into the radiative transfer simulation is straightforward. MODTRAN already has such an option. On the other hand, our case studies have proved that this approximation does not affect the inversion accuracy significantly. The error associated with this approximation is intuitively expected to be much smaller than those due to the failed atmospheric correction over nonvegetated surfaces or the use of a wrong BRDF model during a period of 16 days when a sudden change of the surface conditions affects the current MODIS algorithm.

9.5 DIURNAL CYCLE MODELING

This section is extracted primarily from one section of a previously published paper (Liang et al. 2000) with some new updates.

Radiation budget studies must consider all local times through the diurnal cycle, with the associated changes in solar zenith angle (SZA) and diurnal variation in cloud conditions (Minnis and Harrison 1984). After the simple geometric control of shortwave fluxes by the cosine of the SZA, clouds have the strongest influence on the shortwave radiation budget, and this is the primary driver of the efforts to secure adequate diurnal sampling of the TOA radiation budget by satellites. However, the dependence of surface shortwave reflective properties on SZA also influences the shortwave radiation budget, and this will be easiest to determine under clear skies. In particular, global climate models typically take as input the monthly mean surface albedo, which clearly is strongly influenced by the SZA dependence. While ultimately the all-sky albedo as a function of time is desired, this review largely limits itself to the clear-sky albedo and its diurnal variations.

Variations in SZA of up to 50% in the clear-sky albedo of land surfaces have been demonstrated by observation (Kriebel 1979, Pinker et al. 1980, Minnis and Harrison 1984, Irons et al. 1988, Ranson et al. 1991) and by modeling studies of the interaction of radiation with vegetation and soil (Dickinson 1983, Kimes et al. 1987). The diurnal variations tend to be smaller on cloudy days (Pinker et al. 1980). Dickinson (1983) drew on a mathematical treatment of radiation in plant canopies to develop a simple analytic model of the SZA dependence of apparent albedo, which was presented by Briegleb et al. (1986) in the form

$$\alpha(\mu_0) = \alpha_{0.5} \frac{1+d}{1+2d\mu_0} \tag{9.19}$$

where μ_0 is the cosine of the SZA, α is the shortwave broadband albedo, and the two model parameters are $\alpha_{0.5}$, which is the albedo for $\mu_0 = 0.5$, and d, which controls the strength of the SZA dependence.

Taking the nadir reflectance as the albedo can result in errors of up to 45% (Kimes and Sellers 1985, Irons et al. 1988, Ranson et al. 1991), depending on SZA, land cover, and spectral band. While the error in estimating the albedo from a single directional measurement can be reduced by careful choice of the SZA and view direction, greater accuracy is achievable with multiangle observations.

Geostationary imagers have monitored the globe's surface for many years with good temporal resolution of the diurnal cycle, but only from single view directions and generally with a single visible spectral band much narrower than the 0.2–5.0-μm shortwave band.

Despite the angular and spectral limitations, geostationary satellite observations have been used for albedo estimation (e.g., GOES) (Minnis and Harrison 1984). Polar orbiters that measure radiance in multiple bands from multiple directions are much better suited to estimating instantaneous surface albedo, but a single satellite in a sun-synchronous orbit typically observes any point at one daylight local time per day, and hence even a set of two or three polar orbiters measures diurnal variations poorly. Two approaches are used to estimate the surface albedo at SZAs away from those of the measurements when diurnal sampling is sparse.

1. The first approach is to estimate instantaneous albedo at the time of the satellite overpass and explicitly specify a diurnal variation, typically through a simple parameterization with the coefficients assigned according to a land cover classification (Minnis and Harrison 1984, Brooks et al. 1986). The land cover classification can be an a priori classification, or a scene type identified from the remote sensing data itself as was done with ERBE.

CERES will continue this approach to produce instantaneous TOA short-wave fluxes and correct them to standard time, initially using the ERBE ADMs but eventually better CERES angular distribution models (ADMs) based on a finer land cover classification. CERES will derive 3-hourly surface albedos from 3-hourly TOA fluxes, which are themselves interpolated from the CERES instantaneous measurements (up to six per day) with the aid of geostationary imager data (Wielicki et al. 1998, Young et al, 1998). The TOA-to-surface conversion will be done by two methods: by a simple relation parameterized by atmospheric composition and SZA such as that developed by Li et al. (1993); and by a radiative transfer calculation of the fluxes throughout the atmospheric column. The radiative transfer calculation requires an estimate of the surface spectral albedo, and for cloudy times this will be the previous clear-day value adjusted for SZA with the Dickinson model, with a value of d assigned to land cover classes on the basis of ERBE and ScaRaB (Scanner for Earth Radiation Budget) data. ScaRaB processing follows that of ERBE (Valiente et al. 1995).

In summary, the ERBE, ScaRaB, and CERES processing prescribe the SZA dependence of albedo on the basis of scene type, although for CERES this is just used for initialization of processing and the diurnal behavior of the final result is constrained by the geostationary data.

We note that the ERBE ADMs do not satisfy reciprocity (see discussions in Section 2.2.3). There are, however, efforts to find improved forms of ADMs that do satisfy reciprocity (Manalo-Smith et al. 1998). Both CERES and ScaRaB partly address the need for resolution of the temporal cycle by placing one platform (for ScaRaB, the only platform) in an orbit with a processing local time of equator crossing.

2. A second approach to the treatment of the remote sensing of surface albedo in the face of its SZA dependence is to estimate the BRDF from multiangle observations made at a range of SZAs, and to integrate the BRDF over the hemisphere of surface leaving directions for a specific SZA. This is how albedo is estimated from POLDER (Leroy et al. 1997) and MODIS (Wanner et al. 1997). However, the observations are still made at one or a small number of local times, and so this approach needs to be tested for situations in which the albedo is estimated at SZAs away from those of the measurements. This was considered in the design of the POLDER land surface albedo algorithm, and Lucht (1998) has tested this in detail for the albedo production planned for MODIS/MISR data based on the integration of their BRDF models. From simulations based on modeling of radiative transfer in land cover, for a range of biomes and a range of latitudes and seasons, Lucht found that relative errors of $\geq 30\%$ could arise in the MODIS estimates of spectral albedo at SZAs away from the SZAs of the measurements, although errors of $< 10\%$ were most typical. In general, extrapolation to SZAs larger than those of the measurements was less accurate than extrapolation to smaller SZAs.

Reciprocity is often relied on to fill unmeasured Sun-view-angle bins in the derivation of the angular distribution models (ADMs) or BRDFs from multiangular observations, and it is sometimes suggested that these models be constrained to satisfy reciprocity. Capderou (1998) found reciprocity to be satisfied in a study of ScaRaB data for 1.25° by 1.25° desert areas, but Kriebel (1996) found that airborne measurements of natural surfaces could violate reciprocity. Di Girolamo et al. (1998) discussed applications of reciprocity and examined the basis of the assumption in detail. They noted that in practical remote sensing pure reciprocity is confounded by spatial effects, leading to an apparent breakdown of reciprocity and important implications for remote sensing: namely; (1) directional models should not necessarily be required to satisfy reciprocity, (2) reciprocity can be used to construct ADMs only if the spatial scale is appropriate, and (3) ADMs can strictly be applied only at the spatial resolution at which they were derived.

Grant et al. (2000) examined diurnal variations in clear-sky albedo at three locations separated by up to 750 m across a uniform grassland site. The main conclusions were (1) the gross magnitude of the diurnal variation at the site changed greatly with vegetation state (the ratio of albedo at SZA = 60–30° ranged over at least 1.20–1.41), (2) the albedo differed by up to 0.04 between morning and afternoon times with the same SZA, and (3) variations on the scale of a few hours could give departures of up to 0.02 from a smooth SZA dependence. All of these effects are evident in Fig. 9.5, which shows the albedo through 3 clear days measured at the Uardry grassland site in southeastern Australia (145°18′E, 34°24′S). The mean albedo over time interval T, such as one day or one month, is the ratio of upwelling to downwelling time-integrated energy density:

$$\alpha(T) = \frac{\int_T F_u(t)\,dt}{\int_T F_d(t)\,dt} = \frac{\int_T \alpha(t)F_d(t)\,dt}{\int_T F_d(t)\,dt} \qquad (9.20)$$

Each of the three departures just described, from the kind of simple parameterized diurnal cycle represented by the Dickinson model and used in the processing of data from some polar orbiters, such as ERBE and CERES, can produce errors in the estimation of daily mean albedo, and consequently monthly mean albedo. These errors are at least as large as the errors in the remotely sensed instantaneous albedo. In particular, it is interesting to note that whereas Briegleb et al. (1986) took d in the Dickinson model of SZA dependence of albedo as 0.4 for arable land, grassland, and desert and 0.1 for other land surfaces, Grant et al. found that the best-fit d ranged from 0.1 to 0.7 at their single grassland site. While the albedo changes with SZA due to the varying angular, and even spectral, distribution of downwelling radiance, diurnal surface changes such as dew on vegetation or soil can also influence the diurnal cycle of albedo (Monteith and Szeicz 1961, Minnis et al. 1997).

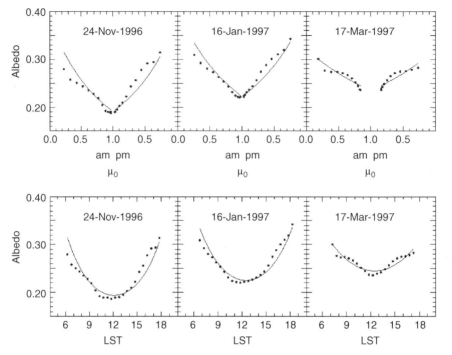

Figure 9.5   Measured land surface broadband albedos as a function of the local time through 3 clear days at the Uardry grassland site in southeastern Australia; the smooth curves are fitted to the data using formula (9.19) [from Liang et al. (2000)].

Preferred azimuthal directions in the land cover induced, for instance, by a prevailing wind direction, will cause diurnal asymmetry in the albedo (Song 1998). Regular diurnal cycles of albedo can be disrupted by rainfall events, and change with trends in vegetation development.

Eventually, attempts should be made to retrieve the surface albedo under cloudy skies. Figure 9.6 shows the downwelling irradiance and albedo through 3 days at Uardry. For one day, clouds have reduced the downwelling irradiance to a roughly constant fraction of the clear-sky time series, while the other 2 days are a few days earlier and later and are clear throughout. The Dickinson model has been fitted to each day's albedos. While the clear-sky albedo has dropped in the 8 days between October 16 and 24, perhaps because of greening of the site, it maintains a strong SZA dependence. But on the heavily clouded day (Oct. 21), the albedo is lower than on either clear day and has almost no SZA dependence. Liang et al. (1999) used simulations to study the effect of the atmosphere on surface albedo, under clear and cloudy skies. Comparison of their Figs. 9.11a and 9.12 suggests that for most nonsnow surfaces the cloudy-sky shortwave albedo is lower than the clear-sky shortwave albedo by an amount comparable to that in Fig. 9.6 here.

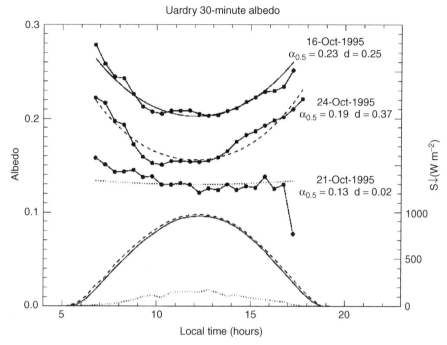

Figure 9.6 Measured downward irradiance (lower three curves) and land surface broadband albedos (upper three curves) under both clear sky and cloudy sky conditions at Uardry grassland site in southeastern Australia; the smooth curves are fitted to the data using formula (9.19) [from Liang et al. (2000)].

Remote sensing of cloudy-sky albedo is probably best done by measuring the surface BRDF under clear conditions and integrating for a cloudy-sky downwelling radiance field. Satellite retrievals of cloudy-sky albedo should, of course, be validated at uniform ground sites such as those of the baseline surface radiation network (BSRN), the atmospheric radiation measurement (ARM) sites, the NOAA integrated surface irradiance study (ISIS) network, global energy balance archive (GEBA) stations, or Australian Continental Integrated Ground-truth Site Network (Prata et al. 1998) (CIGSN) networks.

The highest temporal resolution will continue to come from non-polar-orbiting satellites. The operational geostationary imagers, with their single visible band, continue to monitor each surface point from a fixed single viewing direction for the whole range of SZAs. The Geostationary Earth Radiation Budget (broadband shortwave and longwave) and 12-channel Spinning Enhanced Visible and Infrared Imager are due to monitor the European sector at 15-min temporal resolution from the MSG (Meteosat Second Generation) series of geostationary satellites. Finally, the 10-channel imager on Triana will image the whole dayside Earth disk virtually continuously from the L1 Lagrangian point between Earth and the Sun, and thus monitor every daylight surface point from a changing single view direction at a single local time. These sensors all measure from a single view direction, so their data must be complemented by ADMs or BRDFs derived from polar orbiters in order to derive hemispherical albedos.

The diurnal and seasonal cycles of SZA vary with the latitude of the location purely because of geometry. However, land cover types are generally stratified latitudinally. When studying the global climatology of albedo or the radiation budget, it is important to keep in mind that the influence of SZA and land cover characteristics are coupled by this common dependence on latitude.

A complete dataset for monitoring albedo globally would include measurements at each point of the surface of the surface leaving radiance over all directions of the hemisphere throughout the day. Complete coverage of this multidimensional measurement space is clearly impossible. Current and proposed sensors sample various subsets of this measurement space, and synthesis of observations from sensors is likely to provide the most complete picture. There are, of course, many issues to be solved in order to make this approach useful.

Further work is required to review, analyze old measurements, or collect new measurements on the detailed diurnal cycle of surface albedo at sites representing a range of land cover types. The purpose is to determine limits on accuracy of estimates of daily mean albedo and diurnal variation of albedo from polar orbiters that are imposed by the dynamic nature of land cover, diurnal asymmetry, and other departures from simple parameterizations. In doing this, particular attention must be given to the spatial scale of the remotely sensed data.

9.6 SUMMARY

Absorption of solar radiation and emission of terrestrial radiation drive the general circulation of the atmosphere and are largely responsible for the Earth's weather and climate. The surface of the Earth absorbs about twice as much solar radiation as does the atmosphere. The amount of solar radiation that is absorbed by the surface is modulated by the surface broadband albedo. In this chapter, we have discussed mainly two approaches for estimating land surface broadband albedos from satellite measurements.

The first approach is based on the detailed physical understanding of different components by performing a series of processing steps, which include atmospheric correction that converts TOA radiance to surface directional reflectance, angular modeling for converting directional reflectance to planar albedos, and narrowband–broadband conversions. Since atmospheric correction has been discussed in Chapter 6 and different statistical models were presented in Section 2.2, Section 9.3 has focused on narrowband–broadband conversions based on extensive radiative transfer simulations. The conversion formulas of representative sensors were given, and the validation results were also briefly mentioned. The advantage of this method is that we can incorporate the latest understanding of the physical process into the algorithms by improving the specific processing steps. However, the accuracy of the final albedo products depends on the performance of all these processes. Uncertainties associated with each procedure may be canceled out, but are also very likely to accumulate.

The second approach is to estimate surface broadband albedos directly from TOA radiance without performing all the traditional processing steps. The key idea is to first conduct extensive radiative transfer simulations with representative surface reflectance spectra and atmospheric conditions, and then link the TOA radiance with surface broadband albedos using a nonparametric regression technique. This technique was described in Section 9.4. Some examples were also given to demonstrate the effectiveness of this technique.

The basic characteristics of broadband albedos were discussed in Section 9.2. The temporal integration of surface broadband albedos required for surface energy balance and land surface modeling was discussed in Section 9.5.

Since multiple satellite observations can be used for estimating land surface broadband albedos, it is necessary to compare their characteristics and possibly combine them through data fusion or data assimilation techniques. Extensive ground measurements are also needed over different landscapes under different atmospheric and surface conditions.

REFERENCES

Barkstrom, B. R. and Smith, G. L. (1986), The Earth radiation budget experiment: Science and implementation, *Rev. Geophys.* **24**: 379–390.

Berk, A., Bernstein, L. S., Anderson, G. P., Acharya, P. K., Robertson, D. C., Chetwynd, J. H., and Adler-Golden, S. M. (1998), MODTRAN cloud and multiple scattering upgrades with application to AVIRIS, *Remote Sens. Envir.* **65**: 367–375.

Brest, C. L. and Goward, S. (1987), Deriving surface albedo measurements from narrowband satellite data, *Int. J. Remote Sens.* **8**: 351–367.

Briegleb, B. P., Minnis, P., Ramanathan, V., and E. H. (1986), Comparison of regional clear-sky albedos inferred from satellite observations and model computations, *J. Climate Appl. Meteorol.* **25**: 214–226.

Brooks, D. R., Harrison, E. F., Minnis, P., Suttles, J. T., and S., K. R. (1986), Development of algorithms for understanding the temporal and spatial variability of the earth's radiation balance, *Rev. Geophys.* **24**: 422–438.

Capderou, M. (1998), Confirmation of Helmholtz reciprocity using scarab satellite data, *Remote Sens. Envir.* **64**: 266–285.

Cess, R. D. (1978), Biosphere-albedo feedback and climate modeling, *J. Atmos. Sci.* **35**: 1765–1768.

Chen, T. and Ohring, G. (1984), On the relationship between clear-sky planetary and surface albedos, *J. Atmos. Sci.* **41**: 156–158.

Csiszar, I. and Gutman, G. (1999), Mapping global land surface albedo from NOAA AVHRR, *J. Geophys. Res.* **104**: 6215–6228.

Deschamps, P. Y., Brion, F., Leroy, M., Podaire, A., Bricaud, A., Buriez, J., and Seze, G. (1994), The POLDER mission: Instrument characteristics and scientific objectives, *IEEE Trans. Geosci. Remote Sens.* **32**: 598–615.

Di Girolamo, L., Varnai, T., and Davies, R. (1998), Apparent breakdown of reciprocity in reflected solar radiances, *J. Geophys. Res.* **103**: 8795–8803.

Dickinson, R. E. (1983), Land surface processes and climate-surface albedos and energy balance, *Adv. Geophys.* **25**: 305–353.

Diner, D., Beckert, J. C., Reilly, T. H., Bruegge, C. J., Conel, J. E., Kahn, R. A., Martonchik, J. V. et al. (1998), Multi-angle imaging spectroradiometer (MISR) instrument descrition and experiment overview, *IEEE Trans. Geosci. Remote Sens.* **36**: 1072–1097.

Dirmeyer, P. A. and Shulka, J. (1994), Albedo as a modulator of climate response to tropical deforestation, *J. Geophys. Res.* **99**: 20863–20877.

Dorman, J. L. and Sellers, P. J. (1989), A global climatology of albedo, roughness length and stomatal resistance for atmospheric general circulation models as represented by the simple biosphere model SIB, *J. Appl. Meteorol.* **28**: 833–855.

Duguay, C. R. and LeDrew, E. F. (1992), Estimating surface reflection and albedo from Landsat-5 Thematic Mapper over rugged terrain, *Photogramm. Eng. Remote Sens.* **58**: 551–558.

Garratt, J. R., Krummel, P., and Kowalczyk, E. A. (1993), The surface energy balance at local and regional scales—A comparison of general circulation model results with observations, *J. Climate* **6**: 1090–1109.

Grant, I. F., Prata, A. J., and Cechet, R. P. (2000), The impact of the diurnal variation of albedo on the remote sensing of the daily mean albedo of grassland, *J. Appl. Meteorol.* **39**: 231–244.

Gutman, G. G. (1994), Global data on land surface parameters from NOAA AVHRR for use in numerical climate models, *J. Climate* **7**: 699–703.

Han, W., Stamnes, K., and Lubin, D. (1999), Remote sensing of surface and cloud properties in the arctic from AVHRR measurements, *J. Appl. Meteorol.* **38**: 989–1012.

Hartmann, D. L., Bretherrton, C. S., Charlock, T. P., Chou, M. D., Del Genio, A., Dickinson, R. E., Fu, R., Houze, R. A., King, M. D., Lau, K. M., Leovy, C. B., Sorooshian, S., Washburne, J., Wielicki, B., and Willson, R. C. (1999), Radiation, clouds, water vapor, precipitation, and atmospheric circulation, in *EOS Science Plan*, NASA GSFC, pp. 39–114.

Irons, J. R., Ranson, K. J., and Daughtry, C. (1988), Estimating big bluestem albedo from directional reflectance measurements, *Remote Sens. Envir.* **25**: 185–199.

Jacobowitz, H., Tighe, R. J. et al. (1984), The earth radiation budget derived from the Nimbus-7 ERB experiment, *J. Geophys. Res.* **89**: 4997–5010.

Kandel, R., Viollier, M., Raberanto, P., Duvel, J. P., Pakhomov, L. A., Golovko, V. A., Trishchenko, A. P., Mueller, J., Rashke, E., and Stuhlmann, R. (1998), The scarab earth radiation budget dataset, *Bull. Am. Meteorol. Soc.* **79**: 765–783.

Kaufman, Y. J., Wald, A., Lorraine, L. A., Gao, B. C., Li, R. R., and Flynn, L. (1997), Remote sensing of aerosol over the continents with the aid of a 2.2 μm channel, *IEEE Trans. Geosci. Remote Sens.* **35**: 1286–1298.

Key, J. (2001), *Streamer User's Guide*. Cooperative Institute for Meterological Satellite Studies, University of Wisconsin, 96 pp.

Kiehl, J. T., Hack, J. J., Bonan, G. B., Boville, B. A., Briegleb, P., Williamson, D. L., and Rasch, P. J. (1996), *Description of the NCAR Community Climate Model*, National Center for Atmospheric Research, Boulder, Colorado, NCAR Technical Note NCAR/TN-420 + STR, pp. 1–152.

Kimes, D. S. and Sellers, P. J. (1985), Inferring hemispherical reflectance of the earth's surface for global energy budgets from remotely sensed nadir or directional radiance values, *Remote Sens. Envir.* **18**: 205–223.

Kimes, D. S., Sellers, P. J., and Newcomb, W. W. (1987), Hemispherical reflectance variations of vegetation canopies and implications for global and regional energy budget studies, *J. Climatol. Appl. Meteorol.* **26**: 959–972.

King, M. D., ed. (1999), *EOS Science Plan, the State of Science in the EOS Program*. NASA, Greenbelt, MD.

Knap, W., Reijmer, C., and Oerlemans, J. (1999), Narrowband to broadband conversion of Landsat TM glacier albedos, *Int. J. Remote Sensing* **20**: 2091–2110.

Koepke, P. and Kriebel, K. T. (1987), Improvements in the shortwave cloud-free radiation budget accuracy, part I: Numerical study including surface anisotropy, *J. Climatol. Appl. Meteorol.* **26**: 374–395.

Koster, R. and Suarez, M. (1992), Modeling the land surface boundary in climate models as a composite of independent vegetation stands, *J. Geophys. Res.* **97**: 2697–2715.

Kriebel, K. T. (1979), Albedo of vegetated surfaces: Its variability with differing irradiance, *Remote Sens. Envir.* **8**: 283–290.

Kriebel, K. T. (1996), On the limited validity of reciprocity in measured BRDFs, *Remote Sens. Envir.* **58**: 52–62.

Leroy, M., Deuze, J. L., Breon, F. M., Hautecoeur, O., Herman, M., Buriez, J. C., Tanré, D., Bouffies, S., Chazette, P., and Roujean, J. L. (1997), Retrieval of atmospheric properties and surface bidirectional reflectances over land from POLDER/ADEOS, *J. Geophys. Res.* **102**: 17023–17037.

Li, Z. and Leighton, H. (1992), Narrowband to broadband conversion with spatially autocorrelated reflectance measurements, *J. Appl. Meteo.* **31**: 421–432.

Li, Z. and Garand, L. (1994), Estimation of surface albedo from space: A parameterization for global application, *J. Geophys. Res.* **99**: 8335–8350.

Li, Z., Leighton, H. G., Masuda, K., and Takashima, T. (1993), Estimation of sw flux absorbed at the surface from TOA reflected flux, *J. Climate* **6**: 317–330.

Liang, S. (2001), Narrowband to broadband conversions of land surface albedo, *Remote Sens. Envir.* **76**: 213–238.

Liang, S. (2003), A direct algorithm for estimating land surface broadband albedos from MODIS imagery, *IEEE Trans. Geosci. Remote Sens.* **41**: 136–145.

Liang, S., Strahler, A., and Walthall, C. (1999), Retrieval of land surface albedo from satellite observations: A simulation study, *J. Appl. Meteorol.* **38**: 712–725.

Liang, S., Fallah-Adl, H., Kalluri, S., JaJa, J., Kaufman, Y. J., and Townshend, J. R. G. (1997), An operational atmospheric correction algorithm for Landsat Thematic Mapper imagery over the land, *J. Geophys. Res. Atmos.*, **102**: 17173–17186.

Liang, S., Stroeve, J., Grant, I., Strahler, A., and Duvel, J. (2000), Angular corrections to satellite data for estimating Earth radiation budget, *Remote Sens. Rev.* **18**: 103–136.

Liang, S., Shuey, C., Fang, H., Russ, A., Chen, M., Walthall, C., Daughtry, C., and Hunt, R. (2003a), Narrowband to broadband conversions of land surface albedo: II. Validation, *Remote Sens. Environ.* **84**: 25–41.

Liang, S., Fang, H., Kaul, M., Van Niel, T. G., McVicar, T. R., Pearlman, J., Walthall, C. L., Daughtry, C., and Huemmrich, K. F. (2003b), Estimation of land surface broadband albedos and leaf area index from EO-1 ALI data and validation, *IEEE Trans. Geosci. Remote Sens* **41**: 1260–1268.

Liou, K. N. (1980), *An Introduction to Atmospheric Radiation*, Academic Press.

Lucht, W. (1998), Expected retrieval accuracies of bidirectional reflectance and albedo from EOS-MODIS and MISR angular sampling, *J. Geophys. Res.* **103**: 8763–8778.

Lucht, W. and Roujean, J. L. (2000), Considerations in the parametric modeling of BRDF and albedo from multiangular satellite sensor observations, *Remote Sens. Rev.* **18**: 343–380.

Lucht, W., Schaaf, C. B., and Strahler, A. H. (2000), An algorithm for the retrieval of albedo from space using semiempirical BRDF models, *IEEE Trans. Geosci. Remote Sens.* **38**: 977–998.

Manalo-Smith, N., Smith, G. L., Tiwari, S. N., and Staylor, W. F. (1998), Analytic forms of bidirectional reflectance functions for application to earth radiation budget studies, *J. Geophys. Res.* **103**: 19733–19751.

Martonchik, J. V., Diner, D. J., Kahn, R. A., Ackerman, T. P., Verstraete, M. E., Pinty, B., and Gordon, H. R. (1998), Techniques for the retrieval of aerosol properties over land and ocean using multiangle imaging, *IEEE Trans. Geosci. Remote Sens.* **36**: 1212–1227.

Minnis, P. and Harrison, E. F. (1984), Diurnal variability of regional cloud and clear-sky radiative parameters derived from goes data. Part III: November 1978 radiative parameters, *J. Climatol. Appl. Meteorol.* **23**: 1032–1052.

Minnis, P., Mayor, S., Smith, J. W., and Young, D. F. (1997), Asymmetry in the diurnal variation of surface albedo, *IEEE Trans. Geosci. Remote Sens.* **35**: 879–891.

Monteith, J. L. and Szeicz, G. (1961), The radiation balance of bare soil and vegetation, *Quart. J. Royal Meteorol. Soc.* **87**: 159–170.

Pinker, R., W. , Kustas, W., Laszlo, I., Moran, S., and Huete, A. (1994), Satellite surface radiation budgets on basin scale in semi-arid regions, *Water Resour. Res.* **30**: 1375–1386.

Pinker, R. T. (1985), Determination of surface albedo from satellite. *Adv. Space Res.* **5**: 333–343.

Pinker, R. T., Thompson, O. E., and Eck, T. F. (1980), The albedo of a tropical evergreen forest, *Quart. J. Royal Meteorol. Soc.* **106**: 551–558.

Pinker, R. T., Frouin, R., and Li, Z. (1995), A review of satellite methods to derive surface shortwave irradiance, *Remote Sens. Envir.* **51**: 108–124.

Pinty, B., Roveda, F., Verstraete, M. M., Gobron, N., Govaerts, Y., Martonchik, J. V., Diner, D. J., and Kahn, R. A. (2000a), Surface albedo retrieval from Metrosat. I. Theory, *J. Geophys. Res.* **105**: 18099–18112.

Pinty, B., Roveda, F., Verstraete, M. M., Gobron, N., Govaerts, Y., Martonchik, J. V., Diner, D. J., and Kahn, R. A. (2000b), Surface albedo retrieval from Metrosat. I. Applications, *J. Geophys. Res.* **105**: 18113–18134.

Prata, A. J., Grant, I. F., Cechet, R. P., and Rutter, G. F. (1998), Five years of shortwave radiation budget measurements continental land site in south-eastern Australia, *J. Geophys. Res.* **103**: 266093–26106.

Ranson, K. J., Irons, J., and Daughtry, C. (1991), Surface albedo from bidirectional reflectance, *Remote Sens. Envir.* **35**: 201–211.

Ricchiazzi, P., Yang, S., Gautier, C., and Sowle, D. (1998), SBDART: A research and teaching software tool for plane-parallel radiative transfer in the earth's atmosphere, *Bull. Am. Meteor. Soc.* **79**: 2101–2114.

Russell, M., Nunez, M., Chladil, M., Valiente, J., and Lopez-Baeza, E. (1997), Conversion of nadir, narrowband reflectance in red and near-infrared channels to hemispherical surface albedo, *Remote Sens. Environ.* **61**: 16–23.

Sato, N., Sellers, P. J., Randall, D. A., Schneider, E. K., Shukla, J., Kinter III, J. L., Hou, Y. T., and Albertazzi, E. (1989), Effects of implementing the simple biosphere model (sib) in the general circulation model, *J. Atmos. Sci.* **46**: 2757–2782.

Saunders, R. W. (1990), The determination of broad band surface albedo from AVHRR visible and near-infrared radiances, *Int. J. Remote Sens.* **11**: 49–67.

Schaaf, C., Gao, F., Strahler, A., Lucht, W., Li, X., Tsung, T., Strugll, N., Zhang, X., Jin, Y., Muller, P., Lewis, P., Barnsley, M., Hobson, P., Disney, M., Roberts, G., Dunderdale, M., Doll, C., d'Entremont, R., Hu, B., Liang, S., Privette, J., and Roy, D. (2002), First operational BRDF, albedo nadir reflectance products from MODIS, *Remote Sens. Environ.* **83**: 135–148.

Sellers, P., Hall, F., Kelly, R., Black, A., Baldocchi, D., Berry, J., Ryan, M., Ranson, J., Crill, P., Lettenmaier, D., Margolis, H., Cihlar, J., Newcomer, J., Fitzjarrald, D., Jarvis, P., Gower, S., Halliwell, D., Williams, D., Goodison, B., Wickland, D., and Guertin, F. (1997), BOREAS in 1997: Experiment overview, scientific results, and future directions, *J. Geophys. Res.* **102**: 28731–28769.

Sellers, P. J., Los, S. O., Tucker, C. J., Justice, C. O., Dazlich, D. A., Collatz, G. J., and Randall, D. A. (1996), A revised land surface parameterization (SIB2) for atmospheric GCMs. Part II: The generation of global fields of terrestrial biophysical parameters from satellite data, *J. Climate* **9**: 706–737.

Smith, G. L., Green, R. N., Raschke, E., Avis, L. M., Suttles, J. T., Wielicki, B. A., and Davies, R. (1986), Inversion methods for satellite studies of the Earth's radiation budget: Development of algorithms for the ERBE mission, *Rev. Geophys.* **24**: 407–421.

Song, J. (1998), Diurnal asymmetry in surface albedo, *Agric. Forest Meteorol.*, **92**: 181–189.

Song, J. and Gao, W. (1999), An improved method to derive surface albedo from narrowband AVHRR satellite data: Narrowband to broadband conversion, *J. Appl. Meteor.* **38**: 239–249.

Stroeve, J., Nolin, A., and Steffen, K. (1997), Comparison of AVHRR-derived and in-situ surface albedo over the greenland ice sheet, *Remote Sens. Envir.* **62**: 262–276.

Strugnell, N. and Lucht, W. (2001), Continental-scale albedo inferred from AVHRR data, land cover class and field observations of typical BRDFs, *J. Climate* **14**: 1360–1376.

Townshend, J. G. R. (1994), Global data sets for land applications from the advanced very high resolution radiometer: An introduction, *Int. J. Remote Sen.* **15**: 3319–3332.

Ustin, S. L., Smith, M. O., and Adams, J. B. (1993), Remote sensing of ecological processes: A strategy for developing and testing ecological models using spectral mixture analysis, in *Scaling Physiological Processes: Leaf to Globe*, J. R. Ehleringer and C. B. Field, eds., Academic Press, pp. 339–357.

Valiente, J., Nunez, M., Lopez-Baeza, E., and Moreno, J. (1995), Narrow-band to broad-band conversion for Meteosat-visible channel and broadband albedo using both AVHRR-1 and -2 channels, *Int. J. Remote Sens.* **16**: 1147–1166.

Vermote, E., Tanre, D., Deuze, J. L., Herman, M., and Morcrette, J. J. (1997a), Second simulation of the satellite signal in the solar spectrum: An overview, *IEEE Trans. Geosci. Remote Sens.* **35**: 675–686.

Vermote, E., El Saleous, N., Justice, C., Kaufman, Y., Privette, J., Remer, L., Roger, J., and Tanre, D. (1997b), Atmospheric correction of visible to middle-infrared EOS-MODIS data over land surfaces: Background, operational algorithm and validation, *J. Geophys. Res.* **102**: 17131–17142.

Wielicki, B. A., Barkstrom, B. R., Baum, B. A., Charlock, T. P., Green, R. N., Kratz, D. P., Lee, R. B. I., Minnis, P., Smith, G. L., Wong, T., Young, D. F., Cess, R. D., Coakley, J. A. J., Crommelynck, D. A., Donner, L., Kandel, R., King, M. D., Miller, A. J., Ramanathan, V., Randall, D. A., Stowe, L. L., and Welch, R. M. (1998), Clouds and the earth's radiant energy system (CERES): Algorithm overvie, *IEEE Trans. Geosci. Remote Sens.* **36**: 1127–1141.

Wydick, J. E., Davis, P. A., and Gruber, A. (1987), *Estimation of Broadband Planetary Albedo from Operational Narrowband Satellite Measurements*, NOAA, Technical Report NESDIS 27.

Xue, Y., Sellers, P., Kinter III, J., and Shukla, J. (1991), A simplified biosphere model for global climate studies, *J. Climate* **4**: 345–364.

Yamaguchi, Y., Kahle, A., Tsu, H., Kawakami, T., and Pniel, M. (1998), Overview of advanced spaceborne thermal emission and reflection radiometer (ASTER), *IEEE Trans. Geosci. Remote Sens.* **36**: 1062–1071.

Young, D., Minnis, P., and Wong, I. (1998), Temporal interpolation methods for the clouds and the earth's radiant energy system (CERES) experiment, *J. Appl. Meteorol.* **37**: 572–583.

10

Estimation of Surface Radiation Budget: II. Longwave

Following Chapter 9, this is the second of two parts on surface radiation budget estimation. Similar to Chapter 9 on the shortwave radiation budget, we emphasize the surface variables that are critical in calculating the long-wave radiation budget, which includes land surface temperature (LST) and emissivity. Before discussing how these two variables should be estimated from thermal remote sensing data, Sections 10.2–10.5 introduce thermal radiative transfer theory. This is necessary because previous chapters of this book deal only with solar (shortwave) radiative transfer. Sections 10.6–10.8 introduce various methods for estimating emissivity and LST. Conversion of narrowband to broadband emissivity and surface energy balance modeling are presented in the last two sections.

Section 10.2 introduces the thermal radiative transfer theory and solution at a single wavelength. This is very similar to the solar (shortwave) radiative transfer theory discussed in Chapter 2. It is simpler for multispectral sensors because multiple scattering is usually neglected in the thermal radiative transfer, but there are much stronger wavelength dependences of the gaseous absorption in thermal infrared spectrum.

Section 10.3–10.5 present three different methods for calculating radiative transfer for wavebands with a range of wavelengths. It is most relevant to remote sensing since no bands have a single wavelength. Section 10.6 discusses atmospheric correction methods for thermal IR imagery. Given the atmospheric profiles, atmospheric correction is straightforward using any radiative transfer packages. Section 10.7 introduces various split-window algorithms for estimating land surface temperature using data from two thermal IR bands. Most algorithms require known surface emissivities, which

Quantitative Remote Sensing of Land Surfaces. By Shunlin Liang
ISBN 0-471-28166-2 Copyright © 2004 John Wiley & Sons, Inc.

can be roughly estimated using vegetation indices and land cover information. Section 10.8 introduces various methods for separating emissivity and LST simultaneously from multiple thermal IR bands (more than two). Section 10.9 presents an approach that converts narrowband emissivity to broadband emissivity. This is required in order to calculate surface energy balance. Finally, in Section 10.10, we present the modeling techniques for calculating surface energy balance components that link thermal observations with surface variables.

10.1 INTRODUCTION

The longwave net radiation at the Earth's surface is the difference between the downward flux F_d and the upwelling flux F_u

$$\Delta F = F_d - F_u \tag{10.1}$$

where downward flux depends mainly on the atmospheric profiles of the atmospheric temperature and water vapor. Several sounding sensors provide these profiles operationally, including GOES and MODIS. Given the atmospheric profiles, radiative transfer models can be easily used for calculating downward flux.

Upwelling flux depends mainly on the surface temperature and emissivity:

$$F_u = (1 - \varepsilon) F_d + \varepsilon \sigma T^4 \tag{10.2}$$

where T is the land surface temperature and ε is the surface broadband emissivity. For dense vegetation and water surfaces, broadband emissivity is almost one (0.96–1). For nonvegetated surfaces, it is far below one. Unfortunately, most global circulation (also climate) models (GCMs) and land surface models have assumed unity emissivity, which may lead to errors in net radiation as large as 20 W/m² (Rowntree 1991).

Estimation of both emissivity and land surface temperature simultaneously from thermal infrared remotely sensed data is a very challenging issue. Radiance received by the sensor contains information about atmosphere (e.g., temperature and water vapor profiles) and surface properties (emissivity and land surface temperature). Therefore, the first step for retrieving surface emissivity and LST is to perform atmospheric correction. The second step is to separate emissivity and temperature from the retrieved surface leaving radiance.

For sensors that have two thermal bands, such as AVHRR and GOES, one has to assume a known emissivity (or one inferred from land cover maps or vegetation indices) in order to estimate land surface temperature using the so-called split-window algorithms. Fortunately, the new generation of sensors has multiple thermal bands, such as ASTER and MODIS, which allow us to estimate spectral emissivities and land surface temperature (LST) simultaneously.

In the next section, we introduce thermal radiative transfer formulation at a specific wavelength. Since thermal radiative transfer is similar to the solar radiative transfer presented in Chapters 2–4, the discussion will be brief.

For a waveband, the spectral integration over the range of wavelength is needed. Unlike shortwave solar radiative transfer, atmospheric molecules are highly selective in absorbing thermal radiation and there are a large number of absorption lines within a very small spectral interval. Several techniques are discussed to handle this. The first is the line-by-line method. It is most accurate but computationally very expensive (Section 10.3). The second is the band models that are based on one or more analytic functions with a minimum number of parameters (Section 10.4). They are simple but accuracy is limited. Section 10.5 presents the k distribution and corrected k-distribution methods that represent a good compromise between the accuracy and computational cost.

10.2 MONOCHROMATIC RADIATIVE TRANSFER FORMULATION AND SOLUTIONS

10.2.1 Thermal IR Radiative Transfer Equation

The differential form of the radiative transfer equation in the thermal infrared spectrum in a plane-parallel atmosphere can be written by

$$\mu \frac{dL_v}{d\tau} = L_v - S_v \tag{10.3}$$

where L_v is the radiance in a given line of sight (energy per unit frequency per unit time per steradian), S_v is the source term representing additional energy input into the line of sight, and $d\tau$ is the differential optical depth, defined as

$$d\tau = \rho \kappa_v \, ds \tag{10.4}$$

where ρ is the mass density, κ_v is the opacity (mass extinction coefficient), and ds is the pathlength along the line of sight. From the convention for thermal remote sensing, we have replaced wavelength λ by wavenumber v in all these expressions. As we discussed in Section 1.2.2, wavenumber in (cm^{-1}) can be calculated by $10{,}000/\lambda$, where wavelength λ is expressed in micrometers. In general, L_v is a function of position, frequency, time, and direction (two angles in three dimensions). Considering a stratified, nonscattering atmosphere in thermodynamic equilibrium, the source function equals the Planck function $S_v = B_v = B_v(T)$, and we have

$$\mu \frac{dL_v}{d\tau} = L_v - B_v \tag{10.5}$$

The Planck function for the spectral radiance (W cm^{-2} sr^{-1} μm^{-1}) emitted by a blackbody at a given wavenumber and temperature T can be expressed by

$$B_\lambda(T) = \frac{1.1909561 \cdot 10^{-16} v^5}{e^{1.43879\, v/T} - 1} \qquad (10.6)$$

where v is the wavenumber $(10,000/\lambda$; λ is the wavelength in μm). Quite often the emitted radiance needs to be presented in wavenumber units (W cm^{-2} sr^{-1} cm^{-1}). Thus

$$B_v(T) = \frac{1.1909561 \cdot 10^{-12} v^3}{e^{1.43879\, v/T} - 1} \qquad (10.7)$$

By comparing Eqs. (10.6) and (10.7), it is clear that we have to multiply $v^2/10000$ in order to convert radiance unit from W cm^{-2} sr^{-1} cm^{-1} to W cm^{-2} sr^{-1} μm^{-1}, which is frequently needed in remote sensing data analysis.

Equation (10.3) is the basis of radiative transfer in the thermal infrared spectrum, where scattering effects are often negligible for a clear atmosphere.

10.2.2 Approximations and Numerical Solutions

To determine L_v, we need to specify the boundary conditions:

$$\begin{cases} L_v(\tau_t, -\mu) = 0 \\ L_v(0, \mu) = \varepsilon_v B_v(T_s) + (1 - \varepsilon_v)\dfrac{F_d}{\pi} \end{cases} \qquad (10.8)$$

At the top of the atmosphere (the first boundary condition above with the total optical depth τ_t) we assume that there is no downward thermal radiation. At the surface, upwelling radiance consists of two components: emitted radiance and reflected downward radiation. The total downward flux at the surface can be given as

$$F_d = 2\pi \int_0^1 L_v(0, -\mu)\,\mu\,d\mu \qquad (10.9)$$

Note that we have assumed a Lambertian surface in the lower boundary condition (i.e., emissivity is independent of the viewing angle). The angular dependence of the spectral emissivity is still a premature research issue, and we do not discuss it in detail. Interested readers are referred to the literature (Takashima and Masuda 1987, Rees and James 1992, Prata et al. 1995, Liang et al. 2000) for details.

The formal solutions to (10.5) for upwelling and downward radiances are then obtained by

$$L_v(\tau, +\mu) = e^{-(\tau_t - \tau)/\mu} L_v(0, \mu) + \frac{1}{\mu} \int_0^\tau e^{(\tau' - \tau)/\mu} B_v(\tau') \, d\tau' \quad (10.10)$$

$$L_v(\tau, -\mu) = \frac{1}{\mu} \int_\tau^{\tau_t} e^{-(\tau' - \tau)/\mu} B_v(\tau') \, d\tau' \quad (10.11)$$

The first term on the right side of Eq. (10.10) is the thermal radiation from the surface, attenuated by absorption in the atmosphere until it contributes to the upwelling radiance at level τ. The second term contains the upwelling radiance arriving at τ contributed from the attenuated atmospheric thermal emission from each parcel between the surface and τ. The $1/\mu$ factor in front of the integral accounts for the slant path of the thermally emitting atmosphere. The right side of Eq. (10.11) is similar but contains no boundary contribution since the vacuum above the atmosphere is assumed to emit no thermal radiation. The upwelling and downward radiances in a stratified, thermal atmosphere are fully described by Eqs. (10.10) and (10.11). Let us discuss the practical implementation.

Most radiative transfer codes of the atmosphere calculate the radiance distribution by dividing the atmosphere into many small layers, say, 1-km intervals below 25 km and larger intervals for heights above 25 km. To achieve a high accuracy, we still need to consider the vertical variation of the temperature within each layer. If a linear dependence is assumed, Planck radiance at any location specified by the optical depth τ' can be determined by

$$B_v(\tau') = B_v(T_t) + \frac{\tau'}{\tau} [B_v(T_b) - B_v(T_t)] \quad (10.12)$$

where $B_v(T_t)$ is the Planck function evaluated at the layer top edge temperature T_t and $B_v(T_b)$ is similarly defined at the bottom edge of the layer. Thus, the thermal radiance in the direction (from the entire layer) can be obtained in closed form:

$$E_v(+\mu) = B_v(T_t) - B_v(T_b) + \left\{ B_v(T_b) + \frac{\mu}{\tau} [B_v(T_b) - B_v(T_t)] \right\} \left[1 - \exp\left(-\frac{\tau}{\mu} \right) \right]$$
$$(10.13)$$

for the upwelling radiance and

$$E_v(-\mu) = B_v(T_b) - B_v(T_t) + \left\{ B_v(T_t) - \frac{\mu}{\tau} [B_v(T_b) - B_v(T_t)] \right\} \left[1 - \exp\left(-\frac{\tau}{\mu} \right) \right]$$
$$(10.14)$$

for the downward radiance.

To remove the possible singularity due to small optical depth, Eqs. (10.13) and (10.14) can be written as

$$E_v(+\mu) = \sum (-1)^{n+1} \frac{B_v(T_t) + nB_v(T_b)}{(n+1)!} \left(\frac{\tau}{\mu}\right)^n \qquad (10.15)$$

$$E_v(-\mu) = \sum (-1)^{n+1} \frac{B_v(T_b) + nB_v(T_t)}{(n+1)!} \left(\frac{\tau}{\mu}\right)^n \qquad (10.16)$$

This result is usually referred to as the *linear in τ* approximation.

Another technique is called *Pade approximation.* Let $\Im_v$ be the total transmittance of the layer; then

$$E_v(\tau) = (1 - \Im_v) B_v(\tau) \qquad (10.17)$$

where the simplest Pade approximation for the effective Planck function is of the form

$$B_v(\tau) = \frac{B_v(\bar{T}) + a\tau B_v(T_t)}{1 + a\tau} \qquad (10.18)$$

where $\bar{T}$ is the layer mean temperature and a is a constant ($a = 0.2$). For a two-term Pade approximation, we obtain

$$B_v(\tau) = \frac{B_v(\bar{T}) + (a\tau + b\tau^2) B_v(T_t)}{1 + a\tau + b\tau^2} \qquad (10.19)$$

with $a = 0.193$ and $b = 0.013$.

Given the radiance for each layer, the downward radiance for the stacked layer atmosphere can be calculated layer by layer as follows:

$$\begin{cases} L_v^N(-\mu) = 0 \\ L_v^{N-1}(-\mu) = E_v^{N-1}(-\mu) + L_v^N(-\mu)\exp\left(-\frac{\Delta\tau_N}{\mu}\right) \\ \quad\quad\quad \vdots \\ L_v^0(-\mu) = E_v^0(-\mu) + L_v^1(-\mu)\exp\left(-\frac{\Delta\tau_1}{\mu}\right) \end{cases} \qquad (10.20)$$

where $E_v^i(-\mu)$ is the emitted radiance of layer i specified by Eq. (10.16), and $\Delta\tau_i$ are the optical depth in each layer. It starts from the top of the atmosphere where the solar thermal radiation is effectively zero. Similarly, for the upwelling radiance, it follows that

$$
\begin{cases}
L_v^0(+\mu) = \varepsilon_v B_v(T_s) + (1 - \varepsilon_v) L_v^0(-\mu) \\
L_v^1(+\mu) = E_v^1(+\mu) + L_v^0(+\mu)\exp\left(-\dfrac{\Delta\tau_1}{\mu}\right) \\
\quad\vdots \\
L_v^N(+\mu) = E_v^N(+\mu) + L_v^{N-1}(+\mu)\exp\left(-\dfrac{\Delta\tau_N}{\mu}\right)
\end{cases}
\tag{10.21}
$$

where T_s is surface temperature and ε_v is surface emissivity. The upwelling radiance at the surface layer $L_v^0(+\mu)$ is the sum of the emitted radiance and the reflected downward radiance. In this expression, the surface has been assumed to be Lambertian. $E_v^i(+\mu)$ is specified in (10.15).

It is important to note that formulas (10.20) and (10.21) are not iterative because scattering has been neglected. It is a reasonable approximation for clear-sky conditions. If scattering cannot be neglected, for example, for cloudy atmosphere, iterative formulas similar to (2.85)–(2.87) in Chapter 2 are needed.

Because of the absence of scattering, the angular dependence of radiance becomes much simpler. It depends only on the zenith angle and is assumed to be isotropic in the azimuthal direction. The upwelling flux (F_u) and downward flux (F_d) can be calculated at each layer k using the one-angle Gaussian quadrature

$$
F_d^k = \sum_{j=1}^{J} L^k(-\mu_j)\mu_j w_j
\tag{10.22}
$$

and

$$
F_u^k = \sum_{j=1}^{J} L^k(+\mu_j)\mu_j w_j
\tag{10.23}
$$

where w_j are the Gaussian quadrature weights (see Table 2.6). It is sufficient for small J (say, 3–5). If $J = 1$ and $\mu = 1/1.66$, this is usually called the *diffusivity factor approximation*.

So far we have discussed only the monochromatic radiative transfer (at single frequency or wavelength) in the thermal IR spectrum, but remote sensors detect the Earth environment with several spectral bands. Atmospheric molecules are highly selective in their ability to absorb radiation, and we can observe a large number of absorption lines using an interferometer within a very small spectral interval. The major molecular species (O_2, N_2) of the Earth's atmosphere have essentially no importance for IR absorption, and four of the most important IR-absorbing molecular species are the minor

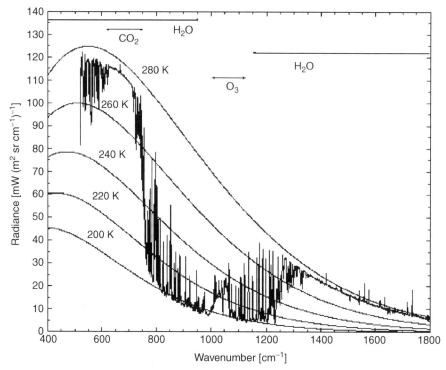

Figure 10.1 Illustration of atmospheric downward radiance relative to values derived from the Planck function for various temperatures. Also noted are the absorption regions for various atmospheric constituents. The spikes in the measured radiance, between 1400 and 1800 cm^{-1}, are a result of water vapor absorption lines that become opaque within the instrument.

polyatmotic moleculars: water vapor (H_2O), carbon dioxide (CO_2), ozone (O_3), and methane (CH_4). They provide the medium for radiative interaction due to absorption of upwelling terrestrial radiation. Subsequently, the energy is isotropically re-emitted at the wavelengths where the absorption occurred.

Figure 10.1 illustrates downwelling spectral radiance as a function of wavenumber. The dominant species are CO_2 (600–800 cm^{-1}, 15 μm), O_3 (1000–1060 cm^{-1}, 9.6 μm), and H_2O. The H_2O absorption signature is of particular interest because of its spectral distribution. The atmospheric window (800–1200 cm^{-1}) is the most transparent region in the spectrum. Measurements between water vapor lines within the atmospheric window yield observations with the least atmospheric contamination. The H_2O absorption lines present an additional problem because the far "wings" of individual H_2O lines combine to form the water vapor continuum. Thus, even the microwindows are not completely transparent.

There are several different approaches to calculating radiance over a spectral range (band), such as line-by-line methods, band models, and correlated k-distribution methods. They are discussed in the following three sections.

10.3 LINE-BY-LINE METHODS

Line-by-line methods are the most accurate but computationally expensive since they calculate radiative transfer at the subwavenumber level. There are many software packages, including GENSPECT (http://www.genspect.com/), FASCODE (fast atmospheric signature code), GENESIS, and GENLN2.

There are basically two types of spectra that are of interest in radiative transfer calculations: line spectra and continua. For line spectra, the positions of the absorbing lines are known and the line shapes can be calculated. Let us first discuss the shape functions of the isolated lines, and spectral sampling in Sections 10.3.1 and 10.3.2. The continuum absorption is discussed in Section 10.3.3.

10.3.1 Lineshapes

To describe the infrared absorption of a radiating molecule, it is necessary to consider the variation of the spectral absorption coefficient for a single line. The observed shape of a single, spectrally isolated line can be described by many different functions, such as rectangular, triangular, Lorentz function, Doppler function, and Voigt (combined Lorentz and Doppler) function. If we do not consider the influence from other lines in the vicinity of the frequency that defines the line center, the extinction coefficient can be expressed as the simple function of frequency. The *Lorentz shape* is defined as

$$k_v = k_a(v) = \frac{S}{\pi} \frac{\alpha_L}{(v - v_0)^2 + \alpha_L^2} \tag{10.24}$$

where S = line strength or line intensity in g^{-1} cm [it is actually a normalization factor $S = \int_{-\infty}^{\infty} k(v)\, dv$]

α_L = the Lorentz half-width in cm^{-1}, defined as half width of the line at half the max value of $k(v)$

v_0 = line center

The *Doppler shape* is defined as

$$k_v = \frac{S}{\alpha_D \sqrt{\pi}} \exp\left[-\left(\frac{(v - v_0)}{\alpha_D}\right)^2\right] \tag{10.25}$$

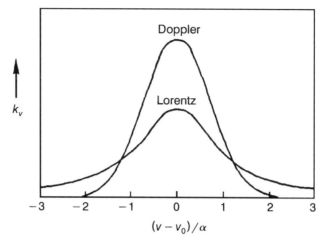

Figure 10.2   Illustration of Lorentz profile and the Doppler profile.

where

$$\alpha_D = \frac{v_0}{c}\sqrt{\frac{2K_BT}{m}} \tag{10.26}$$

where c is the velocity of light; m is the molecular mass; K_B is the Boltzmann constant, 1.38047^{-23} erg/K; and T is the absolute temperature in kelvins.

Figure 10.2 depicts a plot for both the Lorentz profile and the Doppler profile. For comparable intensities and half-widths, the Doppler line has more absorption near the center and less in the wings than the Lorentz line. Lorentzian functions describe the absorption lines subject to collision broadening, and Doppler functions describe Doppler broadening. Broadening is a phenomenon mostly in the lower atmosphere where radiation transitions are sufficiently disturbed by molecular collisions to cause a broadening of spectral lines. Lorentzian shape is dominant in the high-wavenumber and high-pressure regimes, whereas the Doppler shape is mostly in lower-pressure and low-wavenumber regimes where the collision broadening effect is small. In regimes where both broadening effects must be considered, a convolution of these two functions leads to the *Voigt function*:

$$k_v = K(x,y) = K_0\frac{y}{\pi}\int_{-\infty}^{\infty}\frac{\exp(-t^2)}{(x-t)^2+y^2}\,dt \tag{10.27}$$

where

$$x = \frac{(v - v_0)}{\alpha_D} \sqrt{\ln 2} \qquad (10.28)$$

$$y = \frac{\alpha_L}{\alpha_D} \sqrt{\ln 2} \qquad (10.29)$$

and

$$K_0 = \frac{1}{\alpha_D} \sqrt{\frac{\ln 2}{\pi}} \qquad (10.30)$$

For radiative transfer analyses involving gases at low pressure (upper atmo-spheric conditions), the Voigt function is often used since it incorporates the combined influence of the Lorentz and the Doppler broadening.

These parameters can be calculated for a particular gas line from databases, such as the HITRAN'2000 Database (version 11.0) that contains over 1,080,000 spectral lines for 36 different molecules (http://cfa-www.harvard.edu/HITRAN/). The Lorentz width α_L at a specific tem-perature and a total pressure p may be derived from database entries for the self-broadening width α_S and the air-broadening width α_A:

$$\alpha_L = \left(\frac{296}{T}\right)^{T_c} \frac{\alpha_A(p - p') + \alpha_S p'}{1013.25} \qquad (10.31)$$

where T_C is the positive coefficient of temperature dependence, and p' is the partial pressure of the gas. The Doppler broadening width is related only to the mass and the temperature of the molecule and can be computed from Eq. (10.26). The line strength may be computed as

$$S(T) = S_0 Q(T) \exp\left[\frac{c_2 E_L(T - 296)}{296T}\right] \frac{1 - \exp(-c_2 v_0/T)}{1 - \exp(-c_2 v_0/296)} \qquad (10.32)$$

where S_0 is the line strength at the reference temperature (296 K), $Q(T)$ is the internal partition function ratio, E_L is the lower state energy, and $c_2 = 1.438786$.

Given extinction coefficient k_v defined by Eq. (10.24) or Eq. (10.25), the transmittance of the homogeneous path can be calculated by

$$T = \exp(-k_v U) \qquad (10.33)$$

where U is the mass of the absorbing gas per unit area. Figure 10.3 shows a comparison of the transmittances calculated by these three line profiles.

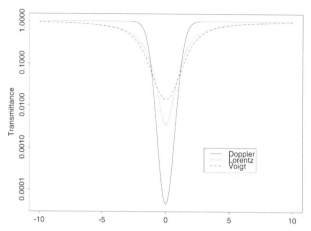

Figure 10.3   Comparison of the transmittance by these three functions where $\alpha_L/\alpha_D = 1$ and $SU/\sqrt{\pi}\,\alpha_D = 10$.

10.3.2   Spectral Sampling

A typical radiative transfer problem involves a huge number of computations of the basic line function over a wide parameter domain, specifically, a number of spectral points, the number of gases, the number of atmospheric layers and the number of lines to be considered. All line-by-line algorithms attempt to reduce the number of times the line function (e.g., Voigt function) must be evaluated. For a specific spectral range, a spectral sampling scheme has to be used for a line-by-line calculation. According to Quine and Drummond (2002), GENSPECT and SEASCAPE use a similar binary-type division, but most other codes adopt one of the following three approaches:

- To decompose the line function into a series of subfunctions (see Fig. 10.4). For example, in FASCODE five domains are chosen for the Voigt lineshape based on the scaled variable $z = (v - v_0)/\alpha_v$ where α_v is the Voigt half-width: $0 \le z \le 4$, $4 \le z \le 16$, $16 \le z \le 64$, $64 \le z \le 25$ cm^{-1}, and beyond 25 cm^{-1}. The first domain extends over the line center, and others over increasingly distant sections of the line. A nominal spectral sampling rate of four points per mean half-width is utilized by FAS-CODE to achieve an accuracy of the order of 0.15%, which corresponds to a sampling interval of 0.02 cm^{-1} for the atmospheric layer at the surface and a sampling interval of 0.000025 cm^{-1} (at 100 cm^{-1}) at the top of the atmosphere (Clough et al. 1992).
- To divide the calculation into two grids (GENESIS, GENLN2, and LINEPACK). The contributions from the line centers where the line function varies relatively rapidly are performed at a fine full-resolution grid, and the contribution from the line function wings (at a distance

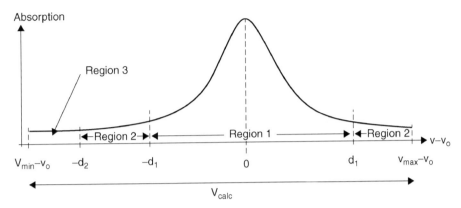

Figure 10.4 Illustration of spectral sampling in the line-by-line computation. Three regions may have different sampling intervals, with the densest sampling in region 1.

from the line center) are performed at a greatly reduced coarse-resolution grid.

- To conduct the calculation in Fourier space, using a Fourier transformed express of the line function (Kobayashi 1999). They could be particularly applicable to problems that are already cast in Fourier space, such as the retrieval of atmospheric constituents from interferograms generated by a Fourier transform spectrometer.

The monochromatic absorption coefficient at a wavenumber can be written as the sum of the absorption contributions from all lines i in the spectral range:

$$k^l(v) = \sum_i k_i(v) \qquad (10.34)$$

All codes include a wing cutoff parameter that limit the extent of this calculation by defining a distance from the line center beyond which the line function contributions can be neglected.

10.3.3 Continuum Absorption

In addition to the line-by-line absorption, the continuum absorption of certain molecules, in particular water, must be taken into account. This is especially important for the thermal IR windows around 4 μm and at 8–12 μm. The continuum absorption coefficient that takes into account both self and foreign components of the continuum absorption is defined as (Clough et al. 1992)

$$k^c(v) = v \tanh\left(\frac{hcv}{2kT}\right) \frac{T_{\text{ref}}}{T} \left[\frac{p_s}{p_{\text{ref}}} C_s^0(v,T) + \frac{p - p_s}{p_{\text{ref}}} C_f^0(v,T)\right] \qquad (10.35)$$

where p_s, p = partial pressure of the absorbing gas and the total gas
pressure, respectively

p_{ref} = is the reference pressure of 1 atm

T = is the path temperature

T_{ref} = reference temperature of 296 K

C_s^0 = continuum absorption parameter for self-component

C_f^0 = parameter for the foreign component in $1/(cm^{-1}\ mole^{-1}\ cm^{-2})$ and stored as the tables in the radiative transfer code

Formulation of the continuum absorption has been used in numerous line-by-line codes, such as FASCODE and GENLN2.

10.3.4 Calculation of Transmittance and Radiance

The total transmittance of all n gases along a path follows Beer's law

$$\Im_v = \exp\left[-\sum_{i=1}^{n} \left(k_i^l + k_i^c \right) u_i \right] \tag{10.36}$$

where the line-by-line absorption coefficients k^l and the continuum absorption coefficients k^c are given by (10.34) and (10.35), and u is the absorber amount. The transmittance is related to optical depth $\Im_v = \exp(-\tau/\mu)$. Thus, the previous formulas (10.20) and (10.21) can be used to calculate upwelling and downward radiances.

The previous lineshapes are valid for constant-pressure and constant-temperature paths. The real atmosphere contains many different absorbers (gases) that are vertically variable, and therefore it is normally treated as a series of layers. The combined absorption coefficient is calculated within each layer. If each layer is very thin and can be approximated by a homogeneous layer, the number of layers will be very large. If the layering grid is very coarse, each layer is no longer a homogeneous layer. To apply the formulas of the homogeneous layer to an inhomogeneous layer, we may develop a scaling method to account for the vertical variations so that the number of layers can be dramatically reduced or the accuracy obtained within each layer can be significantly improved. The simpliest procedure, called the *Curtis–Godsen approximation*, has proved fairly accurate for the infrared radiative transfer calculations involving water vapor and carbon dioxide atmosphere.

The approximation determines the absorption parameters by absorber-weighted averages over the layer. The effective temperature and half-width of the lineshape according to the Curtis–Godsen approximation are

$$\alpha_{\text{eff}} = \frac{\int_{z_1}^{z_2} \rho(z)\,\alpha(z)\,dz}{\int_{z_1}^{z_2} \rho(z)\,dz} \tag{10.37}$$

$$\tau_{\text{eff}} = \frac{\int_{z_1}^{z_2} \rho(z)\,\tau(z)\,dz}{\int_{z_1}^{z_2} \rho(z)\,dz} \tag{10.38}$$

where z_1 and z_2 are the heights of the layer boundary and $\rho(z)$ is the absorber density. Any other desired absorption parameter can be calculated in the same way. This approximation is used in GENLN2 (Edwards 1988). If the Curtis–Godsen approximation is not used, the Pade approximation [see Eq. (10.18) or (10.19)] can be used to calculate radiance flux (Clough et al. 1992).

If the Curtis–Godsen approximation is used, the Planck formula can be used for calculating the layer flux and Eqs. (10.17), (10.20), and (10.21) can be used to calculate the radiance field in both the upwelling and downward directions.

10.4 BAND MODELS

For remote sensing applications with specific wavebands, some degree of integration over the wavenumber domain is required. To simulate the behavior of the radiometer with each band corresponding to a range of wavelengths, and to successfully extract information from the set of radiance measurements, it is therefore necessary to compute integrated quantities, such as radiance in the interval $\Delta\upsilon$

$$L_{\Delta\upsilon} = \frac{\int_{\Delta\upsilon} L_{\upsilon}\tau_{\upsilon}\,d\upsilon}{\int_{\Delta\upsilon} \tau_{\upsilon}\,d\upsilon} \tag{10.39}$$

There are several ways in which the integration can be performed, and they have variable accuracies. Only a brief outline is provided here.

A large number of molecular band models have been developed to represent the very messy transmission problem. They have one or more analytic functions with a minimum number of parameters, such as the Elsasser band model and the random band model. These band model parameters are derived from comparisons with either laboratory data or accurate line-by-line computations. The earlier versions of the MODTRAN code were based on a narrowband model.

In principle, narrowband calculations can be as accurate as line-by-line calculations, provided an "exact" narrowband average can be found. Hartmann et al. (1984) have generated artificial narrowband properties from their (temperature-extrapolated) line-by-line data. Using a resolution of 25 cm^{-1} they observed a maximum 10% error between line-by-line and narrowband absorptivities. Obviously, the assumptions of the applied random statistical model are violated somewhat over a 25-cm^{-1} spectral range. However, using two narrow spectral ranges of H_2O and CO_2, Lacis and Oinas (1991) showed that (for a resolution of 10 cm^{-1}, and for total gas pressures above 0.1 atm) the correlational accuracy can be improved to better than 1% if the model parameters are found through least square fits.

The primary disadvantage of narrowband models is the fact that they provide spectral and band averages for a path-integrated emissivity, with some blackbody (of given temperature) as the emitter. This causes some problems in the case of nonhomogeneous paths, which are successfully overcome by the Curtis–Goodman approximation (Goody and Yang 1989).

10.4.1 Elsasser Model

One of the oldest and most widely recognized models was proposed by Elsasser in which it is assumed that the lines are regularly spaced. The mathematical description of the spectral absorption coefficient for the regular Elsasser model is given based on the Lorentz shape

$$k(v) = \sum_{i=-\infty}^{\infty} \frac{S}{\pi} \frac{\alpha_L}{(v - i\delta)^2 + \alpha_L^2} \tag{10.40}$$

where δ is the (mean) spacing between line centers and $i\delta$ the position of the ith line center.

From the Mittag–Leffler theorem, Eq. (10.40) is equivalent to the following expression

$$k(v) = \frac{S}{\delta} \frac{\sinh \beta}{\cosh \beta - \cos \gamma} \tag{10.41}$$

where

$$\beta = \frac{2\pi\alpha_L}{\delta} \tag{10.42}$$

and

$$\gamma = \frac{2\pi v}{\delta} \tag{10.43}$$

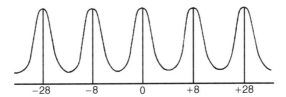

Figure 10.5 Illustration of the Elsasser model.

The transmittance function can be expressed by for a given the amount of absorber u

$$\xi(v) = \frac{1}{\delta} \int_{-\infty}^{\infty} \exp[-k(v)u] \, dv = \frac{1}{2\pi} \int_{-\pi}^{\pi} \exp[-k(\gamma)u] \, d\gamma \quad (10.44)$$

Without providing the details of the mathematical derivation, we only show the final result

$$\xi(v) = 1 - \text{erf}\left(\frac{\sqrt{\pi S \alpha_L u}}{\delta}\right) \quad (10.45)$$

where the error function $\text{erf}(x)$ can be calculated numerically or obtained from a mathematical table. This model is illustrated in Fig. 10.5.

10.4.2 Statistical Model

From observations, it has been found that most molecular lines are not regularly distributed but exhibit a more or less random appearance. Goody (1952) therefore introduced an alternative method of calculating atmospheric transmittance without a priori knowledge of the structure of molecular lines. It is shown in Fig. 10.5. The result is presented here without giving the detailed derivation:

$$\xi(\bar{v}) = \exp\left(-\frac{\sum_{i=1}^{N} U_i/N}{\delta}\right) \quad (10.46)$$

with

$$U_i = \int_{-\infty}^{\infty} \{1 - \exp[-k_i(v)u]\} \, dv \quad (10.47)$$

where $k_i(v)$ is the extinction coefficient of the ith line.

10.5 CORRELATED *k*-DISTRIBUTION METHODS

Another type of algorithm in the 1990s literature is the k distribution and its associated correlated k distribution method (Lacis and Oinas 1991, Fu and Liou 1992, Mlawer et al. 1997). These algorithms provide much better accuracy than the conventional band models, and yet they require two or three orders of less computer time than line-by-line method. Furthermore, they can accommodate multiple scattering in a straightforward manner (Thomas and Stamnes 1999). The latest version of MODTRAN is based on the correlated k-distribution method.

In this method it is observed that, over a narrow spectral range, the rapidly oscillating absorption coefficient k_v attains the same value many times (at slightly different wavenumbers v) each time, resulting in identical radiance I_v and radiative flux F_v. Since the actual wavenumbers are irrelevant in the narrow spectral range, in the correlated k-distribution method the absorption coefficient is reordered, resulting in a smooth dependence of absorption coefficient versus artificial wavenumber g (varying across the given narrow range). This, in turn, makes spectral integration very straightforward.

Let us examine this issue mathematically. The average transmittance $\overline{\Im}$ along a homogeneous ray path through the atmosphere for some spectral interval $\Delta v = v_2 - v_1$ is given by

$$\overline{\Im} = \frac{1}{\Delta v} \int_{v_1}^{v_2} \Im_v \, dv \tag{10.48}$$

where $\Im_v$ is the monochromatic transmittance, defined as

$$\Im_v = \exp(-k_v u) \tag{10.49}$$

where u is the total path absorber amount. Since k_v may have the same values for many times over the spectral range, if we can define a *probability density function* as $f(k)$ the average transmittance can be written as

$$\overline{\Im} = \int_0^1 f(k) \exp(-ku) \, dk \tag{10.50}$$

It is more convenient to use a *cumulative probability function*

$$g(k) = \int_0^k f(k') \, dk' \tag{10.51}$$

such that

$$dg = f(k) \, dk \tag{10.52}$$

This allows the band transmittance integral to be written

$$\overline{\Im} = \int_0^1 \exp[-k(g)u]\, dg \qquad (10.53)$$

where $k(g)$ is the inverse function of $g(k)$, and usually referred to as k *distribution*.

Figure 10.6 shows the example of CO absorption at 4.6 μm for the MOPITT band 1 (Edwards and Francis 2000). Figure 10.6a shows the CO absorption coefficient k for layers at 1000 and 100 mbar with thick and thin lines, respectively. The left panel in Fig. 10.6b shows the probability density function $f(k)$ for the 1000- and 100-mbar layer absorption coefficients, where the spikes are characteristic of local maxima and minima in the k spectra and dependent on the resolution in calculating the density function; the right panel shows the cumulative probability function $g(k)$, Fig. 10.6c shows its inverse, the k distribution $g(k)$. Both functions are seen to be monotoically increasing and smooth.

The average radiance of a layer can be expressed in terms of transmittance:

$$\overline{L} = \frac{1}{\Delta v} \int_{v_1}^{v_2} \left[L_0(v) + \int_{\Im_v}^1 [B(v, T(\Im_v')) - L_0(v)]\, d\Im \right] dv \qquad (10.54)$$

where $L_0(v)$ is the radiance incoming to the layer, $B(v, T)$ is the Planck function at temperature T, and $\Im_v$ is the transmittance of the layer. Assuming that Planck function varies linearly along the absorbing path in the layer, as described by Eq. (10.18); thus (10.54) becomes

$$\overline{L}(\theta) = \frac{1}{\Delta v} \int_{v_1}^{v_2} \left[B_{\mathrm{eff}}(v, T_v) + [L_0(v) - B_{\mathrm{eff}}(v, T_v)] \exp\left(-k_v(P, T)\frac{\rho \Delta z}{\cos \theta}\right) \right] dv$$

$$(10.55)$$

where θ is the zenith angle. Under the mapping from k_v to $g(k)$, it further becomes

$$\overline{L}(\theta) = \int_0^1 \left[B_{\mathrm{eff}}(g, T_g) + [L_0(g) - B_{\mathrm{eff}}(g, T_g)] \exp\left(-k(g, P, T)\frac{\rho \Delta z}{\cos \theta}\right) \right] dg$$

$$(10.56)$$

In actual computation, the domain of the variable g is partitioned into subintervals and the average absorption coefficient for each subinterval k_j is determined. The resulting radiance, weighted by the sizes w_j of their respective subintervals ($\Sigma_j w_j$), are summed to yield an approximation to the average

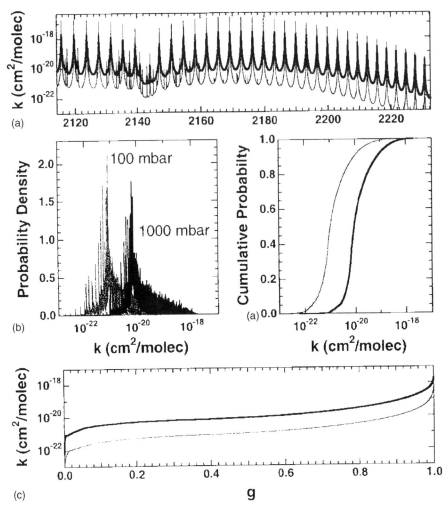

Figure 10.6 Illustration of the correlated k-distribution method for calculating CO_2 band absorptance at 4.7 μm: (a) the absorption coefficient for levels at 1000 mbar (thick line) and 100 mbar (thin line); (b) the left panel shows probability density distributions for the absorption coefficients from the two levels shown in the (a), and the right panel shows the cumulative probability density g distributions; the inverse, the k distributions. [Modified from Edwards and Francis (2001), *J. Geophys. Res.* By permission of American Geophysical Union.]

radiance (10.54):

$$\overline{L}(\theta) \cong \sum_j w_j L_j$$

$$= \sum_j w_j \left[B_{\text{eff},j} + (L_{0,j} - B_{\text{eff},j}) \exp\left(-k_j \frac{\rho \, dz}{\cos \theta}\right) \right] \qquad (10.57)$$

All previous formulas are suitable for a homogeneous atmosphere. However, the real atmosphere has vertical variations, so ideally we need to consider an inhomogeneous path through it. A common approach is to divide the atmosphere into layers, each treated as described above, and use the outgoing radiance at each value of g as the incoming radiance for the same g value for the adjacent layer. This procedure treats each subinterval in the same way as a spectral point is treated in a monochromatic radiative transfer method. Under conditions such that the mapping from k_ν to $g(k)$ is identical for adjacent layers, or equivalently that the k distribution in a given layer is fully correlated in spectral space with the k distribution in the next layer, the extension of the k-distribution method is referred to as the *correlated k-distribution method.*

An important advantage of the correlated k-distribution method is that it retains some of the characteristics of a monochromatic line-by-line calculation, since g may be treated as a pseudowavenumber. This allows layer-by-layer radiative transfer calculations, with the integration over g postponed until the last step. The correlated k-distribution method is potentially as accurate as line by-line calculations, provided "exact" k distributions can be found. Correlated k distributions in general are determined by spectral integration of line-by-line data. The computation time is proportional to the number of layers. In the shortwave, a particular advantage of the correlated k-distribution method over band models is that scattering problems can be addressed directly.

The infrared scattering cross section for molecules is much smaller than the absorption cross section, and extinction in the IR is dominated by absorption such that scattering can be neglected. However, ice crystals and water droplets found in clouds are of similar magnitude or greater in size than are wavelengths associated with IR radiation. Although extinction by absorption continues to dominate, scattering of upwelling terrestrial and atmospheric emission from below the cloud will provide a small contribution to the downwelling measured radiance.

There are a number of issues associated with the development of rapid radiative transfer models using the correlated k-distribution method. The first is to handle the dependence of the absorption coefficient k_ν within a band on pressure and temperature. A common approach is to scale a reference set of $k_\nu(p_r, T_r)$ by an appropriate function of temperature or pressure, where T_r and p_r are the reference temperature and pressure. A linear interpolation is often used in the implementation. It is very similar to the lookup table method, discussed in Section 8.4.

In the previous several sections, we have discussed how to calculate transmittance and radiance given atmospheric properties. In the following sections, we discuss how to apply the thermal IR radiative transfer principles outlined in the previous two sections to estimate land surface emissivity and temperature. Similar to visible and near-IR remote sensing in the previous chapters, atmospheric correction is the first necessary step for quantitatively estimating those surface variables.

10.6 ATMOSPHERIC CORRECTION METHODS

The TOA brightness temperature is usually lower than the surface temperature, but it may be reversed when the atmosphere is warmer than the surface. In the thermal window (10–12-μm) region, the difference between them typically ranges from 1 to 5 K (Prata et al. 1995). Therefore, an accurate atmospheric correction is essential.

If scattering in the thermal IR region under a clear atmosphere is eligible and the radiance is dependent only on the viewing zenith angle, then a simplified expression for spectral radiance at band i reaching at the sensor is the sum of path radiance (L^p) and the surface leaving radiance (L^s) transmitted through the atmosphere

$$L_i = t_i L_i^s + L_i^p \qquad (10.58)$$

where t is the transmittance of the atmosphere between the surface and the sensor (see Fig. 10.7). The upwelling surface radiance is the sum of radiance

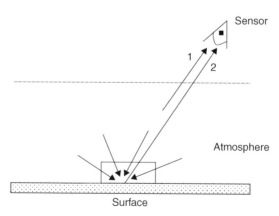

Figure 10.7 The total radiance received by the sensor consists of both surface leaving radiance and atmospheric path radiance.

emitted by the surface with the LST (T_s) and the reflected downward sky radiance (F_d / π):

$$L_i^s = \varepsilon_i B(T_s) + (1 - \varepsilon_i) \cdot \frac{F_d}{\pi} \qquad (10.59)$$

Here we have assumed that surface spectral emissivity ε is angle independent, F_d is downward flux:

$$F_d = 2\pi \int_{-1}^{0} L_i^d(u)\, \mu\, d\mu \qquad (10.60)$$

The key task of the atmospheric correction of thermal IR data for land surface temperature retrieval is to estimate these three quantities: path radiance (L^p), transmittance (t), and the downward flux (F_d) on a pixel basis or on an image basis.

For a single thermal band sensor (e.g., TM/ETM, Meteosat), atmospheric correction requires that the vertical and horizontal distribution of temperature and water vapor in the atmosphere be accurately known. This requirement is almost impossible to meet under most realistic conditions. The solution is to use ancillary information, such as data from numerical weather forecast models, retrieval from sounding instruments on satellites (e.g., TOVS, VISSR (visible and infrared spin-scan radiometer), atmospheric sounder (VAS)), and radiosonde data. For processing massive datasets, we cannot run radiative transfer packages (e.g., MODTRAN) online since it is time consuming. The alternative approach is to create lookup tables using offline radiative transfer calculations, surface temperature and emissivity are interpolated on-line by searching the lookup tables (Schadlich et al. 2001). The lookup table approach is discussed further in Section 8.4.

Another alternative approach is to use an artificial neural network (ANN) rather than a lookup table. The forward radiative transfer computation results are used to train neural networks offline, and ANN can then be used to correct imagery online (Gottsche and Olesen 2002). ANN is discussed in more detail in Section 8.5.1. By *offline calculations* I mean that the calculations are made in advance or using other computers before running the current code. In contrast, by *online calculation* I mean calculations are part of the process in the same computer program.

As we discussed in Chapter 6, the most difficult part in any atmospheric correction algorithm is to estimate the atmospheric parameters from imagery itself. The split-window methods, discussed in the next section, avoid the atmospheric correction issue. Gu et al. (2000) developed a method to estimate atmospheric quantities from the imagery of a hyperspectral sensor SEBASS (spatially enhanced broadband array spectrograph system). The general idea is similar to the differential absorption methods for estimating the total water vapor content in the atmosphere (see Section 6.4.2). Two neighboring bands are selected; one is in the strong water vapor absorption region (e.g., 11.73 μm), and the other is outside the absorption region. If we can assume that surface radiances at these two neighboring bands are almost identical, the difference in the TOA radiances of these two bands is caused mainly by the atmosphere. The empirical relationships were then developed to estimate transmittance and path radiance.

10.7 SPLIT-WINDOW ALGORITHM FOR ESTIMATING LST

From two thermal bands in the 10.5–12.5 μm region, the split-window algorithms have been widely used for estimating LST given surface emissivity. The central part of most split-window algorithms is based on the assumption that LST (T_s) is linearly related to the brightness temperatures of two thermal channels, although some algorithms use a simple nonlinear form. With the assumption that surface emissivities of these two channels are known, the split-window method can largely eliminate atmospheric effects based on the differential absorption between two bands for the LST estimation. A typical split-window algorithm can be written as

$$T_s = a_0 + a_1 T_i + a_2 T_j \tag{10.61}$$

where a_i are the coefficients, and T_i and T_j are the brightness temperature at bands i and j. Coefficients a_i depend mainly on surface spectral emissivity and also the atmospheric conditions, particularly when the atmospheric humidity is very high (Becker and Li 1995).

Before discussing the derivation of the split-window algorithms, we need to note that in all the split-window algorithms, brightness temperature is used rather than band radiance. Given the spectral radiance, the brightness temperature can be derived from Planck formula (10.6):

$$T = \frac{1.43876869 v}{\ln\left(\dfrac{1.1909561 \cdot 10^{-16} v^5}{B_\lambda(T)} + 1\right)} \tag{10.62}$$

This equation will yield small errors when used for a spectral bandpass rather than a single wavenumber, where the bandpass would be represented by the mean wavenumber, v_c. A correction, based on a least squares fit of the measured radiance in a given bandpass over a typical temperature domain, can be applied to Eq. (10.62) such that

$$T(v_c) = \frac{1}{b}\left[\frac{1.43876869 v_c}{\ln\left(\dfrac{1.1909561 \cdot 10^{-16} v^5}{B_\lambda(T)} + 1\right)} - a\right] \tag{10.63}$$

TABLE 10.1   The Central Wavenumbers (v_c) for AVHRR with Two Thermal IR Bands for the Temperature Range 275–320 K (in cm^{-1})

Satellite	Band 4	Band 5
NOAA7	927.220	840.872
NOAA9	929.460	845.190
NOAA11	927.830	842.200

where a and b are the least squares fit y-intercept and slope, respectively. Table 10.1 gives the central wavenumbers of the AVHRR two thermal IR bands, which can be determined from the simple integration

$$v_c = \frac{\int_{v_{min}}^{v_{max}} f(v) v \, dv}{\int_{v_{min}}^{v_{max}} f(v) \, dv} \tag{10.64}$$

where $f(v)$ is the sensor spectral response function that has been discussed in Section 1.3.5.1 and v_{min} and v_{max} specify the band range beyond which $f(v) = 0$.

10.7.1 Theoretical Derivation

Most split-window algorithms were derived from radiative transfer equation with various approximations, although some were developed from measurements. Different split-window algorithms have similar forms but different coefficients that are determined empirically from measurement data or from radiative transfer simulations. To understand why such a simple linear formula can be used to estimate land surface temperature, we must look back to basic radiative transfer theory. The derivation below follows that by Prata (1993).

Equations (10.58) and (10.59) can be combined and written in more detail explicitly for both bands i and j, provided both bands are in the thermal infrared spectrum and very close (usually one within 10–11 μm and another within 11–12 μm):

$$L_i = \left[\varepsilon_i B_i(T_s) + (1 - \varepsilon_i) \cdot L_i^d \right] t_i + (1 - t_i) L_i^a \tag{10.65}$$

$$L_j = \left[\varepsilon_j B_j(T_s) + (1 - \varepsilon_j) \cdot L_j^d \right] t_j + (1 - t_j) L_j^a \tag{10.66}$$

where the last term of the equation is the atmospheric path radiance, L_i^d and L_j^d are downward atmospheric radiance at bands i and j, respectively. L_i^a can be computed by the Planck function with an equivalent atmospheric temperature (T_a):

$$L_i^a = B_i(T_a) \tag{10.67}$$

One of the most important tricks of the derivation is to approximate the Planck function using the first-order Taylor expansion around the central wavenumber (v_i)

$$B_v(T) = B_{v_i}(T) + (v - v_i)\left(\frac{\partial B}{\partial v}\right)_{v_i} \tag{10.68}$$

If we define

$$L_i = B_i(T_i)$$
$$L_i' = B_i(T_j) \tag{10.69}$$

and apply the first-order Taylor expansion to Eq. (10.65), then

$$
\begin{aligned}
L_i = {}& \varepsilon_j t_j B_i(T_s) + (1 - t_j) B_i(T_a) + (1 - t_j) L_j^d t_j \\
& + \varepsilon_j t_j (v_j - v_i)\left(\frac{\partial B_i(T_s)}{\partial v} - \frac{1}{\varepsilon_j}\frac{\partial B_i(T_a)}{\partial v}\right) \\
& + (v_j - v_i)\left(\frac{\partial B_i(T_a)}{\partial v} - \frac{\partial B_i(T_j)}{\partial v}\right)
\end{aligned}
\tag{10.70}
$$

Provided the temperatures T_i, T_j, T_a, and T_s are close in magnitude, the terms involving differences in Planck derivatives are small. After neglecting these terms and eliminating $B_i(T_a)$, the solution for the upwelling surface radiance is

$$B_i(T_s) = \frac{1 + \gamma}{\varepsilon_i}\left(\frac{1}{1 + \gamma t_j \Delta\varepsilon/\varepsilon_i}\right) L_i - \frac{\gamma}{\varepsilon_j}\left(\frac{1}{1 + (1 + \gamma)t_i\Delta\varepsilon/\varepsilon_j}\right) L_i' + \alpha \tag{10.71}$$

where

$$\gamma = \frac{1 - t_i}{t_i - t_j} \tag{10.72}$$

$$\Delta\varepsilon = \varepsilon_i - \varepsilon_j \tag{10.73}$$

$$\alpha = \frac{(1 - t_i)t_j(1 - \varepsilon_j)L_j^d - (1 - t_j)t_i(1 - \varepsilon_i)L_i^d}{\varepsilon_j t_j(1 - t_i) - \varepsilon_i t_i(1 - t_j)} \tag{10.74}$$

If one assumes that a constant value for the downward sky radiance differ-ence can be used, Eq. (10.71) may be written

$$B_i(T_s) = \frac{1+\gamma}{\varepsilon_i}\left(\frac{1}{1+\gamma t_j \Delta\varepsilon/\varepsilon_i}\right)L_i - \frac{\gamma}{\varepsilon_j}\left(\frac{1}{1+(1+\gamma)t_i\Delta\varepsilon/\varepsilon_j}\right)L_i'$$
$$+ \left(\frac{1-\varepsilon_i - \gamma t_j \Delta\varepsilon}{\varepsilon_i + \gamma t_j \Delta\varepsilon}\right)\Delta L^d \tag{10.75}$$

where ΔL^d is a constant downward sky radiance difference that is indepen-dent of atmosphere, but Prata found the use of a mean value of 3.6 mW m^{-2} sr^{-1} cm^{-1} for ΔL^d to be acceptable.

In the ordinary split-window algorithms, brightness temperatures rather than radiance are used. To obtain an algorithm involving brightness tempera-ture, one more approximation using the first-order Taylor expansion around the average temperature $\bar{T}$ is needed:

$$B_v(T) = B_v(\bar{T}) + (T - \bar{T})\left(\frac{\partial B}{\partial T}\right)_{\bar{T}} \tag{10.76}$$

Using this approximation in Eq. (10.76), Eq. (10.75) becomes

$$(T_s - \bar{T})\frac{\partial B_i}{\partial T} + B_i(\bar{T}) = w_1\left[(T_i - \bar{T})\frac{\partial B_i}{\partial T} + B_i(\bar{T})\right]$$
$$+ w_2\left[(T_j - \bar{T})\frac{\partial B_j}{\partial T} + B_j(\bar{T})\right] + w_3\Delta L^d \tag{10.77}$$

where

$$w_1 = \frac{1+\gamma}{\varepsilon_i}\left(\frac{1}{1+\gamma t_j \Delta\varepsilon/\varepsilon_i}\right) \tag{10.78}$$

$$w_2 = -\frac{\gamma}{\varepsilon_j}\left(\frac{1}{1+(1+\gamma)t_i\Delta\varepsilon/\varepsilon_j}\right) \tag{10.79}$$

$$w_3 = \frac{1-\varepsilon_i - \gamma t_j \Delta\varepsilon}{\varepsilon_i + \gamma t_j \Delta\varepsilon} \tag{10.80}$$

Thus, Eq. (10.77) can be written in the split-window form,

$$T_s = w_1 T_i + w_2 T_j + w_4 \tag{10.81}$$

where

$$w_4 = w_3 \Delta L^d \left(\frac{\partial B_i}{\partial T} \right)_{\overline{T}}^{-1} + (1 - w_1 - w_2) \left[\overline{T} - B(\overline{T}) \left(\frac{\partial B_i}{\partial T} \right)_{\overline{T}}^{-1} \right] \quad (10.82)$$

If we further take $\overline{T} = T_4$, the split-window formula (10.81) becomes

$$T_s = a T_i + b T_j + c \qquad (10.83)$$

where

$$a = 1 - w_2 \qquad (10.84)$$

$$b = w_2 \qquad (10.85)$$

$$c = \left[w_3 \Delta L^d + (1 - w_1 - w_2) B_i(T_i) \right] \left(\frac{\partial B_i}{\partial T} \right)_{T_i}^{-1} \qquad (10.86)$$

This resulting formula illustrates the dependence of the split-window algorithms on the surface spectral emissivities and atmospheric transmittance. Any further simplifications always introduce uncertainty in determining the land surface temperature.

The functional forms of these split-window algorithms can be determined this way, but the coefficients need to be further tuned using either ground measurements or radiative transfer simulations. It has been a common practice for people to calibrate the coefficients of the sea surface temperature split-window algorithms using in situ simultaneous measurements and clear satellite imagery. This approach generates the so-called matchup database, which is very representative in both space and time. A statistical regression analysis is used to produce the coefficients.

However, it is very difficult to determine the coefficients of any LST split-window algorithm using such a matchup database from ground measurements because LST varies tremendously in the spatial domain compared to the spatial resolution of satellite data. It is also very time-consuming and expensive to measure the surface temperature of many different land surface cover types (soil, trees, crops, etc.). A common practice is to generate such a matchup database through radiative transfer simulations.

10.7.2 Representative Algorithms

Many split-window formulas are published in the literature. I do not intend to present all of them here. Instead, I will present some representative formulas for estimating land surface temperature (T_s) from AVHRR brightness temperature of bands 4 and 5 (T_4, T_5) that have been widely used. The purpose is to demonstrate the basic principles and provide a handy reference.

TABLE 10.2 ATSR, AVHRR, and GOES Sensors that Have Two Thermal IR Bands for LST Retrieval

Sensor	Thermal Band	Spectral Range (μm)
ATSR	6	10.4–11.3
	7	11.5–12.5
AVHRR	4	10.3–11.3
	5	11.4–12.4
GOES	4	10.2–11.2
	5	11.5–12.5

The general methodology should be suitable for other sensors. Table 10.2 lists three sensors that have two thermal IR bands for LST determination. ASTER and MODIS IR bands for LST determination are listed in Table 10.3.

Price (1984) developed one of the earliest LST split-window algorithms

$$T_s = [T_4 + 3.33(T_4 - T_5)]\left(\frac{5.5 - \varepsilon_4}{4.5}\right) + 0.75T_5(\varepsilon_4 - \varepsilon_5) \quad (10.87)$$

where T_4 and T_5 are the brightness temperature of AVHRR bands 4 and 5 in Celsius, and ε_4 and ε_5 are surface emissivity of the corresponding bands.

More studies have revealed that when the atmosphere is not particularly dry, the traditional split-window algorithm cannot remove the atmospheric effects completely. Therefore, many efforts have been made to incorporate the column water vapor content of the atmosphere into the split-window formulas. Sobrino et al. (1991) suggested a formula that incorporates atmospheric water vapor content (W):

$$T_s = T_4 + A(T_4 - T_5) + B \quad (10.88)$$

$$A = 0.349W + 1.32 + (1.385W - 0.204)(1 - \varepsilon_4)$$
$$+ (1.506W - 10.532)(\varepsilon_4 - \varepsilon_5) \quad (10.89)$$

$$B = \frac{1 - \varepsilon_4}{\varepsilon_4} T_4 u_1 - \frac{1 - \varepsilon_5}{\varepsilon_5} T_5 u_2 \quad (10.90)$$

$$u_1 = -0.146W + 0.561 + (0.575W - 1.966)(\varepsilon_4 - \varepsilon_5) \quad (10.91)$$

$$u_2 = -0.095W + 0.320 + (0.597W - 1.916)(\varepsilon_4 - \varepsilon_5) \quad (10.92)$$

These formulas are linear functions of the brightness temperatures of two AVHRR bands. Sobrino et al. (1994) later proposed a nonlinear function that is believed to improve the LST retrieved when the water vapor content of the atmosphere is high:

$$T_s = T_4 + 1.4(T_4 - T_5) + 0.32(T_4 - T_5)^2 + 0.83 \quad (10.93)$$

Francois and Ottle (1996) suggested a similar form, but the coefficients are given in a table with different conditions. Their formula was developed from radiative transfer simulations with 1761 radiosonde data of the TOVS Initial Guess Retrieval datasets. In fact, the nonlinear functions have been widely used to estimate sea surface temperature from AVHRR data. I further extended this idea and represented the nonlinear relationships using ANN (Liang 1997).

Kerr et al. (1992) proposed an area weighting algorithm for an effective LST based on soil temperature (T_{soil}) and vegetation temperature (T_{veg}) that are calculated from two separate split-window algorithms:

$$T_s = f_v T_{veg} + (1 - f_v) T_{soil} \qquad (10.94)$$

$$T_{veg} = T_4 + 2.6(T_4 - T_5) - 2.4 \qquad (10.95)$$

$$T_{soil} = T_4 + 2.1(T_4 - T_5) - 3.1 \qquad (10.96)$$

$$f_v = \frac{NDVI - NDVI_{min}}{NDVI_{max} - NDVI_{min}} \qquad (10.97)$$

where $NDVI_{min}$ and $NDVI_{max}$ are the minimal and maximal values of NDVI corresponding bare soil and vegetation, respectively.

Becker and Li (1995) developed a split-window algorithm based on extensive radiative transfer simulations with LOWTRAN, which is a low-resolution version of MODTRAN:

$$T_s = A_0 + P\frac{T_4 + T_5}{2} + M\frac{T_4 - T_5}{2} \qquad (10.98)$$

$$A_0 = -7.49 - 0.407W \qquad (10.99)$$

$$P = 1.03 + (0.211 - 0.031 \cos \theta W)(1 - \varepsilon_4) - (0.37 - 0.074W)(\varepsilon_4 - \varepsilon_5) \qquad (10.100)$$

$$M = 4.25 + 0.56W + (3.41 + 1.59W)(1 - \varepsilon_4) - (23.85 - 3.89W)(\varepsilon_4 - \varepsilon_5) \qquad (10.101)$$

where W is the column water vapor content, θ is the viewing zenith angle, and ε_4 and ε_5 are the spectral emissivity of AVHRR bands 4 and 5, respectively. In their simulations, water vapor content values W were taken from radiosonde profiles at different locations. In fact, their formula has the exact same form of a previously developed equation whose coefficients are independent of water vapor (Becker and Li 1990):

$$A_0 = 1.274 \tag{10.102}$$

$$P = 1 + 0.15616\frac{1-\varepsilon}{\varepsilon} - \frac{0.482\Delta\varepsilon}{\varepsilon^2} \tag{10.103}$$

$$M = 6.26 + 3.89\frac{1-\varepsilon}{\varepsilon} - \frac{38.33\Delta\varepsilon}{\varepsilon^2} \tag{10.104}$$

where $\varepsilon = (\varepsilon_4 + \varepsilon_5)/2$ and $\Delta\varepsilon = \varepsilon_4 - \varepsilon_5$.

Wan and Dozier (1996) suggested a generalized split-window algorithm for retrieving land surface temperature from MODIS thermal bands 31 and 32, which are similar to AVHRR bands 4 and 5:

$$
\begin{aligned}
T_s = C &+ \left(A_1 + A_2\frac{1-\varepsilon}{\varepsilon} + A_3\frac{\Delta\varepsilon}{\varepsilon^2}\right)\frac{T_{31} + T_{32}}{2} \\
&+ \left(B_1 + B_2\frac{1-\varepsilon}{\varepsilon} + B_3\frac{\Delta\varepsilon}{\varepsilon^2}\right)\frac{T_{31} - T_{32}}{2}
\end{aligned}
\tag{10.105}
$$

where A_i, B_i, and C are the coefficients, $\varepsilon = (\varepsilon_{31} + \varepsilon_{32})/2$, and $\Delta\varepsilon = \varepsilon_{31} - \varepsilon_{32}$.

Other formulas include those due to Prata and Platt (1991)

$$T_s = 3.45\frac{T_4 - T_0}{\varepsilon_4} - 2.45\frac{T_5 - T_0}{\varepsilon_5} + 40\frac{1-\varepsilon_4}{\varepsilon_4} + T_0 \tag{10.106}$$

where $T_0 = 273.15°$, and those of Ulivieri et al. (1994):

$$T_s = T_4 + 1.8(T_4 - T_5) + 48(1 - \varepsilon) - 75(\varepsilon_4 - \varepsilon_5) \tag{10.107}$$

Note that although these formulas are all for AVHRR, some were developed for earlier NOAA satellites. Because AVHRR thermal bands on different NOAA satellites have different spectral response functions as discussed in Section 1.3.5.1, a correction may be needed (Czajkowski et al. 1998).

A natural question is which one is the best given all these different formulas. To answer such a question, extensive validation and intercomparisons are needed. Limited studies have been reported in the literature (e.g., Prata 1993, Becker and Li 1995, Prata et al. 1995), but it should be a continuing topic. The major limiting factor is the sampling difficulty over heterogeneous landscapes and the associated costs. An appropriate scaling procedure using high-resolution thermal remote sensing data is critical.

10.7.3   Emissivity Specification

Most of the formulas presented above treat spectral emissivity as a variable, but it has to be known. Estimating land surface emissivity accurately at the global scale for use in split-window algorithms is very challenging. "In fact the error due to an error in the emissivity correction is two times larger than that due to an error in the atmospheric correction. ... Certainly this is an area of research that requires much more study" (Prata et al. 1995).

Several methods for estimating emissivity have been reported in the literature on the basis of either NDVI or land cover information. The NDVI-based approach is to link emissivity with the normalized difference vegetation index (NDVI) (Van De Griend and Owe 1993, Olioso 1995, Valor and Caselles 1996). These derived surface emissivity values have been used to produce land surface temperature maps of European, African, and South American areas. Van de Griend and Owe (1993) proposed the following relationship between the measured emissivity (ε) and NDVI based on data collected at Botswana:

$$\varepsilon = 1.0094 + 0.047 \ln(NDVI) \tag{10.108}$$

Valor and Caselles (1996) extended this concept with much more complicated expressions for mixed pixels, which are then applied and extended for different atmospheric conditions and over different areas. But their algorithm requires a priori information about the emissivities of soil and vegetation and the vegetation structure and distributions.

Sobrino et al. (2001) proposed an algorithm for estimating surface emissivities from AVHRR data and found that it can achieve the comparative accuracy to several other algorithms. The algorithm is quite simple, assuming that surface NDVI values be calculated by the first two-band reflectance after atmospheric correction and NDVI mainly vary in the range (0.2–0.5). When $0.2 < NDVI < 0.5$, then

$$\varepsilon_4 = 0.968 + 0.021 f_v \tag{10.109}$$

$$\varepsilon_5 = 0.974 + 0.015 f_v \tag{10.110}$$

where f_v is the fractional proportion of vegetation within the pixel and estimated by NDVI:

$$f_v = \frac{(NDVI - 0.2)^2}{0.09} \tag{10.111}$$

When $NDVI < 0.2$, this pixel is assumed to be a nonvegetated pixel,

$$\varepsilon = \frac{\varepsilon_4 + \varepsilon_5}{2} = 0.98 - 0.042 \rho_1 \tag{10.112}$$

$$\Delta \varepsilon = \varepsilon_4 - \varepsilon_5 = -0.003 - 0.029 \rho_1 \tag{10.113}$$

where ρ_1 is band 1 surface reflectance. When $NDVI > 0.5$, the pixel is assumed to be a pure vegetation pixel, and emissivities of both bands are set to 0.989.

The land-cover-based approach is to associate emissivity with land cover information (Snyder et al. 1998) in which each cover type is assigned one emissivity value. Surfaces are classified into 14 cover types, each of which has fixed emissivity values that are determined using linear BRDF models. The model coefficients are derived from laboratory measurements of material samples and structural parameters of each cover type. The integrated hemi-spherical-directional reflectance is used to calculate emissivity.

Obviously, these two approaches cannot completely capture the tremen-dous amount of variability of surface emissivity, particularly over nonvege-tated regions. An inaccuracy of only 0.01 in emissivity causes errors in LST exceeding those due to atmospheric correction. In the next section, we discuss how to determine the surface emissivity from multiple (> 2) thermal bands.

10.8 MULTISPECTRAL ALGORITHMS FOR SEPARATING TEMPERATURE AND EMISSIVITY

A very practical way for accurately estimating spectral emissivity is from thermal infrared imagery itself. This requires multiple thermal channels. Fortunately, several spaceborne sensors have multiple thermal infrared bands that enable us to estimate LST and emissivity simultaneously, such as MODIS and ASTER. MODIS (Salomonson et al. 1989) has multiple thermal bands in the 3.5–4.2-μm and the 8–13.5-μm ranges, and ASTER (Yama-guchi et al. 1998) has five thermal bands in the 8–12-μm range (see Table 10.3).

TABLE 10.3   ASTER and MODIS Thermal IR Bands
and Their Spectral Ranges for LST Retrieval

System	Band	Spectral Range (μm)
ASTER	10	8.125–8.475
	11	8.475–8.825
	12	8.925–9.275
	13	10.25–10.95
	14	10.95–11.65
MODIS	20	3.660–3.840
	22	3.929–3.989
	23	4.020–0.080
	29	8.400–8.700
	31	10.780–11.280
	32	11.770–12.270
	33	11.185–13.485

The split-window algorithms combine atmospheric correction and LST estimation into one process. However, it is necessary to correct the atmospheric effects before separating temperature and emissivity from multiple thermal infrared imagery. The ASTER science team has proposed correcting the atmospheric effects using the atmospheric profiles produced by the MODIS sounders (Palluconi et al. 1996). The MODIS science team proposed a day/night algorithm (Wan and Li 1997) that adjusts the MODIS atmospheric profiles for each pixel with the surface air temperature and a scaling factor of the total water vapor amount as two unknown variables by solving a 14-equation set.

Assuming that the atmospheric impacts can be effectively removed, we can focus on estimating both LST and emissivity from multispectral thermal infrared imagery. Effects of LST and emissivity on thermal radiance are so closely coupled that their separation from thermal radiance measurements alone is quite difficult. This is because a single multispectral thermal measurement with N bands presents N equations in $N + 1$ unknowns (N spectral emissivities and LST). It is a typical ill-posed inversion problem. With no prior information, it is impossible for us to recover both LST and emissivity exactly. Most LST–emissivity separation studies used one additional empirical equation so that N measurements plus this additional equation can be solved for $N + 1$ unknowns. Various methods are developed to separate temperature and emissivity. The earlier methods are discussed in a special issue of *Remote Sensing of Environment* (No. 42, 1992) and also reviewed by Gillespie et al. (1999). Some of these methods are presented below.

10.8.1 Reference Channel Method

The reference channel method (Lyon 1965) assumes that the value of the emissivity for one of the image channels is constant and known *a priori*, reducing the number of unknowns to the number of equations. Lyon suggested that, for most silicate rocks, the maximum emissivity was commonly about 0.95 and occurred at the long-wavelength end of the 8–14-μm window.

This method is robust and has the virtue of simplicity. It can produce moderately reliably results for a wide range of surface materials. However, it is certainly not a universal algorithm for accurately retrieving surface temperature and emissivity for both vegetation and rocks.

10.8.2 ADE Method

The alpha-derived emissivity (ADE) method (Hook et al. 1992, Kealy and Hook 1993) makes use of the relation between the weighted logarithm values of spectral emissivity and the variance of spectral emissivities. It was developed from an earlier method called the *alpha-residual technique*. The alpha residuals are calculated utilizing Wien's approximation of Planck's law, which neglects the "-1" term in the denominator. If we further neglect the surface reflected term, then

$$L_i = \varepsilon_i \frac{c_1 \lambda_i}{\pi \exp(c_2/\lambda_i T)} \qquad (10.114)$$

where c_1 and c_2 are two constants. This makes it possible to linearize the approximation with logarithms, thereby separating λ and T:

$$\frac{c_2}{T} = \lambda_i \ln(\varepsilon_i) - \lambda_i \ln(L_i) + \lambda_i \ln(c_1) - 5\lambda_i \ln(\lambda_i) - \lambda_i \ln(\pi) \quad (10.115)$$

The next step is to calculate the means for the parameters of the linearized equation, summing over all sensor channels (n):

$$\frac{c_2}{T} = \frac{1}{n} \sum_{i=1}^{n} \lambda_i \ln(\varepsilon_i) - \frac{5}{n} \sum_{i=1}^{n} \lambda_i \ln(\lambda_i) - \frac{1}{n} \sum_{i=1}^{n} \lambda_i \ln(L_i)$$

$$+ \left[\ln(c_1) - \ln(\pi) \right] \frac{1}{n} \sum_{i=1}^{n} \lambda_i \qquad (10.116)$$

The residual is calculated by subtracting the mean from the individual channel values. After some simple algebraic manipulation, we come out with a set of n equations for the residual α_{resid}^i:

$$\alpha_{resid}^i = \lambda_i \ln(\varepsilon_i) - \mu_\alpha = \lambda_i \ln(L_i) - \frac{1}{n} \sum_{i=1}^{n} \lambda_i \ln(L_i) + \kappa_i \quad (10.117)$$

for $i = 1, 2, \ldots, n$, where

$$\kappa_i = 5\lambda_i \ln(\lambda_i) - \sum_{i=1}^{n} \lambda_i \ln(\lambda_i) - \left[\ln(c_1) - \ln(\pi) \right] \left(\lambda_i - \frac{1}{n} \sum_{i=1}^{n} \lambda_i \right) \quad (10.118)$$

which is a constant independent of either temperature or emissivity, and

$$\mu_\alpha = \frac{1}{n} \sum_{i=1}^{n} \lambda_i \ln(\varepsilon_i) \qquad (10.119)$$

From laboratory emissivity spectra, they developed an empirical formula that relates μ_α to the variance of the laboratory spectra v_α:

$$\mu_\alpha = c v_\alpha^{1/x} \qquad (10.120)$$

where c and x are empirically determined coefficients. In their experiments (Kealy and Hook 1993) with different datasets of soils and rocks, x has been set as the integer (3 or 4) in their experiment, and c varies from -1.2983 to -1.8665.

The residual α^i_{resid} is also called the *alpha coefficient* and can be calculated directly from measured radiance using Eq. (10.117). This additional equation and N observations (N equations) allow us to solve for $N+1$ variables.

Gu and Gillespie (2000) commented that since Wiens approximation causes errors of up to 0.025 and 2 K in estimated emissivity and temperature, respectively, it is not suitable for the thermal IR data of either ASTER or MODIS. They suggested approximating the Planck equation by using the first term of the Taylor series at a given temperature.

10.8.3 Temperature-Independent Spectral Indices (TISIs)

Becker and Li (1990, 1995) developed a temperature independent spectral index (TISI) to estimate surface temperature from AVHRR thermal IR observations.

Assume the Planck function can be approximated by the power law:

$$B_\lambda(T) \cong m_\lambda(T_0)T^{n_\lambda(T_0)} \tag{10.121}$$

where m_λ and n_λ depend on the wavelength and the reference temperature T_0 according to

$$n_\lambda(T_0) = \frac{c_2}{\lambda T_0}\left[1 + \frac{1}{\exp(c_2/\lambda T_0) - 1}\right] \tag{10.122}$$

where $c_2 = 1.43879 \cdot 10^{-4}$, and wavelength λ is in micrometers. Assume that thermal data are corrected atmospherically enabling us to obtain the surface leaving radiance for three thermal bands (3–5). This is expressed in Eq. (10.59). Inserting (10.120) into (10.59) yields

$$L_i^s = \varepsilon_i m_i T_s^{n_i} C_i \tag{10.123}$$

where

$$C_i = 1 + \frac{(1 - \varepsilon_i)F_d}{\pi \varepsilon_i B_i(T_s)} \tag{10.124}$$

Using the power-law approximation, Becker and Li (1990) defined the TISI for two channels, i and r from channel surface leaving radiance L

$$\text{TISI} = \frac{C_r^{n_{ir}} m_r^{n_{ir}}}{C_i} \frac{L_i}{m_i} \frac{L_i}{L_r^{n_{ir}}} = \frac{\varepsilon_i}{\varepsilon_r^{n_{ir}}}C_{ir} \tag{10.125}$$

where $C_{ir} = C_i/C_r^{n_{ir}} \approx 1$. Thus, TISI defines the ratio of two band emissivities independent of LST. After calculating TISI_s from AVHRR bands 3, 4 and 5 day and evening data, we can determine the spectral emissivities. The detail is available in their original papers.

10.8.4 The MODIS Day/Night Algorithm

The MODIS science team has proposed making use of seven bands for LST and emissivity estimation (Wan and Li 1997). The spectral ranges of these thermal IR bands are specified in Table 10.3. The generalized split-window algorithm using MODIS bands 31 and 32 have been discussed in Section 10.7.2. An alternative algorithm that uses all seven thermal bands is discussed in this section.

Wan and Li (1997) proposed a physically based algorithm for estimating both spectral emissivity and LST simultaneously based on MODIS observations from both daytime and nighttime. The general idea of this method is outlined below. There are numerous assumptions in this algorithm; the critical ones include

- Surface conditions between day and evening are the same and hence surface spectral emissivities are the same during both day and evening.
- The MODIS retrieved profiles of the atmospheric temperature and water vapor from the sounding channels can represent the actual shapes very well, although the absolute values of their profiles do not. The air temperature at the ground level T_a and a scaling factor α_w are two variables for anchoring the relative shapes of the MODIS sounding profiles;
- For bands at 3.5–4.2 μm (e.g., bands 20, 22 and 23), the downward radiation includes both solar direct radiation and the diffuse sky radiance. The Lambertian surface is assumed to calculate the reflected diffuse solar radiance. The spectral reflectance is $r_i = 1 - \varepsilon_i$ according to Kirchhoffs' law. For calculating the reflected direct solar radiation, it is assumed that the relative shapes of surface BRDF of these bands are determined from the MODIS BRDF product (Schaaf et al. 2002), but its absolute value can be adjusted by a scaling α_{brdf}.

The observation equation for TOA radiance $L(j)$ for band j can be written as

$$L(j) = L_{path}(j) + L_{emitted}(j) + L_{reflected}(j) \qquad (10.126)$$

The path radiance that is not related to surface properties is the sum of the atmospherically scattered $L_s(j)$ and emitted radiance $L_e(j)$:

$$L_{path}(j) = L_e(j) + L_s(j) \qquad (10.127)$$

The surface-emitted radiance that reaches the sensor is the emitted radiance at the ground level multiplied by the atmospheric transmittance $t(j)$:

$$L_{emitted}(j) = t(j)\varepsilon(j)B_j(T_s) \qquad (12.128)$$

The surface-reflected radiance includes three components: direct solar irradiance $E_0(j)$, diffuse solar irradiance $E_d(j)$, and atmospherically emitted downward irradiance $E_t(j)$, which reach to the surface, are then reflected and transmitted to the sensor:

$$L_{\text{reflected}}(j) = \left[\alpha_{\text{brdf}} E_0(j) + E_d(j) + E_t(j) \right] \frac{1 - \varepsilon(j)}{\pi} t(j) \quad (10.129)$$

When we use N-band MODIS from both day and night, we have $2N$ observations [Eq. (10.126)]. The unknown variables include N emissivities, two LSTs ($T_{s,\text{day}}$ and $T_{s,\text{night}}$), and two adjusted parameters (T_a, α_w, and α_{brdf}). To uniquely determine these variables, $2N \geq N + 7$, equivalently, $N \geq 7$. Wan and Li therefore suggested using day and evening imagery of seven MODIS bands to solve 14 unknown variables.

10.8.5 The ASTER Algorithm

The comparisons among different LST/emissivity separation algorithms have been discussed by Gillespie et al. (1999) and Li et al. (1999). On the basis of previous algorithms, the ASTER team (Gillespie et al. 1998, 1999) developed a new temperature and emissivity separation algorithm. The ASTER sensor is aboard Terra and has five thermal bands (see Table 10.3) with a spatial resolution of 90 m and a scanning angle of less than 8.5°. Its temporal resolution is 16 days.

There are three basic modules in the ASTER algorithm: NEM, ratio, and MMD (see Fig. 10.8). A brief discussion on each module is provided below. It assumes that the atmospherically corrected ASTER data is available before running this algorithm.

NEM (normalized emissivity method) module—assumes an initial constant emissivity value (0.97) for all N ($= 5$ for ASTER) bands to provide an initial estimate of LST T_s^0. The reflected sky radiance is removed iteratively to produce the surface emitted radiance ($\tilde{L}_i$).

Ratio module—calculates the ratio of the emissivity (ε_i) from the atmospherically corrected radiance ($\tilde{L}_i$) and the initial temperature (T_s^0) from NEM module and the average emissivity ($\bar{\varepsilon}$):

$$\beta_i = \frac{\varepsilon_i}{\bar{\varepsilon}} = \frac{\tilde{L}_i / B_i(T_s^0)}{\left(\frac{1}{N} \sum_i \tilde{L}_i \right) \bigg/ \left(\frac{1}{N} \sum_i B_i(T_s^0) \right)} \quad (10.130)$$

This module produces the relative shape of the spectral emissivity. The next module determines the absolute spectral emissivity.

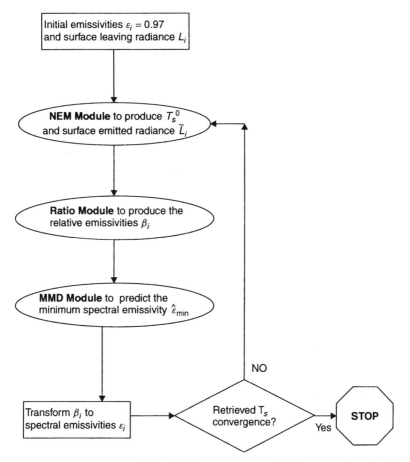

Figure 10.8 Module structure of the ASTER temperature–emissivity separation algorithm.

MMD (*maximum–minimum difference*) *module*—an empirical relationship is used to determine the minimum value ($\hat{\varepsilon}_{\min}$) of the spectral emissivity from the range of the relative emissivity ($\beta_\Delta = \max(\beta_i) - \min(\beta_i)$). It was developed based on laboratory measured emissivity spectra:

$$\hat{\varepsilon}_{\min} = 0.994 - 0.687\beta_\Delta^{0.737} \qquad (10.131)$$

The predicted minimal emissivity $\hat{\varepsilon}_{\min}$ is then used to transform the normalized β_i to the absolute emissivity ε_i

$$\varepsilon_i = \beta_i \frac{\hat{\varepsilon}_{\min}}{\min(\beta_i)}$$

which leads to a newly estimated LST T_s^0. Repeat the procedure from NEM until the difference between the estimated LST of two iterations is smaller than a predefined threshold. This iterative procedure is illustrated in Fig. 10.8.

This algorithm has been used to operationally produce ASTER LST and emissivity products. The numerical simulations show that this algorithm can recover LST to within ~ 1.5 K and emissivity to within 0.015 (Gillespie et al. 1998). However, as Dash et al. (2002) pointed out, this algorithm requires an accurate atmospheric correction. The ASTER atmospheric correction algorithm needs atmospheric input variables (temperature and water vapor profiles) from other sources, which may introduce large uncertainty and eventually impact the accuracy of the final products. Schmugge et al. (1998) applied this algorithm to an airborne multispectral thermal data—TIMS (thermal infrared multispectral scanner) over HAPEX—Sahel and obtained results with typical errors of 3 K. They also applied to the TIMS data over the USDA/ARS Jornada Experimental Range in New Mexico, and excellent results were obtained.

10.8.6 An Optimization Algorithm

I (Liang 2001) proposed an optimization algorithm. Its central idea is very similar to the previous algorithms in the sense that we created a new equation so that $N + 1$ equations are used to solve $N + 1$ unknown variables. However, since any added equations are based on empirical formulas, the solutions are likely to be unstable. Again, this is an ill-posed inversion problem. According to Moritz (1993), the inversion problem is called *properly posed* if the solution satisfies the following requirements: (1) the solution must exist (*existence*); (2) the solution must be uniquely determined by the data (*uniqueness*); and (3) the solution must depend continuously on the data (*stability*). If one or more of these requirements are violated, then we have an *improperly posed*, or *ill-posed*, problem. In our present case, it is an *ill-posed* problem since N observations cannot determine $N + 1$ unknowns uniquely. Additional information is needed to turn it into a *properly posed* problem. One way of providing this additional information is to use appropriate constraints. This process is known as *regularization* of the ill-posed problem. Different regularization methods (Tikhonov and Arsenin 1977, Twomey 1977) were explored to stabilize the solutions in this study.

Nearly 1000 surface emissivity spectra of different cover types were integrated in the ASTER and MODIS bands using their sensor spectral response functions. We developed a new empirical relation to predict the minimum emissivity for the MODIS and ASTER, respectively. This inversion algorithm is quite general and suitable for any multispectral thermal infrared remote sensing. I use MODIS as an example while discussing this new inversion algorithm below, but this method has general applicability.

On the basis of our MODTRAN simulations, we found that band 33 has very low atmospheric transmittance (lower than 0.35 in many cases) under different atmospheric conditions. I therefore considered only six bands.

The additional empirical equation for the MODIS implementation of our method is given by

$$\bar{\varepsilon}_{min} = 0.067 + 0.319\varepsilon_{20} + 0.232\,\varepsilon_{22} + 0.271\varepsilon_{23} + 0.381\,\varepsilon_{29}$$
$$+ 0.289\varepsilon_{31} + 0.261\,\varepsilon_{32} - 0.583\varepsilon_{range} + 0.261\,\varepsilon_{med} \quad (10.132)$$

where ε_{med} and ε_{range} are the median and range of absolute emissivities of six bands. The fitted results were very good, with the R-squared value of 0.999 and the residual standard error (RSE) of 0.0073. If the emissivity spectra are quite smooth, the uncertainty does not affect the inversion results significantly.

Any other prior knowledge might help to determine the solutions more accurately. After examining these spectral emissivity spectra, we found many additional relationships that can be used to constrain the LST and emissivity separation.

To estimate $N + 1$ unknowns from $N + 1$ nonlinear equations, a multidimensional optimization algorithm is needed. For global applications, computational speed is a major concern. On the other hand, the added empirical equation does not fit the data perfectly. If we treat it as a perfect equation, the uncertainty is introduced into the inversion process. Instead, we developed a constrained one-dimensional (1D) inversion procedure. The general idea has been discussed in Section 8.2.

The 1D optimum inversion algorithm for estimating LST is to minimize the merit function that consists of a sum of squares (SS) term and a penalty function:

$$F(\varepsilon_{1,2,\ldots,N}, T) = \text{SS} + \sum f_i(\varepsilon_{1,2,\ldots,N}, T)10^{20} \quad (10.133)$$

The SS term is expressed by the smoothness of the retrieved emissivity curve. Different regularization techniques produce different measures of the smoothness (Tikhonov and Arsenin 1977, Twomey 1977). Four regularization techniques were explored in this study:

Power:

$$\text{SS} = \sum_{i=1}^{N} \varepsilon_i^2 \quad (10.134)$$

Variance:

$$\text{SS} = \sum_{i=1}^{N} (\varepsilon_i - \bar{\varepsilon})^2 \quad (10.135)$$

First-order difference:

$$\text{SS} = \sum_{i=2}^{N} (\varepsilon_{i-1} - \varepsilon_i)^2 \tag{10.136}$$

Second-order difference:

$$\text{SS} = \sum_{i=3}^{N} (\varepsilon_{i-2} - 2\varepsilon_{i-1} + \varepsilon_i)^2 \tag{10.137}$$

where ε_i is the emissivity value at band i and $\bar{\varepsilon}$ is the mean value of the spectral emissivity curve. This first measure corresponds to the power of the emissivity signal. The second technique is closely related to the first one, which eventually is the variance of the spectral emissivity curve. The third and fourth techniques are based on the first-order and second-order differences of the emissivity curve.

The penalty function $f_i(\varepsilon_{1,2,\ldots,N}, T)$ in Eq. (10.133) is designed to force all estimated parameters in reasonable ranges. If the value of a variable is beyond its range, a huge penalty is posed (with the huge weight 10^{20}). This weight was recommended by Siddall (1972).

Estimation of LST and emissivity values from Eq. (10.133) is a typical nonlinear optimization problem that determines the solutions iteratively. The iterative process works as follows. Given an initial LST (T_0), band emissivities are derived from

$$\varepsilon_i = \frac{L_i - L_i^{\text{sky}}}{L_i(T_0) - L_i^{\text{sky}}} \tag{10.138}$$

where L_i and L_i^{sky} respectively are the surface leaving radiance and downward sky radiance from the atmospheric correction procedure and $L_i(T_0)$ is the blackbody radiance that can be calculated by integrating radiance $B_i(T_0)$ using the Planck formula with the sensor band spectral response functions. Note that this algorithm is based on the assumption that observed radiance has been corrected atmospherically. There is therefore no path radiance and transmittance. After determining ε_{min}, $\varepsilon_{\text{range}}$, and ε_{med}, $\bar{\varepsilon}_{\text{min}}$ is predicted from Eq. (10.132). The merit function (10.133) is then calculated. For the next iteration, the iteration length and iteration direction must be found optimally. The iteration continues until the convergence reaches (i.e., $T_{N+1} - T_N$ smaller than a threshold).

To evaluate the performance of this new approach, we conducted a series of numerical inversion experiments. We compared the performance of these four regularization techniques [from (10.134) to (10.137)]. In most cases, the variance measure and the first-order difference measure performed the best.

The results indicated that we could not inverse both LST and emissivities perfectly. However, more than 43.4% inversion results differed from the actual LST within 0.5°, 70.2% within 1°, and 84% within 1.5°.

There are at least four major differences between this new algorithm and the published algorithms in the literature:

- The empirical equations are dramatically different.
- Regularization methods are applied.
- More prior information have been defined and naturally incorporated into the new algorithm in a formal manner.
- The new algorithm uses an optimal inversion algorithm that has solid foundations in computational mathematics. Further tests and validation are needed to make it operational.

10.9 COMPUTING BROADBAND EMISSIVITY

The inversion algorithms discussed above provide us only with the spectral emissivities, but the land surface modeling community needs broadband longwave emissivity. Because of insufficient data, the current GCMs and land surface parameterization schemes usually assume unit emissivity. For vegetated surfaces where emissivity is typically 0.96–0.98, this assumption leads to an error of longwave net radiation as large as 4 W/m². For deserts and other nonvegetated surfaces, the error could be increased to as much as 20 W/m² (Rowntree 1991). The CERES team created an emissivity map based on land cover types and laboratory measured spectra for longwave radiation budget calculations. For seasonally changing surface types, temporal emissivity has to be determined precisely from remotely sensed data.

Following is some work that we recently (at the time of writing) conducted that has not been published elsewhere. We have collected more than 1000 emissivity spectra from different sources (e.g., ASTER spectral library, Salisbury database, USGS spectral library). On the basis of these spectral emissivity spectra, we first calculated broadband emissivity ε

$$\varepsilon = \frac{\sum L_i(T_s)\,\varepsilon_i \Delta \lambda_i}{\sigma T_s^4} \tag{10.139}$$

where T_s is LST, $L_i(T_s)$ is the spectral radiance of the blackbody with temperature T from the Planck equation, and λ is the wavelength. We then integrated these surface spectral emissivity spectra with the sensor spectral response functions to simulate MODIS/AVHRR/ASTER spectral emissivities. On the basis of these findings, simple relationships between the broadband emissivity and those spectral emissivities have been established.

ASTER. A linear formula is fitted:

$$\varepsilon = 0.2966 + 0.1868\,\varepsilon_{10} + 0.0372\,\varepsilon_{12} + 0.2077\varepsilon_{13} + 0.2629\varepsilon_{14} \quad (10.140)$$

where the emissivity of band 11 is not used because of its correlation with other band emissivities. Inclusion of this band does not improve the fitting. The multiple $R^2 = 0.8362$ and the residual standard error (RSE) is 0.03. An effort was made to develop the relationship using a second order polynomial function and then drop the terms that do not make a significant contribution:

$$\varepsilon = 0.1826 - 0.2333\,\varepsilon_{10} + 0.2340\,\varepsilon_{12} + 0.9526\,\varepsilon_{13} + 0.4666\varepsilon_{10}^2$$

$$+ 0.5727\varepsilon_{10}\,\varepsilon_{13} + 0.9265\varepsilon_{10}\,\varepsilon_{14} + 1.4551\,\varepsilon_{12}\,\varepsilon_{13} + 1.1931\,\varepsilon_{12}\,\varepsilon_{14}$$

$$(10.141)$$

with an increased multiple correlation coefficient ($R^2 = 0.8621$) and decreased RSE (0.0276).

AVHRR. A linear equation is fitted

$$\varepsilon = 0.2489 + 0.2386\,\varepsilon_4 + 0.4998\,\varepsilon_5 \qquad (10.142)$$

with $R^2 = 0.815$ and RSE $= 0.0318$. We also found that the second-order polynomial function does not improve the fitting, and the sensor spectral functions of AVHRR on different NOAA satellites do not impact the formulas.

MODIS. If we use only two bands, 31 and 32, then the linear formula is

$$\varepsilon = 0.261 + 0.314\varepsilon_{31} + 0.411\,\varepsilon_{32} \qquad (10.143)$$

Then $R^2 = 0.7884$ and RSE $= 0.0341$. If additional bands are available, then a better formula is

$$\varepsilon = 0.227 + 0.188\,\varepsilon_{29} + 0.217\varepsilon_{31} + 0.359\varepsilon_{32} \qquad (10.144)$$

with $R^2 = 0.8479$ and RSE $= 0.0289$. It was found that inclusion of more bands does not improve the fitting. If we would like to have a nonlinear formula using only bands 31 and 32, the conversion equation becomes:

$$\varepsilon = 0.273 + 1.778\,\varepsilon_{31} - 1.807\varepsilon_{31}\,\varepsilon_{32} - 1.037\varepsilon_{32} + 1.774\varepsilon_{32}^2 \quad (10.145)$$

with $R^2 = 0.8048$ and RSE $= 0.027$.

10.10   SURFACE ENERGY BALANCE MODELING

In the previous sections, we have discussed how to estimate land surface temperature (LST) from thermal IR observations. A further topic is how to link LST with surface environmental variables so that the retrieved LST product can be used for various applications. The bridge is the use of the surface energy balance equation, which is a major interaction point between the land surface models and the atmospheric global circulation models (GCMs). Solving it is the most important component of any land surface scheme as it closes the energy balance at the lower boundary of the atmosphere and determines the temperature of the surface with which the atmosphere is in contact. The surface energy model is also the core of any models that characterize the land-surface processes (ecological, hydrological, biogeochemical, etc.).

The surface energy balance equation for a single layer of the canopy–soil system can be written as

$$R_n = G + H + E \tag{10.146}$$

where R_n is the total net radiation, E is the latent heat flux, commonly called evapotranspiration (ET), G is the soil heat flux, and H is the sensible heat flux. It is illustrated in Fig. 10.9. It is clear that part of the net radiation will be conducted into the surface through G, which may range within 10–50% of the net radiation. Part of the net radiation is used to warm or cool the surface through H. The rest of the net radiation is used to evaporate water from surface to the atmosphere.

Total net radiation can be expressed by

$$R_n = R_n^s + R_n^l = (1 - \alpha) F_d^s + \varepsilon F_d^l - \sigma \varepsilon T^4 \tag{10.147}$$

where R_n^s and R_n^l are the shortwave and longwave components of the net radiation, α is the broadband albedo discussed in Chapter 9, F_d^s and F_d^l are the shortwave and longwave downward fluxes at the ground level that can be

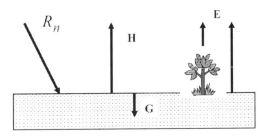

Figure 10.9   Illustration of surface radiation budget.

calculated by a radiative transfer package (e.g., MODTRAN) as long as the atmospheric properties are known, ε is the broadband emissivity discussed in Section 10.9, T is the surface temperature discussed in Sections 10.7 and 10.8, and σ is a constant.

Soil heat flux is usually measured at some depth in the soil

$$G_{z_1} = \int_{z_1}^{z_2} C_s(z) \frac{dT}{dt} dz + G_{z_2} \tag{10.148}$$

where $C_s(z)$ is the volumetric heat capacity of the soil and z_1 is the soil surface. The rate at which heat is transferred to the soil depends on the soil's thermodynamic properties and moisture content, temperature distribution of the soil profile, and the amount of crop cover. One method to calculate soil heat flux is to apply the first law of heat conduction (Fourier's law), which states that heat flux is the direction of and is proportional to the temperature gradient. The formula in one dimension can be written

$$G = k \frac{dT}{dz} \approx k \frac{T_s - T_g}{z} \tag{10.149}$$

where k is thermal conductivity and T_g is the soil temperature at depth z below the soil surface. The thermal conductivity will vary according to soil type and moisture content. Another method is to estimate it as a fraction of the net radiation: $G = \eta R_n$. Unfortunately, η is variable depending on surface conditions, ranging from 0.05 for the mature full-canopy cover to 0.5 for dry soils.

Sensible-heat flux describes the transfer of energy from objects that are warmer than their surroundings to the air or conversely, heat transferred from the air to cooler objects. The heat can be transferred by laminar (conduction or diffusion) or turbulent (wind) processes. A "resistance" approach is often used to calculate sensible heat; that is, similar to Ohm's law in electrical circuits (voltage = current $\times$ resistance $\Rightarrow V = I \times R$ or $I = V/R$), the rate of heat transfer is determined by the driving force (difference in temperature) and the resistance to the movement of heat. Thus, H can be calculated using the bulk transfer equation

$$H = \frac{T_s - T_a}{r_a} \rho C_p \tag{10.150}$$

where T_s is the land surface temperature: T_a is the surface air temperature; ρC_p is the volumetric heat capacity, which can be related to surface elevation; and r_a is the resistance to heat transfer. Estimation of aerodynamic resistance is more complicated. The resistance to heat transfer is a function of wind speed and surface roughness (rougher surfaces create more mixing of the air with the surface). One equation used to calculate resistance (in s/m) is

$$r_a = \frac{\left[\dfrac{\ln(z-d)/z_0}{K}\right]^2}{U} \tag{10.151}$$

where z is the height of wind speed measurements, d is the displacement height (m), z_0 is the roughness length, K is the von Karman constant (~ 0.4), and U is wind speed. For plant canopies with full cover, $z_0 = 0.13\ h$ and $d = 0.63\ h$, where h is canopy height (m). Displacement height is the height at which wind speed becomes essentially zero in the plant canopy, and the roughness length represents the remaining area of the canopy that contributes to turbulent mixing.

Equation (10.151) can create significant errors with low wind speed. A semiempirical equation for resistance that is more robust under low wind speed conditions is

$$r_a = \frac{4.72\left[\dfrac{\ln(z-d)/z_0}{k}\right]^2}{1+0.54U} \tag{10.152}$$

Note that this equation will not tend to infinity as wind speed approaches 0. When trying to apply this approach to sparse cover or bare soil conditions, more rigorous treatment of the resistance term is needed.

Latent heat flux (evapotranspiration) is the component in which we are ultimately interested for agricultural irrigation management and many other applications. Many equations have been developed to predict the rate at which water can be transferred to the atmosphere when water at the surface is not limited (potential evapotranspiration). However, few methods exist to predict actual evapotranspiration (ET), which is the amount of water lost by the plant canopies when water is limited. It can be calculated as

$$E = \frac{\rho C_p}{\gamma}\frac{e_0 - e_a}{r_a + r_0} \tag{10.153}$$

where e_0 is the saturated vapor pressure, e_a is the vapor pressure at the reference height, r_e is the aerodynamic resistance for exchange of latent heat between the surface and the air at reference height, and r_0 is the surface resistance for diffusive transfer of water vapor from the location of vapor source (e.g., leaf stomatal cavities) to the location of the sensible heat exchange (e.g., foliage surface).

The single-layer energy balance models are the simplest. To take into account temperature gradient within canopy, iterative algorithms are needed. For most practical applications, we need to consider at least different canopy and soil temperatures for heterogeneous or sparse vegetation canopies (e.g., Lhomme and Chehbouni 1999, Smith and Golta 1999), and the resulting models are usually referred to *dual-source* or *two-source models*.

There are two different two-source methods. The first is to model two-story canopies as successive layers. The two sources of water vapor and heat are superimposed and coupled. This deals with a typical small-scale heterogeneity. This type of approach has been very successful in interpreting sensible-heat flux over sparse vegetation from radiometric surface temperature (e.g., Lhomme et al. 1994, 1997). Another approach (Norman et al. 1995) is to treat the canopy distribution as a mosaic and to place vegetation patches side by side as they commonly are on a large scale. The latest improvements of their model seem to handle both types of heterogeneity (Kustas and Norman 1999).

One important note is that when we need to consider soil and vegetation canopies separately in the heterogeneous landscape, we also need to distinguish the radiometric temperature measured by the remote sensor and the aerodynamic temperature used in the energy balance equation. A detailed discussion is given in the literature (e.g., Norman et al. 1995).

10.11   SUMMARY

This chapter is composed of three basic parts. The first part discusses longwave radiative transfer theory. The second part is how to estimate land surface temperature and emissivity from thermal IR remote sensing data. The last part is about the conversion of narrowband to broadband emissivity and decomposition of the surface net radiation into latent heat flux (evapotranspiration), the soil heat flux, and the sensible-heat flux. The last part constitutes the bridge to using both LST and emissivity in land surface processes and other applications.

The longwave radiative transfer theory has been discussed in the first several sections, which is complementary to the shortwave radiative transfer of Chapters 2–4. We started with the monochromatic (single wavelength) radiative transfer in Section 10.2, and then moved to radiative transfer calculations over a spectral region from Section 10.3 to Section 10.5. The line-by-line method in Section 10.3 calculates transmittance/radiance in a very narrow spectral region and is an extremely time-consuming numerical technique. Band models and the correlated k-distribution methods are suitable for much larger spectral regions and more suitable for environmental remote sensing problems.

Atmospheric correction of thermal-IR remote sensing data was first discussed. The basic idea is to acquire ancillary information (mainly profiles of atmospheric temperature and water vapor) from other sources and calculate atmospheric quantities using a radiative transfer package. There is an urgent need to develop methods for estimating atmospheric quantities from imagery itself.

Estimation of LST and emissivity from thermal IR data has been discussed in both Sections 10.7 and 10.8. Section 10.7 introduces various methods for estimating LST using the split-window method that requires at least two bands and the known surface emissivity. The methods for estimating both emissivity and LST that require multiple bands were discussed in Section 10.8. The methods that use satellite observations of multiple viewing angles (e.g., ATSR) are not discussed in this book, and the reader should read the related papers (e.g., Prata 1993, Sobrino et al. 1996).

In GCM and most land surface process models, surface broadband emissivity has been set to 1, resulting in significant errors. Section 10.9 presented the formulas for converting narrowband emissivities from several sensors (i.e., AVHRR, ASTER, MODIS) to broadband emissivity.

The last section mainly discussed the energy balance equation and the computation of its components. It demonstrates how LST and broadband emissivity are used in land surface models and outlines the foundation of several applications that will be discussed in Chapter 13.xxx

REFERENCES

Becker, F. and Li, Z. L. (1990), Towards a local split window method over land surface temperature from a satellite, *Int. J. Remote Sens.* **11**: 369–394.

Becker, F. and Li, Z. L. (1995), Surface temperature and emissivity at various scales: Definition, measurement and related problems, *Remote Sens. Environ.* **12**: 225–253.

Clough, S. A., Iacono, M. J., and Moncet, J. L. (1992), Line-by-line calculations of atmospheric fluxes and cooling rates: Application to water vapor, *J. Geophys. Res.* **97**: 15761–15785.

Czajkowski, K. P., Goward, S. N., and Ouaidrari, H. (1998), Impact of AVHRR filter functions on surface temperature estimation from the split window approach, *Int. J. Remote Sens.* **19**: 2007–2012.

Dash, P., Gottsche, F.-M., Olesen, F.-S., and Fischer, H. (2002), Land surface temperature and emissivity estimation from passive data: Theory and practice—current trends, *Int. J. Remote Sens.* **23**: 2563–2594.

Edwards, D. P. (1988), Atmospheric transmittance and radiance calculations using line-by-line computer models, *Proc. SPIE* **928**: 94–116.

Edwards, D. P. and Francis, G. L. (2000). Improvements to the correlated-k radiative transfer method: Application to satellite infrared sounding, *J. Geophys. Res.* **105**: 18135–18156.

Francois, C. and Ottle, C. (1996), Atmospheric corrections in the thermal infared: Global and water vapor dependent split-window algorithms: Applications to ATSR and AVHRR data, *IEEE Trans. Geosci. Remote Sens.* **34**: 457–469.

Fu, Q. and Liou, K. N. (1992), On the correlated k-distribution method for radiative transfer in nonhomogeneous atmospheres, *J. Atmos. Sci.* **49**: 2139–2156.

Gillespie, A., Rokugawa, S., Hook, S. J., Matsunaga, T., and Kahle, A. (1999), *Temperature/Emissivity Separation Algorithm Theoretical Basis Document*, version 2.4.

Gillespie, A. R., Rokugawa, S., Matsunaga, T., Cothern, J., Hook, S., and Kahle, A. (1998), A temperature and emissivity separation algorithm for advanced space-borne thermal emission and reflection radiometer (ASTER) images, *IEEE Trans. Geosci. Remote Sens.* **36**: 1113–1126.

Goody, R. (1952), A statistical model for H_2O absorption, *Quart. J. Royal Meteorol. Soc.* **78**: 165–.

Goody, R. M. and Yang, Y. L. (1989), *Atmospheric Radiation—Theoretical Basis*, 2nd ed., Oxford Univ. Press.

Gottsche, F. M. and Olesen, F. S. (2002), Evolution of neural networks for radiative transfer calculations in the terrestrial infrared, *Remote Sens. Environ.* **80**: 335–353.

Gu, D. and Gillespie, A. R. (2000), A new apporach for temperature and emissivity separation, *Int. J. Remote Sens.* **21**: 2127–2132.

Gu, D., Gillespie, A. R., Kahle, A. B., and Palluconi, F. D. (2000), Autonomous atmospheric compensation (AAC) of high resolution hyperspectral thermal infared remote sensing imagery, *IEEE Trans. Geosci. Remote Sens.* **38**: 2557–2569.

Hartmann, J. M., Levi Di Leon, R., and Taine, J. (1984), Line-by-line and narrow-band statistical model calculations for H_2O, *J. Qant. Spectrosc. Radiat. Transfer* **32**: 119–127.

Hook, S., Gabell, A., Green, A., and Kealy, P. (1992), A comparison of techniques for extracting emissivity information from thermal infrared data for geological studies, *Remote Sens. Envir.* **42**: 123–135.

Kealy, P. and Hook, S. (1993), Separating temperature and emissivity in thermal infrared multispectral scanner data: Implication for recovering land surface temperatures, *IEEE Trans. Geosci. Remote Sens.* **31**: 1155–1164.

Kerr, Y. H., Lagouarade, J. P., and Imbernon, J. (1992), Accurate land surface temperature retrieval from avhrr data with use of an improved split window algorithm, *Remote Sens. Environ.* **41**: 197–209.

Kobayashi, H. (1999), Line-by-line calculation using the Fourier-transformed Voigt function, *J. Quant. Spectrosc. Radiat. Transfer* **62**: 477–483.

Kustas, W. P. and Norman, J. M. (1999), Evaluation of soil and vegetation heat flux predictions using a simple two-source model with radiometric temperatures for a partial canopy cover, *Agric. Forest Meteorol.* **94**: 13–29.

Lacis, A. A. and Oinas, V. (1991), A description of the correlated k distribution method for modeling nongray gaseous absorption, thermal emission, and multiple scattering in vertically inhomogeneous atmospheres, *J. Geophys. Res.* **96**: 9027–9063.

Lhomme, J. P. and Chehbouni, A. (1999), Comments on dual-source vegetation-atmosphere transfer model, *Agric. Forest Meteorol.* **94**: 269–273.

Lhomme, J. P., Monteny, B., and Amadou, M. (1994), Estimating sensible heat flux from radiometric temperature over sparse millet, *Agric. Forest Meteorol.* **68**: 77–91.

Lhomme, J. P., Troufleau, D., Monteny, B., Chehbouni, A., and Bauduin, S. (1997), Sensible heat flux and radiometric surface temperature over sparse Sahelian vegetation, II: A model for the $k_b - 1$ parameter, *J. Hydrol.* 188–189: 839–854.

Li, Z. L., Becker, F., Stoll, M., and Wan, Z. (1999), Evaluation of six methods for extracting relative emissivity spectra from thermal infrared images, *Remote Sens. Envir.* **69**: 197–214.

Liang, S. (1997), Retrieval of land surface temperature and water vapor content from AVHRR thermal imagery using an artificial neural network, *Proc. IGARSS.* **4**: 1959–1961.

Liang, S. (2001), An optimization algorithm for separating land surface temperature and emissivity from multispectral thermal infrared imagery, *IEEE Trans. Geosci. Remote Sens.*, **39**: 264–274.

Liang, S., Strahler, A. H., Barnsley, M. J., Borel, C. C., Diner, D. J., Gerstl, S. A. W., Prata, A. J., and Walthall, C. L. (2000), Multiangle remote sensing: Past, present and future, *Remote Sens. Rev.* **18**: 83–102.

Lyon, R. J. P. (1965), Analysis of rocks by spectral infrared emission (8 to 25 micron), *Econ. Geol.* **60**: 715–736.

Mlawer, E. J., Taubman, S. J., Brown, P. D., Iacono, M. J., and Clough, S. A. (1997), Radiative transfer for imhomogeneous atmospheres: RRTM, a validated correlated-k model for the longwave, *J. Geophys.* **102**: 16663–16682.

Moritz, H. (1993), General considerations regarding inverse and related problems, in *Inverse Problems: Principles and Applications in Geophysics, Technology, and Medicine*, G. Anger, R. Gorenflo, H. Jochmann, H. Moritz, and W. Webers, eds., Akademie Verlag, pp. 11–23.

Norman, J. M., Kustas, W. P., and Humes, K. S. (1995), Source approach for estimating soil and vegetation energy fluxes from observations of directional radiometric surface temperature, *Agric. Forest Meteorol.* **77**: 263–293.

Olioso, A. (1995), Simulating the relationship between thermal emissivity and the normalized difference vegetation index, *Int. J. Remote Sens.* **16**: 3211–3216.

Palluconi, F., Hoover, G., Alley, R., Jentoft-Nilsen, M. and Thompson, T. (1996). *An Atmosphere Correction Method for ASTER Thermal Radiometry over Land*, JPL, NASA EOS ATBD, Revision 2 pp.

Prata, A. and Platt, C. M. R. (1991), Land surface temperature measurements from the AVHRR, *Proc. 5th AVRHH Users Meet.*, 433–438.

Prata, A. J. (1993), Land surface temperature derived from the advanced very high resolution radiometer and the along-track scanning radiometer 1, Theory, *J. Geophys. Res.* **98**: 16689–16702.

Prata, A. J., Caselles, V., Coll, C., Sobrino, J., and Ottle, C. (1995), Thermal remote sensing of land surface temperature from satellites: Current status and future prospects, *Remote Sens. Rev.* **12**: 175–224.

Price, J. C. (1984), Land surface temperature measurements from the split window channels of the NOAA-7/AVHRR, *J. Geophys. Res.* **89**: 7231–7237.

Quine, B. M. and Drummond, J. R. (2002), Genspect: A line-by-line code with selectable interpolation error tolerance, *J. Quant. Spectrosc. Radiat. Transfer* **74**: 147–166.

Rees, W. G. and James, S. P. (1992), Angular variation of the infrared emissivity of ice and water surfaces, *Int. J. Remote Sens.* **13**: 2873–2886.

Rowntree, P. (1991), Atmospheric parameterization schemes for evaporation over land: Basic concepts and climate modeling aspects, in *Land Surface Evaporation: Measurement and Parameterization*, T. Schmugge and J. Andre, eds., Springer-Verlag, pp. 5–34.

Salomonson, V., Barnes, W., Maymon, P., Montgomery, H., and Ostrow, H. (1989), Modis: Advanced facility instrument for studies of the earth as a system, *IEEE Trans. Geosci. Remote Sens.* **27**: 145–153.

Schaaf, C., Gao, F., Strahler, A., Lucht, W., Li, X., Tsung, T., Strugll, N., Zhang, X., Jin, Y., Muller, P., Lewis, P., Barnsley, M., Hobson, P., Disney, M., Roberts, G., Dunderdale, M., Doll, C., d'Entremont, R., Hu, B., Liang, S., Privette, J. and Roy, D. (2002), First operational BRDF, albedo nadir reflectance products from MODIS, *Remote Sens. Envir.* **83**: 135–148.

Schadlich, S., Gottsche, F. M., and Olesen, F. S. (2001), Influence of land parameters and atmosphere on meteosat bightness temperatures and generation of land surface temperature maps by temporally and spatially interpolating atmospheric correction, *Remote Sens. Environ.* **75**: 39–46.

Schmugge, T., Hook, S., and Coll, C. (1998), Recovering surface temperature and emissivity from thermal infrared multispectral data, *Remote Sens. Environ.* **65**: 121–131.

Siddall, J. N. (1972), *Analytical Decision-Making in Engineering Design*. Prentice-Hall.

Smith, J. A. and Golta, S. M. (1999), Simple forest canopy thermal-exitance model, *IEEE Trans. Geosci. Remote Sens.* **37**: 2733–2736.

Snyder, W., Wan, Z., Zhang, Y., and Feng, Y. (1998), Classification-based emissivity for land surface temperature measurement from space, *Int. J. Remote Sens.* **19**: 2753–2774.

Sobrino, J. A., Coll, C., and Casselles, V. (1991), Atmospheric corrections for land surface temperature using AVHRR channels 4 and 5, *Remote Sens. Environ.* **38**: 19–34.

Sobrino, J. A., Raissouni, N., and Li, Z. L. (2001), A comparative study of land surface emissivity retrieval from NOAA data, *Remote Sens. Environ.* **75**: 256–266.

Sobrino, J. A., Li, Z., Stoll, M., and Becker, F. (1994), Improvements in the split window technique for land surface temperature determination, *IEEE Trans. Geosci. Remote Sens.* **32**: 243–253.

Sobrino, J. A., Li, Z. L., Stoll, M. P., and Becker, F. (1996), Multi-channel and multi-angle algorithms for estimating sea and land surface temeprature with ATSR data, *Int. J. Remote Sens.* **17**: 2089–2114.

Takashima, T. and Masuda, K. (1987), Emissivities of quartz and sahara dust powders in the infrared region (7–17 μm), *Remote Sens. Envir.* **23**: 51–63.

Thomas, G. E. and Stamnes, K. (1999), *Radiative Transfer in the Atmosphere and Ocean*, Cambridge Atmospheric and Space Science Series, Cambridge University Press.

Tikhonov, A. and Arsenin, V. (1977), *Solutions of Ill-Posed Problems*, V. H. Winston & Sons, a division of Scripta Technica Publishers.

Twomey, S. (1977), *Introduction to the Mathematics of Inversion in Remote Sensing and Indirect Measurements*, Dover Publications.

Ulivieri, C., Castronuovo, M. M., Francioni, R., and Cardillo, A. (1994), A split window algorithm for estimating land surface temperature from satellites, *Adv. Space Res.* **14**: 59–65.

Valor, E. and Caselles, V. (1996), Mapping land surface emissivity from NDVI: Application to European, African, and American areas, *Remote Sens. Envir.* **57**: 167–184.

Van De Griend, A. A. and Owe, M. (1993), On the relationship between thermal emissivity and the normalized difference vegatation index for natural surface, *Int. J. Remote Sens.* **14**: 1119–1131.

Wan, Z. and Dozier, J. (1996), A generalized split-window algorithm for retrieving land-surface temperature measurement from space, *IEEE Trans. Geosci. Remote Sens.* **34**: 892–905.

Wan, Z. and Li, Z. L. (1997), A physics-based algorithm for retrieving land-surface emissivity and temperatures from EOS/MODIS data, *IEEE Trans. Geosci. Remote Sens.* **35**: 980–996.

Yamaguchi, Y., Kahle, A., Tsu, H., Kawakami, T., and Oniel, M. (1998), Overview of advanced spaceborne thermal emission and reflection radiomater (ASTER), *IEEE Trans. Geosci. Remote Sens.* **36**: 1062–1071.

11

Four-Dimensional Data Assimilation

In the previous few chapters, we discussed various methods published in the literature for estimating land surface variables from the datasets of some specific sensors or their combinations. The values of these variables estimated from different sources may not be physically consistent. Most techniques do not take advantage of observations acquired at different times and cannot handle observations with different spatial resolutions together. In particular, these techniques estimate only variables that significantly affect radiance received by the sensors. In many cases, we also want to estimate some variables that are not related to radiance directly. This four-dimensional (4D) data assimilation is a new technique that can help us address these issues. It has not been widely explored in optical remote sensing so far, but its potential is vast since data assimilation can incorporate heterogeneous, irregularly distributed, and temporally inconsistent observational data

In Section 11.2, we introduce some representative algorithms developed in meteorology/oceanography data assimilation. These begin with some simple algorithms that will be very helpful in understanding the basic principles of data assimilation. Section 11.3 introduces a few typical minimization algorithms that are needed in data assimilation and other applications (e.g., Chapter 8). Sections 11.2 and 11.3 contain many mathematical descriptions. For those who are not interested in the details of the algorithms, the sequence is Section 11.1 and then Sections 11.4 and 11.5. Section 11.4 presents application examples of data assimilation in hydrology. After reporting some individual studies, two jointed activities in the United States and Europe are also briefly mentioned. Section 11.5 discusses the applications of data assimilation using crop growth models.

Quantitative Remote Sensing of Land Surfaces. By Shunlin Liang
ISBN 0-471-28166-2 Copyright © 2004 John Wiley & Sons, Inc.

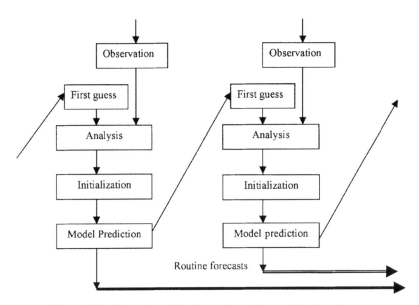

Figure 11.1 Illustration of the intermittent data assimilation scheme.

11.1 INTRODUCTION

Meteorology and oceanography have extensively developed the concept of "adding value" to data, especially remotely sensed data, by combining measurements with the predictions of specially selected dynamic models. This process, which is commonly known as *data assimilation*, is an extension of standard data retrieval techniques. The objective of data assimilation is to provide physically consistent estimates of spatially distributed environmental variables. These estimates are derived from scattered observations taken at a limited number of times and locations, supplemented by any additional "prior information" that may guide the estimation process. The four dimensions we generally refer to are space (x, y, z) and time. As early as the late 1960s Charney et al. (1969) suggested putting current data into a numerical model such that the model's equations provide time continuity and dynamic coupling among the atmospheric field. This concept has proved to be a major advance in numerical weather forecasting (NWF) since the early 1970s. Assimilation of observations is also rapidly developing in the fields of dynamic oceanography and land surfaces.

An assimilation method extensively used in operational meteorology is the analysis–forecast cycle technique, commonly referred to as *intermittent data assimilation* (Fig. 11.1). To get some general ideas about data assimilation, let us examine this method in a bit more detail. This process consists of four steps, which are repeated at each assimilation cycle (typically every 3–12 h). After the observations are checked via quality control, both observations and the background information (first guess) are used in the analysis. The operation that corrects the background field at a given time with new observations is called an *analysis*. If the model state is overdetermined by the observations, then the analysis is reduced to an interpolation problem. In most cases the analysis problem is underdetermined because data are sparse and only indirectly related to the model variables. In order to make it a well-posed problem it is necessary to rely on some *background* information. The background or "first guess" is usually a prior model forecast valid at the analysis time, or it can simply be climatology or a combination of both. Then the analyzed fields are adjusted, or initialized, to conform to some dynamic constraints. The last step is a short-range forecast that generates the background fields for the analysis at the next assimilation cycle. The output can also be used for routine forecasts.

From a temporal perspective, all data assimilation methods can be grouped into two categories: intermittent or continuous in time (Daley 1991). In an *intermittent* method, observations are processed in small batches, which is usually more computationally convenient. In a *continuous* method, observation batches over longer periods are considered, and the correction to the analyzed state is smooth in time, which is physically more realistic. The basic procedures are similar in either case (see Fig. 11.1).

From Fig. 11.1 we can see that in any data assimilation systems there are two basic components: observation data and a dynamic model. What has really spurred the development of data assimilation was the advent of Earth observing satellites that provide observations on the global basis. The observations from remote sensing are in two formats: raw radiance and retrieved model variables through inversion. A dynamic model is physically based and reflects the physical laws governing the basic principles of conservation of mass, energy, and momentum of a dynamic system. In data assimilation, the observed information is accumulated into the model state by taking advantage of consistency constraints with laws of time evolution and physical properties.

For different assimilation systems (meteorology, oceanography, or land surfaces), the dynamic models and observations are different, but the analysis methods are similar. In the next section, we would like to introduce various analysis methods that have been used mainly in meteorology.

11.2   ASSIMILATION ALGORITHMS

11.2.1   Background

Data assimilation is a quantitative, objective method used to infer the state of the dynamic system from heterogeneous, irregularly distributed, and temporally inconsistent observational data with differing accuracies. Since the early 1990s meteorologists and oceanographers have tended to view data assimilation as a model state estimation problem [see Courtier et al. (1993) for a recent annotated bibliography]. The problem may be posed at various levels of generality and abstraction. For the present purposes, suppose that the environmental variables of primary interest (e.g., soil moisture) are "state variables" that evolve over time and space in accordance with partial differential equations based on conservation laws and associated constitutive assumptions.

There are many data assimilation algorithms in meteorology and oceanography that differ in their numerical cost, optimality, and suitability for real-time data assimilation. Land surface data assimilation is relatively new. Most of the research so far concentrates on hydrologic aspects of the land surface process, for example, inferring soil moisture content by assimilating satellite data products and other observations into a land surface model. Assimilation of remotely sensed data into ecosystem or ecological models has attracted increased attention in recent years (Nouvellon et al. 2001, Weiss et al. 2001).

From a purely algorithmic perspective, all current data assimilation algorithms can be grouped into two categories: sequential assimilation and variational assimilation (Talagrand 1997). In *sequential assimilation*, the dynamic model is integrated over the time interval over which the observations are available. Whenever the new observations are available, the system states predicted by the model are used as a background and then "updated" or "corrected" with the new observations. The integration of the model is then restarted from the updated state, and the process is repeated until all the available observations have been utilized. One appealing feature of sequential assimilation is its constant updating of the state fields predicted by the model on the basis of new observations, which makes sequential assimilation well adapted to numerical weather prediction (NWP). Sequential assimilation only considers observation made in the past until the time of analysis, which is the case of real-time assimilation systems, and each individual observation is used only once. The information contained in the observations are propagated only from the past into the future, but not from the future into the past. This certainly is a disadvantage for a posteriori reassimilation of past observations, where it seems preferable to use algorithms capable of carrying information both forward and backward in time.

Variational assimilation, on the other hand, aims at globally adjusting a model solution to all the observations available over the assimilation period. The adjusted states at all times are influenced by all the observations. This idea has been discussed in Section 8.2 in terms of optimization methods, but the time dimension was not considered there. We start with some simple methods that were used in the earlier data assimilation systems. Although they are rarely used now in an operational forecast system in meteorology or oceanography, presentation of these methods will be helpful for us to gain the perspectives on the development of different methods and to understand more advanced and sophisticated methods. The methods presented in Sec-

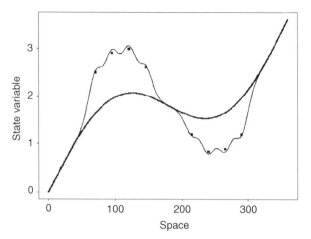

Figure 11.2 Direct insertion scheme. The thick line represents the background information (first guess); points are observations. The thin line is the output from the analysis.

tions 11.2.2 and 11.2.3 are often called *objective analysis* in the data assimilation literature. The sequential and variational assimilation methods will be discussed in Sections 11.2.4–11.2.6.

11.2.2 The Method of Successive Correction

We start with a discussion of the direct insertion and local polynomial fitting methods, and then move to successive correction techniques.

One of the earliest algorithms is the *direct-insertion* technique, which sets the observed values as the values of the model variables at the observation location. The values of the model variables at other locations are interpolated. This is illustrated in Fig. 11.2, where it is assumed the model state is univariate (single independent variable). If we denote $\mathbf{x}_b$ by the first guess of the model state (background), and by $\mathbf{x}_o(\mathbf{r}_j)$, $j = 1, 2, \dots, n$, a set of observations of the same parameter, where $\mathbf{r}$ defines the spatial location in a one-, two-, or three-dimensional domain. The model state $\mathbf{x}_a$ defined at each gridpoint i can be determined by

$$\mathbf{x}_a(\mathbf{r}_i) = \mathbf{x}_b(\mathbf{r}_i) + \frac{\sum_{j=1}^{n} w(\mathbf{r}_i, \mathbf{r}_j)[\mathbf{x}_o(\mathbf{r}_j) - \mathbf{x}_b(\mathbf{r}_j)]}{\sum_{j=1}^{n} w(\mathbf{r}_i, \mathbf{r}_j)} \tag{11.1}$$

where $w(\mathbf{r}_i, \mathbf{r}_j)$ is the weighting function dependent on the distance $d_{i,j}$

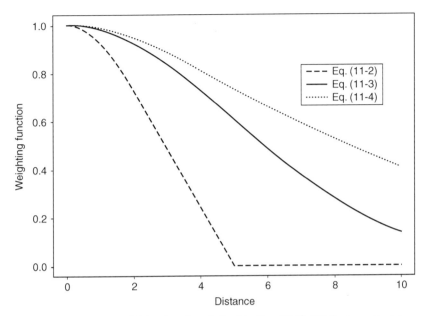

Figure 11.3   Three weighting functions defined by Eqs. (11.2)–(11.4) where $R = 5.0$.

between points $\mathbf{r}_i$ and $\mathbf{r}_j$

$$w(\mathbf{r}_i, \mathbf{r}_j) = \max\left(0, \frac{R^2 - d_{i,j}^2}{R^2 + d_{i,j}^2}\right) \tag{11.2}$$

or

$$w(\mathbf{r}_i, \mathbf{r}_j) = \exp\left(-\frac{d_{i,j}^2}{2R^2}\right) \tag{11.3}$$

or

$$w(\mathbf{r}_i, \mathbf{r}_j) = \left(1 + \frac{d_{i,j}}{R}\right)\exp\left(-\frac{d_{i,j}}{R}\right) \tag{11.4}$$

These three weighing functions are compared in Fig. 11.3. It is evident that these three functions differ quite dramatically as the distance becomes larger.

In Eq. (11.1), $\mathbf{x}_b(\mathbf{r}_j)$ is the background state interpolated to point $\mathbf{r}_j$. The weight function equals one [i.e., $\mathbf{x}_a(\mathbf{r}_j) = \mathbf{x}_o(\mathbf{r}_j)$] if the gridpoint $\mathbf{r}_i$ is collocated with observation $\mathbf{r}_j$, and zero [i.e., $\mathbf{x}_a(\mathbf{r}_j) = \mathbf{x}_b(\mathbf{r}_j)$] if the distance is greater than a predefined range R, which is usually called the *radius of influence*. This technique implicitly assumes that all observed values are error-free and represent the "true" values of the dynamic system.

Another algorithm is *local polynomial fitting*, in which all observations in each neighborhood are used to determine the coefficients of a low-order polynomial function through the least squares estimation. The model state x_a may not be equal to the observed value at the observation location but represents a smooth surface of the variable and resists the random errors of the observed values. This method has been used widely in other disciplines for surface trend analysis. Further development of this concept forms the *successive correction method* (SCM). Let us first consider the basic formulation for a single correction. The objective is to determine the adjustment magnitude.

At the analysis gridpoint and observation location, the background estimates contain errors denoted $\varepsilon_b(\mathbf{r}_i)$ and $\varepsilon_b(\mathbf{r}_j)$, the observation error is denoted $\varepsilon_o(\mathbf{r}_j)$. Assume that both the background error and observation error are homogeneous and spatially uncorrelated and that the expected background error variance $E_b^2 = \langle \varepsilon_b^2(\mathbf{r}) \rangle$ and observation error variance $E_o^2 = \langle \varepsilon_o^2(\mathbf{r}) \rangle$ are independent of location. First consider the case of a single observation at location $\mathbf{r}_j$. Two estimates can be used for the estimate of the analysis: (1) the background value at the gridpoint $\mathbf{r}_i$ [i.e., $x_b(\mathbf{r}_i)$] and (2) the difference between the observation $x_o(\mathbf{r}_j)$ and the background $x_b(\mathbf{r}_j)$ plus $x_b(\mathbf{r}_i)$ [i.e., $x_b(\mathbf{r}_i) + x_o(\mathbf{r}_j) - x_b(\mathbf{r}_j)$]. Bergthorsson and Doos (1955) developed the formulas by weighing these two estimates in an optimal way:

$$x_a(\mathbf{r}_i) = \frac{E_b^{-2} x_b(\mathbf{r}_i) + E_o^{-2} w(\mathbf{r}_i, \mathbf{r}_j)\{x_b(\mathbf{r}_i) + [x_o(\mathbf{r}_j) - x_b(\mathbf{r}_j)]\}}{E_b^{-2} + E_o^{-2} w(\mathbf{r}_i, \mathbf{r}_j)} \quad (11.5)$$

where the distance weighing function $w(\mathbf{r}_i, r_j)$ is defined in Eq. (11.2) or (11.3). It is clear that the analysis value is the linear combination of these two estimates (see Fig. 11.4), the weight for background value is its inverse of error variance, and the weight for the observation is both its inverse of error variance and the distance weighting function. If $\mathbf{r}_i$ is far away from $\mathbf{r}_j$, then $w(\mathbf{r}_i, \mathbf{r}_j) = 0$, and the analysis value equals the background value [i.e., $x_a(\mathbf{r}_i) = x_b(\mathbf{r}_i)$]. If both error variances are very small (i.e., $E_o^2 \ll 0$ and $E_b^2 \ll 0$), then (11.5) is equivalent to (11.1) with one observation. If $\mathbf{r}_i = \mathbf{r}_j$, then (11.5) becomes

$$x_a(\mathbf{r}_i) = \frac{E_b^{-2} x_b(\mathbf{r}_i) + E_o^{-2} x_o(\mathbf{r}_i)}{E_b^{-2} + E_o^{-2}} \quad (11.6)$$

It is more useful if we rewrite Eq. (11.5) as

$$x_a(\mathbf{r}_i) - x_b(\mathbf{r}_i) = \frac{E_o^{-2} w(\mathbf{r}_i, \mathbf{r}_j)}{E_b^{-2} + E_o^{-2} w(\mathbf{r}_i, \mathbf{r}_j)} [x_o(\mathbf{r}_j) - x_b(\mathbf{r}_j)] \quad (11.7)$$

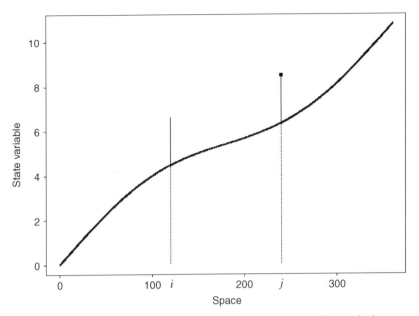

Figure 11.4 Illustration of Eq. (11.5) in the successive correction method.

In particular, if we define a posterior weight $W_{i,j}$ as

$$W_{i,j} = \frac{E_b^2 w(\mathbf{r}_i, \mathbf{r}_j)}{E_o^2 + E_b^2 w(\mathbf{r}_i, \mathbf{r}_j)} \qquad (11.8)$$

then Eq. (11.7) becomes a much simple form:

$$x_a(\mathbf{r}_i) - x_b(\mathbf{r}_i) = W_{i,j}\big[x_o(\mathbf{r}_j) - x_b(\mathbf{r}_j)\big] \qquad (11.9)$$

The discussions above pertain to one single observation. Suppose that we have N observations; then Eq. (11.7) can be extended to become

$$x_a(\mathbf{r}_i) - x_b(\mathbf{r}_i) = \frac{\displaystyle\sum_{j=1}^{N} w(\mathbf{r}_i, \mathbf{r}_j)\big[x_o(\mathbf{r}_j) - x_b(\mathbf{r}_j)\big]}{E_o^2/E_b^2 + \displaystyle\sum_{j=1}^{N} w(\mathbf{r}_i, \mathbf{r}_j)} \qquad (11.10)$$

So far we have discussed only a single correction of the background field. Now let us consider multiple corrections, which might result in a better analysis. Let us rewrite Eq. (11.9) as the first correction

$$x_a^1(\mathbf{r}_i) = x_b(\mathbf{r}_i) + W_{i,j}\left[x_o(\mathbf{r}_j) - x_b(\mathbf{r}_j)\right] \qquad (11.11)$$

then the second correction uses the estimated analysis value as the background value:

$$x_a^2(\mathbf{r}_i) = x_a^1(\mathbf{r}_i) + W_{i,j}\left[x_o(\mathbf{r}_j) - x_a^1(\mathbf{r}_j)\right] \qquad (11.12)$$

This iteration proceeds according to the following general formula:

$$x_a^{n+1}(\mathbf{r}_i) = x_a^n(\mathbf{r}_i) + W_{i,j}\left[x_o(\mathbf{r}_j) - x_a^n(\mathbf{r}_j)\right] \qquad (11.13)$$

There are many possible variants to the method discussed above, one of which is to vary the posterior weight in each iteration. Additionally, the conditions under which this iteration process converges, what it converges to, and how fast this iteration process is are all variable. Detailed discussions of these issues are beyond the scope of this book, but the reader is referred to the related references for more information.

11.2.3 Optimal Interpolation

Statistical interpolation is a powerful and widely used technique in meteorology data assimilation. One of these interpolation algorithms is called *optimal interpolation*, which significantly contributed to the progress of numerical weather prediction during the 1970s and 1980s. It is actually a minimum variance method that is closely related to the Kriging technique widely used in various Earth science disciplines.

Let us first consider the *linear case*. Assume that the state of the model is a random variable denoted by $\mathbf{x}$. The observation variable is denoted by $\mathbf{y}$ and related to the state variable linearly

$$\mathbf{y} = \mathbf{H}\mathbf{x} + \boldsymbol{\varepsilon} \qquad (11.14)$$

where $\mathbf{H}$ is the linear observation operator, and $\boldsymbol{\varepsilon}$ represents the observation error. The observations are assumed to be unbiased ($\langle\boldsymbol{\varepsilon}\rangle = 0$) and have a known error covariance $\langle\boldsymbol{\varepsilon}\boldsymbol{\varepsilon}^T\rangle = \mathbf{R}$. An unbiased background $\mathbf{x}_b$ has the same mean value as the state variable of the model ($\langle\mathbf{x}_b\rangle = \langle\mathbf{x}\rangle$). The background error covariance is characterized by $\langle(\mathbf{x}_b - \langle\mathbf{x}\rangle)(\mathbf{x}_b - \langle\mathbf{x}\rangle)^T\rangle = \mathbf{B}$. The best unbiased estimator is then given by

$$\mathbf{x}_a = \mathbf{x}_b + \mathbf{K}(\mathbf{y} - \mathbf{H}\mathbf{x}_b) \qquad (11.15)$$

where

$$\mathbf{K} = \mathbf{B}\mathbf{H}^T(\mathbf{H}\mathbf{B}\mathbf{H}^T + \mathbf{R})^{-1} = (\mathbf{B}^{-1} + \mathbf{H}^T\mathbf{R}^{-1}\mathbf{H})^{-1}\mathbf{H}^T\mathbf{R}^{-1} \quad (11.16)$$

The covariances of analysis error follow

$$\langle (\mathbf{x}_a - \langle \mathbf{x} \rangle)(\mathbf{x}_a - \langle \mathbf{x} \rangle)^T \rangle = (\mathbf{I} - \mathbf{K}\mathbf{H})\mathbf{B} = (\mathbf{B}^{-1} + \mathbf{H}^T\mathbf{R}^{-1}\mathbf{H})^{-1} \quad (11.17)$$

In numerical weather forecasting, Eqs. (11.15) and (11.16) are actually never used because of the dimension of the model variables. For example, $\mathbf{B}$ is a matrix of $10^7 \times 10^7$. This is far beyond current computer archiving capabilities. Efforts have been made to simplify Eqs. (11.15) and (11.16) to an affordable amount of computer resources, while preserving some of the essential characteristics.

If the observation operator $\mathbf{H}$ is weakly linear (*quasilinear case*), the *tangent linear approximation* can be used to generalize the previous results. Assume that $\mathbf{H}(\mathbf{x})$ is not linear, but the tangent linear approximation is valid

$$\mathbf{H}(\mathbf{x}) - \mathbf{H}(\mathbf{x}_b) = \mathbf{H}'(\mathbf{x} - \mathbf{x}_b) \quad (11.18)$$

where the prime sign represents the first derivative. Thus

$$\mathbf{y} = \mathbf{H}(\mathbf{x}) + \varepsilon = \mathbf{H}(\mathbf{x}_b) + \mathbf{H}(\mathbf{x}) - \mathbf{H}(\mathbf{x}_b) + \varepsilon$$
$$= \mathbf{H}(\mathbf{x}_b)\mathbf{H}(\mathbf{x} - \mathbf{x}_b) + \varepsilon \quad (11.19)$$

Equations (11.15) and (11.16) become

$$\mathbf{x}_a = \mathbf{x}_b + \mathbf{K}(\mathbf{y} - \mathbf{H}(\mathbf{x}_b)) \quad (11.20)$$

with

$$\mathbf{K} = \mathbf{B}\mathbf{H}'^T(\mathbf{H}'\mathbf{B}\mathbf{H}'^T + \mathbf{R})^{-1} \quad (11.21)$$

and

$$\langle (\mathbf{x}_a - \langle \mathbf{x} \rangle)(\mathbf{x}_a - \langle \mathbf{x} \rangle)^T \rangle = (\mathbf{I} - \mathbf{K}\mathbf{H}')\mathbf{B} \quad (11.22)$$

The quasilinear case suffers the same problem as the linear case. The optimal interpolation methods use the same equation as Eqs. (11.15)–(11.17) with several important simplifications. The basic one is to use only the neighboring observations for a given location. The second approximation is the use of an analytical model for the forecasting of error covariances that need to be computed at the observation locations. In most implementations, the spatial correlations can be expressed as a product of two functions, one of which depends on the horizontal distance only and the other on the vertical distance only. Spatial correlation functions have been discussed in Section 8.1.3 in terms of variograms and the autocorrelation function.

11.2.4 Variational Analysis Algorithms

The variational analysis scheme attempts to minimize the objective function that represents the difference between the observations and the model predictions. The general ideas are similar to that used for estimating land surface biophysical parameters from remotely sensed data in Section 8.2 on the model inversion, although we did not use this notation. Similar techniques have been used for retrieving atmospheric profiles from satellite sounding such as TOVS radiances or microwave measurements, and retrieving surface wind fields from a collection of scatterometer ambiguous wind measurements, or for the analysis of land surface properties in a numerical weather prediction model. In atmospheric data assimilation, algorithms performing a global analysis of the meteorological fields are often called *three-dimensional* (3D) or *four-dimensional* (4D) *variational analysis*, but the number of variables is much larger.

11.2.4.1 3D Variational Analysis
The 3D variational analysis (3D-Var) scheme attempts to minimize the objective function J

$$J(\mathbf{x}) = \tfrac{1}{2}(\mathbf{x} - \mathbf{x}_b)^T \mathbf{B}^{-1}(\mathbf{x} - \mathbf{x}_b) + \tfrac{1}{2}(\mathbf{H}(\mathbf{x}) - \mathbf{y})^T \mathbf{R}^{-1}(\mathbf{H}(\mathbf{x}) - \mathbf{y}) = J_b + J_o \tag{11.23}$$

where $\mathbf{y}$ is the observation vector (reflectance/radiance and/or the retrieved variables), $\mathbf{x}$ the model state variables (i.e., soil properties), $\mathbf{x}_b$ the background field (or first guess), $\mathbf{H}$ the model operator (e.g., crop growth model), $\mathbf{R}$ the observation error covariance matrix, and $\mathbf{B}$ the background error covariance matrix.

In Eq. (11.23), the first term, J_b, is to force the optimal parameters as close as possible to background fields, and the second term J_o is to adjust parameters so that model outputs will be as close to the observations as possible. Specifying $\mathbf{O}$ and $\mathbf{B}$ depends on the relative accuracy of background information and remote sensing data products.

Starting with an initial guess field $\mathbf{x}^0$, normally $\mathbf{x}^0 = \mathbf{x}_b$, the final analysis $\mathbf{x}^{final} = \mathbf{x}^\infty$ can be obtained using a minimization algorithm. The simplest method is the so-called steepest descent, which can be expressed as

$$\mathbf{x}^{n+1} = \mathbf{x}^n - \alpha \nabla_x J \tag{11.24}$$

where the superscript denotes the iteration cycle number, α is a constant ($\alpha > 0$), and $\nabla_x J$ is the gradient of the cost function J with respect to $\mathbf{x}$:

$$\nabla_x J = B^{-1}(\mathbf{x} - \mathbf{x}_b) + \mathbf{H}^T \mathbf{R}^{-1}[\mathbf{H}(\mathbf{x}) - \mathbf{y}] \tag{11.25}$$

where **H** is the tangent linear operator of *H*. A detailed discussion of the steepest-descent method is given in Section 11.3.1.

Given *N* model gridpoints and *M* observation locations, **x** and $\mathbf{x}_b$ are vectors of length *N*, **y** is a vector of length *M*, **B** is an $N \times N$ matrix, and **R** is an $M \times M$ matrix.

If *N* is very large, calculating $\mathbf{B}^{-1}$ is very time-consuming. To avoid the inversion of **B**, a new vector **v** is introduced (Huang 2000), defined as

$$\mathbf{v} = \mathbf{B}^{-1}(\mathbf{x} - \mathbf{x}_b) \tag{11.26}$$

and the cost function *J* can now be written as

$$J(\mathbf{v}) = J_b(\mathbf{v}) + J_0(\mathbf{v}) = \tfrac{1}{2}\mathbf{v}^T B^T \mathbf{v} + \tfrac{1}{2}\left[H(B\mathbf{v} + \mathbf{x}_b) - \mathbf{y} \right]^T$$
$$\times \mathbf{R}^{-1}\left[H(B\mathbf{v} + \mathbf{x}^b) - \mathbf{y} \right] \tag{11.27}$$

Using **v** as the control variable, the minimization scheme is now rewritten as

$$\mathbf{v}^{n+1} = \mathbf{v}^n - \alpha \nabla_v J \tag{11.28}$$

where ∇_v is the gradient of *J* with respect to **v**:

$$\nabla_v J = B^T \left\{ \mathbf{v} + H^T R^{-1}\left[H(B\mathbf{v} + \mathbf{x}_b) - \mathbf{y} \right] \right\} \tag{11.29}$$

Starting with an initial guess field $\mathbf{v}^0$, we obtain

$$\mathbf{v}_0 = \mathbf{B}^{-1}\left(\mathbf{x}^0 - \mathbf{x}_b \right) \tag{11.30}$$

The minimum of *J* is reached at $\mathbf{v}^\infty$, and the final analysis $\mathbf{x}^{\text{final}}$ becomes

$$\mathbf{x}^{\text{final}} = \mathbf{x}_b + B\mathbf{v}^\infty \tag{11.31}$$

In this scheme $\mathbf{B}^{-1}$ is needed only to provide the initial guess value for the control variable. Consequently, if we assume $\mathbf{v}^0 = 0$ we have obtained a variational analysis scheme that does not need the inversion of **B**.

In meteorology data assimilation the number of the variables is very large and the minimization of the cost function becomes computationally very expensive. A more common practice is to implement this minimization procedure using an incremental algorithm that has two different resolutions: coarse resolution for minimizing the cost function and high resolution for the analysis. For each iteration *n* in the incremental method, an approximate analysis $\mathbf{x}_n$ is calculated from an initial approximation $\mathbf{x}_{n-1}$ by

$$\mathbf{x}_n = \mathbf{x}_{n-1} + S^* \delta \mathbf{x}_n \tag{11.32}$$

where S^* denotes the pseudoinverse of an operator S that reduces the resolution of the model fields to the resolution used for the cost function. The increment $\delta \mathbf{x}_n$ is calculated by minimizing the cost function:

$$J(\delta \mathbf{x}_n) = \tfrac{1}{2}(S\mathbf{x}_{n-1} + \delta \mathbf{x}_n - S\mathbf{x}_b)^T \mathbf{B}^{-1}(S\mathbf{x}_{n-1} + \delta \mathbf{x}_n - S\mathbf{x}_b)$$

$$= \tfrac{1}{2}\left[H(\mathbf{x}_{n-1}) + H'(\delta \mathbf{x}_n) - \mathbf{y}\right]^T O^{-1}\left[H(\mathbf{x}_{n-1}) + H'(\delta \mathbf{x}_n) - \mathbf{y}\right]$$

$$(11.33)$$

Note in particular that the observation operator H acts on the high-resolution fields $\mathbf{x}_{n-1}$ whereas the operator H' acts on the low-resolution increment.

The incremental method is described above as an iterative procedure. However, there is generally no guarantee that the iterations will converge. For this reason, a typical implementation of the method performs only a few iterations (less than 100).

A correct specification of observation and background error covariances $\mathbf{R}$ and $\mathbf{B}$ is crucial to the quality of the analysis, because they determine the extent to which background fields will be corrected to match the observations. The essential parameters are the variances, but the correlations are also very important. In general, the only way to estimate statistics is to assume that they are stationary over a period of time and uniform over a domain so that one can take a number of error realizations and make empirical statistics. In a sense this is a climatology of errors. Another empirical way to specify error statistics is to take them to be a fraction of the climatological statistics of the fields themselves.

Specification of the observation error statistic $\mathbf{R}$ relies on the knowledge of instrumental characteristics, which can be estimated using collocated observations. The method currently used at ECMWF (European Centre for Medium-Range Weather Forecasts) to estimate background error statistics is to run an ensemble of independent analysis experiments. For each experiment, the observations are perturbed by adding random noise drawn from the assumed distribution of observation error. The effect of the perturbations is to generate differences in the analysis for each experiment. These are propagated to the next analysis cycle as differences in backgrounds. After a few days of assimilation, the statistics of differences between background fields for pairs of members of the ensemble come to equilibrium, and in principle become representative of the true statistics of background error.

11.2.4.2 *4D Variational Analysis*

Four-dimensional variational analysis (4D-Var) is a simple generalization of 3D-Var for observations that are distributed in time. The equations are the same, provided the observation operators are generalized to include a forecast model that will allow a comparison between the model state and the observations at the appropriate time.

The 4D-Var cost function is

$$J(\mathbf{x}(t_0)) = \tfrac{1}{2}[\mathbf{x}(t_0) - \mathbf{x}_b]^T \mathbf{B}^{-1}[\mathbf{x}(t_0) - \mathbf{x}_b]$$

$$+ \tfrac{1}{2}\sum_{i=1}^{n}[\mathbf{H}_i(\mathbf{x}(t_i)) - \mathbf{y}_i]^T \mathbf{R}_i^{-1}[\mathbf{H}_i(\mathbf{x}(t_i)) - \mathbf{y}_i] \quad (11.34)$$

Note that the cost function is regarded as a function of $\mathbf{x}(t_0)$, whereas the observation is at time t_i. We can eliminate $\mathbf{x}(t_i)$ by noting that $\mathbf{x}(t_i)$ is the result of a model integration with initial conditions $\mathbf{x}(t_0)$. Let us write this as

$$\mathbf{x}(t_i) = \mathbf{M}_{t_0 \to t_i}(\mathbf{x}(t_0)) \quad (11.35)$$

The cost function is then

$$J(\mathbf{x}(t_0)) = \tfrac{1}{2}[\mathbf{x}(t_0) - \mathbf{x}_b]^T \mathbf{B}^{-1}[\mathbf{x}(t_0) - \mathbf{x}_b]$$

$$+ \tfrac{1}{2}\sum_{i=1}^{n}\left[\mathbf{H}_i\left(\mathbf{M}_{t_0 \to t_i}(\mathbf{x}(t_0))\right) - \mathbf{y}_i\right]^T \mathbf{R}_i^{-1}\left[\mathbf{H}_i\left(\mathbf{M}_{t_0 \to t_i}(\mathbf{x}(t_0))\right) - \mathbf{y}_i\right]$$

$$(11.36)$$

As in the 3D case, the incremental formulation can be introduced under the assumption of quasilinearity of the model and the observation operators.

$$J(\delta\mathbf{x}) = \tfrac{1}{2}\delta\mathbf{x}^T \mathbf{B}^{-1}\delta\mathbf{x} + \tfrac{1}{2}\sum_{i=0}^{n}[\mathbf{H}_i'\delta\mathbf{x}(t_i) - \mathbf{d}_i]^T \mathbf{R}_i^{-1}[H_i'\delta\mathbf{x}(t_i) - \mathbf{d}_i] \quad (11.37)$$

where $\delta\mathbf{x}(t_i) = \mathbf{M}_{t_0 \to t_i}'\delta\mathbf{x}$, the innovation vector is given at each timestep t_i, by $\mathbf{d}_i = \mathbf{y}_i - \mathbf{H}_i(\mathbf{x}_b(t_i))$ and $\mathbf{x}_b(t_i)$ can be calculated by integrating the nonlinear model $\mathbf{M}$ from time t_0 to t_i.

The incremental 4D-Var works in this way. First, the background trajectory (full model high resolution trajectory whose initial condition is the background $\mathbf{x}_b$) is compared with the observations producing the innovation vectors d_i. Then, the incremental 4D-Var minimization problem is solved for a simplified problem. Denoting the simplification operator by S, we introduce the control variable of the simplified problem that is incremented at time t_0, $\delta\mathbf{w} = S\delta\mathbf{x}(t_0)$. Integrating the simplified dynamics L over time produces a simplified trajectory for the increments $\delta\mathbf{w}(t_i)$. G_i is the simplified operator $(G_i \approx H_i'S^*)$. The 4D-Var algorithm consists of minimizing the simplified problem:

$$J(\delta\mathbf{w}) = \tfrac{1}{2}\delta\mathbf{w}^T \mathbf{B}^{-1}\delta\mathbf{w} + \tfrac{1}{2}\sum_{i=0}^{n}[G_i\delta\mathbf{w}(t_i) - \mathbf{d}_i]^T \mathbf{O}_i^{-1}[G_i\delta\mathbf{w}(t_i) - \mathbf{d}_i]$$

$$(11.38)$$

where $\delta\mathbf{w}(t_i) = L(t_i, t_0)\delta\mathbf{w}$.

The analysis increments at time t_0, the result of the above minimization problem, are added to the background $\mathbf{x}^a$ to provide the analysis $\mathbf{x}^a$

$$\mathbf{x}^a = \mathbf{x}^b + S^* \delta \mathbf{w} \tag{11.39}$$

11.2.5 Physically Spaced Statistical Analysis Scheme

The physically spaced statistical analysis scheme (PSAS) described by Cohn et al. (1998) is another variational analysis scheme avoiding the inversion of $\mathbf{B}$. The PSAS solution is given as

$$\mathbf{x}^{\mathrm{PSAS}} = \mathbf{x}^b + \mathbf{B}\mathbf{H}^T \mathbf{w}^\infty \tag{11.40}$$

where $\mathbf{w}^\infty$ is the minimization result from the following cost function:

$$J(\mathbf{w}) = \tfrac{1}{2}\mathbf{w}^T(\mathbf{H}\mathbf{B}\mathbf{H}^T + \mathbf{R})\mathbf{w} - \mathbf{w}^T(\mathbf{y} - \mathbf{H}\mathbf{x}^b) \tag{11.41}$$

Although PSAS does not invert $(\mathbf{H}\mathbf{B}\mathbf{H}^T + \mathbf{R})$, this $M \times M$ matrix could still be a computational burden for meteorology data assimilation. In the implementation of Cohn et al. (1998), simplifications were used to make the matrix sparse.

It is still unclear whether PSAS is superior to the conventional variational formulations, 3D-Var and 4D-Var. More comments were made by Courtier (1997). Here are some pros and cons:

- PSAS is only equivalent to 3D/4D-Var if $\mathbf{H}$ is linear, which means that it cannot be extended to weakly nonlinear observation operators.
- Most implementations of 3D/4D-Var are incremental, which means that they do rely on a linearization of $\mathbf{H}$ anyway; they include nonlinearity through incremental updates, which can be used identically in an incremental version of PSAS.
- Background error models can be implemented directly in PSAS as the $\mathbf{B}$ operator. In 3D/4D-Var they need to be inverted (unless they are factorized and used as preconditioners).
- The size of the PSAS cost function is determined by the number of observations P instead of the dimension of the model space N. If $P \ll N$, then the PSAS minimization is done in a smaller space than 3D/4D-Var. In a 4D-Var context, P increases with the length of the minimization period whereas N is fixed, so this apparent advantage of PSAS may disappear.
- The conditioning of a PSAS cost function preconditioned by the square root of $\mathbf{O}$ is identical to that of 3D/4D-Var preconditioned by the square root of $\mathbf{B}$. However, the comparison may be altered if more sophisticated preconditionings are used, or if one square root or the other is easier to specify.

- Both 3D/4D-Var and PSAS can be generalized to include model errors. In 3D/4D-Var this means increasing the size of the control variable, which is not the case in PSAS, although the final cost of both algorithms is about the same.

11.2.6 Extended Kalman Filter (EKF)

For a perfect, linear model and linear observation operators, both 4D-Var and the Kalman filter give the same values for the model variables at the end of the 4D-Var assimilation window, provided both systems start with the same covariance matrices at the beginning of the window. The fundamental difference between the Kalman filter and 4D-Var is that the former explicitly evolves the covariance matrix, whereas the covariance evolution in 4D-Var is implicit. This means that when we come to perform another cycle of analysis, the Kalman filter provides us with both a model state (background) and its covariance matrix. 4D-Var does not provide the covariance matrix.

The Kalman filter is a sequential assimilation method. Assume that the state variable x is governed by an equation

$$\mathbf{x}(i+1) = \mathbf{M}_i[\mathbf{x}(i)] + \mathbf{\varepsilon}(i) \tag{11.42}$$

where $\mathbf{\varepsilon}$ is a noise process with zero mean and covariance $\mathbf{Q}$. The EKF consists of a forecast step and analysis step. For forecasting, we have

$$\mathbf{x}_f(i) = \mathbf{M}_i \mathbf{x}_a(i-1) \tag{11.43}$$

and the associated uncertainties can be computed through the following equation

$$\mathbf{P}_f(i) = \mathbf{M}_{i-1}\mathbf{P}_a(i-1)\mathbf{M}_{i-1}^T + \mathbf{Q}_{i-1} \tag{11.44}$$

where $\mathbf{P}_f(i)$ is the model error covariance matrix at time i, calculated from the covariance matrix $\mathbf{P}_a(i-1)$ at time i, the Jacobian matrix $\mathbf{M}_{i-1}$ of the dynamic model m, and the sequential error matrix $\mathbf{Q}_{i-1}$. $\mathbf{P}_a$ will be updated later in Eq. (11.47).

The analysis step in which the observation available at time i is blended with the previous information, carried forward by the forecast step:

$$\mathbf{x}_a(i) = \mathbf{x}_f(i) + \mathbf{K}_i[\mathbf{y}(i) - \mathbf{H}_i(\mathbf{x}_f(i))] \tag{11.45}$$

where $\mathbf{y}(i)$ are the observations

$$\mathbf{y}(i) = \mathbf{H}_i(\mathbf{x}(i)) + \text{noise} \tag{11.46}$$

and the term between brackets represents the innovation vector, and

$$\mathbf{P}_a(i) = (\mathbf{I} - \mathbf{K}_i\mathbf{H}_i)\mathbf{P}_f(i-1) \tag{11.47}$$

The equation used to compute the Kalman gain $\mathbf{K}$ whenever observations are available is

$$\mathbf{K}_i = \mathbf{P}_f(i)\mathbf{H}_i^T\left(\mathbf{H}_i\mathbf{P}_f(i)\mathbf{H}_i^T + \mathbf{R}_i\right)^{-1} \tag{11.48}$$

where $\mathbf{H}$ is the Jacobian of the observation model h and $\mathbf{R}$ is the error covariance of the observations.

If the system is linear, the algorithm provided above is the standard Kalman filter algorithm. If $\mathbf{H}$ and/or $\mathbf{M}$ are nonlinear, the algorithm written above is called the *extended Kalman filter* (EKF). The term *extended* in EKF means that partial derivatives are taken to form the model matrix $\mathbf{M}$ and the observation model $\mathbf{H}$.

There are many similarities between 4D-Var and the EKF, and it is important to understand the fundamental differences between them:

- 4D-Var can be run for assimilation in a realistic NWP framework because it is computationally much cheaper than the KF or EKF.
- 4D-Var is more optimal than the (linear or extended) KF inside the time interval for optimization because it uses all the observations at once, that is, it is not sequential and just smooths all data points.
- Unlike the EKF, 4D-Var relies on the hypothesis that the model is perfect (i.e., $\mathbf{Q} = 0$).
- 4D-Var can be run only for a finite time interval, especially if the dynamic model is nonlinear, whereas the EKF can be run forever in principle.
- 4D-Var itself does not provide an estimate of $\mathbf{P}_f$; a specific procedure to estimate the quality of the analysis must be applied, which costs as much as running the equivalent EKF.

The ensemble Kalman filter (Evensen 1994, Houtekamer and Mitchell 1998) takes a statistical approach to the solution of the Kalman filter equations for the covariance matrices of analysis and background error. The idea of the method is to generate a statistical sample of analyses. This is done by running the analysis system several times for a given date, each time using backgrounds that differ by an amount characteristic of background error, and observations that have been perturbed by adding random noise drawn from the distribution of observation error. (The backgrounds are produced by running short forecasts from each member of the preceding ensemble of analyses.)

11.3 MINIMIZATION ALGORITHMS

From Section 11.2, we can see that minimization is a critical part of the data assimilation implementation. The discussions below are extracted largely from the Website at http://www.gothamnights.com/trond/Thesis /hoved.html. The computer codes of these algorithms are widely available (e.g., Press et al. 1989), but it is worth introducing some basic principles and concepts of minimization algorithms.

Let us begin with a brief introduction of the minimization problem. Assume that the function for which we are trying to find the minima is referred to as $f(\mathbf{x})$, where $\mathbf{x}$ are the unknown variables. An unconstrained optimization procedure starts by choosing a starting point, that is, an initial guess for the values of the unknown parameters in $f(\mathbf{x})$, $\mathbf{x}_0$. Once the initial point is chosen, we must make two decisions before the next point can be generated:

1. Determine a direction along which the next point is to be chosen.
2. Decide on a step size to be taken in that chosen direction.

We then have the following iterative picture:

$$\mathbf{x}_{k+1} = \mathbf{x}_k + \alpha_k \mathbf{d}_k, \qquad (k = 0, 1, \dots) \qquad (11.49)$$

where $\mathbf{d}_k$ is the direction and $|\alpha_k d_k|$ is the step size. The different optimization methods differ in the choice of $\mathbf{d}_k$ and α_k.

Roughly speaking, all methods can be grouped into two categories: (1) methods using the function only and (2) methods using the derivitive. In general, the algorithms using the derivitive are more powerful, and we should try to use them if we can calculate derivitives. In the multidimensional case, the derivitive is the gradient. Since the methods using the function only are briefly discussed in Section 8.2, we will discuss a few gradient methods.

For illustrative purposes, a quadratic function will be used in the presentation of these optimizing methods, due in some part to its simplicity, but mainly because several of these methods where originally designed to solve problems equivalent to minimizing a quadratic function. The n-dimensional quadratic function used here is of the form

$$f(\mathbf{x}) = \tfrac{1}{2}\mathbf{x}^T \cdot \mathbf{Q} \cdot \mathbf{x} + \mathbf{b} \cdot \mathbf{x} + c \qquad (11.50)$$

where $\mathbf{Q}$ is a positive definite matrix, $\mathbf{x}$ and $\mathbf{b}$ are vectors, and c is a scalar constant.

11.3.1 Steepest-Descent Method

The method of *steepest descent* is the simplest of the gradient methods. The choice of direction is where f decreases most quickly, which is in the direction opposite to $\nabla f(\mathbf{x}_i)$. The search starts at an initial point $\mathbf{x}_0$ and then slides down the gradient, until we are close enough to the solution. In other words, the iterative procedure is

$$\mathbf{x}_{k+1} = \mathbf{x}_k - \alpha_k \nabla f(\mathbf{x}_k) = \mathbf{x}_k - \alpha_k \mathbf{g}(\mathbf{x}_k) \qquad (11.51)$$

where $\mathbf{g}(\mathbf{x}_k)$ is the gradient at one given point. Now, the question is how large the step taken in that direction should be, that is, what the value of α_k is. What we have here is actually a minimization problem along a line, where the line is given by (11.51) for different values of α_k. This is usually solved by doing a *linear search* (also referred to as a *line search*): searching for a minimum point along a line. Some of these methods are discussed by Press et al. (1989). The next step is then taken in the direction of the negative gradient at this new point [i.e., $-\nabla f(\mathbf{x}_k)$]. This iteration continues until the global minimum has been determined within a chosen accuracy ε. This implementation of the steepest-descent method is often referred to as the *optimal gradient method*.

As seen, the method of *steepest descent* is simple and easy to apply, and each iteration is fast. It is also very stable. If a minimum point exists, the method is guaranteed to locate it after an infinite number of iterations. But even with all these positive characteristics, the method has one very important drawback—it generally has slow convergence. The steepest-descent method can be used where one has an indication of where the minimum is, but is generally considered to be a poor choice for any optimization problem. It is used mostly only in conjunction with other optimizing methods.

11.3.2 Conjugate Gradient Methods

The conjugate gradients method is a special case of the method of conjugate directions, where the conjugate set is generated by the gradient vectors. It is also called the *Fletcher–Reeves algorithm*. This seems to be a sensible choice since the gradient vectors have proved their applicability in the steepest-descent method. For a quadratic function, as given in Eq. (11.50), the procedure is as follows.

The initial step is in the direction of the steepest descent:

$$\mathbf{d}_0 = -\mathbf{g}(\mathbf{x}_0) = -\mathbf{g}_0 \qquad (11.52)$$

Subsequently, the mutually conjugate directions are chosen so that

$$\mathbf{d}_{k+1} = -\mathbf{g}_{k+1} + \beta_k \mathbf{d}_k \qquad (11.53)$$

where the coefficient β_k is given by, for example, the Fletcher–Reeves formula:

$$\beta_k = \frac{\mathbf{g}_{k+1}^T \cdot \mathbf{g}_{k+1}}{\mathbf{g}_k^T \cdot \mathbf{g}_k} \qquad (11.54)$$

The step length along each direction is given by

$$\alpha_k = \frac{\mathbf{d}_k^T \cdot \mathbf{g}_k}{\mathbf{d}_k^T \cdot (\mathbf{Q} \cdot \mathbf{d}_k)} \tag{11.55}$$

The iteration formula is the same as (11.51).

Even though the conjugate gradient method is designed to find the minimum point for simple quadratic functions, it also does the job well for any continuous function $f(\mathbf{x})$ for which the gradient $\nabla f(\mathbf{x})$ can be computed. Equation (11.54) cannot be used for nonquadratic functions, so the step length has to be determined by linear searches. An alternative to the Fletcher–Reeves formula is the Polak–Ribière formula:

$$\beta_k = \frac{\mathbf{g}_{k+1}^T \cdot (\mathbf{g}_{k+1} - \mathbf{g}_k)}{\mathbf{g}_k^T \cdot \mathbf{g}_k} \tag{11.56}$$

There is not much difference in the performance of these two formulas, but the Polak–Ribière formula is known to perform better for nonquadratic functions.

The conjugate gradient method is not only an optimization method but also one of the most prominent iterative methods for solving sparse systems of linear equations. It is fast and uses small amounts of storage since it needs only the calculation and storage of the first derivative at each iteration. The latter becomes significant when the number of variables n is so large that problems of computer storage arise. However, in the worst case this method may not converge at all. Nevertheless, it is generally to be preferred over the steepest-descent method.

11.3.3 Newton–Raphson Method

The Newton–Raphson method differs from the steepest-descent and conjugate gradient methods in that the information of the second derivative is used to locate the minimum of the function $f(\mathbf{x})$. This results in faster convergence, but not necessarily less computing time. The computation of the second derivate and the handling of its matrix can be very time-consuming, especially for large systems.

The idea behind the *Newton–Raphson method* is to approximate the given function $f(\mathbf{x})$ in each iteration by a quadratic function, as given in Eq. (11.50), and then move to the minimum of this quadratic. The quadratic function for a point $\mathbf{x}$ in a suitable neighborhood of the current point $\mathbf{x}_k$ is given by a truncated Taylor series:

$$f(\mathbf{x}) \approx f(\mathbf{x}_k) + (\mathbf{x} - \mathbf{x}_k)^T \cdot \mathbf{g}_k + \tfrac{1}{2}(\mathbf{x} - \mathbf{x}_k)^T \cdot \mathbf{H}_k \cdot (\mathbf{x} - \mathbf{x}_k) \tag{11.57}$$

where both the gradient $\mathbf{g}_k$ and the Hessian matrix $\mathbf{H}_k$ are evaluated at $\mathbf{x}_k$. The derivative of (11.57) is

$$\nabla f(\mathbf{x}) = \mathbf{g}_k + \tfrac{1}{2}\mathbf{H}_k(\mathbf{x} - \mathbf{x}_k) + \tfrac{1}{2}\mathbf{H}_k^T \cdot (\mathbf{x} - \mathbf{x}_k) \qquad (11.58)$$

The Hessian matrix is always symmetric (i.e., $\mathbf{H}_k = \mathbf{H}_k^T$) if the function $\mathbf{x}_k$ is twice as continuously differentiable at every point, which is the case here. Hence, Eq. (11.58) reduces to

$$\nabla f(\mathbf{x}) = \mathbf{g}_k + \mathbf{H}_k(\mathbf{x} - \mathbf{x}_k) \qquad (11.59)$$

If we assume that $f(\mathbf{x})$ takes its minimum at $\mathbf{x} = \mathbf{x}^*$, the gradient is zero

$$\mathbf{g}_k + \mathbf{H}_k(\mathbf{x} - \mathbf{x}_k) = 0 \qquad (11.60)$$

which is nothing more than a linear system. The Newton–Raphson method uses the $\mathbf{x}^*$ as the next current point, resulting in the iterative formula

$$\mathbf{x}_{k+1} = \mathbf{x}_k - \mathbf{H}_k^{-1} \cdot \mathbf{g}_k \qquad (k = 0, 1, \ldots) \qquad (11.61)$$

where $-\mathbf{H}_k^{-1} \cdot \mathbf{g}_k$ is often referred to as the *Newton direction*. If the approximation in (11.57) is valid, the method will converge in just a few iterations. For example, if the function to be optimized, $f(\mathbf{x})$, is a n-dimensional quadratic function, it will converge in only one step from any starting point.

Even though the convergence may seem to be fast, judging by the number of iterations, each iteration does include the calculation of the second derivative and handling of the Hessian. The size of the Hessian can be crucial to the effectiveness of the Newton–Raphson method. For systems with a large number of dimensions, where the function $f(\mathbf{x})$ has a large number of variables, both the computation of the matrix and calculations that includes it will be very time-consuming. This can be mended by either just using the diagonal terms in the Hessian, ignoring the cross-terms, or simply not recalculating the Hessian at each iteration (this can be done due to slow variation of the second derivative). Another serious disadvantage of the Newton–Raphson method is that it is not necessarily globally convergent, meaning that it may not converge from any starting point. If it does converge, it is not unusual for it to expend significant computational efforts in getting close enough to the solution where the approximation in Eq. (11.57) becomes valid. A fix to this problem is to adjust the step size in the Newton direction, assuring that the function value decreases. Equation (11.61) is then as follows:

$$\mathbf{x}_{k+1} = \mathbf{x}_k - \alpha_k \mathbf{H}_k^{-1} \cdot \mathbf{g}_k \qquad (k = 0, 1, \ldots) \qquad (11.62)$$

Despite these problems, the Newton–Raphson method enjoys a certain amount of popularity because of its fast convergence in a sufficiently small neighborhood of the stationary value. The convergence is quadratic, which, loosely speaking, means that the number of significant digits doubles after each iteration. In comparison, the convergence of steepest-descent method is at best linear and for the conjugate gradient method it is superlinear.

As shown previously, the method of steepest descent starts out having a rather rapid convergence, while the Newton–Raphson method has just the opposite quality; starts out slowly and ends with a very rapid convergence. It is too obvious to ignore the fact that these two methods can also be combined, resulting in a very efficient method. One starts out with the steepest-descent method, and switches to the Newton–Raphson method when the progress by the former gets slow, and enjoys the quadratic convergence of the latter. This is the subject of the next section.

11.3.4 Quasi-Newton Methods

Comparing the iterative formula of steepest descent [Eq. (11.51)] with (11.62) shows that the steepest-decent method is identical to the Newton–Raphson method when $\mathbf{H}_k^{-1} = \mathbf{I}$ where $\mathbf{I}$ is the unit matrix. This is the basis for the *secant method*, also known as the *variable-metric method*.

As seen in the previous subsection, the Hessian matrix needed in the Newton–Raphson method can be both laborious to calculate and invert for systems with a large number of dimensions. The idea of the secant method is not to use the Hessian matrix directly, but rather start the procedure with an approximation to the matrix, which, as one gets closer to the solution, gradually approaches the Hessian.

The earliest Secant methods were formulated in terms of constructing a sequence of matrices B_k that builds up the *inverse* Hessian:

$$\lim_{k \to \infty} \mathbf{B}_k = \mathbf{H}^{-1} \tag{11.63}$$

It is even better if the limit is achieved after n iterations instead of infinitely many iterations. The iterative formula is then

$$\mathbf{x}_{k+1} = \mathbf{x}_k - \alpha_k \mathbf{B}_k \cdot \mathbf{g}_k \qquad (k = 0, 1, \dots) \tag{11.64}$$

where $\mathbf{g}_k$ is the gradient. The inverse Hessian approximation might seem to offer an advantage in terms of the number of arithmetic operation required to perform an iteration, since one needs to inverse the matrix for the Hessian update.

For an n-dimensional quadratic function, the inverse Hessian will be achieved after at most n iterations, and the solution will then be located in only one iteration. Hence, the solution will be arrived at after at most n iterations. In other words, this method resembles the method of steepest descent near the starting point of the procedure, while near the optimal point, it morphs into the Newton–Raphson method.

The secant method is quite similar to the conjugate gradients, in some ways. They both converge after n iterations for an n-dimensional quadratic function, they both accumulate information from successive iterations, and they both require the calculation of only the first derivative. The main difference between the two methods is the way they store and update the information; the secant method requires an $n \times n$ matrix, while the conjugate gradient method only needs an n-dimensional vector. Usually, the storage of the matrix in the secant method is not disadvantageous, at least not for a moderate number of variables. So how are the $\mathbf{B}_k$ being updated at each iteration? There are two main schemes: Davidon–Fletcher–Powell (DFP) and Broyden–Fletcher–Goldfarb–Shanno (BFGS). Let us first consider the DFP algorithm. At a point $\mathbf{x}_k$, the approximated inverse Hessian at the subsequent point is given by Press as

$$\mathbf{B}_{k+1} = \mathbf{B}_k + \frac{\mathbf{p}_k \times \mathbf{p}_k}{\mathbf{y}_k \cdot \mathbf{p}_k} - \frac{(\mathbf{B}_k \cdot \mathbf{y}_k) \times (\mathbf{B}_k \cdot \mathbf{y}_k)}{\mathbf{y}_k (\mathbf{B}_k \cdot \mathbf{y}_k)} \tag{11.65}$$

where $\mathbf{p}_k = \mathbf{x}_{k+1} - \mathbf{x}_k$ and $\mathbf{y}_k = \mathbf{g}_{k+1} - \mathbf{g}_k$, where $\mathbf{g}$ is the gradient. Even though the deduction of Eq. (11.65) is in terms of a quadratic function, it basically applies for all kinds of equations and systems. Another property of (11.65) is that if $\mathbf{B}_0$ is positive definite, it will remain so for each k; thus $\mathbf{B}_k$ is a sequence of positive definite matrices.

The BFGS algorithm was later introduced to improve some unbecoming features of the DFP formula, concerning, for example, rounding error and convergence tolerance. The BFGS updating formula is exactly the same, apart from one additional term

$$\cdots + \left| \mathbf{y}_k \cdot (\mathbf{B}_k \cdot \mathbf{y}_k) \right| \mathbf{u} \times \mathbf{u} \tag{11.66}$$

where $\mathbf{u}$ is defined by

$$\mathbf{u} \equiv \frac{\mathbf{p}_k}{\mathbf{y}_k \cdot \mathbf{p}_k} - \frac{\mathbf{B}_k \cdot \mathbf{y}_k}{\mathbf{y}_k \cdot (\mathbf{B}_k \cdot \mathbf{y}_k)} \tag{11.67}$$

The BFGS formula is usually to be preferred.

The step length α_k in Eq. (11.64) is determined in the same way as for the Newton–Raphson method; that is, either by linear search, backtracking, or by deciding on $0 < \alpha \le 1$ and gradually changing it to unity as the solution is being approached.

The secant method has become very popular for optimization; it converges fast (at best superlinear), is stable, and spends relatively modest computing time on each iteration. The method is often preferred over the conjugate gradient method, but the latter will have an advantage over the former for systems with large number of dimensions (e.g., 10,000). Since both the DFP and the BFGS updating formulas are guaranteed to uphold the positive definiteness of the initial approximation of the inverse Hessian (assuming that there are no roundoff errors that results in loss of positive definiteness), the secant method will only be able to locate minimum points.

11.4 DATA ASSIMILATION IN HYDROLOGY

Data assimilation concepts have been applied to a number of problems relating to the estimation of land–atmosphere fluxes and related state variables (e.g., soil moisture). We will first outline some individual studies reported in the literature, and then introduce two major activities in the United States and Europe.

An earlier review of this subject is provided by McLaughlin (1995). Entekhabi et al. (1994) used an extended Kalman filter to estimate time-dependent near-surface soil moisture and temperature profiles from synthetic measurements of microwave and infrared radiation. Houser et al. (1998) applied some of the earlier data assimilation techniques to hydrological issues, such as direct insertion as described in Section 11.2.2 (where model states predicted by the forward hydrological model are replaced by observations when available) and nudging techniques (where predicted model states are relaxed toward observations). Boni et al. (2001) used a variational land data assimilation technique to provide estimates of components of the surface energy balance and land surface control on evaporation from satellite remotely sensed surface temperature.

Considerable research effort has been spent in deriving land surface characteristics from satellite data, as presented in the previous chapters. An important aspect of the land surface is its partitioning of available radiative energy over sensible and latent heat, which is determined by the amount of vegetation and available soil moisture. The relation between evaporation and soil moisture content can be used to derive regional or global soil moisture estimates from satellite data. Since these satellite data alone do not contain all the necessary information for calculation of surface flux budgets, merging these data with operational meteorological model predictions and data assimilation possibly provides an important step forward in realistic surface flux and soil moisture estimations.

Many authors (e.g., Ottle and Vidal-Madjar 1994, Houser et al. 1998) have used remotely sensed soil moisture data for assimilation in hydrological and soil physical models. Walker et al. (2000) described an assimilation approach in which a physical model describing soil moisture movement is linked with a data assimilation technique that uses (irregularly spaced) remotely sensed near-surface soil measurements. Calvet et al. (1998) proposed estimating bulk soil water content from surface soil moisture or temperature estimates. These procedures illustrate how data assimilation can extend remotely sensed measurements of surface conditions into the underlying soil column.

Van den Hurk et al. (1997) described how evaporation maps derived from satellite data may be used to assimilate (initial) soil moisture fields for use in an operational NWP model. Jones et al. (1998a) used GOES-derived surface temperature changes in time to update the soil moisture condition in a regional atmospheric model. They adjusted the soil moisture content in the model, forcing the model to exhibit a surface heating rate comparable to that from the satellite observations. Jones et al. (1998b) described a case study showing that the method was able to pick up a soil moisture distribution that is compatible with a record of antecedent rainfall in the Great Plains area of the United States. A technique has recently been described by Suggs et al. (1999) for assimilating GOES-IR skin temperature tendencies into the surface budget equation of the mesoscale model (MM5) developed by Penn State University and the National Center for Atmospheric Research. The assimilation of GOES data aims at minimizing the difference between the simulated rate of temperature change and GOES satellite observations of skin temperature through adjusting the soil moisture availability and hence the latent heat flux. Van den Hurk (2001) gives an overview of some recent attempts in using remotely sensed thermal infrared data for adjusting soil moisture evolution in an NWP model environment. The results imply that satellite-based information on the spatial variability of land surface temperature and shortwave reflectance may be used in partitioning the available net radiant energy into latent-heat and sensible-heat fluxes from the land surface.

A few investigators have also used data assimilation techniques to characterize horizontal spatial variability at the land–atmosphere interface. For example, Bouttier et al. (1993a, 1993b) used a three-dimensional mesoscale soil moisture model to assimilate data from the HAPEX-MOBILHY field experiment in southern France. Reichle et al. (2001) applied a data assimilation technique to estimate the soil moisture profiles at the resolution of a few kilometers from brightness imagery with a resolution of tens of kilometers, provided micrometeorological, soil texture, and land cover inputs are available. Surface heterogeneity has also been addressed by Shuttleworth (1998) by assimilating remote sensing into the land surface models.

Besides these individual activities, there are two major ongoing research projects in the United States and Europe. In the United States, a joint collaboration composed of dozens of research institutes, federal agencies, and universities is undertaking the development of the north American land data assimilation system (NLDAS) and the global land data assimilation system (GLDAS) aiming at more accurate reanalysis and forecast simulations by numerical weather prediction (NWP) models. (http://ldas.gsfc. nasa.gov/index.shtml). These systems will reduce the errors in the stores of soil moisture and energy that are often present in NWP models and that degrade the accuracy of forecasts. NLDAS is currently running retrospectively and in near real time on a $\frac{1}{8}°$ grid while GLDAS is running at $\frac{1}{4}°$ resolution. The systems are currently forced by terrestrial (NLDAS) and space based (GLDAS) precipitation data, space-based radiation data and numerical model output. In order to create an optimal scheme, the projects involve several land surface models (LSMs), many sources of data, and several institutions.

The European community is also working on a project called "Development of a European land data assimilation system to predict floods and droughts (ELDAS) (http://www.knmi.nl/samenw/eldas/). ELDAS is designed to develop a general data assimilation infrastructure for estimating soil moisture fields on the regional (continental) scale, and to assess the added value of these fields for the prediction of the land surface hydrology in models used for NWP and climate studies. The goals of this project are briefly summarized below:

- To combine current (European) expertise in soil moisture data assimilation, and design and implement a common flexible and practical data assimilation infrastructure at a number of European NWP centers
- To validate the assimilated soil moisture fields using independent observations
- To assess the added value of soil moisture data assimilation for prediction of the seasonal hydrological cycle over land (associated with drought prediction) and for the risk of flooding.
- To build a demonstration database covering at least one seasonal cycle in the European continent.
- To anticipate on the use of data expected from new satellite platforms, in particular METEOSAT Second Generation (MSG) and the ESA Soil Moisture/Ocean Salinity (SMOS) mission.
- To provide a European contribution to the global land data assimilation system (GLDAS), a U.S. initiative for generating near-realtime information on land surface characteristics on a global scale.

11.5   DATA ASSIMILATION WITH CROP GROWTH MODELS

11.5.1   Background

Precision farming relies on detailed information about the spatial and temporal variability within fields to modulate cultural practices (Moran et al. 1997). Remote sensing is the only means to monitor this variability at various scales. However, only the primary biophysical variables of the canopy or soil can be derived directly from remotely sensed data (Baret et al. 2000), and cannot be used directly for crop management decision making in many cases. It has been well recognized (e.g., Moran et al. 1997, Brisco et al., 1998, Stafford 2000) that the current satellite sensors have limited direct applications because of the few spectral bands, coarse spatial resolution, and inadequate repeat coverage. Although the current and upcoming commercial Earth observation satellites will overcome some of these limitations, it is still almost impossible to acquire frequent satellite imagery over every farm in the United States in the near future because of numerous factors (e.g., the overall costs, and the weather conditions).

Remotely sensed variables have to be integrated with ancillary data, such as soil, weather, and the past management practices, to provide higher level information that is pertinent to making all strategic, tactic, and operational decisions at a field level through a crop decision support system (DSS).

The core of any crop DSS is its crop growth models. As Jones and Luyten (1998) summarized, crop growth models have been developed for a wide range of crops and for a variety of applications, such as irrigation management, nutrient management, pest management, land use planning, crop rotation, climate change assessment, and yield forecasting. The simulation-assisted decisions using crop growth models result in greater economic returns as a consequence of better information about the crop. However, most of these applications have been in the research mode. One major reason is a limited knowledge on model inputs. It is very time-consuming and extremely expensive to measure the model inputs in the field. For the best site-specific applications, crop growth models must be calibrated with realistic inputs. The most obvious and important role that remote sensing has is to improve the capacity and accuracy of DSS and crop growth models by providing accurate input information or as a means of model calibration or validation.

Various methods for integrating a crop growth model with remote sensing data were initially described by Maas (1988a, 1988b), then revisited by Delecolle et al. (1992), and also reviewed by Fischer et al. (1997) and Moulin et al. (1998). In the following discussions, we group these methods into two groups: forcing and assimilation.

The forcing approach is to drive the crop model using remote sensing products as the values of some variables. The crop model runs at a constant time step (e.g., daily), but it is very difficult to acquire remote sensing imagery at such high-resolution temporal steps. The common practice is to fit an empirical curve to the estimated values from remote sensing observations and then interpolate according to the model timestep. The concept is depicted in Fig. 11.5.

The assimilation approach minimizes the difference between values from remote sensing and the predicted ones from crop models by adjusting model parameters or the initial conditions. We can assimilate either the estimated variables from remote sensing data, or the observed spectral reflectance/radiance at the top of atmosphere directly. If the direct assimilation of the observed radiance/reflectance is used, the radiative transfer models of the atmosphere/crop/soil are needed to couple with the crop growth models (see Fig. 11.6). This has become a recent trend in the European community (Guerif and Duke. 1998, 2000; Baret et al. 2000). The combination of these two methods together by minimize the differences between both radiance/reflectance and retrieved variables (e.g., LAI) might provide us with the best solution to practical problems.

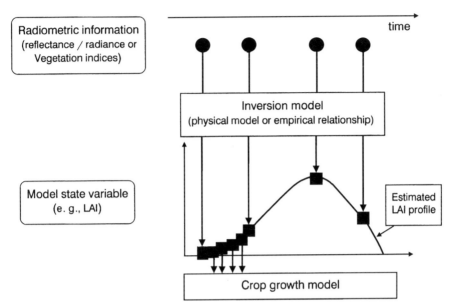

Figure 11.5 Forcing technique that generates the parameter fields to drive the crop growth models. [From Moulin et al. (1998), *Int. J. Remote Sens.* Reproduced by permission of Taylor & Francis, Ltd.]

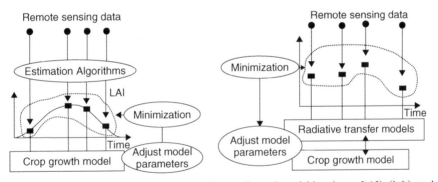

Figure 11.6 Data assimilation schemes using estimated variables (e.g., LAI) (left) and reflectance/radiance (right).

11.5.2 Case Studies

Maas (1988a, 1993) used ground radiometric measurements over maize crops to constrain the simulations of the crop growth model (GRAMI). LAI values were derived from optical measurements and stress index values, from thermal measurements. The initial LAI value at crop emergence and three shape coefficients of the LAI seasonal curve were estimated. The correct retrieval of these parameters is necessary to obtain a consistent estimation of crop production.

Carbone et al. (1996) applied remote sensing and geographic information system (GIS) technologies to drive SOYGRO crop model for examine the spatial variability in county soybean yield. This is a simple example of the forcing method. Remote sensing data were used to classify land cover and to identify agricultural regions with the county, while the GIS allows the spatial organization of soil and weather data input to the crop model.

Clevers (1997) applied the data assimilation scheme to estimate a series of parameters in a crop growth model for predicting the sugarbeet yield. FPAR is first estimated from remotely sensed data using a vegetation index: WDVI (see Section 8.1.1.4). An optimum inversion procedure was used to estimate four parameters in a simple crop growth model (SUCROS): sowing date, relative growth rate, light use efficiency, and maximum leaf area. This study found that FPAR near the end of June or the beginning of July and the final beet yield have a highly linear relationship.

Guerif and Duke (1998) combined the SUCROS crop model with the SAIL canopy reflectance model for accurate estimation of sugarbeet yield. Ground-measured reflectance spectra were used to match the predicted reflectance from the combined SUCROS + SAIL model by adjusting the parameters characterizing the crop emergence and early growth, such as the final number of plants that emerged, the temperature sum needed for 80% of the final number of plants to emerge, and LAI value at emergence.

In a follow-up study, Guerif and Duke (2000) found that this procedure is very sensitive to the canopy radiative transfer models, such as soil reflectance and leaf optical properties. Instead of using the original spectral reflectance, they found that using the vegetation index TSAVI (see Section 8.1.1.3) is a better choice since this index provides more consistent results for the estimates of the sowing date and emergence parameters, which remained poorer than yield estimates. Over a wide range of unknown crop situations corresponding to extreme sowing dates and emergence conditions, the proposed method allowed to estimate the sugarbeet yield with relative errors varying from 0.6 to 2.6%.

More studies have been reported in the literature that discuss how to couple crop growth models with canopy radiative transfer models, although more experiments on data assimilation are needed. Moulin et al. (1995) simulated the temporal profiles of satellite data for an agricultural area in the field (20 m, SPOT HRV) and regional (4 km, NOAA AVHRR) scales, using a wheat crop growth model (AFRCWHEAT2) and a canopy radiative transfer model (SAIL). Weiss et al. (2001) coupled a crop growth model with the SAIL canopy radiative transfer model through several state variables: leaf area index, leaf chlorophyll content, organ dry matter, and relative water content. The relationships between the crop model outputs (agronomic variables) and canopy radiative transfer model inputs (biophysical variables) were defined using experimental datasets corresponding to wheat crops under different climatic and stress conditions. The coupling scheme is then tested on the data set provided by the Alpilles–ReSeDA (remote sensing data assimilation) campaign. Results show a good fitting between the simulated reflectance data at top of canopy and the measured ones provided by SPOT images corrected from atmospheric and geometric effects, with a root mean square error lower than 0.05 for all the wavebands.

11.6 SUMMARY

In previous chapters, we discussed various models and algorithms that can be used to estimate various land surface variables. The limitations of these estimation methods are twofold: (1) only those variables that directly affect the radiometric signals detected and recorded by remote sensors can be effectively estimated—many important variables, such as deep-soil properties or agricultural management practice from application perspectives cannot be well estimated from remote sensing; and (2) much ancillary information cannot be effectively incorporated into these estimation algorithms. Four-dimensional data assimilation is an advanced estimation technique that can overcome these limitations. However, studies on land surface data assimilation are still relatively new. The purpose of including this chapter in the book is to encourage more researchers to work on this area. It represents an important research direction in quantitative remote sensing.

Several major assimilation algorithms have been discussed in Section 11.2. Those methods are quite representative, but I did not attempt to introduce all algorithms extensively. Interested readers should consult with the related books or review articles (e.g., Daley 1991).

One major component in data assimilation is the use of minimization algorithms. Most minimization algorithms can be roughly grouped into two categories according to whether they use the derivative. Section 11.3 briefly introduced some representative derivative methods that are used mainly in data assimilation. Nonderivative algorithms are briefly outlined in Section 8.2.

Sections 11.4 and 11.5 introduced many different examples of applying data assimilation to hydrology and crop growth model assessment. Assimilating remote sensing data into other land-surface-related models has been widely reported in the literature (e.g., Olioso et al. 1999, Cayrol et al. 2000, Nouvellon et al. 2001).

REFERENCES

Baret, F., Weiss, M., Troufleau, D., Prevot, L., and Combal, B. (2000), Maximum information exploitation for canopy characterization by remote sensing, *Aspects Appl. Biol.* **60**.

Bergthorsson, P. and Doos, B. (1955), Numerical weather map analysis, *Tellus* **7**: 329–340.

Boni, G., Entekhabi, D., and Castelli, F. (2001), Land data assimilation with satellite measurements for the estimation of surface energy balance components and surface control on evaporation, *Water Resour. Res.* **37**: 1713–1722.

Bouttier, F., Mahfouf, J.-F., and Noilhan, J. (1993a), Sequential assimilation of soil moisture from atmospheric low-level parameters, part I: Sensitivity and calibration studies, *J. Appl. Meteorol.* **32**: 1335–1351.

Bouttier, F., Mahfouf, J.-F., and Noilhan, J. (1993b), Sequential assimilation of soil moisture from atmospheric low-level parameters, part II: Implementation in a mesoscale model, *J. Appl. Meteorol.* **32**: 1352–1364.

Brisco, B., Brown, R. J., Hirose, T., McNairn, H., and Staenz, K. (1998), Precision agriculture and the role of remote sensing: A review, *Can. J. Remote Sens.* **24**: 315–327.

Calvet, J.-C., Noilhan, J., and Bessemoulin, P. (1998), Retrieving the root-zone soil moisture from surface soil moisture or temperature estimates: A feasibility study based on field measurements, *J Appl. Meteorol.* **37**: 371–386.

Carbone, G. J., Narumalani, S., and King, M. (1996), Application of remote sensing and gis technologies with physiological crop models, *Photogramm. Eng. Remote Sens.* **62**: 171–179.

Cayrol, P., Dedieu, G., Mordelet, P., Nouvellon, Y., Chehbouni, A., and Kergoat, L. (2000), Grassland modeling and monitoring with SPOT-4 vegetation instrument during the 1997–1999 salsa experiment, *Agric. Forest Meteorol.* **105**: 91–115.

Charney, J., Halem, M., and Jastrow, R. (1969), Use of incomplete historical data to infer the present state of the atmosphere, *J. Atmos. Sci.* **26**: 1160–1163.

Clevers, J. G. P. W. (1997), A simplified approach for yield prediction of sugar beet on optical remote sensing data, *Remote Sens. Envir.* **61**: 221–228.

Cohn, S. E., Da Silva, A., Guo, J., Sienkiewicz, M., and Lamich, D. (1998), Assessing the effects of data selection with the DAO physical-space statistical analysis system, *Monthly Weather Rev.* **126**: 2913–2926.

Courtier, P. (1997), Dual formulation of four-dimensional variational assimilation, *Quart. J. Royal Meteool. Soc.* **123**: 2449–2461.

Courtier, P., Derber, J., Errico, R., Louis, J.-F., and Vukicevic, T. (1993), Important literature on the use of adjoint, variational methods and the kalman filter in meteorology, *Tellus* **45A**: 342–357.

Daley, R. (1991), *Atmospheric Data Analysis*. Cambridge University Press, 457 pp.

Delecolle, R., Maas, S. J., Guerif, M., and Baret, F. (1992), Remote sensing and crop production models: Present trends, *ISPRS J. Photogramm. Remote Sens.* **47**: 145–161.

Entekhabi, D., Nakamura, H., and Njoku, E. G. (1994), Solving the inverse problem for soil moisture and temperature profiles by sequential assimilation of multifrequency remotely sensed observations, *IEEE Trans. Geosci. Remote Sens.* **32**: 438–448.

Evensen, G. (1994). Sequential data assimilation with a nonlinear quasi-geostropic model using Monte Carlo methods to forecast error statistics, *J. Geophys. Res.* **99**: 10143–10162.

Fischer, A., Kergoat, L., and Dedieu, G. (1997), Coupling satellite data with vegetation functional models: Review of different approaches and perspectives suggested by the assimilation strategy, *Remote Sens. Rev.* **15**: 283–303.

Guerif, M. and Duke, C. L. (1998), Calibration of the sucros emergence and early growth module for sugar beet using optical remote sensing data assimilation, *Eur. J. Agron.* **9**: 127–136.

Guerif, M. and Duke, C. L. (2000), Adjustment procedures of a crop model to the site specific characteristics of soil and crop using remote sensing data assimilation, *Agric. Ecosyst. Envir.* **81**: 57–69.

Houser, P. R., Shuttleworth, W. J., Famiglietti, J. S., Gupta, H. V., Syed, K. H., and Goodrich, D. C. (1998), Integration of soil moisture remote sensing and hydrologic modeling using data assimilation, *Water Resour. Res.* **34**: 3405–3420.

Houtekamer, P. L. and Mitchell, H. L. (1998), Data assimilation using an ensemble kalman filter technique, *Monthly Weather Rev.* **126**: 796–811.

Huang, X. (2000), Variational analysis using spatial filters, *Monthly Weather Rev.* **128**: 2588–2600.

Jones, A. S., Guch, I. C., and VonderHaar, T. H. (1998a), Data assimilation of satellite derived heating rates as proxy surface wetness data into a regional atmospheric mesoscale model. Part I: Methodology, *Monthly Weather Rev.* **126**: 634–645.

Jones, A. S., Guch, I. C., and VonderHaar, T. H. (1998b), Data assimilation of satellite derived heating rates as proxy surface wetness data into a regional atmospheric mesoscale model. Part II: Case study, *Monthly Weather Rev.* **126**: 646–667.

Jones, J. W. and Luyten, J. C. (1998), Simulation of biological processes, in *Agricultural Systems, Modeling and Simulation*, R. M. Peart and R. B. Curry, eds., Marcel Dekker, pp. 19–62.

Maas, S. J. (1988a), Use of remotely-sensed information in agricultural crop growth models, *Ecol. Model.* **41**: 247–268.

Maas, S. J. (1988b), Use satellite data to improve model estimates of crop yield, *Agron. J.* **80**: 655–662.

Maas, S. J. (1993), Parameterized model of gramineous crop growth. II: Within-season simulation calibration, *Agron. J.* **85**: 354–358.

McLaughlin, D. B. (1995), Recent developments in hydrologic data assimilation, *U.S. Natl. Rep. Int. Union Geod. Geophys., Rev. Geophys. 1991–1994* **33**: 995–1003.

Moran, M. S., Inoue, Y., and Barnes, E. (1997), Opportunities and limitations for image-based remote sensing in precision crop management, *Remote Sens. Environ.* **61**: 319–346.

Moulin, S., Bondeau, A., and Delecolle, R. (1998), Combining agricultural crop models and satellite observations: From field to regional scales, *Int. J. Remote Sens.* **19**: 1021–1036.

Moulin, S., Fischer, A., Dedieu, G., and Delecolle, R. (1995), Temporal variations in satellite reflectances at field and regional scales compared with values simulated by linking crop growth and sail models, *Remote Sens. Envir.* **54**: 261–272.

Nouvellon, Y., Moran, M. S., Lo Seen, D., Bryant, R., Rambal, S., Ni, W. M., Begue, A., Chehbouni, A., Emmerich, W. E., Heilman, P., and Qi, J. G. (2001), Coupling a grassland ecosystem model with landsat imagery for a 10-year simulation of carbon and water budgets, *Remote Sens. Envir.* **78**: 131–149.

Olioso, A., Chauki, H., Courault, D., and Wigneron, J. P. (1999), Estimation of evapotranspiration and photosynthesis by assimilation of remote sensing data into SVAT models, *Remote Sens. Envir.* **68**: 341–356.

Ottle, C. and Vidal-Madjar, D. (1994), Assimilation of soil moisture inferred from infrared remote sensing in a hydrological model over the Hapex-mobilhy region, *J. Appl. Meteorol.* **38**: 1352–1369.

Press, W. H., Flannery, B. P., Teukolsky, S. A., and Vetterling, W. T. (1989), *Numerical Recipes. The Art of Scientific Computing* (*Fortran Version*), Cambridge Univ. Press.

Reichle, R. H., Entekhabi, D., and McLaughlin, D. B. (2001), Downscaling of radio brightness measurements for soil moisture estimation: A four-dimensional variational data assimilation approach, *Water Resour. Res.* **37**: 2353–2364.

Shuttleworth, W. J. (1998), Combining remotely sensed data using aggregation algorithms, *Hydrol. Earth System Sci.* **2**: 149–158.

Stafford, J. V. (2000), Implementing precision agriculture in the 21th century, *J. Agric. Eng. Res.* **76**: 267–275.

Suggs, R. J., Jedlovec, G. J. and Guillory, A. R., (1998). Retrieval of geophysical parameters from GOES: Evaluation of a split-window technique, *J. Appl. Meteor.* **37**: 1205–1227.

Talagrand, O. (1997), Assimilation of observations, an introduction, *J Meteorol. Soc. Japan* **75**: 191–209.

van den Hurk, B. (2001), Energy balance based surface flux estimation from satellite data, and its application for surface moisture assimilation, *Meteorol. Atmos. Phys.* **76**: 43–52.

van den Hurk, B. J. J. M., Bastiaanssen, W. G. M., Pelgrum, H., and Van Meijgaard, E. (1997), A new methodology for assimilation of initial soil moisture fields in weather prediction models using Meteosat and NOAA data, *J. Appl. Meteorol.* **36**: 1271–1283.

Walker, J. P., Willgoose, G. R., and Kalma, J. D. (2001), One-dimensional soil moisture profile retrieval by assimilation of near-surface observations: A comparison of retrieval algorithms, *Adv. Water Resour.* **24**: 631–650.

Weiss, M., Troufleau, D., Baret, F., Chauki, H., Prevot, L., Olioso, A., Bruguier, N., and Brisson, N. (2001), Coupling canopy functioning and radiative transfer models for remote sensing data assimilation, *Agric. Forest Meteorol.* **108**: 113–128.

12

Validation and Spatial Scaling

We have discussed various models for estimating land surface variables in the previous chapters. At the early stages of model development, most models need to be calibrated (or tuned) under different conditions. Derived biophysical/geophysical products also need to be validated before being distributed to the user. Both model calibration and product validation require the field measurements on the ground. As spatial resolution becomes coarser, validation is simply not equivalent to field measurements, and advanced methodologies (e.g., scaling techniques) must be developed in order to determine the reliable estimates of product accuracy. The relevant issue is about spatial scaling.

In Section 12.1, we provide some background information and discuss the need for validation. Section 12.2 is the core of this chapter, in which different validation methodologies for land surface products are discussed. This starts with the direct field measurements in which (1) instrumentation is briefly outlined, (2) spatial sampling principles are presented, and (3) various operational observation networks worldwide will be briefly introduced. Comparison of different algorithms and products is another way to evaluate the accuracy of the data products that will be discussed in the last part of Section 12.2.2. Section 12.2.3 introduces the basic structure and elements of the NASA EOS validation Program.

Because of its significant importance to remote sensing beyond validation, spatial scaling is discussed in a separate section, Section 12.3. Spatial scaling (both upscaling and downscaling) principles and methods are also discussed. Various downscaling methods are discussed in more detail.

Quantitative Remote Sensing of Land Surfaces. By Shunlin Liang
ISBN 0-471-28166-2 Copyright © 2004 John Wiley & Sons, Inc.

12.1   RATIONALE OF VALIDATION

Validation involves the specification of the transformations required to extract estimates of high-level geo/biophysical quantities from calibrated digital numbers (DNs) and specification of the uncertainties in these high-level products. Validation requires detailed knowledge of the relationships between DNs and bio/geophysical quantities of interest over the full range of possible conditions.

Besides providing radiance data products to the user, there has been a general trend in modern remote sensing that a set of high-level products are also generated as the standard data products for the end users. The emphasis of validation activities should vary before and after the satellite launch. *Prelaunch activities* include determination of algorithms and characterization of uncertainties resulting from parameterizations and their algorithmic implementation. *Postlaunch activities* include refinement of algorithms and uncertainty estimates based on near–direct comparisons with correlative data and selected, controlled analyses.

The importance of accurately assessing the products of remotely sensed data is universally recognized. In fact, the accuracy of a remotely sensed data product is equally important as the information presented in the product. Without a known accuracy, the product cannot be used reliably and, therefore, has limited applicability. Fenstermaker (1994) edited a compendium on accuracy assessments of image classification, including papers published mostly before the early 1990s. According to the Working Group on Calibration and Validation (WGCV) of the international Committee on Earth Observation Satellites (CEOS), validation is the process of assessing by independent means the quality of the data products derived from the system outputs. Validation of products involves three steps:

- Acquisition of remote sensing imagery and derivation of information estimates from the imagery using product algorithms from the instrument science teams
- Acquisition of independent measures of the information product quantity, ideally from in situ measurements at ground level, using independent methods
- Comparison of the image-based information product estimate with the independent measure for identical (or similar) locations and times

As such, validation is not simply a pass/fail issue. It is the issue that determines the accuracy with which the image-based methods are expected to perform. There is a long processing chain for turning remotely sensed, calibrated DNs to high-level bio/geophysical products. Level 1 products usually represent TOA radiance in a specific map projection. Both level 0 and level 1 are often treated as *engineering products*. Validation of level 1 is primarily associated with vicarious calibration (see Chapter 5). Validation campaigns involve ground-based, airborne, and on-orbit satellite sensors making simultaneous radiometric measurements of spatially and spectrally homogeneous Earth targets for purposes of validating the on-orbit satellite radiometric calibration. These campaigns provide an effective check of the operation and reliability of the satellite onboard calibration systems. The campaigns also enable the comparison and validation of the on-orbit calibration of follow-on sensors that may employ different optical designs and on-board calibration systems.

Generating higher-level products of land surface bio/geophysical variables requires a great deal of scientific understanding of the remote sensing system, as discussed briefly in Section 1.3 and in detail in the previous chapters. Data products at level 2 and above are often treated as the *scientific products*. Validation of the scientific products is primarily the subject of this chapter.

Quality assessment (QA) is an integral part of the data product generation chain, although it is rarely treated as part of validation. The objective of QA is to evaluate and document the scientific quality of the high-level products with respect to their intended performance. The QA results can be made available on a routine basis and are formally stored as product metadata and as per pixel information. Users are encouraged to check these QA results when they order and use individual products to ensure that they were generated without error or artifacts.

12.2 VALIDATION METHODOLOGY

To adequately cover the broad range of surface–atmosphere systems that will be encountered around the world, multiple validation methods applicable to different temporal and spatial scales are necessary. Because of the complexity of the land surface conditions and the validation data itself is not error-free, the validation approach has to be comprehensive and at times redundant. In the following, we will start with direct measurements in the field, including instrumentation, spatial sampling and observation networks. It will be followed by introductions of the inter-comparison methods and the NASA EOS Validation Program.

12.2.1 Direct Correlative Measurements

Direct correlative measurements have to be the essential part of any validation program. Comparisons with in situ data collected over a distributed set of validation test sites may generate the uncertainty information directly.

According to the MODIS Land Discipline (MODLAND) validation plans, field data collection will include instantaneous measures of spectral reflectances and thermal IR radiance, and various bio/geophysical variables at various test sites, first as a calibration activity and second as validation of our radiometric variables. Second will be the establishment of a semipermanent array of test sites, usually including a flux tower, for extended temporal measurement of terrestrial biophysical dynamics over a range of land cover types.

12.2.1.1 Instrumentation and Measurement Techniques
We briefly introduce some representative instruments commonly used for ground measurements of a series of variables.

Radiometric (Radiation) Variables

- *Reflectance Spectra*. Spectroradiometers are often used to measure reflectance as the function of wavelength. Under natural conditions, a reference panel, which is usually white and has spectrally stable, high reflectance, is used to convert the readings of a spectroradiometer to

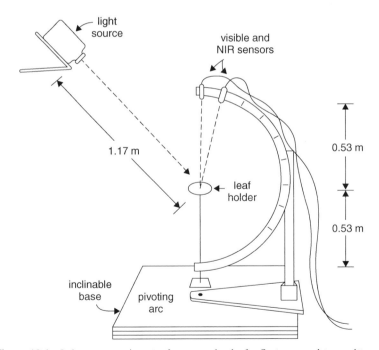

Figure 12.1 Laboratory goniometer for measuring leaf reflectance and transmittance.

spectral reflectance at the specific viewing direction. Most cheaper spectroradiometers such as the SE590 (Spectron Engineering, 1987) cover the spectral range of only 400–1100 nm, while some expensive ones can cover from 350–2500 nm at the spectral resolution of 1 nm [e.g., analytical spectral devices (ASD) spectroradiometer]. For canopy reflectance calculations, leaf optics must also be measured. One way is to take them into the lab for the measurement. A laboratory goniometer is illustrated in Fig. 12.1 (Walter-Shea et al. 1989).

- *Directional Reflectance*. Radiometers or spectroradiometers can measure surface reflectance at a specific direction at a specific time. With or without a mounting device, it might be possible to measure multiangular

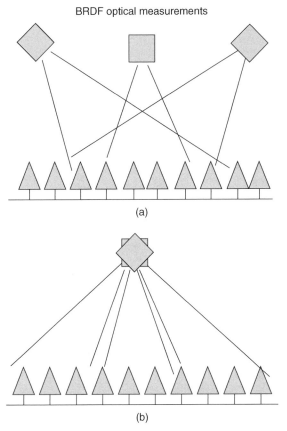

BRDF optical measurements

(a)

(b)

Figure 12.2 Two protocols for directional reflectance measurements in the field (Walthall et al. 2000). Method (a) has the platform moving over the target while maintaining the target in view. Method (b) has the platform at a stationary point with the instrument looking away from the point. This method assumes a homogeneous target area.

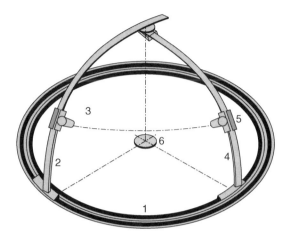

Figure 12.3 European goniometer facility.

reflectance during a short period of time. Some special devices have been developed to measure surface directional reflectance. There are two protocols for directional reflectance measurements as shown in Fig. 12.2 (Walthall et al. 2000). The first one views the same target while changing the viewing zenith angle by moving the platform. Various goniometers, including the European Goniometer Facility (see Fig. 12.3) (Rothkirch et al. 2000) and ASAS are in this category. MISR and Air-MISR employ the similar strategy. The second one looks off in different directions away from a central point. The targets within the instantaneous field of view (IFOV) at different directions are assumed to have the same properties. PARABOLA (see Fig. 12.4) (Bruegge et al. 2002) belongs to the second category.

- *Hemispheric Albedos.* Albedometers are usually used to measure the broadband albedo. An albedometer usually consists of two pyranometers, one pointing up and another pointing down (see Fig. 12.5). A pyranometer has been routinely used in weather stations for measuring solar radiation. Different filters can be used so that the solar radiation in narrow spectral regions (say, visible or near-IR) can be measured. These can be further used to calculate broadband albedos (Liang et al. 2003).

- *Radiation Fluxes.* Examples are PAR and shortwave fluxes. Quantum sensors can measure photosynthetically active radiation (PAR) in the 0.4–0.7-μm waveband and produce an analog voltage response proportional to the scene irradiance. Pyranometers are the common instruments for measuring shortwave radiation flux. A filter can be used for measuring PAR flux only.

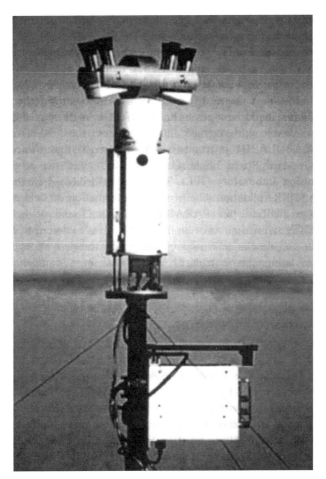

Figure 12.4   Photograph of PARABOLA III in a tower.

- *Skin Temperature.* There are both contact and noncontact sensors. Temperature transducers are contact sensors. They are very useful because of their very low cost, accuracy (0.1°C), and repeatability (<0.1°C). They can be distributed spatially to represent areal measurements. Various IR radiometers are noncontact sensors and available at different price ranges (Kannenberg 1998).

- *Thermal Emissivity.* The spectral emissivity can be reliably measured in the laboratory (e.g., Salisbury and D'Aria 1992). In the field, the so-called gold-box method (Buettner and Kern 1965) is probably the most reliable.

Figure 12.5   Two albedometers at the two ends of a tripod in the field.

Biophysical / Biochemical Variables

- *LAI/FPAR*: These two variables can be measured through a variety of techniques and instruments, such as using allometric relationships determined through destructive canopy sampling for LAI; using indirect measurements from a light-sensitive instrument such as LICOR's LAI-2000 (Norman and Cambell 1989), through the TRAC technique (Chen and Cihlar 1995), or by analyzing hemispherical photos (Rich 1990). However, measuring LAI is still a challenging issue for small leaves and coniferous needles. The fraction of photosynthetically active radiation intercepted by vegetation (FPAR) at the flux stations can be measured using three-point PAR quantum sensors or a long line PAR quantum sensor. Theoretically, most of these instruments can be used for either LAI or FPAR.

- *Leaf Angle Distribution* (LAD). The spatial coordinate apparatus (SCA) (Lang 1973) has been widely used. Others are also available (Welles and Cohen 1996).
- *Chlorophyll and Other Biochemical Concentrations.* They are usually measured in the lab. Handheld chlorophyll meters have been widely used, but its reliability and accuracy have been often questioned.

12.2.1.2 *Spatial Sampling Design*

The highly heterogeneous nature of land systems, in contrast to most atmospheric and ocean systems, makes in situ measurement of coarse resolution parameters extremely difficult. An appropriate spatial sampling scheme is important in environmental monitoring, model calibration, and product validation. Spatial sampling is actually highly relevant to the design of monitoring network. Optimal collection of ground "truth" for remote sensing validation is an area that needs more exploration. In this section we will present some general principles rather than specific examples that have not been widely reported in the literature.

Let us start with the introduction of a few basic concepts. Some people may think of accuracy and precision as the same thing, or be confused by their difference. *Accuracy* is the degree to which data products match true or accepted values. The level of accuracy required for particular applications varies greatly. Highly accurate data can be very difficult and costly to produce and compile. *Precision* refers to the level of measurement and exactness of description of the data products. Two additional terms are frequently used as well. *Data quality* refers to the relative accuracy and precision of a particular data product. These facts are often documented in data quality reports. *Error* encompasses both the imprecision of data and its inaccuracies.

To ensure that the ground data are representative for the spatial population, a suitable sample design has to be chosen (Levy and Lemeshow 1999). Designing a sample scheme includes a number of consideration about the relations between study area, sample site, and subplot (Fig. 12.6), such as the

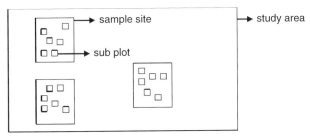

Figure 12.6 Sketch of the relations between the study area, sample site, and subplot.

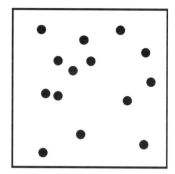

Figure 12.7   Random spatial sampling.

spatial distribution and the number of sample sites required within a study area, the required size of the individual sample site, and the number and size of subplots required within one sample site.

By *sample site* we mean the place where plants or soils are measured or collected for lab analysis. When designing an appropriate sampling scheme for collecting ground data, the primary consideration is the choice of the sample site's distribution. Each sample design must account for the area being studied and the cover type being associated. Here are a few sampling methods that have been widely used in various scientific disciplines.

1. *Random Spatial Sampling.* The simplest spatial sampling method probably is *random spatial sampling*, where one selects a sample location by using two random numbers, one for each direction [defined as either (x, y) coordinates, or as east–west and north–south location]. The result is a pattern of randomly chosen points that can be shown as a set of dots on a map, as in the example in Fig. 12.7. This method is very simple and straightforward to implement and analyze the collected data, but the sampling variance is usually larger than other methods.

2. *Stratified Spatial Sampling.* The concept of stratification can readily be applied to spatial sampling by redefining the subpopulation as a subarea, which is called a *stratum*. To do this we break the total area to be surveyed into subunits, either as a set of regular blocks (as shown in Fig. 12.8) or into 'natural' areas based on factors such as soil type or land cover type. The result is a pattern of (usually randomly-chosen) points within each subarea. The advantage of stratifying is that it is potentially more efficient. We can achieve the same variance with lower cost than with random sampling when a stratified scheme is used. The disadvantage is that there could be a loss in efficiency with inappropriate stratification or a suboptimal allocation of sample sizes.

3. *Systematic Spatial Sampling.* All forms of *systematic spatial sampling* produce a regular grid of points, although the structure of the grid may vary.

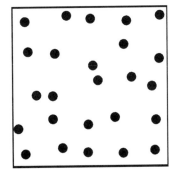

Figure 12.8   Stratified spatial sampling.

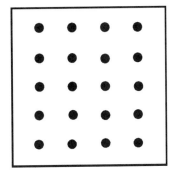

Figure 12.9   Systematic spatial sampling.

It may be square (as shown in Fig. 12.9), rectangular, hexagonal, or any other appropriate geometric system. The advantage of systematic sampling is the uniform spread of the sampled observations over the entire population. The major disadvantage, on the other hand, is that the selection procedure implies that each unit in the population does not have an equal chance of being included in the sample. Systematic sampling can ether be random systematic or stratified systematic.

There are also many other sampling schemes, such as two-stage sampling, cluster sampling, and compound sampling (De Gruijter 1999). These sampling techniques will be implemented differently depending on the availability of the prior information on the spatial variability. In general, systematic sampling is more efficient than most random counterparts, which can be explained by geostatistics (Burgess et al. 1981, Atkinson 1991).

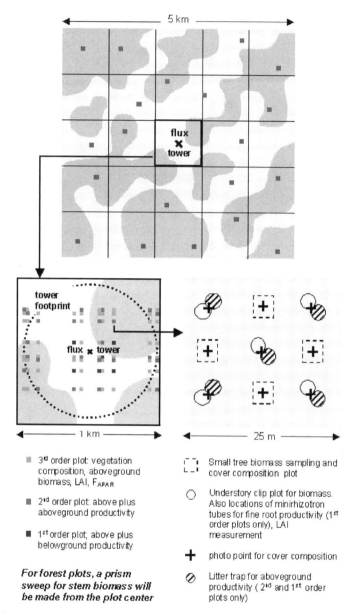

Figure 12.10 Spatial sampling scheme of the BigFoot project (from www.fsl.orst.edu/ larse/bigfoot/).

The spatial sampling method used depends on which variables we measure in the field. For reflectance spectra measurements, we often carry radiometers along several transects and randomly within a plot (Liang et al. in press a, in press b). Because it is relative easy to measure reflectance for many points during a short period of time, the number of samples has not been an issue. For some measurements that are very time-consuming (e.g., soil samples or biomass measurements in the field), sampling design is an important issue to consider. A carefully designed spatial sampling scheme utilized by the BigFoot project (see Section 12.2.1.3) is presented in Fig. 12.10. The BigFoot sample design calls for 100 ground validation measurements of land cover, LAI, FAPAR, and NPP at each site. Plot size is 25 × 25 m, chosen to roughly correspond to the pixel size of ETM + data and neatly nesting at various increments up to 1 km. Between 60 and 80 plots will be concentrated within a 1-km cell centered on the site's eddy flux tower, with the balance of the 100 plots located outside the tower cell, but within the 5 × 5 km BigFoot footprint. This density of plots within the tower footprint ensures adequate characterization of the vegetation properties within that footprint, a critical accomplishment if flux data are to be properly interpreted and used to assess scaled carbon and water flux estimates from biogeochemical models. The 20–40 plots outside the tower footprint (i.e., within the 24 external cells) are apportioned within basic land cover strata to enable independent validation of BigFoot surface products over the full BigFoot footprint.

12.2.1.3 Observation Networks

Consistent and well-registered time-series satellite and field measurements of specific, well-characterized sites are critical to developing an understanding of seasonal and interannual changes of various bio/geophysical variables. This strategy will require consistent observations, obtained repeatedly, of specific ground targets by multiple Earth-observing satellites. In addition, it requires continued acquisition of the standard in situ ecological data sets that have been developed in the major field campaigns.

Selecting globally distributed sites to evaluate these high-level data products and to establish current land surface conditions is more difficult than using the bare-earth desert sites often selected for vicarious sensor calibration. Phenological changes and spatial complexity require a significant long-term commitment of resources at test sites. Great strides have been made in identifying and organizing global sites through the EOS Pathfinder program, the International Geosphere–Biosphere Program (IFBP), and national and international long-term ecological programs.

In the following paragraphs, we present some major observation networks that are related to land surfaces, including (1) AERONET, (2) FLUXNET, (3) EOS Land Core Validation Sites, (4) BigFoot, (5) BSRN, (6) Oklahoma Mesonet, (7) SURFRAD, (8) ISIS, (9) LTER, and (10) ILTER (see Acronyms list in book front matter for definitions). Although their Web addresses may change as time passes, readers should be able to locate them by searching the World Wide Web (www).

1. *AERONET* (aerosol robotic network). The AERONET program (http://aeronet.gsfc.nasa.gov:8080/) is an inclusive federation of ground-based remote sensing aerosol networks established by AERONET and PHOTON and greatly expanded by AEROCAN (the Canadian Sun-Photometer Network) and other agency, institute, and university partners. The goal is to assess aerosol optical properties and validate satellite retrieval of aerosol optical properties. It is very important for us to develop and test atmospheric correction algorithms in quantitative remote sensing of land surface. The network imposes standardization of instruments, calibration, and processing. Data from this collaboration provides globally distributed observations of spectral aerosol optical depths, inversion products, and precipitable water in geographically diverse aerosol regimes. Three levels of data can be downloaded from their Website.

2. *FLUXNET* (http://www-eosdis.ornl.gov/FLUXNET). FLUX-NET is a global network of micrometeorological tower sites that use eddy covariance methods to measure the exchanges of carbon dioxide (CO_2), water vapor, and energy between terrestrial ecosystem and atmosphere. At present, over 150 tower sites are operating on a long-term and continuous basis. Researchers also collect data on site vegetation, soil, hydrological, and meteorological characteristics at the tower sites. FLUXNET builds on regional networks of tower sites: South and North America (AmeriFlux), Europe (CarboEurope), Asia (AsiaFlux), Australia and New Zealand (OzFlux), and independent tower sites.

3. *EOS Land Core Validation Sites* (http://modisland.gsfc.nasa.gov/val/coresite_gen.asp). The EOS land validation core sites are intended as a focus for land product validation over a range of biome types. Nearly 30 sites represent a consensus among the EOS instrument science teams and validation investigators, developed through a number of meetings and discussions. Most of the sites build on an existing program of long-term measurements and have an infrastructure to support in situ measurements. Each site has a point of contact responsible for overall validation coordination at the sites. Although these sites are not intended to meet all EOS test site needs, they will provide a focus for satellite, aircraft, and ground data collection of land product validation, and will provide sites for which scientists can readily access in situ and EOS instrument data. The detailed descriptions of this network are available in the literature (Justice et al. 1998, Morisette et al. 1999).

4. *BigFoot* (http://www.fsl.orst.edu/larse/bigfoot/). The overall goal of BigFoot is to provide validation of MODLAND (MODIS Land science team) science products, including land cover, leaf area index (LAI), fraction absorbed photosynthetic active radiation (FPAR), and net primary production (NPP). Ground measurements, remote sensing data, and

ecosystem process models at sites representing different biomes have been used. BigFoot sites are 5×5 km in size and surround the relatively small footprint (≈ 1 km^2) of CO_2 flux towers. At each site we make multiyear in situ measurements of ecosystem structure and functional characteristics that are related to the terrestrial carbon cycle. Our sampling design allows us to explicitly examine scales of fine-grained spatial patterns in these properties, and provides for a field-based ecological characterization of the flux tower footprint. Multiyear measurements ensure that interannual validity of MOD-LAND products can be assessed.

5. *BSRN* (baseline surface radiation network) (`http://bsrn.ethz. ch/`). BSRN is a project of the World Climate Research Programme (WCRP) aimed at detecting important changes in the Earth's radiation field which may cause climate changes. At a small number of stations (fewer than 40) in contrasting climatic zones, covering a latitude range from 80°N–90°S, solar and atmospheric radiation is measured with instruments of the highest available accuracy and at a very high frequency (minutes). The radiation data are stored together with collocated surface and upper-air observations and station "metadata" in an integrated database. High-accuracy BSRN radiation measurements are already used to validate the radiation schemes in climate models and to calibrate satellite algorithms.

6. *Oklahoma Mesonet* (`http://www.mesonet.ou.edu/`). The Oklahoma Mesonet is a world-class network of environmental monitoring stations. The network was designed and implemented by scientists at the HYPERLINK http://www.ou.edu University of Oklahoma and at HYPERLINK http://www.okstate.edu Oklahoma State University. The Oklahoma Mesonet consists of 114 automated stations covering Oklahoma. There is at least one Mesonet station in each of Oklahoma's 77 counties. At each site, the environment is measured by a set of instruments located on or near a 10-m-tall tower. The measurements are packaged into "observation" every 5 min, then the observations are transmitted to a central facility every 15 min, 24 h per day year-round. The Oklahoma Climatological Survey at the University of Oklahoma receives the observations, verifies the quality of the data, and provides the data to Mesonet customers. It takes only 10–20 minutes from the time the measurements are acquired until they become available to customers, including schools.

7. *SURFRAD* (surface radiation budget network) (`http://www.srrb. noaa.gov/surfrad/index.html`). Accurate and precise ground-based measurements of surface radiation budget in differing climatic regions are essential to refine and verify the satellite-based estimates, as well as to support specialized research. To fill this niche, SURFRAD was established in 1993 through the support of NOAA's Office of Global Programs. Its primary objective is to support climate research with accurate, continuous, and long-term measurements of the surface radiation budget over the United States. Currently (2003) six SURFRAD stations are operating in climatologically diverse regions: Montana, Colorado, Illinois, Mississippi, Pennsylvania,

and Nevada. This represents the first time that a full surface radiation budget network has operated across the U.S. Independent measures of upwelling and downward, solar, and infrared are the primary measurements; ancillary observations include direct and diffuse solar, photosynthetically active radiation, UVB (ultraviolet B), spectral solar, and meteorological parameters. Data are downloaded, quality controlled, and processed into daily files that are distributed in near real time by anonymous File Transfer Protocol ftp and the www (http://www.srrb.noaa.gov). Observations from SURFRAD have been used for evaluating satellite-based estimates of surface radiation, and for validating hydrological, weather prediction, and climate models. Quality assurance built into the design and operation of the network, and good data quality controls ensure that a continuous, high-quality product is released.

8. *ISIS* (Integrated Surface Irradiance Study) (http://www.srrb. noaa.gov/isis/index.html). The Integrated Surface Irradiance Study (ISIS) is a continuation of earlier NOAA surface-based solar monitoring programs, in the visible and ultraviolet wavebands. ISIS provides basic surface radiation data with repeatability, consistency, and accuracy based on reference standards maintained at levels better than 1% to address questions of spatial distributions and time trends, at sites selected to be (1) regionally representative, (2) long-term continuous, and (3) strategic foci for the research that is now needed.

9. *LTER* (long term ecological research) (http://lternet.edu/). The LTER network is a collaborative effort involving more than 1100 scientists and students in the United States investigating ecological processes operating at long timescales and over broad spatial scales. The network promotes synthesis and comparative research across sites and ecosystems and among other related national and international research programs. The U.S. National Science Foundation established a program in 1980 to support research on long-term ecological phenomena in the United States. The network now consists of 24 sites representing diverse ecosystems and research emphases. A network office coordinates communication, network publications, and planning activities.

10. *ILTER* (international long-term ecological research) (http://www. ilternet.edu/). The International LTER is closely associated with the global terrestrial observing system (GTOS) and related efforts. As of May 2000, 21 countries had established formal national LTER programs and joined the ILTER network and 10 more were actively pursuing the establishment of national networks and many others had expressed interest in the model.

12.2.2 Intercomparisons of Algorithms and Products

The collection of ground data for a particular data product from a large number of representative locations and conditions is the most important

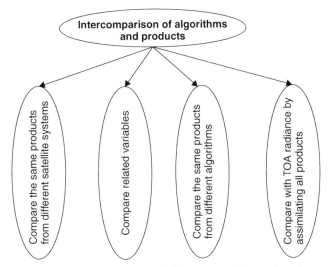

Figure 12.11 Intercomparison of different algorithms and products.

process in validation. However, several other approaches also allow us to evaluate the random and systematic errors of the data products. The concept is illustrated in Fig. 12.11, and the details are as follows:

- The *first approach* is to compare the data in question with the same type of data and products from other spaceborne sensors. If we examine the spectral coverage of some typical sensors in Table 9.3 and Fig. 1.10, there is clearly spectral overlap. The retrieved surface products that are wavelength-dependent, such as reflectance, albedo, and emissivity, can be compared directly with the aid of statistical analysis. The other products independent of wavelength (e.g., LAI, temperature, FPAR) are probably more amenable to direct comparisons. The spectral response of different sensors and spatial resolution may not be the same, but spatial and temporal correlation analyses can give us some clues about product accuracy and any possible anomalies.

- The *second approach* is to compare trends derived from independently obtained reference data and other types of satellite products. For example, LAI values over snow-covered surfaces or desert should not be significantly larger than zero. Surfaces with large LAI values should have much smaller visible albedo. The skin temperature of the snow/ice areas should not exceed that in the desert regions. All this human knowledge can be transformed into a set of rules and numerous threshold values, which will help monitor the production of various products.

- The *third approach* is to compare a particular data product (and its associated errors), with the same product (and its errors) obtained from a different algorithm. Ideally, several independent retrieval methods are used. In this manner systematic errors associated with a satellite product have often been identified, leading to an improvement of the accuracy of that particular data product. Validation is therefore an iterative approach.

- The *last approach* is through data assimilation methods. The concepts and basic principles of four-dimensional data assimilation methods have been discussed in Chapter 11. Assimilating all satellite products and other measurements (ground-based or other satellite measurements) into a validated radiative transfer model enables continuous, collocated comparison with the TOA observations. This is an effective way of using forward calculations to validate the inversion products (see Fig. 1.6).

12.2.3 NASA EOS Validation Program

The EOS (Earth Observing System) is one of the NASA's Earth Science Enterprise long-term research programs dedicated to understanding how human-induced and natural changes affect our global environment. A series of satellites have been or will soon be launched and dozens of high-level geo/biophysical products are being generated. The EOS Validation Program defines a clear hierarchical responsibility structure.

The EOS Validation Program defines these responsibilities in a hierarchy: Project Science Office, instrument science teams, and interdisciplinary science teams.

- The responsibilities of the EOS Project Science Office include
  Organizing and coordinating projectwide validation program
  Providing direct and supplemental support for team validation tasks
- The responsibilities of the instrument science teams include
  Developing scientifically sound and comprehensive data product validation plans, emphasizing basic remote sensing products (levels 1–3)
  Implementing validation tasks, archive data, and disseminating results
- The primary responsibility of the interdisciplinary science teams is concentrating on high-order (level 3 and 4) data products and their validation.

Independent science validation teams were also formed to validate some of the key products. Successful validations are accomplished if timely and accurate product uncertainty information becomes routinely available to the product users within a few years (say, 2–3) after launch. During this period, EOS instrument science teams can adjust algorithms as necessary to provide better performance and consistency with validation data. Previously derived products will be reprocessed in a timely manner following changes to the operational algorithms.

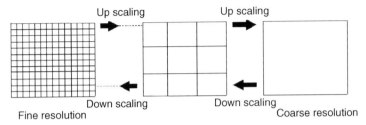

Up scaling   Up scaling

Down scaling   Down scaling
Fine resolution   Coarse resolution

Figure 12.12 Illustration of both upscaling and downscaling schemes.

Scaling is an important part of validation methodology. Because of its significant importance in quantitative remote sensing, we discuss various scaling methods in the following separate section.

12.3 SPATIAL SCALING TECHNIQUES

One basic characteristic of a remote sensing image is its spatial resolution. A vast amount of remotely sensed data with variable spatial resolutions is being acquired operationally (see Section 1.3.5.2). Scales and scaling have been one of the central issues in remote sensing. People have addressed this issue from many different perspectives (e.g., Woodcock and Strahler 1987, Townshend and Justice 1988, Raffy 1992, Curran et al. 1997, Pax-Lenney and Woodcock 1997, Quattrochi and Goodchild 1997; Raffy and Gregoire 1998).

There are two scaling processes: (1) *upscaling*, the conversion from a fine resolution to a coarse resolution, and (2) *downscaling* from the coarse resolution to the fine resolution. It is illustrated in Fig. 12.12. Downscaling and upscaling respectively are often referred to as *disaggregation* and *aggregation* in the literature, although some people tend to distinguish them for some specific applications.

12.3.1 Upscaling Methods

Since land surface is very heterogeneous, upscaling is probably more important than downscaling in product validation. To validate coarse-resolution products (e.g., 1 km), it is difficult to find a large homogeneous region of at least 2–3 pixels in size as the sampling site. Additionally, there are sometimes large variations even within the same land cover type. We often need to acquire fine resolution imagery in the validation process. Thus, we have to address the scaling issue one way or another. A natural question is whether we should derive bio/geophysical variables from fine-resolution imagery and then aggregate to the coarse-resolution or aggregate fine-resolution imagery into coarse resolution first and then derive these variables. This question is illustrated in Fig. 12.13. Mathematically, suppose that the derived variable

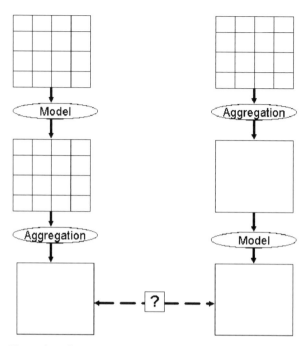

Figure 12.13 Illustration of two upscaling schemes: to run the model at fine scale and then upgrade the results into the coarse resolution and to aggregate the input data from the fine scale to the coarse scale and the run the mode at the coarse scale.

(e.g., LAI, albedo) Y is functionally related to spectral variable x at the fine scale,

$$Y = f(x) \qquad (12.1)$$

The question is whether the following equation is valid

$$\overline{f(x)} = f(\bar{x}) \qquad (12.2)$$

where $f(\cdot)$ is the inversion algorithm that converts spectral variables to bio/geophysical variables Y. If $f(\cdot)$ is linear, the answer is affirmative. However, it is nonlinear in most cases, as we can see from Chapters 8–10.

We explored the upscaling laws of directional reflectance, albedo, and LAI from 30 m to 1 km on the basis of extensive atmospheric and canopy radiative transfer simulations (Liang 2000). Two questions were posed in our study (Fig. 12.14): (1) whether the retrieved LAI values from coarse-resolution remotely sensed data are equivalent to the ground "true" values and (2). whether we should aggregate BRDF and then calculate albedo at the coarse resolution or calculate albedo from BRDF at the fine scale and then aggregate albedo to the coarse scale. From the numerical experiments using

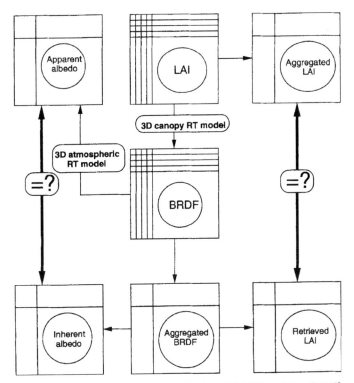

Figure 12.14   Upscaling studies on albedo and LAI [from Liang (2000)].

both three-dimensional atmospheric and canopy radiative transfer models, we found, using summer data from Beltsville, MD, that (1) BRDF upscaling is linear; (2) "effective" LAI values from coarse-resolution remotely sensed data could be quite different from the "true" values if the surface is quite heterogeneous, and the difference is linearly related to the LAI variance; and (3) the upscaling law of spectral albedos is basically linear from 30 m resolution to coarser resolutions (200, 500, and 1000 m) and not significantly subject to the variations of the atmospheric conditions.

These findings on BRDF and albedo scaling are very significant to remote sensing validation. This implies that if we can map land surface BRDF and albedo at fine resolutions (e.g., 30 m) in the summer, these fine-resolution BRDF and albedo values can be linearly aggregated to the coarser resolutions (e.g., 200, 500, and 1000 m). If ground "point/plot" measurements are used to calibrate these fine-resolution products, we are essentially able to carry surface measurement information to the validation of the coarser resolution satellite products. The conclusion that "effective" LAI values retrieved from coarse-resolution remotely sensed data are not always equal to "true" values can be explained from Fig. 12.15. Suppose that we have two pixels that correspond to a small LAI value (LAI_1) and a large LAI value

Reflectance

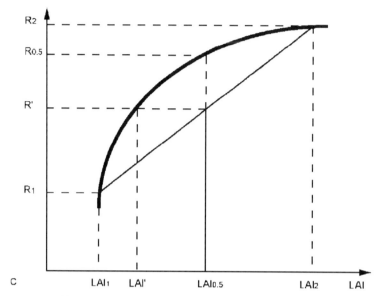

Figure 12.15 Illustration of nonlinear upscaling of LAI.

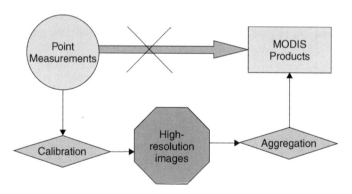

Figure 12.16 Validation methodology for validating MODIS reflectance and albedo products. [From Liang et al. (2002), *Remote Sens. Environ.* Copyright © 2002 with permission from Elsevier.]

(LAI_2). The average reflectance of the mixed pixel corresponds to a LAI value smaller than the average LAI value $[(LAI_1 + LAI_2)/2]$. This indicates that the LAI validation sites need to be more homogeneous. It also raises the question as to how the retrieved LAI product from satellite observations in heterogeneous regions can be effectively used in various land surface models.

In validating MODIS albedo/BRDF products, We (Liang et al. in press a) implemented a validation scheme illustrated in Fig. 12.16. We cannot compare ground point measurements to MODIS pixel values because of the scale mismatch. Unless the surface is large and perfectly homogeneous or a sufficient number of point measurements can be made during the satellite overpass, "point" measurements may not be sufficient to validate the 1-km MODIS products if direct comparison is employed. Therefore, upscaling from "ground" point measurements to the MODIS resolutions using high-resolution remotely sensed imagery is a necessary and critical step. The ground measurements were used to "calibrate" the products from high-resolution imagery, which are then aggregated to the MODIS resolutions.

12.3.2 Downscaling Methods

Because of surface heterogeneity, there are always some pixels at any spatial resolutions that contain multiple cover types. When the spatial resolution becomes coarser and coarser, the number of the mixed pixels increases dramatically. Determining the subpixel information is considered a downscaling process. We will discuss several approaches in this section.

12.3.2.1 Linear Unmixing Methods
Linear unmixing methods have been successfully applied in a number of remote sensing applications (Anser and Lobell 2000, Quarmby et al. 1992, Gong et al. 1994, Bajjouk et al. 1998). If we assume that the signal on a pixel level is a linear combination of the signals of all components that are called *endmembers*, the pixel reflectance for a specified spectral band i can be written as follows:

$$R_i = \sum_{j=1}^{n} p_{ij} r_j + \varepsilon_i \qquad (12.3)$$

where

$$\sum_{j=1}^{n} p_{ij} = 1 \qquad (12.4)$$

and

$$p_{ij} > 0 \qquad (12.5)$$

where R_i = average reflectance at band i
p_{ij} = the proportion of component (end-member) j at band I $(j = 1, \ldots, n)$ that is called *fractional abundances*
r_j = average reflectance of endmember j,
ε_i = random error term

The purpose of the unmixing algorithms is to estimate endmember reflectance r_j and fractional abundances p_{ij} from a group of pixels at multiple wavebands. Some studies assumed that the endmembers are already known, while others first seek endmembers, and others still estimate both quantities at the same time. An unmixing algorithm typically consists of three sequential procedures: dimension reduction, endmember determination, and inversion (Keshava and Mustard 2002).

Dimension reduction of the data in the scene is optional; it is invoked by some algorithms only to reduce the computational load of subsequent steps. The typical algorithm is based on the *principal-components analysis* (PCA), which determines orthogonal axes by performing an eigendecomposition of the sample convariance matrix of the data

$$\Gamma = \frac{1}{N} \sum_{i=1}^{N} (R_i - \mu_R)(R_i - \mu_R)^T \tag{12.6}$$

where μ_R is the mean vector of the pixel reflectance R_i. The covariance matrix can be decomposed into eigenvector matrix U and eigenvalues $\Gamma = U\Sigma U^T$, where Σ is the diagonal matrix of eigenvalues. After selecting the first few largest eigenvalues and the corresponding eigenvectors, we can significantly reduce the dimension of the dataset and yet ensure that it retains the requisite information for successful unmixing in the lower dimension. This is traditional PCA. There are many other variants, such as maximum noise fraction or noise adjusted PCA (Green et al. 1988, Lee et al. 1990). Details on these variations are available in the literature and beyond the scope of this book.

Endmember determination is a very important step in any unmixing algorithm. Theoretically, we can identify the number of endmembers as equal to the number of bands plus one. However, the number of endmembers that may be practically identified and used is far lower, typically ranging from three to seven. Determining the endmembers may be interactive or automated. In *interactive endmember determination*, endmember selection is achieved through an educated trial-and-error approach (Bateson and Curtiss 1996). An analyst has some knowledge of the field site or dataset, and determines the endmembers on the basis of the data structure and/or using library or reference endmembers. This is illustrated in Fig. 12.17, where open circles are endmembers and the fractional calculation for each data point (filled circle) corresponds to the relative distance to these open circles. In automated endmember determination, various algorithms have been developed to determine the essential endmembers automatically, including clustering analysis (Bezdek et al. 1984, Foody and Cox 1994), parametric methods (Tompkins et al. 1997), and geometric approaches (Craig 1994). Bateson et al. (2000) incorporated endmember variability into mixture analysis by representing each endmember by a set or bundle of spectra, each of which could reasonably be the reflectance of an instance of the endmember. Endmember bundles are constructed from the data themselves by an extension to a previously described method of manually deriving endmembers

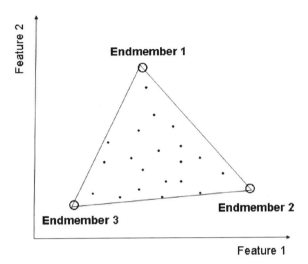

Figure 12.17 Determination of endmembers in two-dimensional space manually.

from remotely sensed data. Applied to remotely sensed images, bundle unmixing produced maximum and minimum fraction images bounding the correct cover fractions and specifying error due to endmember variability.

Inversion is the last step and the core in any unmixing algorithm. It includes the least squares methods (Strang 1988, Shimabukuro and Smith 1991), the regularization method (Settle and Drake 1993), minimum variance methods (Manolakis et al. 2000), and variable endmember methods (Ramsey and Christensen 1998, Roberts et al. 1998). Detailed discussion of mathematical expressions is beyond the scope of this book; readers should consult with cited references.

Keep in mind that linear mixture is an approximation of reality. If the mixture is nonlinear, the calculated fractions using the linear unmixing algorithms will be significantly in error. If the endmembers are mixed on spatial scales smaller than the pathlength of photons in the mixture, such as sand grains of different composition in a beach deposit, light typically interacts with more than one component because of multiple scattering. Thus, such a mixture is nonlinear. Radiative transfer methods have been developed to address this issue (see Chapters 3, 4, and 8). Nonlinear statistical unmixing methods are also available (e.g., Borel and Gerstl 1994, Mustard et al. 1998, Li and Mustard 2000).

12.3.2.2 *Methods for Generating Continuous Fields*

Characterization of terrestrial vegetation from the coarse-resolution imagery (e.g., AVHRR) on the global to regional scales has traditionally been accomplished using classification schemes with discrete numbers of vegetation classes. Representation of vegetation into a limited number of homogeneous classes does not account for the variability within land cover, nor does the portrayal recognize transition zones between adjacent cover types. An alternative paradigm to describing land cover as discrete classes is to represent land cover as continuous fields of vegetation characteristics using a linear mixture model approach (DeFries et al. 1997, 2000), as discussed in the previous section.

The procedure for deriving the continuous fields of vegetation characteristics is fully explained by DeFries et al. (2000) and utilizes a linear mixture model approach applied to 1-km AVHRR data. A set of 156 Landsat multispectral scanner (MSS) scenes were used to train the linear models for vegetation characteristics permitting estimation of endmember values (DeFries et al. 1998). The spectral response of the AVHRR data is then unmixed using the endmembers, and estimates of leaf longevity (percent evergreen and percent deciduous), leaf type (percent broadleaf and percent needleleaf), and percent tree cover are identified. A separate model was developed for each continent to determine the mixture of broadleaf evergreen, broadleaf deciduous, needleleaf evergreen, and needleleaf deciduous woody vegetation depending on which forest types are present in each continent. The approach is based on the annual phenological cycle of vegetation derived from 30 metrics acquired from the AVHRR. These metrics are the annual maximum, minimum, mean, and amplitude for the annual NDVI time series, and channels 1–5 of the AVHRR. The 24 metrics were calculated from April 1992 to April 1993) to account for the full growing season in both hemispheres, but only the 8 months with the highest NDVI are used to describe green vegetation. Six metrics based on surface temperature were also derived from channel 4 of the AVHRR to account for snow cover at higher latitudes. Linear discriminates or linear combinations of the weighted metrics were then made to reduce the statistical complexity and error associated with using 30 metrics in the linear model. The resulting data set thus represent a percentage map where each cell is composed of between 10–80% of the respective vegetation characteristic.

Hansen et al. (in press) improved their procedure using a regression tree method rather than the unmixing approach. Figure 12.18 outlines the prototype methodology (DeFries et al. 2000) and the improved technique presented in their new paper. The two approaches share one feature, the use of annual phenological metrics as the independent variables to predict tree cover. They differ in the following ways:

- The new technique is fully automated, which the authors claimed was their greatest achievement.
- The new training dataset is a continuous variable, not discrete class labels. In the new approach, the classification results from high-resolution imagery are aggregated to coarser scales by labeling each stratum with a mean cover value (0%, 25%, 50%, and 80% for the aforementioned classes) and then averaging over the coarser output cells. In this way a continuous tree cover training dataset is created.

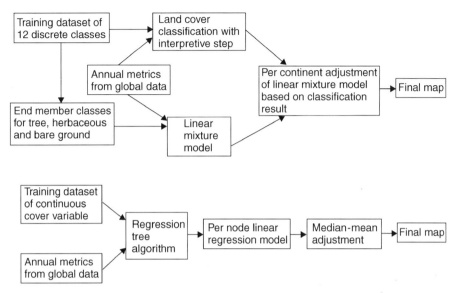

Figure 12.18   Comparison of two methods for creating continuous forest cover. [From Hansen et al. (2002), *Remote Sens. Environ.* Copyright © 2002 with permission from Elsevier.]

- The new algorithm is a regression tree as opposed to a linear mixture model.
- Modified by a land cover classification. The regression tree method has been discussed in Section 8.5.3.
- The new approach operates globally, without per continent adjustments of the mixture model.

12.3.2.3   Decomposition of NDVI Temporal Profiles

This is the approach of spatial downscaling using temporal information. The idea is to decompose the temporal NDVI profile of a mixed pixel into the temporal NDVI profiles of several specific covers within the mixed pixel.

Fisher (1994a, 1994b) described the NDVI temporal profile using an empirical statistical model for agriculture crops, where each crop corresponds to a set of coefficients. The formula is a double logistic function with five coefficients (k, c, p, d, q) and two constant values (vb and ve):

$$\mathrm{NDVI}(t) = vb + \frac{k}{1 + \exp[-c(t-p)]} - \frac{k + vb - ve}{1 + \exp[-d(t-q)]} \quad (12.7)$$

where t is the time variable representing the day of the year, January 1 is set zero, k is related to the asymptotic value of NDVI, c and d (day^{-1}) denote the slopes at the first and second inflection points, p and q (day) are the dates of these two points, and vb and ve are the NDVI values at the

TABLE 2.1   Fitted Parameters to the Double Logistic Model from the Ground-Measured NDVI Profiles of Four Crops at Field Level

	Wheat	Spring Barley	Corn	Sugarbeet
k	0.65	0.80	0.64	0.71
c	0.093	0.119	0.140	0.138
p	93.8 (April 3)	118.2 (April 28)	174.5 (June 23)	164.8 (June 14)
d	0.083	0.078	0.082	0.108
q	202.3 (July 21)	194.5 (July 13)	282.9 (Oct. 10)	290.4 (Oct. 17)
RMSE	0.025	0.031	0.017	0.024

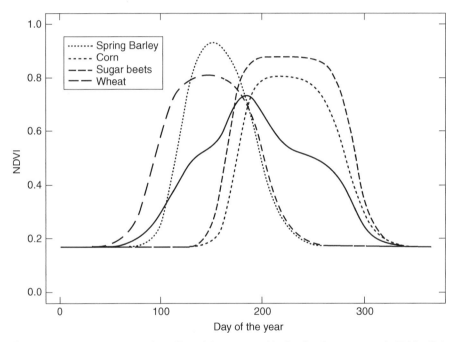

Figure 12.19   NDVI temporal profiles of four crops with the fitted parameters in Table 13.1 and their composite NDVI temporal profile as a mixed pixel.

beginning and the end of the growing season. Fisher (1994a) set $vb = 0.3$ and $ve = 0.17$ for winter wheat and $vb = ve = 0.17$ for summer crops. The fitted values from the measured NDVI profiles of several crops at the field level are given in Table 12.1. The root mean square error (RMSE) is also given for each crop. For demonstration purposes, the fitted curves using the fitted parameters in Table 12.1 are shown in Fig. 12.19. Obviously, different crops have different shapes.

It is assumed (Fisher 1994a) that the regional NDVI profiles are the linear combination of the NDVI profiles of the n crops within this region

$$\text{NDVI}(t) = \sum_{i=1}^{n} \sigma_i \text{NDVI}_i(t) \tag{12.8}$$

where σ_i is the percentage area of crop i ($\sigma_i \geq 0$ and $\sum_{i=1}^{n} \sigma_i = 1$). Note that this assumption is an approximation because surface reflectances are linearly upscaled if surface adjacency is neglected, but NDVI is a nonlinear transformation of the spectral reflectance. The composite NDVI profile from these four crops with equal area (i.e., $\sigma_i = \frac{1}{4}$) is also demonstrated in Fig. 12.19.

Because it is a linear mixture problem, we might be able to estimate the component NDVI profile for the specific crops from the observed regional NDVI profile given the relative areas are known. Similarly, if we know the component NDVI profiles, we might be able to estimate the relative areas of those crops. In the first case, the inversion is a nonlinear problem because we eventually need to estimate their parameters. But it is a linear inversion problem for the second case.

12.3.2.4   *Multiresolution Data Fusion*

Fusion of multiresolution imaging data has been a very active research area. The fusion that we are interested in is a process that merges coarse-resolution images with fine-resolution images so that the coarse-resolution imagery can therefore obtain the details at the scale of fine resolution. The process is considered to be a downscaling procedure. The fine-resolution image may come from the same sensor or from another sensor, but the images must be registered geometrically. The process has been widely used for sharpening lower-resolution images in digital image processing because it merges the spatial-information from a "high-resolution image" with the radiometric information from a "low-resolution image." Most algorithms can be roughly divided into two groups (Zhukov et al. 1999): spectral substitute methods and spatial domain methods.

Spectral component substitution techniques have been developed principally to fuse multispectral and panchromatic images where the panchromatic band usually has a much higher spatial resolution. They are based on replacing a spectral component of the low-resolution multispectral image by the radiometrically adjusted panchromatic image. The most frequently used spectral component substitution techniques include hue–intensity–saturation (HIS) methods, PCA methods, and regression methods.

Spatial domain techniques transfer high-resolution information from a high-resolution image to all the low-resolution spectral bands using various deterministic or statistical predictors. In order to preserve the available radiometric information of the low-resolution image, only the excess high-spatial-frequency components have to be transferred to the low-resolution bands. This can be achieved by highpass filtration in the image domain or by using various multiresolution representations: the wavelet decomposition, the Laplacian pyramid, or the Fourier decomposition.

In the following text, we will briefly introduce some of these techniques, including hue–intensity–saturation (HIS) methods, PCA methods, regression methods, classification-based methods, and wavelet methods.

HIS Method. With three multispectral bands, we can easily produce red-green-blue (RGB) color composite imagery. The HIS methods transfer three multispectral bands from the RGB color space to the HIS color space, the intensity component of the HIS space is then replaced by the panchromatic band, and finally the revised HIS space is transformed back to RGB space. First, the three multispectral bands denoted by RGB are transformed to HIS space with the formulae suggested by Pratt (1991) and used by Zhou et al. (Zhou et al. 1998)

$$
\begin{pmatrix} I \\ V_1 \\ V_2 \end{pmatrix} = \begin{pmatrix} \frac{1}{3} & \frac{1}{3} & \frac{1}{3} \\ -\frac{1}{\sqrt{6}} & -\frac{6}{\sqrt{6}} & \frac{2}{\sqrt{6}} \\ \frac{1}{\sqrt{6}} & -\frac{1}{\sqrt{6}} & 0 \end{pmatrix} \begin{pmatrix} R \\ G \\ B \end{pmatrix} \tag{12.9}
$$

$$
H = \tan^{-1}\left(\frac{V_1}{V_2}\right) \tag{12.10}
$$

$$
S = \sqrt{V_1^2 + V_2^2} \tag{12.11}
$$

Second, the panchromatic band can be then used to replace component I after being linearly stretched to match the mean and variance of the intensity component, which can be denoted as I'. The last step is to transform the revised IHS space into the RGB space explicitly:

$$
\begin{pmatrix} R \\ G \\ B \end{pmatrix} = \begin{pmatrix} 1 & -\frac{\sqrt{6}}{6} & \frac{\sqrt{6}}{2} \\ 1 & -\frac{\sqrt{6}}{6} & -\frac{\sqrt{6}}{2} \\ 1 & \frac{\sqrt{6}}{3} & 0 \end{pmatrix} \begin{pmatrix} I' \\ H \\ S \end{pmatrix} \tag{12.12}
$$

Note that in Eq. (12.9) intensity I is the average of the three multispectral bands:

$$
I = \frac{(R + G + B)}{3} \tag{12.13}
$$

There are also several other algorithms, for example

$$I = \max(R, G, B) \tag{12.14}$$

or

$$I = \frac{\max(R, G, B) + \min(R, G, B)}{2} \tag{12.15}$$

PCA Method. The multispectral bands (X_i) can be represented by the principal components (Y_i) using the linear PCA transformation

$$\begin{pmatrix} Y_1 \\ Y_2 \\ \dots \\ Y_n \end{pmatrix} = \begin{pmatrix} a_1 & a_2 & \cdots & a_n \end{pmatrix} \begin{pmatrix} X_1 \\ X_2 \\ \dots \\ X_n \end{pmatrix} \tag{12.16}$$

where a_i are the eigenvectors that orthogonalize the covariance matrix of **X** such that the covariance matrix of **Y** is a diagonal matrix. The result of the principal component is a set of decorrelated images whose variances are ordered in amplitude. Similar to the HIS method, the first principal component (Y_1) that corresponds to the largest eigenvalue is replaced by the panchromatic band that is linearly stretched with the same mean and variance as the first principal component (Chavez et al. 1991, Shettigara 1992). The last step is to transform the revised principal components back to the original space. More details on PCA are given in Section 12.3.2.1. Software for performing PCA is widely available.

Regression Methods. Price (1999) applied the regression method to add the details of high-resolution imagery to coarse-resolution imagery. This procedure is based on the statistical relationship between imagery of both the fine resolution and the coarse resolution. The idea is quite simple. The fine resolution is first aggregated to the coarse resolution, and the pixel values of both the coarse-resolution images and the aggregated fine-resolution images within each local window (say, 30×30) are used to establish a linear regression equation. If the established regression equation at the coarse scale is assumed to be valid at the fine scale, the subpixel values of the coarse-resolution imagery can be predicted by the fine-resolution imagery. This is illustrated in Fig. 12.20. In this case, the pixel size of the coarse-resolution imagery is twice that of the fine-resolution imagery. The upper panel of the figure shows a training process that establishes a regression equation within a window of 3×3 pixels, and the lower panel depicts prediction of the subpixel value of the coarse-resolution image. In a previous study Price (1987) used the same equation for the whole imagery, which results in ambiguous results when different spectral signatures $Y_i = f(X_i)$ were inferred from a single higher-resolution spectral measurement.

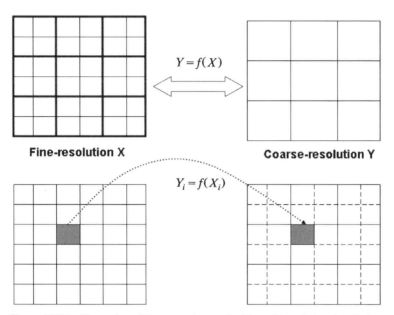

Figure 12.20 Illustration of the regression method in multiresolution data fusion.

Classification-Based Methods. Zhukov et al. (1999) and Zege and Kokhanovskiy (1989) proposed a classification-based method, and Minghelli-Roman et al. (2001) applied a similar procedure to merge Landsat TM imagery (30 m) with MERIS imagery (250 m) so that MERIS images have the details at the resolution of 30 m. This algorithm consists of several steps:

1. The first step is to register coarse- and fine-resolution imagery. This step is required by all algorithms.
2. The second step is to classify fine-resolution imagery into a finite number of classes. The classification could be supervised or unsupervised, but the only requirement is that all pixels be assigned to one specific class, with no pixel remaining unclassified.
3. The third step is to determine class spectra. Within a coarse-resolution pixel, there are many different classes; the pixel value (L) is assumed to be the linear combination of these classes:

$$L = \sum_{i=1}^{N} \alpha_i L_i \tag{12.17}$$

where α_i is the areal proportion of each class, which can be easily determined from the classification from the fine-resolution imagery. Given the pixel values of the coarse-resolution imagery (L), we can easily determine the radiance of each class from Eq. (12.17) using the least squares method.

4. After determining the radiance of each class, we actually have determined the subpixel value of the coarse-resolution pixels at the fine-resolution scale.

This approach is very similar in principal to the linear unmixing method described in Section 12.3.2.1. The linear unmixing methods assume that the spectral values of the endmember can be determined manually or automatically. The classification-based method in this section computes the proportion of each class from the fine-resolution imagery. Obviously, the classification-based method not only produces a much higher number of "pure" classes (only very few endmembers can be determined in the linear unmixing methods) but also determines their locations at the subpixel level.

Wavelet Methods. Wavelet decomposition is being increasingly used in remote sensing. The method is based on the decomposition of the image into multiple channels on the basis of their local frequency content. The wavelet transform provides a framework to decompose images into a number of new images, each one of them with a different degree of resolution. The following texts are largely extracted from the presentation by Nuñez et al. (1999).

The wavelet transform of a distribution $f(t)$ can be expressed as

$$W(f)(a,b) = |a|^{-(1/2)} \int_{-\infty}^{+\infty} f(t) \psi\left(\frac{t-b}{a}\right) dt \qquad (12.18)$$

where a and b are scaling and translational parameters, respectively. Each base function $\psi[(t-b)/a]$ is a scaled and translated version of a function called "mother wavelet." These base functions satisfy $\int \psi[(t-b)/a] = 0$.

The discrete approach of the wavelet transform can be done with several different algorithms. Nuñez et al. (1999) applied the discrete wavelet transform known as *à trous* ("with holes") algorithm to decompose the image into wavelet planes. Given an image P, we construct the sequence of approximations:

$$\begin{cases} F_1(P) = P_1 \\ F_2(P_1) = P_2 \\ F_3(P_2) = P_3 \\ \qquad \cdots \end{cases} \qquad (12.19)$$

To construct the sequence, this algorithm performs successive convolutions with a filter obtained from an auxiliary function named *scaling function*. Nuñez et al. (1999) employed a scaling function that has B_3 cubic spline profile. The use of a B_3 cubic spline leads to a convolution with a mask of 5×5:

$$\frac{1}{256} \begin{pmatrix} 1 & 4 & 6 & 4 & 1 \\ 4 & 16 & 24 & 16 & 4 \\ 6 & 24 & 36 & 24 & 6 \\ 4 & 16 & 24 & 16 & 4 \\ 1 & 4 & 6 & 4 & 1 \end{pmatrix} \qquad (12.20)$$

The wavelet planes are computed as the differences between two consecutive approximations P_{i-1} and P_i. Letting $w_i = P_{i-1} - P_i$ and $P_0 = P$, we can write the reconstruction formula

$$P = \sum_{i=1}^{n} w_i + P_r \qquad (12.21)$$

In this representation, the images P_i are decomposed from the original image at increasing scales (decreasing resolution levels) forming the multiresolution wavelet planes, and P_r is a residual image. If a dyadic decomposition scheme is used, the original image ($P_0 = P$) has twice the resolution of P_1; the image P_1 double the resolution of P_2; and so on. If the resolution of image P_0 is, for example, 10 m, the resolution of P_1 would be 20 m, the resolution of P_2 would be 40 m, and so forth. Note that all the consecutive approximations (and wavelet planes) in this process have the same number of pixels as the original image. Wavelet fusion can be carried out using either the substitution method or the additive method.

   The *substitution method* involves replacing the first wavelet plane of the coarse-resolution imagery with the first wavelet plane of the fine-resolution imagery and then reconstructing the coarse-resolution imagery. Both coarse- and fine-resolution imagery first need to be registered and normalized. The wavelet decomposition of these coarse- and fine-resolution images can be implemented by Eq. (12.21). Finally, the first wavelet plane of the coarse-resolution imagery is replaced by that of the fine-resolution imagery. The reconstructed coarse-resolution imagery will then have the details of the fine-resolution scale.

   The *additive methods* involve adding the wavelet planes of the high-resolution image directly to the coarse-resolution imagery. The details are not given here.

2.3.2.5 Methods for Statistical Downscaling of GCM Outputs

Finally, we will introduce several downscaling methods used by the climate modeling community. They are used in a different discipline, but might be helpful for us to develop corresponding downscaling algorithms in quantitative remote sensing.

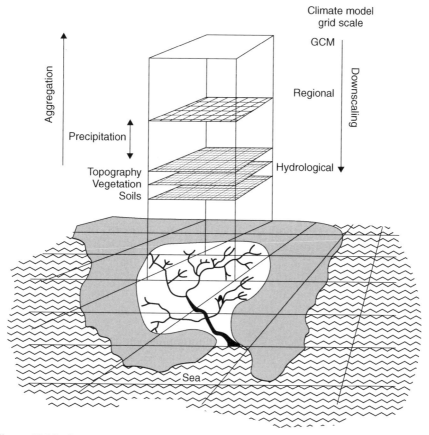

Figure 12.21   Downscaling of GCM outputs. [From Wilby and Wigley (1997), *Progress Phys. Geogr.* Reproduced by permission of Arnold Journals.]

The current GCM outputs have very coarse spatiotemporal resolution (typically $0.25-1°$ degree spatially). People quite often need to assess the impacts of global change on land surfaces at the local scale and require the regional-scale atmospheric parameters. Downscaling techniques can bridge the gaps between what climate modelers are currently able to provide and what impact assessors require. The general idea is illustrated in Fig. 12.21. Wilby and Wigley (1997) grouped all statistical downscaling techniques into four categories: regression methods, weather-pattern-based methods, stochastic weather generators, and limited-area modeling methods. Xu (1999) further discussed the methods for scaling hydrological variables.

Regression methods generally involve establishing linear or nonlinear relationships between subgrid-scale (e.g., single site) parameters and coarse-resolution GCM grid predictor variables. For example, Wigley et al. (1990) regressed site values of temperature and precipitation against spatial area averages of temperature, precipitation, mean sea-level pressure, 700 mb (millibars) geopotential heights, and zonal/meridional components of the GCM predictor variables. The slight variants of this approach involve regressing the same parameter from a regional to local scale, or across several scales. Carbone and Bramante (1995) regressed spatially averaged monthly maximum and minimum temperatures against the same variables at multiple stations across the southeastern United States.

The *weather pattern approach* typically involves statistically relating observed station or area-average meteorological data to a given weather classification scheme. For a given weather pattern, we can condition the local surface variables, such as precipitation. This is accomplished by deriving conditional probability distributions for observed data such as probability of a wet day following a wet day or the mean wet-day amount associated with a given atmospheric circulation pattern (Hughes and Guttorp 1994).

Stochastic weather generators typically downscale weather variables in the time dimension. Given monthly or annually observations, these models generate the series of subtimescale. Most models are based on Markov renewal processes, in which, for each successive day, the precipitation occurrences and amount are governed by outcomes on previous days. Many crop growth models incorporate weather generators since they require daily inputs of weather variables, but many users only have climatic data (monthly or yearly average data). Mearns et al. (1996) used this type of model to investigate the direct effect of changes in daily and interannual variability of temperature and precipitation on crop yields in the central Great Plains of the United States.

The final downscaling method is use of the *limited-area climate models*, where a higher-resolution limited-area climate model is embedded within the GCM, using the GCM to define the time-varying boundary conditions (Giorgi 1990). Limited-area climate models can simulate fine-scale atmospheric features and provide atmospheric data for impact assessment that reflect the natural heterogeneity of the climate at regional scales.

We have discussed four major statistical downscaling methods used by the climate community. Some of these techniques can be easily applied to remote sensing. For example, AVHRR data has high-temporal resolution but low spatial resolution. The regional average NDVI series can be used for predicting NDVI series at a specific site or small field. The spatial pattern of a large region might be possibly used for predicting the proportional percentage of certain cover types within a much smaller area.

12.4   SUMMARY

Validation is a very important process in quantitative remote sensing. An inspection by the human eyes really cannot reveal the accuracy of the products derived from remote sensing observations. The core of any validation activity is the correlative ground measurements. Some critical issues associated with ground measurements include the instrumentation, measurement methodology, and data management. Since field data collections are usually very time-consuming and expensive, data sharing from any field experiments is absolutely necessary. Thus, collaborations must be warranted between scientists conducting the field measurements and those developing information extraction algorithms from remotely sensed data.

The EOS land community has adopted core land validation sites representing a range of biomes and atmospheric conditions that must be augmented and strengthened to validate future land data products. Coordinated and periodic ground-based measurement of such variables as surface reflectance, LAI/FPAR, canopy structure, land cover, fire-burned area, and net primary productivity will provide a critical component for assessing product accuracy. A research challenge is to tie measurements to the equivalent satellite products. In particular, continuous tower-based (point) measurements will have to be related to satellite and model-derived products.

Ground measurements are most likely at the point/plot scale, but pixel size of remote sensing data is usually much greater. Since most land surfaces are heterogeneous, a scaling up from point/plot scale to pixel-scale scale is needed. Scaling is not just the issue associated with validation; it is actually a common one in quantitative remote sensing in general.

Future remote sensing systems generating higher-order data products should include an explicit validation component to enable science users to characterize error budgets in their analysis. Automated and reliable in situ instrumentation is highly desirable. The validation community should continue to develop protocols for collecting data for validation and for instrument calibration that allow a distributed and international validation system to be developed.

REFERENCES

Asner, G. P. and Lobell, D. B. (2000), A biogeophysical approach for automated SWIR unmixing of soils and vegetation, *Remote Sens. Envir.* **74**: 99–112.

Atkinson, P. M. (1991), Optimal ground-based sampling for remote sensing investigations: Estimating the regional mean, *Int. J. Remote Sens.* **12**: 559–567.

Bajjouk, T., Populus, J., and Guillaumont, B. (1998), Quantification of subpixel cover fractions using PCA and a linear programming method: Application to the coastal zone of Roscaff (France), *Remote Sens. Envir.* **64**: 153–165.

Bateson, A. and Curtiss, B. (1996), A method for manual endmember selection and spectral unmixing, *Remote Sens. Envir.* **55**: 229–243.

Bateson, C. A., Asner, G. P., and Wessman, C. A. (2000), Endmember bundles: A new approach to incorporating endmember variability into spectral mixture analysis, *IEEE Trans. Geosci. Remote Sens.* **38**: 1083–1094.

Bezdek, J. C., Ehrlich, R., and Full, W. (1984), FCM: The fuzzy c-means clustering algorithm, *Comput Geosci.* **10**: 191–203.

Borel, C. C. and Gerstl, S. A. W. (1994), Nonlinear spectral mixing models for vegetative and soil surfaces, *Remote Sens. Envir.* **47**: 403–416.

Bruegge, C. J., Helmlinger, M. C., Conel, J. E., Gaitley, B. J., and Abdou, W. A. (2002), Parabola III: A sphere-scanning radiometer for field determination of surface anisotropic reflectance functions, *Remote Sens. Rev.* **19**: 75–94.

Buettner, K. and Kern, C. (1965), The determination of emissivities of terrestrial surfaces, *J. Geophys. Res.* **70**: 1329–1337.

Burgess, T. M., Webster, R., and McBratney, A. B. (1981), Optimal interpolation and isarithmic mapping of soil properties iv. Sampling strategy, *J. Soil Sci.* **32**: 643–659.

Carbone, G. J. and Bramante, P. D. (1995), Translating monthtly temperature from regional to local scale in the southeastern United States, *Climate Res.* **5**: 229–242.

Chavez, P. S., Sides, S. C., and Anderson, J. A. (1991), Comparison of three different methods to merge multiresolution and multispectral data: Landsat TM and SPOT panchromatic, *Photogramm. Eng. Remote Sens.* **57**: 295–303.

Chen, J. M. and Cihlar, J. (1995), Quantifying the effect of canopy architecture on optical measurements of leaf area index using two gap size analysis method, *IEEE Trans. Geosci. Remote Sens.* **33**: 777–787.

Craig, M. D. (1994), Minimum-volume transforms for remotely sensed data, *IEEE Trans. Geosci. Remote Sens.* **32**: 99–109.

Curran, P. J., Foody, G. M., and van Gardingen, P. R. (1997), Scaling up, in *Scaling-up: From Cell to Landcsape*, P. R. van Gardingen, G. M. Foody, and P. J. Curran, eds., Cambridge Univ. Press, pp. 1–5.

DeFries, R. S., Hansen, M., Townshend, J. R. G., and Sohlberg, R. (1998), The use of training data derived from landsat imagery in decision tree classifiers, *Int. J. Remote Sens.* **19**: 3141–3168.

DeFries, R. S., Hansen, M. C., Townshend, J. R. G., Janetos, A. C., and Loveland, T. R. (2000), A new global 1-km dataset of percentage tree cover derived from remote sensing, *Global Change Biol.* **6**: 247–254.

DeFries, R. S., Hansen, M., Steininger, M., Dubayah, R., Sohlberg, R., and Townshend, J. R. G. (1997), Subpixel forest cover in central Africa from multisensor, multitemporal data, *Remote Sens. Envir.* **60**: 228–246.

de Gruijter, J. (1999), Spatial sampling schemes for remote sensing, in *Spatial Statistics for Remote Sensing*, A. Stein, F. van der Meer, and B. Gorte, eds., Kluwer Academic Publishers, pp. 211–242.

Fenstermaker, L. K. (1994), *Remote Sensing Thematic Accuracy Assessment: A Compendium*, The American Society for Photogrammetry and Remote Sensing.

Fisher, A. (1994a), A model for the seasonal variations of vegetation indices in coarse resolution data and its inversion to extract crop parameters, *Remote Sens. Envir.* **48**: 220–230.

Fischer, A. (1994b), A simple model for the temporal variations of NDVI at regional scale over agricultural countries. Validation with ground radiometric measurements, *Int. J. Remote Sens.* **15**: 1421–1446.

Foody, G. and Cox, D. (1994), Sub-pixel land cover composition estimation using a linear mixture model and fuzzy membership model and fuzzy membership functions, *Int. J. Remote Sens.* **15**: 619–631.

Giorgi, F. (1990), Simulation of regional climate using a limited area model nested in a general circulation model, *J. Climate* **3**: 941–963.

Gong, P., Miller, J. R., and Spanner, M. (1994), Forest canopy closure from classification and spectral unmixing of scene components—multisensor evaluation of an open canopy, *IEEE Trans. Geosci. Remote Sens.* **32**: 1067–1080.

Green, A. A., Berman, M., Switzer, P., and Craig, M. D. (1988), A transformation for ordering multispectral data in terms of image quality with implications for noise removal, *IEEE Trans. Geosci. Remote Sens.* **26**: 65–74.

Hansen, M., DeFries, R. S., Townshend, J. R. G., Sohlberg, R., Dimiceli, C., and Carroll, M. (2002), *Remote Sens. Envir.* **83**: 303–319.

Hughes, J. P. and Guttorp, P. (1994), A class of stochastic models for relating synoptic atmospheric patterns to regional hydrologic phenomena, *Water Resour. Res.* **30**: 1535–1546.

Justice, C., Starr, D., Wickland, D., Privette, J., and Suttles, T. (1998), EOS land validation coordination: An update, *Earth Observer* **10**: 55–60.

Kannenberg, B. (1998), IR instrument comparison workshop at the rosentiel school of marine and amospheric science, *Earth Observer* **10**: 51–54.

Keshava, N. and Mustard, J. F. (2002), Spectral unmixing, *IEEE Signal Process. Mag.* **19**: 44–57.

Lang, A. R. G. (1973), Leaf orientation of a cotton plant, *Agric. Meteorol.* **11**: 37–51.

Lee, J. B., Woodyatt, S., and Berman, M. (1990), Enhancement of high spectral resolution remote-sensing data by a noise-adjusted principal component transform, *IEEE Trans. Geosci. Remote Sens.* **28**: 295–304.

Levy, P. S. and Lemeshow, S. (1999), *Sampling of Populations: Methods and Applications*, 3rd ed., Wiley.

Li, L. and Mustard, J. F. (2000), Compositional gradients across mare-highland contacts: The importance and geological implications of lateral mixing, *J. Geophys. Res.* **105**: 20431–20450.

Liang, S. (2000), Numerical experiments on spatial scaling of land surface albedo and leaf area index, *Remote Sens. Rev.* **19**: 225–242.

Liang, S., Fang, F., Chen, M., Shuey, C., Walthall, C., Daughtry, C., Morisette, J., Schaaf, C., and Strahler, A. (2002), Validating MODIS land surface reflectance and albedo products: Methods and preliminary results, *Remote Sens. Envir.* **83**: 149–162.

Liang, S., Shuey, C., Fang, H., Russ, A., Chen, M., Walthall, C., Daughtry, C., and Hunt, R. (2003), Narrowband to broadband conversions of land surface albedo: II. Validation, *Remote Sens. Environ.* **84**: 25–41.

Manolakis, D. G., Ingle, V. K., and Kogon, S. M. (2000), *Statistical and Adaptive Signal Processing*, McGraw-Hill.

Mearns, L. O., Rosenzweig, C., and Goldberg, R. (1996), The effect of changes in daily and interannual climatic variability on ceres-wheat: A sensitivity study, *Climatic Change* **32**: 257–292.

Minghelli-Roman, A., Mangolini, M., Petit, M., and Polidori, L. (2001), Spatial resolution improvement of meris images by fusion with TM images, *IEEE Trans. Geosci Remote Sens.* **39**: 1533–1536.

Morisette, J., Privette, J., Justice, C., Olson, D., Dwyer, J., Davis, P., Starr, D., and Wickland, D. (1999), The EOS land validation core sites: Background information and current status, *Earth Observer* **11**: 11–26.

Mustard, J. F., Li, L., and He, G. (1998), Nonlinear spectral mixtue modleing of lunar multispectral data: Implications for lateral transport, *J. Geophys. Res.* **103**: 419–425.

Norman, J. M. and Cambell, G. S. (1989), Canopy structure, in *Plant Physiological Ecology; Field Methods and Instrumentation*, J. R. E. R. W. Pearcy, H. A. Mooney, and P. W. Rundel, eds., Chapman & Hall, pp. 301–325.

Nuñez, J., Otazu, X., Fors, O., Prades, A., Pala, V., and Arbiol, R. (1999), Multiresolution-based image fusion with additive wavelet decomposition, *IEEE Trans. Geosc. Remote Sens.* **37**: 1204–1211.

Pax-Lenney, M. and Woodcock, C. E. (1997), The effect of spatial resolution on the ability to monitor the status of agricultural lands, *Remote Sens. Envir.* **61**: 210–220.

Pratt, W. K. (1991), *Digital Image Processing*, Wiley.

Price, J. C. (1987), Combining panchromatic and multispectral imagery from dual resolution satellite instruments, *Remote Sens. Envir.* **21**: 119–128.

Price, J. C. (1999), Combining multispectral data of differing spatial resolution, *IEEE Trans. Geosci. Remote Sens.* **37**: 1199–1203.

Quarmby, N. A., Townshend, J. R. G., Settle, J. J., White, K. H., Milnes, M., Hindle, T. L., and Silleos, N. (1992), The use of multitemporal ndvi measurements from avhrr data for crop yield estimation and prediction, *Int. J. Remote Sens.* **13**: 415–425.

Quattrochi, D. A. and Goodchild, M. F., eds. (1997), *Scale in Remote Sensing and GIS*, Lewis Publishers.

Raffy, M. (1992), Change of scale in models of remote sensing: A general method for spatialization of models, *Remote Sens. Envir.* **40**: 101–112.

Raffy, M. and Gregoire, C. (1998), Semi-empirical models and scaling: A least square method for remote sensing experiments, *Int. J. Remote Sens.* **19**: 2527–2541.

Ramsey, M. S. and Christensen, P. R. (1998), Mineral abundance dtermination: Quantitative deconvolution of thermal emission spectra, *J. Geophys. Res.* **103**: 577–596.

Rich, P. M. (1990), Characterizing plant canopies with hemispherical photographs. *Remote Sens. Envir.* **5**: 13–29.

Roberts, D. A., Gardner, M., Church, R., Ustin, S., Scheer, G., and Green, R. O. (1998), Mapping chapparal in the santa monica mountains using multiple endmember spectral mixture models, *Remote Sens. Envir.* **65**: 267–279.

Rothkirch, A., Meister, G., Hosgood, B., Spitzer, H., and Bienlein, J. (2000), BRDF measurements on urban materials using laser light equipment characteristics and estimation of error sources, *Remote Sensing Reviews*, **19**: 21–36.

Salisbury, J. W. and D'Aria, D. M. (1992), Emissivity of terrestrial materials in the 8-14 um atmospheric window, *Remote Sens. Envir.* **42**: 83–106.

Settle, J. J. and Drake, N. A. (1993), Linear mixing and the estimation of ground cover proportions, *Int. J. Remote Sens.* **14**: 1159–1177.

Shettigara, V. K. (1992), A generalized component substitution technique forspatial enhancement of multispectral images using a higher resolution data set, *Photogramm. Eng. Remote Sens.* **58**: 561–567.

Shimabukuro, Y. E. and Smith, J. A. (1991), The leasts mixing models to generate fraction images derived from remote sensing multispectral data, *IEEE Trans. Geosci. Remote Sens.* **29**: 16–20.

Strang, G. (1988), *Linear Algebra and Its Applications*, Harcourt Brace Jovanovich.

Tompkins, S., Mustard, J. F., Pieters, C. M., and Forsyth, D. W. (1997), Optimization of endmembers for spectral mixture analysis, *Remote Sens. Envir.* **59**: 472–489.

Townshend, J. R. G. and Justice, C. O. (1988), Selecting the spatial resolution of satellite sensors required for global monitoring of land transformations, *Int. J. Remote Sens.* **9**: 187–236.

Walter-Shea, E. A., Norman, J. M., and Blad, B. L. (1989), Leaf bidirectional reflectance and transmittance in corn and soybean, *Remote Sens. Envir.* **29**: 161–174.

Walthall, C., Jean-Louis, R., and Morisette, J. (2000), Field and landscape BRDF optical wavelength measurements: Experience, techniques and the future, *Remote Sens. Rev.* **18**: 503–531.

Welles, J. M. and Cohen, S. (1996), Canopy structure measurement by gap fraction analysis using commercial instrumentation, *J. Exp. Botany* **47**: 1335–1342.

Wigley, T. M. L., Jones, P. D., Briffa, K. R., and Smith, G. (1990), Obtaining sub-grid scale information from coarse resolution general circulation model output, *J. Geophys. Res.* **95**: 1943–1953.

Wilby, R. L. and Wigley, T. M. L. (1997), Downscaling general circulation model output: A review of methods and limitations, *Progress Phys. Geogr.* **21**: 530–548.

Woodcock, C. E. and Strahler, A. H. (1987), The factor of scale in remote sensing, *Remote Sens. Envir.* **21**: 311–332.

Xu, C. (1999), From gcm to river flow: A review of downscaling methods and hydrologic modelling approaches, *Progress Phys. Geogr.* **23**: 229–249.

Zege, E. P. and Kokhanovskiy, A. A. (1989), Approximation of the anomalous diffraction of coated spheres, *Atmos. Oceanic Phys.* **25**: 883–887.

Zhou, J., Civco, D. L., and Silander, J. A. (1998), A wavelet transform method to merge landsat tm and spot panchromatic data, *Int. J. Remote Sens.* **19**: 743–757.

Zhukov, B., Oertel, D., Lanzl, F., and Reinhackel, G. (1999), Unmixing-based multisensor multiresolution image fusion, *IEEE Trans. Geosci. Remote Sens.* **37**: 1212–1226.

13

Applications

Earth observing satellite missions are always driven by scientific questions and practical applications. In the previous chapters, we examined various techniques for estimating land surface variables. In this chapter, I would like to demonstrate how these derived products can be used to solve practical real-world problems. The application areas range from environment and resource monitoring (agriculture and urban heat island effect) to global change related issues (carbon cycle and atmosphere-land surface interactions).

Section 13.1 discusses how to integrate remote sensing with ecological process models. Different strategies are illustrated. The methodology should be general and suitable for any land surface dynamic process models. Section 13.2 discusses various applications in agriculture. It starts with precision agriculture, then moves to some specific examples using remote sensing information in decision support systems, methods for agricultural drought monitoring, and various approaches for crop yield estimation over large areas. Section 13.3 presents some historical and more recent studies on detecting and assessing urban heat island effects using remote sensing techniques. Section 13.4 introduces how remote sensing products discussed in previous chapters can be used for assessing global carbon cycle. Two specific models are also presented to demonstrate the importance of remote sensing techniques. Section 13.5 discusses how remote sensing products can be used for improving land–atmosphere interaction studies.

13.1 METHODOLOGIES FOR INTEGRATING REMOTE SENSING WITH ECOLOGICAL PROCESS MODELS

In the previous chapters, we discussed various techniques for estimating land surface variables. A majority of the work on developing estimation methods has been attempted without reference to the needs of various applications. In fact, I believe that the future development of quantitative remote sensing of land surface relies on a successful linkage with various land surface process models.

Several review articles summarize the methods of linking remotely sensed data with various models (e.g., Maas 1988a, Bouman 1992, Delecolle et al. 1992, Fischer et al. 1997, Moulin et al. 1998, Plummer, 2000). We will closely follow a discussion by Plummer (2000) in the following text.

In this section, ecological process models comprise mechanistic simulation of physiological processes in natural vegetation, although many of the following observations derive from experience in developing similar models of crop growth and surface-vegetation-atmosphere transfer (SVAT).

There are a number of approaches for linking remotely sensed data and ecological processing models at a range of spatial and temporal scales, which can be roughly grouped into four strategies:

- To use remotely sensed data to provide estimates of variables required to drive ecological processing models. This is also called the "forcing" strategy.
- To use remotely sensed data to test, validate or verify predictions of ecological process models. This is called the "calibration" strategy.
- To use remotely sensed data to update or adjust ecological process model predictions. This is called the "assimilation" strategy.
- To use ecological process models to understand remotely sensed data.

It will be demonstrated that remote sensing is not just a source of spatially comprehensive driving variables but can also be used to constrain and validate model behavior. It is also important to note that the process is not unidirectional and the terrestrial remote sensing community can benefit by using ecological models to aid the interpretation of remotely sensed observations. In the following, we discuss these strategies in detail.

13.1.1   Strategy 1: Remotely Sensed Data for Driving Ecological Process Models

The most common rationale for interfacing remote sensing and any land surface dynamic models is to use remotely sensed data to generate model initialization products (Fig. 13.1). These input data correspond to forcing functions or state variables in ecological modeling. The approach is frequently applied to extract measurements of surface atmospheric conditions and vegetation related information. Variables characterizing the atmospheric conditions at the surface level and estimated from remote sensing include incident and reflected photosynthetically active (PAR) and shortwave radiation, cloud cover, air temperature, and atmospheric precipitation. Vegetation-related variables include leaf area index, land surface temperature, soil

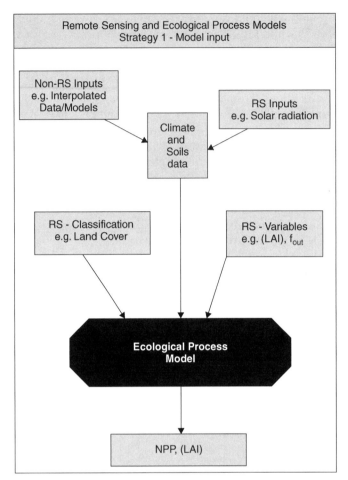

Figure 13.1   Linking remote sensing with ecological process models: strategy 1—the use of remote sensing for data input into ecological process models. [From S. E. Plummer (2000), *Ecol. Model.* Copyright © 2000 with permission from Elsevier.]

moisture, albedo, and land cover. Land cover maps are traditionally used as surrogates to which vegetation attributes can be applied or to define a geographic segmentation. Increased efforts are being made to estimate vegetation variables (e.g., LAI, FPAR, foliar biochemical quantities) directly from remote sensing as discussed in the previous chapters, rather than using mean values as assigned to land cover classes.

From a remote sensing perspective, we need to define the most appropriate spatial and temporal scales for these variables, and conduct methodological research directed at specifying and matching the required estimation accuracy. A bigger and more difficult issue is that we need to revisit ecological models to define the most appropriate variables that remote sensing should be used to provide. For example, LAI has been used as an intermittent variable by ecological models in representing photosynthetic processes and estimating canopy radiation absorption. One obvious alternative strategy is to provide direct estimation of radiation interactions from remote sensing and reformulate the ecological models to accept these remote sensing products. Of course, LAI is used in the calculation of many other process rates from canopy gas exchange to litterfall nutrient return. It remains an essential variable for most ecological models.

13.1.2 Strategy 2: Remotely Sensed Data to Test Predictions of Ecological Models

The second use of remotely sensed data in ecological models is to test and validate whether they are able to produce reasonable results (see Fig. 13.2).

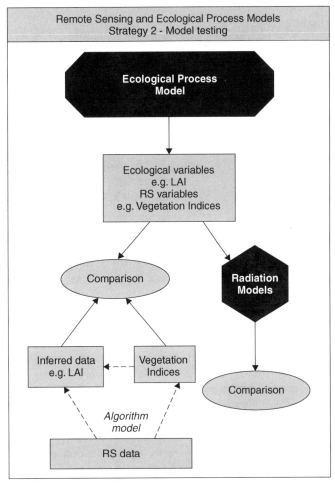

Figure 13.2 Linking remote sensing with ecological process models: strategy 2—the use of remote sensing for calibration/validation of ecological process models. [From S. E. Plummer (2000), *Ecol. Model.* Copyright © 2000 with permission from Elsevier.]

Comparisons can be made at several levels. In direct comparison, the output of the ecological model is simply compared against the estimated variable from remote sensing (e.g., LAI, skin temperature, albedo). If an ecological model is coupled with a canopy radiative transfer model, we can compare the model predicted canopy reflectance with the retrieved surface reflectance from remote sensing. If an atmospheric radiative transfer model is also coupled, we can compare the top-of-atmosphere (TOA) reflectance or vegetation indices with the direct measurements of remote sensing.

There is a need to determine whether it is necessary to validate models globally or at specific sites representing the key global biomes and whether there is a need to ensure that the spatial scales of model and observation are comparable. From an ecological modeling perspective, we need to focus on the adaptation of ecological process models so that they generate more of the required variables needed to model reflectance or estimate reflectance accurately as a model output.

13.1.3 Strategy 3: Remotely Sensed Data to Constrain Ecological Process Models through Data Assimilation

This is the so-called data assimilation strategy discussed in Section 11.5. The development of this approach has been pioneered in crop growth modeling but has wider relevance in ecological modeling. This is illustrated in Fig 13.3.

The simplest approach is direct insertion. Whenever remotely sensed values are available, model state variables are replaced. It works well only when we have regular observations through the growth cycle. This method was called "updating" by Maas (1988a).

A more advanced approach is to adjust the initial model conditions to "best fit" a wide range of observed data that could be the retrieved values (e.g., LAI) or direct measurements (e.g., TOA reflectance). It is related to reinitialization and reparameterization discussed earlier (Maas 1988a). Reinitialization endeavors to determine the values of initial variable conditions that minimize the error between model and observation. Reparameterization is similar to reinitialization except that the model parameters are adjusted rather than the variables.

Since there are generally more parameters than initial conditions, reparameterization is usually more complex than reinitialization and often requires an iterative numerical approach to find the "most acceptable" value from the multiple solutions. There is no guarantee that this will lead to the "correct" set of parameters. Whether by reinitialization or reparameterization, assimilation provides an added advantage in that it can be used to understand the sensitivities of the model system to various parameters as well as uncertainties in the estimates. Many sophisticated assimilation algorithms have been discussed in Chapter 11.

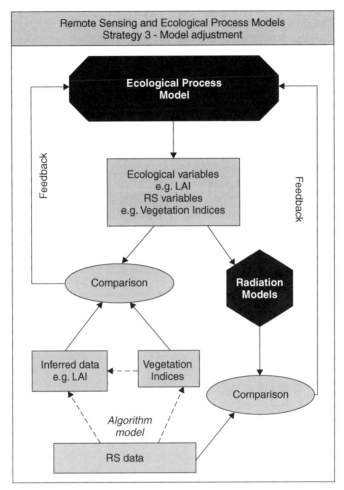

Figure 13.3 Linking remote sensing with ecological process models: strategy 3—the assimilation of remotely sensed data products into ecological process models. [From S. E. Plummer (2000), *Ecol. Model.* Copyright © 2000 with permission from Elsevier.]

13.1.4 Strategy 4: Remotely Sensed Data to Aid the Interpretation of Ecological Models

In the three previous strategies the objective was to use remotely sensed data to support ecological modeling. However, there is value in inverting the approach, such that the ecological model is used to constrain, validate, or understand remotely sensed data. The fourth strategy suggested here concerns the use of ecological models to aid the interpretation of remotely sensed data. Figure 13.4 shows two suggested pathways for this process. In the first, ecological models are deployed to constrain reflectance model inversion, and secondly, ecological models are used to assess the predictive or diagnostic capacity of remotely sensed data.

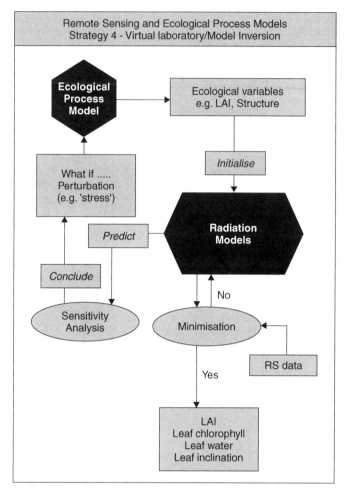

Figure 13.4 Linking remote sensing with ecological process models: strategy 4—the use of ecological process models to understand remote sensing. [From S. E. Plummer (2000), *Ecol. Model.* Copyright © 2000 with permission from Elsevier.]

Because of the complexity of the natural environment, it is difficult to develop models that are accurate and invertible. As we discussed in Section 8.2, the problem with traditional optimization inversion is that there are usually multiple minima, which implies instability. "Successful" inversion can therefore be satisfactorily resolved only by predetermining many parameters. Currently the initialization of a model for numerical inversion is conducted using local knowledge or from experience gleaned from the literature. In theory, ecological models could provide the initialization data, based on local climatic and soil conditions. However this will require adaptation of ecological models and development of long-term monitoring sites to act as tests for the inversion procedure. Another aspect is that most inversion methods discussed in Chapter 8 are based on spectral and angular signatures of remote sensing, and ecological models can provide the valuable information to inversion in the time dimension.

The second approach where ecological models could prove useful is in rigorous assessment of remote sensing observations and algorithms. An example of this is the use of remote sensing for detection of stress in vegetation. The response of vegetation to stress depends on the timing, duration, and toxicity of the contamination, and it will be manifest spectrally through a change in leaf area, projected leaf area (wilting), or the photosynthetic apparatus (leaf color). A more mechanistic approach to the use of remote sensing for "stress" detection must account for dependence on local conditions. The physiological behavior of a plant canopy, constrained by the local meteorology, hydrology, and pedology, can be used to examine rigorously the remotely sensed response of the plant to different pollution scenarios through the coupling of mechanistic models of (1) plant physiology, (2) pollutant pathways, and (3) radiative transfer in a vegetation canopy. The interpretation of "stress" is then through comparison of the observed biophysical variables (LAI, chlorophyll) against the expected performance predicted by the physiological model run in the absence of contamination.

13.2 AGRICULTURAL APPLICATIONS

13.2.1 Remote Sensing and Precision Agriculture

According to The U.S. National Research Council (NRC 1997), "Precision agriculture is a management strategy that uses information technology to bring data from multiple sources to bear on decisions associated with crop production." According to Bouma et al. (1999), management decisions made by farmers can be classified into three groups: strategic, tactical, and operational. *Strategic decisions* concern long-term issues (10 years or more), such as the selection of a farming system (mixed, organic or integrated). *Tactical decisions* concern medium-term issues (2–5 years) regarding crop rotation. *Operational decisions* concern short-term issues on a daily basis in the growing season, including the selection and timing of management operations, such as planting, harvesting, fertilizer application, and crop protection measure. Remote sensing has not been used to assist strategic and tactic management decisionmaking extensively, but it has been used for operational management in a number of ways. Some specific examples of applying remote sensing to decision support systems (DSSs) are described in Section 13.2.2.

13.2.1.1 Methodology Overview
There are several approaches to the application of remote sensing to precision agriculture:

- In *the first approach*, remote sensing images are used for anomaly detection. By comparing the imagery of the current season with those from previous seasons, or from one field to another, farmers can obtain useful information for guiding their management practice. However, anomaly detection does not provide quantitative recommendations that can be directly applied to precision farming.

- *A second approach* involves correlating remote sensing signals to specific variables such as soil properties or nitrogen deficiency. Various methods have been discussed in Chapter 8. For example, in the case of nitrogen deficiency, once site-specific relationships have been developed, remote sensing images can then be translated directly to maps of fertilizer application rates. Additional examples about soil properties, litter, and weed mapping are given in the following two subsections. The related information is also available in Section 13.2.3.

- *The third approach* is to convert remotely sensed data quantitatively into various biophysical variables (such as LAI or temperature) and integrate this information into physically based crop growth models. We have discussed this topic in Chapter 11 in depth. For example, Moran et al. (1995) utilized remotely sensed estimates of LAI and evapotranspiration as inputs to a simple alfalfa growth model. The remotely sensed estimates were used to adjust the model's parameters throughout the season and resulted in improved predictions (Mougin et al. 1995, Carbone et al. 1996). Using remotely sensed inputs to crop growth models also provides a means to obtain crop yield predictions over large areas (see also Section 13.2.4). Jones and Barnes (2000) presented an approach to precision farming decisionmaking. The additional use of remote sensing data provides an efficient method to describe spatial variability in terms that can be related to a crop model, making the decisionmaking approach feasible for precision farming applications. The crop model provides information that can be used by the decision model, and the remote sensing data are used to fine-tune the calibration of the crop model, maximizing the accuracy of its results.

The latter two approaches have potential for incorporating remote sensing into decision support systems in a geographic information system (GIS) environment (e.g., Brown and Steckler 1995). Further developments could ultimately allow farm managers to make informed decisions about site-specific applications of farm materials.

It has been widely recognized (e.g., Moran et al. 1997, Brisco et al. 1998, Stafford, 2000), however, that the current satellite sensors have limited applications directly because of the low number of spectral bands, coarse spatial resolution, and inadequate repeat coverage. Although the current and upcoming commercial Earth observation satellites will overcome some of these limitations, it will still be almost impossible to acquire satellite imagery very frequently over every farm in the United States in the near future because of numerous factors (e.g., overall costs and weather conditions). Moreover, remote sensing can be used only for estimating certain biophysical variables, and cannot be used directly for crop management decisionmaking in many cases. The remotely sensed variables have to be integrated with ancillary data, such as soil, weather, and the past management practices, to provide higher-level information that is pertinent for making all strategic, tactic, and operational decisions at a field level through the crop decision support systems. One of the best approaches for achieving this goal is use of data assimilation, which is a new area and has been discussed in Section 11.5.

Advances in precision farming technology (geographic information systems, global positioning systems, and variable-rate equipment) provide the tools needed to apply information from multispectral images to management problems. There is still considerable work to be done before the full benefits of remotely sensed data can be realized, but there are applications that can benefit from those data at the present time. Following are a few examples of how remote sensing can currently meet some of the information needs in precision agriculture.

13.2.1.2 *Soil Properties*
Many studies have related surface reflectance to various soil properties, such as texture, organic matter, iron oxide content, and soil nutrient (Irons et al. 1989, Ben-Dor et al. 1999, Barnes and Baker 2000). Soil properties are among those key variables that affect crop growth. Because it is very time-consuming and expensive to take soil samples from the field and conduct lab analysis, information available to researchers on soil properties is seldom sufficient. Using remote sensing techniques to characterize soil variables is definitely an area that deserves more study.

Soil physical properties such as organic matter have been correlated to specific spectral responses (Dalal and Henry 1986, Shonk et al. 1991). Therefore, multispectral images have shown potential for the automated classification of soil mapping units (Leone et al. 1995). Such direct applications of remote sensing for soil mapping are limited because several other variables can impact soil reflectance, such as tillage practices and moisture content. However, bare soil reflectance could have an indirect application in interpolating the results of gridded soil samples. The nitrogen status of crops has also been estimated using remotely sensed data (Filella et al. 1995). Palacios-Orueta and Ustin (1998) demonstrated that hyperspectral remote sensing is very useful for estimating some soil properties, such as organic matter, iron content, and texture. Liu et al. (2002) and Muller and Decamps (2001) found that surface reflectance can be well related to soil moisture content, which is consistent with some of earlier conclusions (e.g., Bedidi et al. 1992). Related discussions of the dependence of reflectance on soil moisture have been given in Section 4.5.6.

Chang et al. (2001) evaluated the ability of near-IR reflectance spectroscopy to predict diverse soil properties: 33 chemical, physical, and biochemical properties were studied for 802 soil samples collected from four major land resource areas (MLRAs). Total C, total N, moisture, cation-exchange capacity (CEC), 1.5 MPa water, basal respiration rate, sand, silt, and Mehlich III extractable Ca were successfully predicted ($r^2 > 0.8$). Some Mehlich III extractable metals (Fe, K, Mg, Mn) and exchangeable cations (Ca, Mg, and K), sum of exchangeable bases, exchangeable acidity, clay, potentially mineralizable N, total respiration rate, biomass C, and pH were also estimated but with less accuracy. The results indicate that near-IR reflectance spectroscopy can be used as a rapid analytic technique to simultaneously estimate several soil properties with acceptable accuracy in a very short time.

Shepherd and Walsh (2002) developed a scheme for development and use of soil spectral libraries for rapid nondestructive estimation of soil properties based on analysis of diffuse reflectance spectroscopy. A diverse library of over 1000 archived topsoils from eastern and southern Africa was used to test the approach. They showed the response of prediction accuracy to sample size and demonstrated how the predictive value of spectral libraries can be iteratively increased through detection of spectral outliers among new samples. The spectral library approach opens up new possibilities for modeling, assessment and management of risk in soil evaluations in agricultural, environmental, and engineering applications.

13.2.1.3 *Litter Cover*

Litter is senescent (or dead) plant material that gradually decomposes into soil. The decay of litter adds nutrients to the soil, improves soil structure, and reduces soil erosion (Aase and Tanaka 1991). The annual loss of 1.25 billion tons of soil in the United States could be reduced by leaving litter on bare soil (USDA 1991). Litter also affects water infiltration, evaporation, porosity, and soil temperatures. In agriculture systems, quantifying crop residue cover is also necessary to evaluate the effectiveness of conservation tillage practices. Quantifying plant litter cover is important for interpreting vegetated landscapes and for evaluating the effectiveness of conservation tillage practices. Current methods of measuring litter cover are subjective, requiring considerable visual judgment. Reliable and objective methods are needed.

Remote sensing techniques have been used in quantifying litter cover. The spectral reflectance curves of plant litter and soils have similar, generally featureless shapes in the visible and near infrared (0.4–1.1 μm) wavelength ranges (Aase and Tanaka 1991; Daughtry et al. 1995, 2001), but the slope of the reflectance spectra at the VIS–NIR transition (i.e., 680–780 nm) is generally greater for litter than for soils depending on moisture conditions and litter decomposition age (Goward et al. 1994, Daughtry 2001). Several studies have noted the unique spectral features of dried litter and soils in the shortwave infrared (1.1–2.5 μm) region (Elvidge 1990, Daughtry et al. 1995, Daughtry 2001). Nagler et al. (2000) observed an absorption feature associated with cellulose and lignin at 2.1 μm in the spectra of dry plant litter, which was not present in the spectra of soils. A new spectral variable, *cellulose absorption index* (CAI), was defined using the relative depth of the reflectance spectra at 2.1 μm. The CAI of dry litter was significantly greater than CAI of soils. CAI generally decreased with age of the litter. Water absorption dominated the spectral properties of both soils and plant litter and significantly reduced the CAI of the plant litters. Nevertheless, the CAI of wet litter was significantly greater than CAI of wet soil.

13.2.1.4 Weed and Insect Mapping

Weed growth and reproduction in agricultural fields is one of the most consistent production problems facing farmers. Weeds can influence remote sensing and modeling of crop production in agricultural fields in two ways. First, they become mixed with crop vegetation in agricultural fields and, therefore, can interfere with accurate measurement of crop parameters by remote sensing. New algorithms for spectral and textural analysis will be needed to accurately discriminate weed patches from crops (Radhakrishnan et al. 2002). Second, weeds compete with crops for light, moisture, and nutrients, often leading to reduced crop productivity. Recently developed competition models (Kropff and van Laar 1994) can be incorporated into production models to estimate yield losses from weeds. Remote sensing can provide the necessary information to accurately map weed population distributions across fields that, in turn, will be required to take full advantage of precision weed management technologies. By combining accurate weed maps derived from remote sensing with the power of modeling to project potential yield losses due to weeds, site-specific weed management strategies can be developed that maximize economic gains and minimize environmental contamination.

Extensive reviews of the spectral properties of plants for potential applications in crop/weed discrimination (Zwiggelaar 1998) and site-specific weed management (Radhakrishnan et al. 2002) suggest that it should be possible to use high spatial, temporal, and spectral resolution satellite or airborne images to accurately map weed spatial distributions at the field level for incorporation into decision support systems. Sprayer-mounted sensors have been found useful for the control of herbicide applications (Shearer and Jones 1991). Brown and Steckler (1995) developed a method to use digitized color-infrared photographs to classify weeds in a no-till cornfield. The classified data were placed in a GIS, and a decision support system was then used to determine the appropriate herbicide and amount to apply.

A related topic is the assessment of insect damage. Peñuelas et al. (1995) used reflectance measurements to assess mite effects on apple trees. Powdery mildew has also shown to be detectable with reflectance measurements in the visible portion of the spectrum (Lorenzen and Jensen 1989). Using remote sensing imagery for weed and pest detection in this manner can lead to focused application of herbicide and pesticide. This has obvious financial and environmental advantages.

13.2.2 Examples of Decision Support in Agriculture

Diak et al. (1998) developed a suite of products for agriculture that are based in satellite and conventional observations, as well as state-of-the-art forecast models of the atmosphere and soil canopy environments. These products include

- An irrigation scheduling product for potato crops based on satellite estimates of daily solar energy
- A frost protection product that relies on prediction models and satellite estimates of clouds
- A product for the prediction of foliar disease that is based in satellite net radiation, rainfall, and a detailed model of the soil canopy environment

Two computer simulation models were used. One is a mesoscale model of the atmosphere above the United States that provides 1- and 2-day forecasts. The second is a detailed plant–environment model to assist the development of simple surface models that are interfaced with the mesoscale model to provide site-specific information. We will briefly describe these three products, but more detailed descriptions are also available from their Website at `http://cimss.ssec.wisc.edu/intrdisc/ag/ag.html`.

13.2.2.1 Irrigation Management

The purpose of this tool is to assist potato farmers in determining how much and when to irrigate. This minimizes the amount of water used and fertilizer reaching the groundwater beneath their crops.

Irrigation is an essential practice on sandy soils, and serves to stabilize yield fluctuations on other soils. In the north central United States, water resources are plentiful and irrigation is a sustainable practice (in contrast to more arid regions). Excessive irrigation water application, however, leads to leaching of nitrogen and pesticides from the soil and into the groundwater. Additionally, the seasonal demand for electricity to pump irrigation water coincides with that for air conditioning demand, creating large peak power requirements.

A decision support system (DSS) for irrigation scheduling is in common use, but the required daily evaporation estimate input is derived from sparse ground-based observations and is not always available to end users on demand. Attempts have been made to link this DSS with spatially complete data inputs provided by satellites and disseminated by commercial services, assuring immediate access. Reduced groundwater contamination and energy usage will be realized at the farm level through wider adoption of scheduling that more closely matches crop needs.

Potential evapotranspiration (ET) is calculated from satellite-derived measurements of solar radiation (Diak et al. 1996) and air temperatures at regional airports. The ET values are a reasonable estimate of daily crop water use for most crops that have reached at least 80% coverage of the ground. Prior to 80% or greater coverage, ET will be a fraction of the full ET value in proportion to the amount of crop coverage. This can also be estimated from remote sensing imagery. The topic of ET estimation discussed further in Section 13.2.3.

13.2.2.2 *Potato Foliar Disease Prediction*

This tool predicts development of leaf blight disease on potatoes, allowing growers to apply fungicide sprays only when necessary. Related technical details are discussed by Anderson et al. (2001).

Development of plant diseases that can be ruinous to high-value crops such as potatoes, onions, or green beans depends strongly on temperature and duration of foliage wetness. In potato crops (a \$120–\$150 million/year crop in Wisconsin), repeated fungicide sprays are needed to control devastating foliar diseases. In the past, applications were made at regular time intervals (5–7 days).

The prediction of foliar wetness in canopies is so difficult that when plant pathologists developed a potato blight prediction model, they required growers to measure humidity in potato canopies rather than attempt prediction. The inconvenience and expense associated with this humidity measurement discourages growers from participating in the potato blight prediction program.

Foliage wetness can arise from rainfall, irrigation, or dew formation. Canopy wetness can be predicted from a model that uses data from grower-measured irrigation and rainfall; GOES satellite radiation estimates; measurements of temperature, humidity, and wind speed from weather service locations and agricultural measurement networks; and forecasts from a mesoscale model of the atmosphere.

The predictions of conditions leading to leaf wetness have been combined by farmers with local information about irrigation and rain, to derive an estimate of how long leaves have been wet. This wetness duration estimate can then be used with existing disease prediction models to determine if a fungicide spray is needed.

13.2.2.3 *Cranberry Frost Prediction*

The purpose of this tool is to advise cranberry growers of the chance of nighttime frost in their marshes, both early in the day and as the evening develops. Ready access to real-time data and mesoscale model predictions of minimum temperatures will reduce unnecessary applications of water for frost protection.

Cranberries are cultivated on modified wetlands where frost occurs throughout the summer growing season. Reduction of frequent (depending on the year) and heavy water applications required for frost protection will reduce off-site movement of agricultural production chemicals in surrounding natural wetlands and reduce environmental concerns.

Cranberry beds tend to be several degrees cooler than nearby higher ground during nights of high radiant energy loss. This is due to the dense cover of vines and leaves, which blocks radiant transfer from the soil, and the low, level beds interspersed with berms, conditions that eliminate cold-air drainage.

The tools used to develop these decision aids are data sources and research models. The primary data sources are estimates of solar energy at the Earth's surface based on GOES satellite images (Diak et al. 1996, 2000) and surface weather observations.

13.2.3 Drought Monitoring

Drought is a complex natural event and one of the most damaging environmental phenomena. It originates from a deficiency of precipitation that results in a water shortage on the land surface. Droughts can be considered the opposite of flooding. Floods result from an overabundance of rainfall; droughts, from a lack of rainfall. Drought can have a greater impact, financially and on human health, than floods. Floods usually affect only a small portion of the region, whereas droughts can have widespread effects. The impact of the a large-area severe drought in 1988 on the U.S. economy has been estimated at \$40 billion, which is 2–3 times the estimated losses from the 1989 San Francisco (Loma Prieta) earthquake (Riebsame et al. 1990). Unlike floods and blizzards, whose impacts are felt in a matter of hours or days, a severe drought may not really start to impact an area for months or years since droughts usually last longer and develop more slowly. The insidious nature of drought makes it very difficult to assess its severity early on.

A universally accepted definition of drought does not exist, but four major types of droughts are broadly defined and agreed on in the literature (McVicar and Jupp 1998): meteorological drought, agricultural drought, hydrological drought, and socioeconomic drought. In this section, we discuss mainly how to monitor and assess agricultural drought using remote sensing technique. Agricultural drought occurs when available plant water falls below that required by the plant community during a critical growth stage. This leads to below-average yields.

An excellent review on drought monitoring using remote sensing technique has been provided by McVicar and Jupp (1998). They grouped all methods for monitoring agronomic conditions into four categories:

1. Vegetation condition monitoring with reflective remote sensing
2. Environmental condition monitoring with thermal remote sensing
3. Soil moisture monitoring with microwave remote sensing
4. Environmental stress monitoring with the combined thermal and reflective remote sensing

We discuss mainly methods 1, 2, and 4 in the rest of this section.

13.2.3.1 *Vegetation Condition: Monitoring with Reflective Imagery*

AVHRR data have been widely used for monitoring and assessing the global drought (Kogan 1997). There are a number of parameters that can be used to characterize and assess vegetation condition. One basic approach is to examine multitemporal NDVI profiles and detect any anomaly. Since the AVHRR data are usually quite noisy, normalization processes have made such procedures more effective. For example, NDVI has been converted into the *vegetation condition index* (VCI)

$$\mathrm{VCI}_i = \frac{\mathrm{NDVI}_i - \mathrm{NDVI}_{min}}{\mathrm{NDVI}_{max} - \mathrm{NDVI}_{min}} \tag{13.1}$$

where NDVI_{min} and NDVI_{max} are the minimum and maximum NDVI values of the whole time sequence. The primary purpose of developing VCI was to assess changes in the NDVI signal through time due to weather conditions, by reducing the influence of geographic or ecosystem variables, such as climate, soils, vegetation type, and topography (Kogan 1997).

Relating NDVI with meteorological parameters is a further step in addressing this issue. Efforts have been made to link time-series analysis of an NDVI signal with a moisture balance model to explain and assist in monitoring of the NDVI signal. For example, Nicholson and Farrar (1994) illustrated that while rainfall is the main determinant of a response to NDVI, there are also other environmental variables to consider. They found that soil type is more important than vegetation type in Botswana. They also explored the generation of available plant soil moisture as a function of soil porosity, through the use of a water balance model. They also found that moisture use efficiency, defined as the slope of the linear regression between NDVI and soil moisture, is influenced more strongly by soil type than vegetation type.

Currently, the USGS Famine Early Warning System Network (FEWS-NET) employs data from two operational remote sensing products to monitor agricultural areas for signs of drought on a near-real-time, spatially continuous basis. These include AVHRR NDVI data from NASA and rainfall estimate images prepared by the Climate Prediction Center of the National Oceanic and Atmospheric Administration (NOAA).

13.2.3.2 *Environmental Condition: Monitoring with Thermal Imagery*

Similar to the VCI, the retrieved land surface temperature T from AVHRR thermal data is also normalized to form the *temperature condition index* (TCI)

$$\mathrm{TCI}_i = \frac{T_i - T_{min}}{T_{max} - T_{min}} \tag{13.2}$$

where T_{min} and T_{max} are the minimum and maximum T values of the whole time sequence. Drier surfaces tend to have higher skin temperature. To take advantage of both VCI and TCI, a linear combination of these two indices generates a new index for more effective monitoring. Kogan (1997) suggested 0.5 as the weight for each.

Rising leaf temperatures is a good indicator of plant moisture stress and precedes the onset of drought. Plant moisture stress occurs when the demand for water exceeds the available soil moisture level. The most established method for detecting crop water stress remotely is through the measurement of a crop's surface temperature (Jackson 1982). The correlation between surface temperature and water stress is based on the assumption that as a crop transpires, the evaporated water cools the leaves so that their temperatures are below air temperature. As the crop becomes water stressed, transpiration will decrease, and thus the leaf temperature will increase. Other factors need to be accounted for in order to obtain a good measure of actual stress levels, but leaf temperature is one of the most important.

In the following paragraphs we mainly introduce two indices: *crop water stress index* (CWSI) and *normalized difference temperature index* (NDTI). Both empirical CWSI and theoretical CWSI are described.

Empirical CWSI. Idso et al. (1981) and Jackson (Jackson et al. 1981, Jackson 1982) pioneered a method using the daytime canopy temperature called the *crop water stress index* (CWSI), which is now used routinely for assessing crop health and establishing irrigation scheduling at the field scale.

CWSI is expressed as (Idso et al. 1981)

$$\text{CWSI} = \frac{\Delta T_{ca} - \Delta T_l}{\Delta T_u - \Delta T_l} \tag{13.3}$$

where ΔT_{ca} is the difference between crop canopy and air temperature, ΔT_u is the upper limit of canopy minus air temperature (nontranspiring crop), and ΔT_l is the lower limit of canopy minus air temperature (well-watered crop). A CWSI of 0 indicates no water stress, and a value of 1 represents maximum water stress. The crop water stress that signals the need for irrigation is crop-specific and should consider factors such as yield response to water stress, probable crop price, and water cost. Reginato and Howe (1985) found that cotton yield showed the first signs of decline when the CWSI average during the season was greater than 0.2. For corn, a CWSI value of 0.3–0.4 might be a conservative timing parameter to avoid excess irrigation (Yazar et al. 1999).

Several methods are used to determine the upper and lower limits in the CWSI equation. One method, developed by Idso et al. (1981), accounts for changes in the upper and lower limits due to variation in *vapor pressure deficit* (VPD). VPD is calculated as

$$\text{VPD} = \text{VP}_{sat} - \text{VP} \tag{13.4}$$

where VP_{sat} is the maximum vapor pressure for a given air temperature and

TABLE 13.1 Baseline Parameters for Various Crops—Sunlit Conditions

Crop	Intercept (α)	Slope (β)
Alfalfa	0.51	−1.92
Barley (preheading)	2.01	−2.25
Barley (postheading)	1.72	−1.23
Bean	2.91	−2.35
Beet	5.16	−2.30
Corn (no tassels)	3.11	−1.97
Cotton	1.49	−2.09
Cowpea	1.32	−1.84
Cucumber	4.88	−2.52
Lettuce, leaf	4.18	−2.96
Potato	1.17	−1.83
Soybean	1.44	−1.34
Tomato	2.86	−1.96
Wheat (preheading)	3.38	−3.25
Wheat (postheading)	2.88	−2.11

Source: Idso (1982).

pressure (the maximum water vapor that the air can hold) and VP is the actual vapor pressure (partial pressure of water vapor in the atmosphere). Therefore, a VPD of 0 indicates that the air is holding as much water vapor as possible (this also corresponds to a relative humidity of 100%). The lower limit in the CWSI will change as a function of vapor pressure because at lower VPDs, moisture is removed from the crop at a lower rate; thus the magnitude of cooling is decreased. Idso (1982) demonstrated that the lower limit of the CWSI is a linear function of VPD for a number of crops and locations. The canopy air temperature difference for a well-watered crop (lower limit) and severely stressed crop (upper limit) can be calculated from Eq. (13.4) as

$$\Delta T_l = \alpha + \beta \cdot \text{VPD} \tag{13.5}$$

$$\Delta T_u = \alpha + \beta \left(\text{VP}_{\text{sat}}(T_a) - \text{VP}_{\text{sat}}(T + \alpha) \right) \tag{13.6}$$

where $\text{VP}_{\text{sat}}(T_a)$ is the saturation vapor pressure at air temperature T_a and $\text{VP}_{\text{sat}}(T_a + \alpha)$ is the saturation vapor pressure at air temperature plus the intercept value α for the crop of interest. Slope α and intercept β have been determined for a number of crops as shown in Table 13.1.

Thus, with a measure of humidity (relative humidity, wet bulb temperature, etc.), air temperature, and canopy temperature, it is now possible to determine CWSI. The slope and intercept parameters in Table 13.1 were determined from experimental data, so the CWSI calculated using this method is often referred to as the "empirical" CWSI.

There have been many studies using the empirical CWSI to time irrigations of field crops, some of which were very successful. For example, Anconelli et al. (1994) found that tomato crops yield was about 35 tons/ha with no irrigation, and increased to 51, 57, and 60 tons/ha irrigated using CWSI thresholds of 0.6, 0.35, and 0.1. But there were also some studies reporting negative results. For example, Al-Faraj et al. (2001) stated that CWSI is not a reliable index because the empirically determined coefficients could not take into account many other factors.

Theoretical CWSI. Another method is the so-called "theoretical" CWSI. It has the same equation as (13.3), but the coefficients are determined by combining the Penman–Monteith equation with a one-dimensional energy balance equation discussed in Section 10.10. By rearranging terms of the surface energy balance equation in Section 10.10, Jackson et al. (1981) were able to develop an equation to predict the canopy minus air temperature difference ($T_c - T_a, °C$):

$$T_c - T_a = \frac{r_a(R_n - G)}{YC_p}\left[\frac{K(1 + r_c/r_a)}{\gamma + K(1 + r_c/r_a)}\right] - \frac{\text{VPD}}{\gamma + K(1 + r_c/r_a)} \quad (13.7)$$

where all the terms are as defined in Section 10.10 and γ is the slope of the saturated vapor pressure–temperature relation. Jackson et al. (1981) found that γ could be sufficiently represented by

$$\gamma = 45.03 + 3.014\overline{T}_{\text{ca}} + 0.05345\overline{T}_{\text{ca}}^2 + 0.00224 \times 10^{-3}\overline{T}_{\text{ca}}^3 \quad (13.8)$$

where $\overline{T}_{\text{ca}}$ is the average of the canopy and air temperature (°C). Even with the simplification of Eq. (13.8), air and canopy temperature still appear on both sides of Eq. (13.7), so iteration is needed to obtain a solution. Equation (13.7) can now be used to obtain the upper and lower bounds for the CWSI. In the case of the upper limit (nontranspiring crop) canopy resistance will approach infinity, and Eq. (13.7) reduces to

$$\Delta T_u = \frac{r_a(R_n - G)}{(YC_p)} \quad (13.9)$$

In the case of a non water stressed crop, assuming r_c is essentially 0:

$$\Delta T_l = \frac{r_a(R_n - G)}{YC_p}\left[\frac{K}{\gamma + K} - \frac{\text{VPD}}{\gamma + K}\right] \quad (13.10)$$

Now Eqs. (13.9) and (13.10) can be used to determine the CWSI as given in

Eq. (13.3), which in fact can be expressed as

$$\text{CWSI} = 1 - \frac{\text{ET}}{\text{ET}_p} = 1 - m_{\text{ad}} \tag{13.11}$$

where ET and ET_p are the daily actual evapotranspiration and the daily potential evapotranspiration, respectively and m_{ad} is the *moisture availability* (McVicar and Jupp 1998).

CWSI has been used extensively for such important farm applications as irrigation scheduling, predicting crop yields, and detecting certain plant diseases. Since it requires a measurement of foliage temperature and most remote sensing systems measure the surface temperature of the mixed medium (both canopy and soil), application has generally been limited to fully vegetated sites, such as agricultural fields (Moran et al. 1996). The extension to partially vegetated surfaces is discussed in Section 13.2.3.3.

Normalized Difference Temperature Index (NDTI). The NDTI is defined as (McVicar and Jupp 1998):

$$\text{NDTI} = \frac{T_\infty - T_s}{T_\infty - T_0} \tag{13.12}$$

where T_∞ is the modeled surface temperature if there is an infinite surface resistance (i.e., ET = 0), T_0 is the modeled surface temperature if there is zero surface resistance (i.e., ET = ET_p), and T_s is the estimated LST from remote sensing data. According to McVicar and Jupp (1998), there are a number of advantages to driving the NDTI rather than the CWSI:

- NDTI is a very close approximation of moisture availability. NDTI can be developed as a standard product, which is an easily computable surrogate for moisture availability. The required modeling techniques to determine NDTI for the time and date of a satellite overpass have been developed.
- When NDVI is high, NDTI is also high. This is of benefit for the intercomparisons of the remotely sensed products.

13.2.3.3 *Environmental Stress: Monitoring with Combined Thermal and Reflective Imagery*

Combining thermal and reflective data to determine either soil water status or surface water availability has been reported in numerous studies (Nemani and Running 1993, Carlson et al. 1994, Moran et al. 1994, Gillies et al. 1997). A scatterplot of vegetation cover and surface temperature results in a characteristic triangular or "trapezoid" envelope of pixels. The surface temperature response is a function of varying vegetation cover and surface

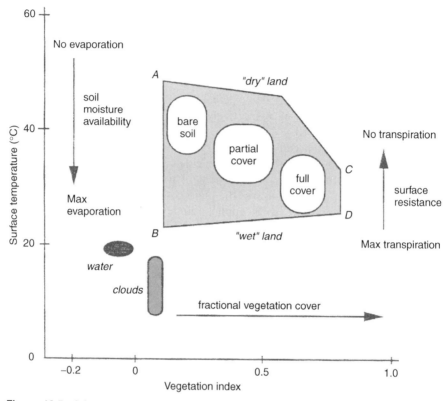

Figure 13.5 Schematic plot of surface temperature and vegetation index space, and the conceptual relationships with evaporation, transpiration, and fractional vegetation coverage. [From Lambin and Ehrlich (1996), *Int. J. Remote Sens.* Reproduced by permission of Taylor & Francis, Ltd.]

soil water content (which can be defined as a surface moisture availability). Lambin and Ehrlich (1996) gave an excellent explanation of the vegetation index (VI) and LST space in terms of evaporation, transpiration and fractional vegetation coverage (see Fig. 13.5) based on previous studies. As they explained, variations in surface brightness temperature are highly correlated with variations in surface water content over base soil. Thus, points *A* and *B* in Fig. 13.5 represent dry bare soil (low VI, high LST) and moist bare soil, respectively. As the fractional vegetation cover increases, surface temperature decreases as a result of several biophysical mechanisms. Point *C* corresponds to continuous vegetation canopies with a high resistance to evapotranspiration (high VI, relatively high LST), which may result from a low soil water availability. Point *D* corresponds to continuous vegetation canopies with low resistance to evapotranspiration (high VI, low LST), which may occur on well-water surfaces. The upper envelope of observations in the

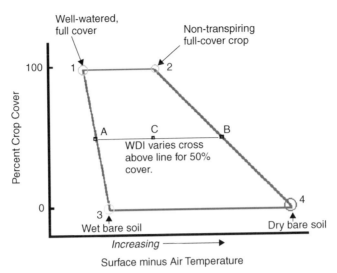

Figure 13.6 Illustration of the water deficit index trapezoid. [From Moran et al. (1994), *Remote Sens. Environ.* Copyright © 1994 with permission from Elsevier.]

VI-LST space, *A–C*, represents the low-evapotranspiration line (i.e., dry conditions). The lower envelop, *B–D*, represents the line of potential evapotranspiration (wet conditions).

In the following paragraphs, we discuss the *water deficit index* (WDI) (Moran et al. 1994) that was based on a similar "VI/temperature trapezoid" defined by the fractional vegetation cover and surface minus air temperature.

The previous discussions of the CWSI have assumed that a measure of canopy temperature was available or that the crop completely covered the soil surface. To consider the situation of bare soils, Moran et al. (1994) developed the WDI that uses both surface minus air temperature and a vegetation index to estimate the relative water status of a field. The concept of the WDI is illustrated in Fig. 13.6.

The distribution of surface minus air temperature at a particular time was found to form a trapezoid when plotted against percent cover. Vegetation indices can be used to calculate percent cover, as discussed in Chapter 8. The upper left of the trapezoid corresponds to a well-watered crop at 100% cover and the upper right to a nontranspiring crop at 100% cover (points 1 and 2, respectively). These two points are the same as those as the upper and lower limits for the standard CWSI and can be estimated using the same techniques. The lower portion of the trapezoid (bare soil) is bound by a wet and dry soil surface. These points can also be calculated using the energy balance concepts previously discussed in Section 10.10; however, the calculation of the resistance terms needs to be more rigorous, particularly in the case of a dry bare soil, as the assumption of atmospheric stability is seldom valid and can introduce large errors. Details on calculating the soil corners are not presented here, but are explained by Moran et al. (1994).

With corners of the trapezoid, the WDI for a measure percent cover becomes

$$\text{WDI} = \frac{\Delta T - \Delta T_{L13}}{\Delta T_{L24} - \Delta T_{L13}} \qquad (13.13)$$

where ΔT is the measure of surface minus air temperature at a particular percent cover, ΔT_{L13} is the surface minus air temperature determined by the line from points 1 to 3 for the percent cover of interest ("wet" line), and ΔT_{L24} is the temperature difference on the line formed between points 2 and 4 ("dry" line). Graphically, the WDI can be viewed as the ratio of the distances AC to AB in Fig. (13.6).

In some cases, the trapezoid resembles a triangle. When vegetation cover and surface temperature measurements are combined in a scatterplot, the data show a characteristic triangle-shaped envelope of pixels. The surface temperature response along the abscissa (horizontal, x axis) is a function of varying vegetation cover and surface soil water content (defined as the surface moisture availability). The so-called "triangle method" has been described in the literature (Nemani and Running 1993, Carlson et al. 1994, Gillies et al. 1997, Moran et al. 1997). A similar concept has been used to develop a new approach called *temperature/vegetation index* (TVX) that has been widely used for estimating soil moisture condition and surface air temperature for the NPP model (Goward et al. 2002) (also see Section 13.4). Jiang and Islam (2001) developed another vegetation index (VI) and LST (T_s) method by linear decomposition of the triangular distribution of the VI–T_s diagram and estimated the α parameter of Priestley-Taylor's equation. This method has clear advantages of simplicity and consistency. It does not require any surface meteorology data. Nishida et al. (2003) developed an algorithm for estimating evaporation fraction (EF), expressed as the ratio of actual evapotranspiration (ET) to available energy, from MODIS data. In their algorithm, a simple two-source model of ET was used and the VI–T_s diagram is used for estimation of EF of bare soil. They have proposed to use this algorithm for producing global distribution of EF every 8 days operationally.

13.2.4 Crop Yield Estimation

Crop yields are strongly influenced by climate, genetics, crop management, and the physical and chemical properties of soils. They can vary considerably from field to field on any farm, and from year to year. Accurate, objective, reliable, and timely predictions of crop yields over large areas are critical for national food security through policymaking on import/export plans and prices, and for the farmers for making their plans on utilization of fertilizer, other chemicals, and irrigation. Yield prediction at the farm level is also very important to the farmers for on-site precision management that have significant economical and environmental implications.

In the following sections, we focus on estimation of crop yields over a large area. We start from the traditional agricultural statistics method Section 13.2.4.1) for estimating and forecasting crop yields, then discuss different methods that are related to remote sensing technique (Sections 13.2.4.2–13.2.4.5). The final two sections are devoted to two operational systems in the United States and Europe.

Note that crop production estimation and forecasts have two components: acres to be harvested and expected yield per acre. Most methods that we discuss are for estimating the crop yield per area. Estimates of acreage rely on high-resolution remote sensing imagery in conjunction with sampling techniques.

13.2.4.1 *Agricultural Statistics*
Agricultural statistics are reliable in estimating crop yields by sampling the field measurements of standing crops. The U.S. Department of Agriculture (USDA) National Agricultural Statistics Survey (NASS) currently uses two major survey techniques for crop yield forecasting and estimation (Haboudane et al. 2002). The first is the use of list or multiple frame-based sample surveys of farm operators. Farmers selected in the statistically based sample are asked to report their final harvested yield or their best evaluation of potential yield, based on current conditions, during the forecast season. The second is objective yield surveys that utilize plant counts and fruit measurements from random plots in selected fields. Because of the high cost of in-field observations, objective yield surveys are normally used only in major producing states. Data from multiple years are used to build models that relate preharvest counts and measurements to the final postharvest yield. A detailed document of USDA/NASS crop yield estimation methods using agricultural statistics (Vogel and Bange 1999) is also available on the Internet (http://www.usda.gov/nass/).

Agricultural statistics methods are time-consuming and very costly. Agricultural information should have the qualities of objectivity, reliability, timeliness, adequacy of coverage, efficiency, and effectiveness. Agricultural statistics in many important agricultural countries do not meet any of these standards. Remote sensing is a promising alternative approach.

13.2.4.2 *Agrometeorological Models*
Early crop yield estimation mainly relied on agrometeorological models (e.g. Thompson 1969, Rudorff and Batista 1990, Ray et al. 1999) that are based on multiple linear regressions of historical yields and monthly averages of temperature and precipitation. The agrometeorological model used in the large area crop inventory experiment (LACIE) (Macdonald and Hall 1980) was a zone-specific polynomial model based on multiple linear regression of

the form:

$$Y = \alpha X + \beta + \varepsilon \tag{13.14}$$

where α and β are coefficients, X is composed of a list of meteorological variables, and ε is the random error. In fitting the model to the data, three factors were considered. Conceptually, this is

$$Y = A + B + C \tag{13.15}$$

where the crop yield (Y) per area depends on three factors: preceding year yield for average weather (A), yearly adjustment for technology trend (B), and effects of current weather (C). Landsat MSS imagery through a stratified random sampling scheme were used to identify crops and estimate their hectage for harvest, and the global weather data were used to forecast crop yield per hectage area. Their method was successfully tested in the United States, Canada, and the former Soviet Union. In a 1977 quasioperational test, the LACIE in-season forecast of a 30% shortfall in the 1977 Soviet spring wheat crop came within 10% of official Soviet figures released several months after harvest (MacDonald and Hall 1980).

As pointed out by Horie et al. (1992), weather variables are highly correlated, but multiple regression requires independent predictors. This makes agrometeorological models unstable. This problem can be partially solved by employing other indices as the predictors. For example, the Palmer moisture anomaly index for assessing drought was used to predict wheat yield in Australia (Sakamoto 1978). Boatwright and Whitehead (1986) presented an empirical model for yield y (kg/ha) as the function of the accumulated CWSI (x):

$$y = 4290 + 3.97x - 2.91x^2 \tag{13.16}$$

CWSI can be estimated from the retrieved daytime temperature from remote sensing, as discussed in Section 13.2.3.2. Many other drought indices might serve this purpose as well. This method is ultimately a statistical model and cannot predict the time-dependent processes of growth and field formation that are critical for real-time yield forecasting.

13.2.4.3 AVHRR Indices

Many studies have utilized empirical regression methods to relate crop yield to surface reflectance and their combinations (i.e., vegetation indices) from AVHRR data. Hayes and Decker (1996) used the 16-km AVHRR global vegetation index (GVI) data as the input in the VCI defined in (13.1) for 42 crop reporting districts in 11 crop producing states in the mid-West of the United States. Tucker et al. (1980) regressed yield to an NDVI time integral at specific times during the growing season. Quarmby et al. (1992, 1993) integrated the area under the AVHRR NDVI signal to determine crop yields. The prediction formula looks like

$$\text{Yield}_{(t)} = A_{(1986)} + B_{(1986)} \sum \text{NDVI}_{(t)} \qquad (13.17)$$

where $\text{Yield}_{(t)}$ is the yield for a particular crop for a particular year t, $\sum \text{NDVI}_{(t)}$ is the integrated NDVI, the coefficients A and B are determined from the official statistical yield data collected in 1986.

Maselli et al. (1992) proposed a new standardization procedure of NDVI for predicting crop yields after discussing the potential problems associated with the use of VCI:

$$\text{NDVI}_s = \frac{\text{NDVI}_{30th, July} - \overline{\text{NDVI}}}{\sigma_{\text{NDVI}}} \qquad (13.18)$$

where $\text{NDVI}_{30th, July}$ is the NDVI value around the critical phonological date, July 30, each year; $\overline{\text{NDVI}}$ is the multiannual mean value of NDVI, and σ_{NDVI} is the multiannual NDVI standard deviation. The normalized NDVI_s values are then related to precipitation and crop yield.

The derived coefficients from NDVI may not be universally suitable for all cultivars, crop growth stages, environments, and agronomic technology. More recent studies (Unganai and Kogan 1998, Dabrowska-Zielinska et al. 2002, Liu and Kogan 2002) have revealed that combining both AVHRR vegetation indices in terms of VCI and TCI is better for predicting crop yields.

Manjunath et al. (2002) developed the so-called spectrometeorological model for estimating wheat yield over the districts or agroclimatic zones. Their validation results indicated that the incorporation of monthly rainfall in the regression yield models in addition to NDVI improves the model performance significantly. Observations indicate that the rainfall prior to sowing has a very large influence on the yields, as it is responsible for the efficiency of germination and is a deciding factor for the root zone soil moisture availability for the crop during its growth.

To date, research shows that AVHRR-based crop yield estimates are not nearly as precise as NASS's existing survey system for U.S. crop yield data. However, in areas with no survey or ground data systems, AVHRR-based data can provide limited year-to-year interpretations of changes in yield levels.

13.2.4.4 *Physiological Models*

Another method is to rely on physiology-based crop models by estimating crop-absorbed PAR from remote sensing (Asrar et al. 1985, Wiegand and Richardson 1990; Clevers 1997). Growth of a crop is the result of absorbed PAR (APAR) accumulation. The total aboveground phytomass production (Y) can be related to the integrated daily APAR (Monteith 1972, Monteith and Unsworth 1990)

$$Y = \int_o^t e \cdot \text{APAR}(t) \, dt \qquad (13.19)$$

where e is the efficiency factor, and $e = e_c e_i e_s$, e_c, e_i, and e_s are the photochemical, interception, and stress efficiency factors, respectively. Monteith 1972) found that e_c was relatively constant for given species throughout the growing season. Asrar et al. (1984) found that its values computed from the aboveground fraction of phytomass for winter wheat were affected by stages of physiological development and some management practices. Diurnal and seasonal variations of e_i were evaluated by Hipps et al. (1983) using direct measurements of components of PAR in winter wheat canopies. e_s is the stress factor including the effects of nutrient deficiencies, temperature, and water stress. For water stress, CSWI defined in Eq. (13.3) can be used to represent its value. Since daily APAR can be calculated by FPAR and the total incoming PAR, dry-matter accumulation between emergency and harvest can be written as

$$Y = e_c \int_{\text{emergency}}^{\text{harvest}} e_i \cdot e_s \cdot \text{FPAR}(t) \cdot \text{PAR}(t) \, dt \qquad (13.20)$$

where FPAR can be estimated from remote sensing observations (see various methods in Chapter 8).

It can be further assumed that the final crop yield is linearly related to the absorbed PAR (Clevers 1997)

$$\text{Yield} = \alpha + \beta \int_{\text{emergence}}^{\text{harvest}} \text{FPAR}(t) \cdot \text{PAR}(t) \, dt \qquad (13.21)$$

where α and β are coefficients. In practice, it may be difficult to obtain APAR throughout the growing season. However, it was found that the APAR at the end of June and early July is highly related to the final sugarbeet yield.

13.2.4.5 Mechanical Crop Growth Models

The final approach we want to discuss in this section is the use of mechanical crop growth models. Jones and Luyten (1998) summarized data indicating that crop growth models have been developed for a wide range of crops and for a variety of applications, such as irrigation management, nutrient management, pest management, land use planning, crop rotation, climate change assessment, and yield estimation.

Duchon (1986) used the CERES-Maize model to predict maize yield during the growing season for large area yield prediction. Hodges et al. (1987) also used the CERES-Maize model to estimate annual fluctuations in maize production in the U.S. Cornbelt for the years 1982–1985. Johnson et al. (1994) described how to apply the results from the yield prediction process to the development of a triparties system of land evaluation based on crop yield prediction, expert systems for biophysical and economical land suitability–risk analysis. They used the AUSCANE model, which is a version of EPIC (erosion–productivity impact calculator) (Williams et al. 1983) that has been modified to simulate the growth and yield of sugarcane in Australia. They simulated the crop yields for 69 core soil types at the farm scale and 69 core soil types at the regional scale over a 25-years period. Haskett et al. (1995) examine the performance of the GLYCIM soybean crop model in modeling regional soybean yields in Iowa at the state and county levels over a 20-year period. Chipanshi et al. (1999) simulate wheat yields using the CERES wheat model at the crop district level containing hundred farms with different soils, climates, and management practices. DSSAT (Decision Support System for Agrotechnology Transfer) is a microcomputer software program combining crop soil and weather databases with crop models and application programs. It has been widely used for crop yield simulations and estimation (Hoogenboom et al. 1999).

Traditionally, crop simulation models are point-based systems, in that they predict growth and development of a plant for a confined space rather than a region (Schulze 2000). Most input data for crop simulation models are also point-based, such as weather data input from a weather station and soil information from a vertical soil profile. Crop management normally relates to local crop inputs at a field level. Crop simulation models are not very well equipped to handle extrapolation over space (Olesen et al. 2000). Many studies have explicitly used GIS in the process of crop yield prediction (e.g., Engel et al. 1997, Hartkamp et al. 1999, Priya and Shibasaki 2001).

At an individual plot level it is relatively easy to determine all required inputs for the crop models, such as the water holding capacity of the soil profile, initial soil water and nitrogen, as well as crop management. On larger scales this becomes much more problematic. For example, it is difficult to determine the crop management information at a county level, so determining which crops have been planted and at what time is rather critical, especially for yield forecasting during the current growing season. Remote sensing can play a key role in determining the various input parameters and initial conditions required by the crop simulation models. Maas (1988b) inverted LAI from Landsat TM imagery and then used the retrieved LAI to estimate the initial LAI value at emergence, which strongly determines the canopy behavior and the production. Different approaches and application examples for assimilating remotely sensed products into crop growth models for the application of crop yield estimation have been discussed in Section 11.5.

In the following paragraphs, we introduce two operational systems in the United States and Europe.

13.2.4.6 *USDA / FAS Program*
The Production Estimates and Crop Assessment Division (PECAD) of the USDA Foreign Agricultural Service (FAS) has an operational system for providing information on crop growth conditions and crop yields of several major crops over the most regions of the world (Reynolds 2001). PECAD is the operational outgrowth of the LACIE and AgRISTARS (Agriculture and Resources Inventory Surveys through Aerospace Remote Sensing) program

which began in 1974 and 1980, respectively (Macdonald and Hall 1980, Boatwright and Whitehead 1986). Its core is based on a database management system called *crop condition data retrieval and evaluation* (CCDRE).

PECAD relies on several different data sources to monitor weather anomalies that affect crop production and quality of agricultural commodities. Two main agrometeorological input data sources include the WMO (World Meteorological Organization) meteorological measurements of more than 6000 stations and the gridded weather data (1/8-mesh grid reference system) from the U.S. Air Force Weather Agency (AFWA) that combines the WMO ground station data with satellite imagery, such as GOES, SSM/I, Meteosat, and GMS.

CCDRE contains several crop models and data reduction algorithms that are executed for these input agrometeorological data. These models include

- Crop calendars models that simulate the crop growth incrementally based on growing degree-days (or thermal unit)
- Crop stress models that combine soil moisture, crop calendar, and a hazard algorithm for alerting analysts of abnormal temperature or moisture stresses that may affect yields
- Several different crop yield reduction models (e.g., Sinclair soybean and CERES wheat models) (Sinclair et al. 1991, Richtie et al. 1998)

Most of these models were developed by the AgRISTARS program, but new models are constantly reviewed for possible integration into the CCDRE operational system.

CCDRE generates more agrometerological variables and other information on crop stage and crop stress for wheat and corn, relative yield reductions for corn, wheat, and soybeans, vegetation indices, and anomalies. Regional PECAD analysts then utilize the agrometeorological data sets output from CCDRE, other high-resolution satellite imagery (e.g., TM/ETM+, SPOT), FAS attach crop reports, in-country sources, wire services, and personal knowledge to estimate national crop production before the 12th of each month. Their estimates of foreign production, use and trade are submitted for interagency review and clearance. Other analysts separately propose changes in foreign use and trade data. This division of responsibilities provides a check and balance system to ensure accuracy and integrity of crop production assessment in the supply and demand estimates. Official crop statistics of other nations, where available, are critical in forming current estimates. But many major producing and trading countries do not publish crop reports until well after the crop has been harvest. In the interim, the USDA must rely on the historical record compared with current conditions. The final estimates are published in the monthly *Crop Production* report and WASDE (*World Agricultural Supply and Demand Estimates*) report.

13.2.4.7 *European MARS Project*

The MARS (monitoring agriculture through remote sensing) project of the Joint Research Center of European Union (EU) has operationally predicted the yields of many crops for the EU countries. The crop growth monitoring system (CGMS) is the MARS agro-meteorological modeling system based on the WOFOST model (Van Diepen et al. 1989) that uses data on soil and meteorological conditions to simulate crop growth. In 1994, it was expanded to cover 10 different crops: winter wheat, soybean, field beans, barley, maize, sunflower, sugarbeet, rice, potato, and grape. The system is used on a regular basis to produce 10-day outputs and monthly summaries of

- Meteorological condition indicators
- Crop state indicators, including development stage, leaf area index, soil moisture index, total biomass, and storage organ biomass for either water limited or irrigated conditions

The outputs are prepared for each of these 10 crops in cartographic format on a 50×50 km grid square basis. The operational version of CGMS, was enhanced by introducing the capacity to directly forecast national yields employing the EUROSTAT's CRONOS database and by adding a number of user friendly facilities for the production of maps.

Basically, CGMS contains the following three modules:

- The processing of daily meteorological data, replacement of missing values; calculation of derived parameters, such as solar radiation (from cloud cover or sunshine duration), vapor pressure, and potential evapotranspiration; interpolation to a regular grid of 50×50 km; and production of output maps of the meteorological conditions during a given 10-day period, month, or season, both as actual values and as departures from the climatological average conditions.
- Agrometeorological crop growth simulation for each of the major annual crop types that, according to a crop knowledge bases, are likely to grow in a given 50×50-km grid. Since various soil types and crop varieties coexist in a grid, the output of a basic square is produced for each major soil type and available water profile capacity, so as to reach a representation of approx. 80% of the suitable soil coverage.
- A statistical module relating the model outputs through a regression analysis and possibly in combination with a technological time trend function drawn from historical yield data, to the series of regional and national yields available in EUROSTAT's REGIO and CRONOS databases. The regression analysis of past years is used only if it gives satisfactory results in terms of significance of the multiple determination coefficients, the partial correlation coefficients, the stability of the regression coefficients, and the error analysis; if not, only the time trend function or previous year's yield is used as predicted values.

The outputs of the system are basically the mapped outputs of indicators on the quality of the agricultural season, including

- *Biomass and grain production*, under actual rainfall conditions and as if all required moisture were available, estimated actual soil moisture reserve, differences as compared to the previous decade or month, and state of development stage of the crop during a given decade
- *Alarm warning*—detection of abnormal weather conditions during a given decade or accumulated since the start of the season
- *Tables with calculated yield forecasts*—information on the quality of the regression equations such as the coefficient of multiple determination, the stability of the regression coefficients, and the errors of the one-year-ahead predictions obtained from previous years

Within the MARS framework, Supit (1997) gave an example by using the crop growth model WOFOST (Van Diepen et al. 1989) to predict operationally the total wheat yield for the EU.

13.3 "URBAN HEAT ISLAND" EFFECTS

The development of temperature differences between cities and their surrounding rural regions has been well documented. Urbanization within the United States has had the greatest influence on minimum (compared to maximum or mean) temperature records (Karl et al. 1988). Higher temperature in the summer affects energy consumption for air conditioning. The urban heat island may also increase cloudiness and precipitation in the city, as a thermal circulation sets up between the city and surrounding region.

The reason why the city is warmer than the country comes down to a difference between the energy gains and losses of each region. A number of factors contribute to the relative warmth of cities:

- During the day in rural areas, the solar energy absorbed near the ground causes water to evaporate from the vegetation and soil. Thus, while there is a net solar energy gain, this is compensated to some degree by evaporative cooling. In cities, where there is less vegetation, the buildings, streets, and sidewalks absorb the majority of solar energy input.
- Because the city has less water, runoff is greater in the cities because pavement is largely nonporous. Thus, evaporative cooling is less contributing to higher air temperatures.
- Waste heat from city buildings, automobiles, and trains is another factor contributing to the warm cities. Heat generated by these objects eventually makes its way into the atmosphere.

- The thermal properties of buildings add heat to the air by conduction. Tar, asphalt, brick, and concrete are better conductors of heat than is the vegetation of rural areas.
- The "canyon" structure that tall buildings create enhances the warming. During the day, solar energy is trapped by multiple reflections off the buildings while infrared heat losses are reduced by absorption.
- The urban heat island effects can also be reduced by weather phenomena. The temperature difference between the city and surrounding areas is also a function of winds. Strong winds reduce the temperature contrast by mixing together the city and rural air.

Satellite-derived surface temperature data have been utilized for urban climate analyses in several studies up until 1990 (Rao 1972, Carlson et al. 1977, Matson et al. 1978, Price 1979, Roth et al. 1989, Carnahan and Larson 1990). Urban increases in temperature and changes in cloud and precipitation patterns are local examples of what many expect to occur on a global scale in future decades. Methods and techniques currently being developed to study local phenomena will provide an increase in the skills and knowledge necessary to deal with possible future global climate change.

Rao (1972) was the first to demonstrate that urban areas could be identified from analyses of thermal data acquired by a satellite. Matson et al. (1978) utilized AVHRR thermal data (10.5–12.5 μm) acquired at night to examine urban and rural surface temperature differences. Price (1979) utilized Heat Capacity Mapping Mission (HCMM) data (10.5–12.5 μm) to assess the extent and intensity of urban surface heating in the northeastern United States. Roth et al. (1989) utilized AVHRR thermal data (10.5–11.5 μm) to assess the urban heat island (UHI) intensities of several cities on the west coast of North America. Daytime thermal patterns of surface temperature were associated with land use; higher surface temperatures were observed in industrial areas than in vegetated regions.

The greatest urban and rural air temperature differences are observed at night, while the greatest differences in surface radiant temperature are observed during midday (e.g., Roth et al. 1989). Additional land surface properties, beyond surface temperature, may provide more information related to the factors that contribute to the observed UHI effect. The reduced amount of heat stored in the soil and surface structures in rural areas covered by transpiring vegetation, compared to the relatively unvegetated urban areas, has been cited as a significant contributor to the UHI effect (Carnahan and Larson 1990). Thus, a measure of the difference in density of urban and rural vegetation may be an indicator of the magnitude of the differences observed in the minimum air temperatures of urban and rural areas. Vegetation indices computed from remotely sensed data have been demonstrated as useful estimators of the amount of leaf area and related variables associated with agricultural. In the following paragraphs we briefly introduce several studies on this subject that have been published since 1998.

Gallo and Owen (1999) compared the monthly and seasonal relationships between urban–rural differences in minimum, maximum, and average temperatures measured at surface based observation stations to satellite-derived AVHRR estimates of NDVI and surface radiant temperature for correcting the UHI bias. They found that the urban–rural differences in air temperature were linearly related to urban–rural differences in the NDVI and surface temperature. They concluded that the use of satellite-derived data may contribute to a globally consistent method for analysis of the UHI bias.

Quattrochl et al. (2000) described an information support system to assess urban thermal landscape characteristics as a means for developing more robust models of the UHI effect. The remote sensing data for four cities in the United States have been used to generate a number of products for use by "stakeholder" working groups to convey information on what the effects are of the UHI and what measures can be taken to mitigate them. In turn, these data products are used to both educate and inform policymakers, planners, and the general public about what kinds of UHI mitigation strategies are available.

Weng (2001) integrated remote sensing and GIS for detecting urban growth and assessing its impact on surface temperature in the Zhujiang Delta of southern China. Multitemporal TM imagery were used to detect land use/cover changes and a GIS-based modeling approach was used to evaluate the urban growth patterns. Their results revealed a notable and uneven urban growth in the study area, and the urban development had raised surface radiant temperature by 13.01 K in the urbanized area.

Streutker (2002) derived radiative surface temperature maps of Houston, TX from AVHRR data acquired over a 2-year period. The urban heat island was modeled as a two-dimensional Gaussian surface superimposed on a planar rural background for characterizing the complete urban heat island in magnitude and spatial extent without the use of in situ measurements and determining whether a correlation exists between heat island magnitude and rural temperature. The UHI magnitude was found to be inversely correlated with rural temperature, while the spatial extent was found to be independent of both heat island magnitude and rural temperature.

13.4 CARBON CYCLE STUDIES

13.4.1 Background

As a result of human activities, primarily fossil fuel burning and changing land use, atmospheric carbon concentration has increased by more than 28% over the past 150 years (Watson et al. 2000). As atmospheric concentrations of carbon dioxide continue to increase, the Earths climate is expected to

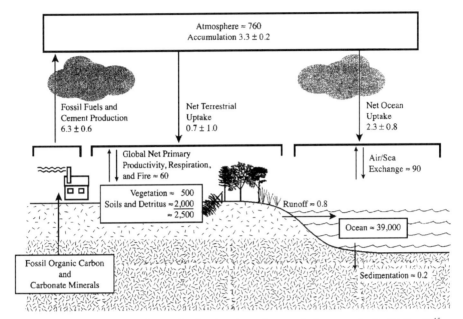

Figure 13.7   The global carbon cycle, showing the carbon stocks in reservoirs (in Gt C = 10^{15} g C) and carbon flows (in Gt C per year) relevant to the anthropogenic perturbation as annual averages during 1989–1998 [from Watson et al. (2000)].

change significantly over the next several decades. To prevent the further increase of carbon and other greenhouse gases, an accurate knowledge of global carbon cycle is urgently needed.

The global carbon cycle connects the major three components of the Earth system: the atmosphere, oceans, and land, as shown in Fig. 13.7 (Watson et al. 2000)]. Carbon stocks in reservoirs [in gigatonnes (metric tonnes) (Gt) C = 10^{15} g C) and carbon flows (in Gt C/year) relevant to anthropogenic perturbation as annual averages over the decade from 1989 to 1998 are also displayed. There are large uncertainties in the global carbon budget, the largest of which is due to the terrestrial ecosystem amounts to 1.3 Gt C/year (IPCC 2000). There is a strong need to quantify terrestrial carbon stocks and fluxes by developing global, systematic observations of carbon cycle. Quantitative remote sensing has a key role to play in this process.

In a workshop organized by the global terrestrial observing system (GTOS) and the integrated global observing strategy partnership (IGOSP), two approaches have been proposed to determine the change of the terrestrial carbon distribution in three main pools (atmosphere, plants, and soils) (Cihlar et al. 2000):

- The first is the "bottom-up approach," which starts with a specific parcel of land to account for the various pathways of carbon exchange

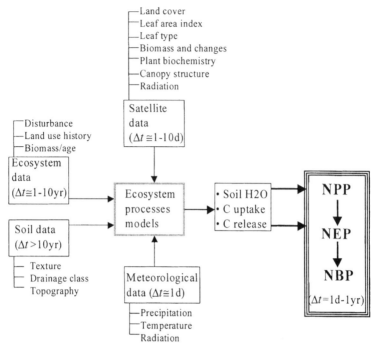

Figure 13.8 Typical data flow for calculation of global carbon cycle using the bottom-up approach [From Cihlar et al. (2000)].

between the ecosystem and the atmosphere and then scales up to much larger regions.

- The second one is the "top-down approach," which begins with measured changes in atmospheric gas concentrations and attempts to infer the spatial distribution and magnitude of the net exchange.

In the following, we discuss mainly the bottom-up approach.

13.4.2 "Bottom-up" Approach

The bottom-up approach relied on both ecosystem process models and spatial datasets. This model can be developed at the local scale and validated using some conventional measurements. Satellite observations provide spatial distribution and up-to-date information frequently, in relation to the rate of change of the variables that drive the model. Figure 13.8 gives an overview of some important variables and the data flow involved in the bottom-up approach. In this figure, carbon is represented by net primary productivity (NPP), net ecosystem productivity (NEP) and net biome productivity (NBP).

NPP is the net biomass increase through photosynthesis, and a parameter for quantifying the net carbon absorption rate by living plants. NPP is a measure of plant growth, and it provides highly synthesized, quantitative information for sustainable resource management. NPP is also an important component of the biosphere carbon cycle. NEP involves net carbon exchange with the atmosphere after accounting for soil respiration and organic matter decomposition. Net carbon flow to/from terrestrial ecosystems (NEP) = NPP − (soil respiration) (grams carbon m^{-2} $year^{-1}$). NBP involves the net production of organic matter in a region containing a range of ecosystems (a biome) and includes, in addition to heterotrophic respiration, other processes leading to loss of living and dead organic matter (harvest, forest clearance, fire, etc). NBP is appropriate for the net carbon balance of large areas (100–1000 km^2) and longer periods of time (several years and longer) (Watson et al. 2000).

The information required for global carbon cycle study that can be estimated from optical and thermal remote sensing are briefly summarized below (Cihlar et al. 2000, 2001). We then introduce two specific models that utilize some of this information.

- *Land Cover and Land Use*. A primary characterization of land cover and land cover change is fundamental for carbon cycle observation and assessment, and also to nearly every aspect of land management. Land use is a critical carbon cycle parameter. The knowledge of present and historical (decades to centuries) land use is essential to properly present carbon exchanges in models. Satellite mapping is an ideal means of characterizing land cover because it spans a range of spatial and temporal scales. Although we did not discuss this subject in the book directly, some chapters are very relevant, such as sensor calibration (Chapter 5), atmospheric correction (Chapter 6), and topographic correction (Chapter 7). Readers are referred to some related papers (Defries et al. 1995, Friedl et al. 1999, Liang 2001).

- *Aboveground and Belowground Biomass*. Estimates of above/belowground biomass provide fundamental information on the size and changes of the terrestrial carbon pool as land use and associated land management practice change. At present, remote sensing provides high-resolution global coverage of land cover and cover changes that are relevant to the estimation of aboveground biomass, but there is no reliable satellite-based capability to estimate biomass directly. Future LiDAR satellite programs, such as VCL and ICESat GLAS, will provide information on forest height and structure, thus allowing detailed and robust biomass estimates globally. Multiangle remote sensing (e.g., MISR, POLDER) also has the potential to infer this type of information (Liang and Strahler 2000).

- *Seasonal Growth Cycle*. Seasonal growth characteristics such as leaf area, growing season duration, and timing of growth (onset and senescence) provide strong constraints on carbon sequestration. Many ecological and carbon models require leaf area index (LAI) and information on vertical and horizontal leaf distribution to account for different light-use efficiencies of sunlit and shaded leaves. Both multispectral (e.g., Landsat, MODIS) and multiangular (e.g., MISR, POLDER, EPIC) remotely sensed data can be used for deriving these information. We have discussed vegetation indices and LAI estimation in detail in Chapter 8 and several related chapters.

- *Fire*. Fire causes the strongest disturbance of vegetation in many regions of the world. Worldwide information about fire is necessary to calculate net carbon sinks, and fire may be a major cause of the large observed interannual variations in carbon emissions from ecosystems. Fire is also a very important factor influencing ecosystem succession and land use. Remote sensing is the most effective means for burned-area detection. Small fires can be detected and mapped using high-resolution imagery, such as TM/ETM + and SPOT, and the area burned in large fires can be accurately mapped using coarse-resolution imagery, such as AVHRR and MODIS. Chapters 5–7 are related to this topic, although no direct discussions are provided in this book. Readers are referred to the related papers (e.g., Kasischke and Stocks 2000, Justice and Korontzi 2001)

- *Solar Radiation*. Global solar radiation [shortwave (SW)] and its photosynthetically active radiation (PAR) component are major drivers of surface processes such as photosynthesis and evapotranspiration. For carbon uptake modeling, daily PAR estimates are needed at a resolution of 50 km (minimum) to 10 km (preferred). Several methods have been developed to derive SW from satellite radiance measurements (Charlock and Alberta 1996). The products generated so far are the result of various research programs using different satellite data, such as ERBE and GOES. Estimates of PAR averages at weekly and monthly timescales can be derived as a constant fraction of SW. The limitations of these products are coarse spatial and temporal resolution, and their availability for a limited period (back to 1995). Other satellite data are eventually useful for generating SW and PAR products. Chapters 2 and 6 are particularly relevant to this topic.

- *Ecosystem Productivity*. Ecosystem productivity quantities are determined using satellite-derived products, soil and meteorological databases, and bio/geochemical models. These biogeochemical processes mimic the ecosystem processes involving carbon uptake and transformations within the ecosystem as well as exchange with the atmosphere. To date, regional or ecosystem productivity products using

AVHRR data have been generated by various groups at global and regional levels, using top-down and/or bottom-up approaches.

13.4.3 Case Demonstrations

For demonstration purposes, we present two models for calculating NPP and ecosystem productivity with numerous inputs of remote sensing products below.

1. *GloPEM.* The global production efficiency model (GloPEM; also abbreviated GLO-PEM) is the first attempt to utilize the production efficiency concept globally, in which the canopy absorption of photosynthetically active radiation (APAR) is used with a conversion "efficiency" or carbon yield of APAR in terms of gross primary productivity (GPP). See Eq. (13.19) for details. The GloPEM model is thus based on physiological principles, in particular the amount of carbon fixed per unit APAR is modeled rather than fitted using field observations. The details of this model and some applications are given by Prince and colleagues (Prince and Goward 1995, Goetz 1997, Goward and Dye 1997, Prince et al. 1998, Goetz and Prince 1999, Goetz et al. 1999, Goward et al. 2002). The modeling components include

- The amount of incident photosynthetically active radiation (PAR) is derived from total ozone mapping spectrometer (TOMS) ultraviolet observations of cloud cover, which are used to modify incident PAR as derived from a clear-sky model. This is the only variable used in the model that is not directly inferred from the AVHRR instrument.
- Surface reflectance properties in the visible and infrared wavelengths are converted to spectral vegetation index (SVI) values that are linearly related to the fraction of incident PAR absorbed by terrestrial vegetation (FPAR). This topic is discussed in Section 8.1. When combined with incident PAR, this provides a measure of canopy light absorption (APAR).
- The minimum value of visible reflectance in the annual observation period is related to the amount of standing aboveground biomass, taking into account the effects of solar zenith angle and cloud shadows.
- Surface radiometric temperature (T_s) and atmospheric column precipitable water vapor amount (U) are derived from thermal measurements in different spectral wavelength bands (using the "split-window" approach). This topic is discussed in Sections 6.4 and 10.7.
- The regression relationship between a moving-window array of SVI and T_s values [termed *temperature/vegetation index* (TVX)] is used to derive an estimate of ambient air temperature (T_a) by extrapolating to a high SVI value (~ 0.9) that represents an infinitely thick canopy. It is assumed that canopy and air temperature are equivalent at this point.

- The atmospheric water vapor amount (U) is extrapolated to the surface and used to estimate surface humidity and dewpoint temperature. When combined with T_a, this is used to calculate vapor pressure deficit (VPD).
- An index of soil surface moisture is derived from changes in the slope of the TVX relationship through time. This topic is discussed in Section 13.2.3.3.
- The potential amount of carbon fixation per unit APAR is calculated from the quantum yield of photosynthesis and a climatological mean air temperature to differentiate between photosynthetic pathways (C3, C4).
- Potential carbon fixation is reduced by "stress" factors related to plant physiological control (i.e., Ta, VPD, soil moisture) to derive actual carbon fixation in the form of gross primary productivity (GPP).
- Respiration related to the growth and maintenance of biomass is subtracted from GPP to derive global net primary productivity (NPP).

2. *BEPS and InTEC.* BEPS (boreal ecosystem productivity simulator) was developed by the Canadian Center for Remote Sensing (CCRS) (Liu et al.

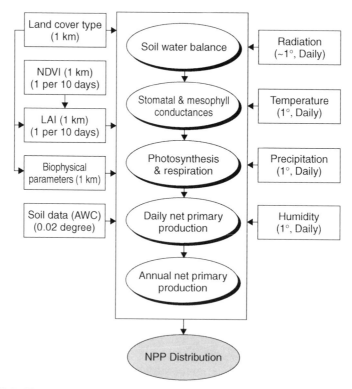

Figure 13.9 Framework of BEPS showing the major modeling steps and the required input data; some of the key variables can be estimated from remote sensing. [From Liu et al. (1997), *Remote Sens. Environ.* Copyright © 1997 with permission from Elsevier.]

1997) to assist in natural resources management and to estimate the carbon budget over the Canadian landmass. It uses principles of FOREST biogeochemical cycles (FOREST-BGC) (Running and Coughlan 1988) for quantifying the biophysical processes governing ecosystems productivity with an improvement of the canopy radiation process. The major output of BEPS is net primary productivity (NPP). The inputs to BEPS and the major processes within BEPS are shown in Fig. 13.9. It requires daily inputs and calculates daily NPP. Data sources from remote sensing include NDVI, LAI, land cover, and other biophysical parameters.

On the basis of BEPS on NPP calculation, an integrated terrestrial ecosystem C-budget (InTEC) model was also developed to estimate the carbon budget of Canada's forests (Chen et al. 2000a, 2000b). The InTEC model estimates the carbon budgets of forests from atmospheric, climatic and biotic changes since the pre-industrial period. Remote sensing derived data of leaf area and NPP, network measurements of nitrogen deposition, and historical data of climate and disturbance rates are used in the InTEC model. InTEC results indicate that in the past 100 years, Canada's forests as a whole were a small carbon source of ~ 30 Mt C/year in the period 1895–1905 as a result of large disturbances near the end of the nineteenth century; a large carbon sink of ~ 170 Mt C/year (1930–1970) due to forest regrowth in

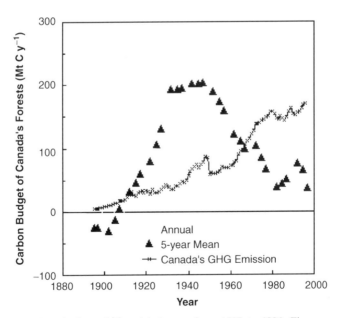

Figure 13.10 Carbon budget of Canada's forests from 1895 to 1996. Five-year means are plotted; the last point is only the average from 1995 to 1996. Also included is a record of Canada's greenhouse gas carbon emission rate.

previously burnt areas; and moderate carbon sinks of ~ 50 Mt C/year (1980–1996) (Fig. 13.10).

13.5 LAND–ATMOSPHERE INTERACTION

Since the early 1980s, the role of the land surface in modulating the global climate and influencing data-to-day weather events has gained increased recognition. The basic elements of land–atmosphere interaction are the exchange of moisture and energy between these two systems. As summarized by Dickinson (1995), many of the important aspects of this interaction have been treated in areas related to micrometeorology, agriculture and forest meteorology, planetary boundary layer, hydrology, biochemical cycling, climate, mesoscale meteorology, and numerical weather prediction.

Land has been recognized as a major element of the climate system. As a lower boundary condition, land may be less important than the oceans in some ways. For example, it provides much less storage and negligible horizontal transport of thermal energy. On the other hand, it has been well known that the ocean surface conditions represented by the sea surface temperature (SST) play a major role in forcing the atmosphere and SST can be retrieved from daily satellite data accurately. In contrast, land is more variable and exchangeable than the oceans for many of the important coupling processes. The temperature and moisture of the soil over land are not routinely observed or input daily as a boundary condition into global forecast model. Koster and Suarez (1995) used a series of general circulation model simulations to quantify the relative contributions of land surface and ocean variability to variability in precipitation. The simulations show that land surface processes contribute significant to the variance of annual precipitation over continents. In contrast, ocean processes acting alone have a much smaller effect, particularly in the midlatitudes.

A land surface model has to be used by solving a full system of parameterized equations that involve surface radiation budget and surface energy budget. The radiation budget is strongly influenced by the solar diurnal cycle and the cloud field. The surface energy balance equation is a major interaction point between the land surface model and GCMs. Solving this equation is the most important component of any land surface model as it closes the energy balance at the lower boundary of the atmosphere and determines the temperature of the surface with which the atmosphere is in contact. Both surface radiation budget and surface energy budget have been discussed in Chapters 9 and 10 and previous sections of this chapter.

A variety of land surface models have been developed, incorporating varying levels of complexity of the land-atmosphere interaction (Xue et al. 1991, Sellers et al. 1996, Rosenzweig and Abramopoulos 1997, Zeng et al. 2002). Sellers et al. (1997) reviewed all land surface models that estimate the exchanges of energy, heat, and momentum between the land surface and the atmosphere in three generations. The first, developed in the late 1960s and 1970s, was based on simple aerodynamic bulk transfer formulas and often uniform prescriptions of surface parameters (albedo, aerodynamic roughness, and soil moisture availability) over continents. In the early 1980s, a second generation of models explicitly recognized the effects of vegetation in the calculation of the surface energy balance (e.g., BATS, SiB). The third generation models use modern theories relating photosynthesis and plant water relations to provide a consistent description of energy exchange, evapotranspiration, and carbon exchange by plants (e.g., SiB-2, CLM)

Land surface modeling is cited as one of the major causes of the uncertainties in current climate change predictions (Houghton et al. 1996). The Project for Intercomparison of Land Surface Parametrization Scheme (PILPS) (Henderson-Sellers et al. 1996) is dedicated to evaluating the relative accuracy of existing land surface models. The project was set up in order to improve the understanding of the interaction between the atmosphere and the continental surface to make improvements to GCM.

In fact, land surface model development is constrained to a large extent by the availability of data acquired to initialize, parameterize, and test the models. The most promising approach to obtaining these data is the development of algorithms to predict the state of the land surface and the atmosphere from remote sensing platforms (Running et al. 1999).

Molders (2001) evaluated the impacts of the uncertainty of plant and soil parameters (albedo, evaporative conductivity, roughness length, soil volumetric heat capacity, field capacity, capillarity, emissivity), soil type, sub-grid-scale heterogeneity, and inhomogeneity on mesoscale model results (e.g., fluxes, variables of state, cloud, and precipitation formation) based on extensive simulations. The distribution of the precipitation fields is shifted upstream by ~ 4 km when roughness length is doubled, while it shifts downstream by ~ 4 km when a uniformly sandy loam is assumed in the entire model domain. Albedo affects the extension of the precipitation fields to a high degree, as does the intensity of precipitation. Surface emissivity, roughness length, thermal diffusivity, volumetric heat capacity, field capacity, capillarity, heterogeneity, and inhomogeneity appreciably influence the statistical behavior of the surface fluxes, variables of state, as well as cloud, and precipitating particles. Bounoua et al. (2000) investigated the sensitivity of global and regional climate to changes in vegetation density using a coupled biosphere–atmosphere model. This kind of sensitivity study has been conducted extensively (e.g., Buermann et al. 2001, Zeng et al. 2002).

In June 1992, an interdisciplinary Earth science workshop was convened in Columbia, Maryland, to assess recent progress in land–atmosphere research, specifically in the areas of models, satellite data algorithms, and field experiments. At the workshop, representatives of the land-atmosphere modeling community stated that they had a need for global datasets to prescribe boundary conditions, initialize state variables, and provide near-surface meteorological and radiative forcings for their models. The Initiative I CDs have been available to the public for initializing, forcing, and validating all three

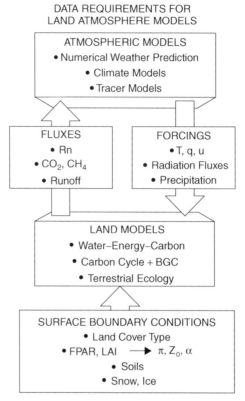

DATA REQUIREMENTS FOR
LAND ATMOSPHERE MODELS

Figure 13.11 Schematic showing relationships between different kinds of atmospheric and land models and the associated data sets in ISLSCP. The "surface boundary conditions box includes vegetation-dependent parameters that are derived from the fraction of photosynthetically active radiation absorbed by the green portion of the canopy (FPAR) or leaf area index (LAI). These parameters are the canopy PAR use parameter (p), the roughness length (z_0), and the albedo (a). (Information available at http://daac.gsfc.nasa.gov/CAMPAIGN_DOCS/ ISLSCP/islscp_i1.html.)

classes of land models identified by this workshop (see Fig. 13.11). Some of these variables were produced from remote sensing. Initiative II is also under way.

Before finishing this section, I'd like to briefly mention a few more recent studies on this topic that use some of these variables derived from remote sensing data. Sellers et al. (1996) used AVHRR data to specify the time-varying phenological properties of FPAR, leaf area index, and canopy greenness fraction as the inputs to SiB2. Hall et al. (1995) provided an assessment of the state of the art of surface state remote sensing algorithms to infer land surface parameters on a global basis. They first considered the modeling requirements for land cover parameters (e.g., vegetation community composition), biophysical parameters (LAI, biomass density, FPAR), and soil moisture. Then they reviewed the status of remote sensing algorithms for obtaining these parameters and examined a number of issues involved in the global implementation and testing of these algorithms.

Prigent et al. (2001) presented an evaluation and comparison of visible, near-infrared, passive, and active microwave observations for vegetation characterization, on a global basis, for a year, with spatial resolution compatible with climatological studies. The capacity to discriminate between vegetation types and to detect the vegetation phenology was assessed in the context of a vegetation classification obtained from in situ observations.

Muchoney and Strahler (2002) reported on the integrated analysis of site and remote sensing data to directly predict and map land surface biophysical parameters relating to structure, morphology, phenology, and physiognomy, as an alternative to using intermediate classification categorizations to relate these site variables and parameters to remote sensing features. Given the ability to define vegetation and land cover at the site level on the basis of attributes such as physiognomy, horizontal and vertical structure, vegetation phenology, and leaf morphology, direct parameterization and mapping using remotely sensed data can enhance the ability to characterize and monitor these important biogeophysical parameters.

13.6 SUMMARY

This chapter has demonstrated some of the benefits of quantitative remote sensing. The intent is to shed some light on how the estimated land surface variables discussed in previous chapters can be used for solving the real-life problems. The list of application areas is by no means inclusive, and the choice was largely arbitrary but perhaps reflects the authors current research interests to some degree.

This chapter brings an end to one cycle starting from understanding remote sensing data (the first part of the book), land surface information extraction (the second part), and their applications. The new requirements of the applications motivate the development of new sensor systems, new models for understanding the data acquired by the new sensors, and new algorithms for estimating surface variables, leading to another cycle. The future of quantitative remote sensing over land surfaces depends largely on how well the remote sensing data and the derived products are integrated with various land surface process and application models.

REFERENCES

Aase, J. K. and Tanaka, D. L. (1991), Reflectance from four wheat residue cover densities as influenced by three soil backgrounds, *Agron. J.* **83**: 753–757.

Al-Faraj, A., Meyer, G. E., and Horst, G. L. (2001), A crop water stress index for tall fescue irrigation decision-making—a traditional method, *Comput. Electron. Agric.* **31**: 107–124.

Anconelli, S., Mannini, P., and Battilani, A. (1994), CWSI and baseline studies to increase quality of processing tomatoes, *Acta Horticult.* **376**: 303–306.

Anderson, M. C., Bland, W. L., Norman, J. M., and Diak, G. D. (2001), Canopy wetness and humidity prediction using satellite and synoptic-scale meteorological observations, *Plant Disease* **85**: 1018–1026.

Asrar, G., Fuchs, M., Kanemasu, E., and Hatfield, J. (1984), Estimating absorbed photosynthetic radiation and leaf area index from spectral reflectance in wheat, *Agron. J.* **76**: 300–306.

Asrar, G., Kanemasu, E. T., Jackson, R. D., and Pinter, P. J. J. (1985), Estimation of total above-ground phytomass production using remotely sensed data, *Remote Sens. Envir.* **17**: 211–220.

Barnes, E. M. and Baker, M. G. (2000), Multispectral data for mapping soil texture: Possibilities and limitations, *Appl. Eng. Agric.* **16**: 731–741.

Bedidi, A., Cervelle, B., Madeira, J., and Pouget, M. (1992), Moisture effects on visible spectral characteristics of lateritic soils, *Soil Sci.* **153**: 129–141.

Ben-Dor, E., Irons, J., and Epema, G. (1999), Soil reflectance, in *Manual of Remote Sensing*, 3rd ed., A. Rencz, ed., Wiley, pp. 111–188.

Boatwright, G. O. and Whitehead, V. S. (1986), Early warning and crop condition assessment research, *IEEE Trans. Geosci. Remote Sens.* **24**: 54–64.

Bouma, J., Stoorvogel, J., van Alphen, B. J., and Booltink, H. W. G. (1999), Pedology, precision agriculture, and the changing paradigm of agricultural research, *Soil Sci. Am. J.* **63**: 1763–1768.

Bouman, B. A. M. (1992), Linking physical remote sensing models with crop growth simulation models, applied for sugar beet, *Int. J. Remote Sens.* **13**: 2565–2581.

Brisco, B., Brown, R. J., Hirose, T., McNairn, H., and Staenz, K. (1998), Precision agriculture and the role of remote sensing: A review, *Can. J. Remote Sens.* **24**: 315–327.

Brown, R. B. and Steckler, J.-P. G. A. (1995), Prescription maps for spatially variable herbicide application in no-till corn, *Trans. ASAE (Am. Soc. Agric. Eng.)* **38**: 1659–1666.

Buermann, W., Dong, J. R., Zeng, X. B., Myneni, R. B., and Dickinson, R. E. (2001), Evaluation of the utility of satellite-based vegetation leaf area index data for climate simulations, *J. Climate* **14**: 3536–3550.

Carbone, G. J., Narumalani, S., and King, M. (1996), Application of remote sensing and GIS technologies with physiological crop models, *Photogramm. Eng. Remote Sens.* **62**: 171–179.

Carlson, T., Gillies, R., and Perry, E. (1994), A method to make use of thermal infrared temperature and NDVI measurements to infer surface soil water content and fractional vegetation cover, *Remote Sens. Rev.* **9**: 161–173.

Carlson, T. N., Augustine, J. A., and Boland, F. E. (1977), Potential application of satellite temperature measurements in the analysis of land use over urban areas, *Bull. Am. Meteorol. Soc.* **58**: 1301–1303.

Carnahan, W. H. and Larson, R. C. (1990), An analysis of urban heat sink, *Remote Sens. Envir.* **33**: 65–71.

Chang, C. W., Laird, D. A., Mausbach, M. J., and Hurburgh, C. R. (2001), Near-infrared reflectance spectroscopy-principal components regression analyses of soil properties, *Soil Sci, Am, J.* **65**: 480–490.

Charlock, T. P. and Alberta, T. L. (1996), The CERES/ARM/GEWEX experiment (CAGEX) for the retrieval of radiative fluxes with satellite data, *Bull. Am. Meteorol. Soc.* **77**: 2673–2683.

Chen, J. M., Chen, W., Liu, J., and Cihlar, J. (2000a), Annual carbon balance of Canada's forests during 1895–1996, *Global Biogeochem. Cycle* **14**: 839–850.

Chen, W., Chen, J. M., and Cihlar, J. (2000b), Integrated terrestrial ecosystem carbon-budget model based on changes in disturbance, climate, and atmospheric chemistry, *Ecol. Model.* **135**: 55–79.

Chipanshi, A. C., Ripley, E. A., and Lawford, R. G. (1999), Large-scale simulation of wheat yields in a semi-arid environment using a crop-growth model, *Agric. Syst.* **59**: 57–66.

Cihlar, J., Denning, A. S., and Gosz, J. (2000), *Global Terrestrial Carbon Observation: Requirements, Present Status, and Next Steps*, CCRS, Ottawa, Canada.

Cihlar, J., Denning, S., Ahern, F., Arino, O., Belward, A., Bretherton, F., Cramer, W., Dedieu, G., Field, C., Francey, R., Gommes, R., Gosz, J., Hibbard, K., Igarashi, T., Kabat, P., Olson, R., Plummer, S., Rasool, I., Raupach, M., Scholes, R., Townshend, J., Valentini, R., and Wickland, D. (2001), *IGOS-p Carbon Cycle Observation Theme: Terrestrial and Atmospheric Components*, CCRS, Ottawa, Canada.

Clevers, J. G. P. W. (1997), A simplified approach for yield prediction of sugar beet on optical remote sensing data, *Remote Sens. Envir.* **61**: 221–228.

Dabrowska-Zielinska, K., Kogan, F., Ciolkosz, A., Gruszczynska, M., and Kowalik, W. (2002), Modelling of crop growth conditions and crop yield in poland using AVHRR-based indices, *Int. J. Remote Sens.* **23**: 1109–1123.

Dalal, R. C. and Henry, R. J. (1986), Simultaneous determination of moisture, organic carbon and total nitrogen by near infrared reflectance spectrophotometry, *Soil Sci. Soc. Am. J.* **50**: 120–123.

Daughtry, C. (2001), Discriminating crop residues from soil by shortwave infrared reflectance, *Agron. J.* **93**: 125–131.

Daughtry, C., McMurtrey III, J. E., Chappelle, E. W., Dulaney, W. P., Irons, J. R., and Satterwhite, M. B. (1995), Potential for discriminating crop residues from soil by reflectance and fluorescence, *Agron. J.* **87**: 165–171.

Daughtry, C., Hunt, R., Walthall, C., Gish, T., Liang, S., and Kramer, E. (2001), Assessing the spatial distribution of plant litter, paper presented at AVIRIS Earth Science and Applications Workshop, Pasadena, CA.

DeFries, R., Hansen, M., and Townshend, J. R. G. (1995), Global discrimination of land cover from metrics derived from AVHRR pathfinder data sets, *Remote Sens. Envir.* **54**: 209–222.

Delecolle, R., Maas, S. J., Guerif, M., and Baret, F. (1992), Remote sensing and crop production models: Present trends, *ISPRS J. Photogramm. Remote Sens.* **47**: 145–161.

Diak, G. R., Bland, W. L., Mecikalski, J. R. et al. (1996), A note on first estimates of surface insolation from GOES-8 visible satellite data, *Agric. Forest Meteorol.* **82**: 219–226.

Diak, G. R., Bland, W. L., Mecikalski, J. R., and Anderson, M. C. (2000), Satellite-based estimates of longwave radiation for agricultural applications, *Agric. Forest Meteorol.* **103**: 349–355.

Diak, G. R., Anderson, M. D., Bland, W. L., Norman, J. M., Mecikalski, J. M., and Aune, R. M. (1998), Agricultural management decision aids driven by real-time satellite data, *Bull. Am. Meteorol. Soc.* **79**: 1345–1355.

Dickinson, R. E. (1995), Land atmosphere interaction, *Rev. Geophys.*, (Suppl.): 917–922.

Dorman, J. L. and Sellers, P. J. (1989), A global climatology of albedo, roughness length and stomatal resistance for atmospheric general circulation models as represented by the simple biosphere model sib, *J. Appl. Meteorol.* **28**: 833–855.

Duchon, C. E. (1986), Corn yield prediction using climatology, *J. Climate Appl. Meteorol.* **25**: 581–590.

Elvidge, C. D. (1990), Visible and near infrared reflectance characteristics of dry plant materials, *Int. J. Remote Sens* **10**: 1775–1795.

Engel, T., Hoogenboom, G., Jones, J. W., and Wilkens, P. W. (1997), AEGIS/win: A computer program for the application of crop simulation models across geographical areas, *Agron. J.* **89**: 919–928.

Filella, I., Serrano, L., Serra, J., and Penuelas, J. (1995), Evaluating wheat nitrogen status with canopy reflectance indices and discriminant analysis, *Crop Sci.* **35**: 1400–1405.

Fischer, A., Kergoat, L., and Dedieu, G. (1997), Coupling satellite data with vegetation functional models: Review of different approaches and perspectives suggested by the assimilation strategy, *Remote Sens. Rev.* **15**: 283–303.

Friedl, M. A., Brodley, C. E., and Strahler, A. H. (1999), Maximizing land cover classification accuracies at continental to global scales, *IEEE Trans. Geosci. Remote Sens.* **37**: 969–977.

Gallo, K. P. and Owen, T. W. (1999), Satellite-based adjustments for the urban heat island temperature bias, *J. Appl. Meteorol.* **38**: 806–813.

Gillies, R., Cui, J., Carlson, T., Kustas, W., and Humes, K. (1997), Verification of the triangle method for obtaining surface soil water content and energy fluxes from remote measurements of ndvi and surface radiant temperature, *Int. J. Remote Sens.* **18**: 3145–3166.

Goetz, S. J. (1997), Multi-sensor analysis of NDVI, surface temperature and biophysical variables at a mixed grassland site, *Int. J. Remote Sens.* **18**: 71–94.

Goetz, S. J. and Prince, S. D. (1999), Modeling terrestrial carbon exchange and storage: The evidence for and implications of functional convergence in light use efficiency, *Adv. Ecol. Res.* **28**: 57–92.

Goetz, S. J., Prince, S. D., Goward, S. N., Thawley, M. M., Small, J., and Johnston, A. (1999), Mapping net primary production and related biophysical variables with remote sensing: Application to the Boreas region, *J. Geophys. Res.* **104**: 27719–27733.

Goward, S. N. and Dye, D. (1997), Global biospheric monitoring with remote sensing, in *The Use of Remote Sensing in Modeling Forest Productivity*, H. L. Gholtz, K. Nakane, and H. Shimoda, eds., Kluwer Academic Publishers.

Goward, S. N., Huemmrich, K. F., and Richard, R. H. (1994), Visible-near infrared spectral reflectance of landscape components in western Oregon, *Remote Sens. Envir.* **47**: 190–203.

Goward, S. N., Xue, Y., and Czajkowski, K. P. (2002), Evaluating land surface moisture conditions from the remotely sensed temperature/vegetation index measurements: An exploration with the simplified simple biosphere model, *Remote Sens. Envir.* **79**: 225–242.

Haboudane, D., Miller, J. R., Tremblay, N., Zarco-Tejada, P. J., and Dextraze, L. (2002), Integrated narrow-band vegetation indices for prediction of crop chlorophyll content for application to precision agriculture, *Remote Sens. Envir.* **81**: 416–426.

Hall, F. G., Townshend, J. R. G., and Engman, E. T. (1995), Status of remote sensing algorithms for estimation of land surface state parameters, *Remote Sens. Envir.* **51**: 138–156.

Hartkamp, A. D., Whit, J. W., and Hoogenboom, G. (1999), Interfacing geographic information systems with agronomic modeling: A review, *Agron. J.* **91**: 761–772.

Haskett, J. D., Pachepsky, Y. A., and Acock, B. (1995), Estimation of soybean yields at county and state levels using GLYCIM: A case study for Iowa, *Agron. J.* **87**: 926–931.

Hayes, M. J. and Decker, W. L. (1996), Using NOAA AVHRR data to estimate maize production in the United States corn belt, *Int. J. Remote Sens.* **17**: 3189–3200.

Henderson-Sellers, A., McGuffie, K., and Pitman, A. (1996), The project for intercomparison of land surface parameterization schemes (PILPS): 1992–1995, *Climate Dynam.* **12**: 849–859.

Hipps, L. E., Asrar, G., and Kanemasu, E. T. (1983), Assessing the intercept of photosynthetically active radiation in winter wheat, *Agric. Forest Meteorol.* **28**: 253–259.

Hodges, T., Botner, B., Sakamoto, C., and Hays, H. J. (1987), Using the CERES-maize model to estimate production for the US corn belt, *Agric. Forest Meteorol.* **40**: 293–303.

Hoogenboom, G., Wilkens, P. W., and Tsuji, G. Y., eds. (1999), *Dssat v3*, University of Hawaii, 286 pp.

Horie, T., Yajima, M., and Nadagawa, H. (1992), Yield forecasting, *Agric. Syst.* **40**: 211–236.

Houghton, J. T., Filho, L. G. M., Callander, B. A., Harris, N., Kattenberg, A. et al., eds. (1996), *Climate Change 1995*, Cambridge Univ. Press.

Idso, S. B. (1982), Non-water-stressed baselines: A key to measuring and interpreting plant water stress, *Agric. Meteorol.* **27**: 59–70.

Idso, S. B., Jackson, R. D., Pinter, P. J., Reginato, R. J., and Hatfield, J. L. (1981), Normalizing the stress-degree-day parameter for environmental variability, *Agric. Meteorol.* **24**: 45–55.

Irons, J. R., Weismiller, R. A., and Petersen, G. W. (1989), Soil reflectance, in *Theory and Applications of Optical Remote Sensing*, G. Asrar, ed., Wiley, pp. 66–106.

Jackson, R. D. (1982), Canopy temperature and crop water stress, in *Advances in Irrigation*, D. I. Hillel, ed., Academic Press, pp. 43–85.

Jackson, R. D., Idso, S. B., Reginato, R. J., and Pinter, J. P. J. (1981), Crop temperature as a crop water stress indicator, *Water Resour. Res.* **17**: 1133–1138.

Jiang, L. and Islam, S. (2001), Estimation of surface evaporation map over southern great plains using remote sensing data, *Water Resour. Res.* **37**: 329–340.

Johnson, A. K. L., Cramb, R. A., and Wegener, M. K. (1994), The use of crop yield prediction as a tool for land evaluation studies in northern Australia, *Agric. Syst.* **46**: 93–111.

Jones, D. and Barnes, E. M. (2000), Fuzzy composite programming to combine remote sensing and crop models for decision support in precision crop management, *Agric. Syst.* **65**: 137–158.

Jones, J. W. and Luyten, J. C. (1998), Simulation of biological processes, in *Agricultural Systems, Modeling and Simulation*, R. M. Peart and R. B. Curry, eds., Marcel Dekker, pp. 19–62.

Justice, C. O. and Korontzi, S. A. (2001), A review of satellite fire monitoring and the requirements for global environmental change research, in *Global and Regional Vegetation Fire Monitoring from Space: Planning a Coordinated International Effort*, F. G. Ahern and C. O. Justice, eds., SPB Academic Publishing, pp. 1–18.

Karl, T. R., Diaz, H. F., and Kukla, G. (1988), Urbanization: Its detection and effect in the United States climate record, *J. Climate* **1**: 1099–1123.

Kasischke, E. S. and Stocks, B. J., eds. (2000), *Fire, Climate Change and Carbon Cycling in the Boreal Forest*, Vol. 138, Ecological Studies Series, Springer-Verlag.

Kogan, F. (1997), Global drought watch from space, *Bull. Am. Meteorol. Soc.* **78**: 621–636.

Koster, R. D. and Suarez, M. J. (1995), Relative contributions of land and ocean processes to precipitation variability, *J. Geophys. Res.* **100**: 13775–13790.

Kropff, M. and van Laar, H. H., eds. (1994), *Modeling Crop-weed Interactions*. Wallingford, England: CAB International, 274 pp.

Lambin, E. F. and Ehrlich, D. (1996), The surface temperature–vegetation index space for land cover and land-cover change analysis, *Int. J. Remote Sens.* **17**: 463–487.

Leone, A. P., Wright, G. G., and Corves, C. (1995), The application of satellite remote sensing for soil studies in upland areas of southern Italy, *Int. J. Remote Sens.* **16**: 1087–1105.

Liang, S. (2001), Land cover classification methods for multiyear AVHRR data, *Int. J. Remote Sens.* **22**: 1479–1493.

Liang, S. and Strahler, A. eds. (2000), Land surface bidirectional reflectance distribution function (BRDF): Recent advances and future prospects, *Remote Sens. Rev.* **18**: 83–551.

Liu, J., Chen, J. M., Cihlar, J., and Park, W. M. (1997), A process-based boreal ecosystem productivity simulator using remote sensing inputs, *Remote Sens. Envir.* **62**: 158–175.

Liu, W., Baret, F., Gu, X., Tong, Q., and Zheng, L. (2002), Relating soil surface moisture to reflectance, *Remote Sens. Envir.* **81**: 238–246.

Liu, W. T. and Kogan, F. (2002), Monitoring Brazilian soybean production using NOAA/AVHRR based vegetation condition indices, *Int. J. Remote Sens.* **23**: 1161–1179.

Lorenzen, B. and Jensen, A. (1989), Changes in leaf spectral properties induced in barley by cereal powdery mildew, *Remote Sens. Envir.* **27**: 201–209.

Maas, S. J. (1988a), Use of remotely-sensed information in agricultural crop growth models, *Ecol. Model.* **41**: 247–268.

Maas, S. J. (1988b), Use satellite data to improve model estimates of crop yield, *Agron. J.* **80**: 655–662.

MacDonald, R. and Hall, F. (1980), Global crop forecasting, *Science* **208**: 670–679.

Manjunath, K. R., Potdar, M. B., and Purohit, N. L. (2002), Large area operational wheat yield model development and validation based on spectral and meteorological data, *Int. J. Remote Sens.* **23**: 3023–3038.

Maselli, F., Conese, C., Petkov, L., and Gilabert, M. (1992), Use of NOAA-AVHRR NDVI data for environmental monitoring and crop forecasting in the Sahel. Preliminary results, *Int. J. Remote Sens.* **13**: 2743–2749.

Matson, M., McClain, E. P., McGinnis, J., D. F., and Pritchard, J. A. (1978), Satellite detection of urban heat islands, *Monthly Weather Rev.* **106**: 1725–1734.

McVicar, T. R. and Jupp, D. L. B. (1998), The current and potential operational uses of remote sensing to aid decisions on drought exceptional circumstances in Australia: A review, *Agric. Syst.* **57**: 399–468.

Molders, N. (2001), On the uncertainty in mesoscale modeling caused by surface parameters, *Meteorol. Atmos. Phys.* **76**: 119–141.

Monteith, J. L. (1972), Solar radiation and productivity in tropical eco-systems, *J. Appl. Ecol.* **9**: 747–766.

Monteith, J. L. and Unsworth, M. H. (1990), *Principles of Environmental Physics*, Edward Arnold Publishers.

Moran, M. S., Inoue, Y., and Barnes, E. (1997), Opportunities and limitations for image-based remote sensing in precision crop management, *Remote Sens. Envir.* **61**: 319–346.

Moran, S. M., Maas, S. J., and Pinter, P. J. J. (1995), Combining remote sensing and modeling for estimating surface evaporation and biomass production, *Remote Sens. Rev.* **12**: 335–353.

Moran, S. M., Clarke, T. R., Inoue, Y., and Vidal, A. (1994), Estimating crop water deficit using the relationship between surface-air temperature and spectral vegetation index, *Remote Sens. Envir.* **49**: 246–263.

Mougin, E., Lo Seen, D., Rambal, S., Gaston, A., and Hiernaux, P. (1995), A regional sahelian grassland model to be coupled with multispectral satellite data. II: Toward the control of its simulations by remotely sensed indices, *Remote Sens. Envir.* **52**: 194–206.

Moulin, S., Bondeau, A., and Delecolle, R. (1998), Combining agricultural crop models and satellite observations: From field to regional scales, *Int. J. Remote Sens.* **19**: 1021–1036.

Muchoney, D. and Strahler, A. (2002), Regional vegetation mapping and direct land surface parameterization from remotely sensed and site data, *Int. J. Remote Sens.* **23**: 1125–1142.

Muller, E. and Decamps, H. (2001), Modeling soil moisture-reflectance, *Remote Sens. Envir.* **6**: 173–180.

Nagler, P. L., Daughtry, C. S. T., and Goward, S. N. (2000), Plant litter and soil reflectance, *Remote Sens. Envir.* **71**: 207–215.

Nemani, R. and Running, S. (1993), Developing satellite-derived estimates of surface moisture status, *J. Appl. Meteorol.* **32**: 548–557.

Nicholson, S. E. and Farrar, T. J. (1994), The influence of soil type on the relationships between NDVI, rainfall and soil moisture in semiarid botswana. I. Ndvi response to rainfall, *Remote Sens. Envir.* **50**: 107–120.

Nishida, K., Nemani, R. R., Running, S. W., and Glassy, J. M. (2003), Remote sensing of land surface evaporation (i) theoretical basis for an operational algorithm, *J. Geophy. Res.* **108**.

NRC (1997), *Precision Agriculture in the 21st Century*, National Academy Press.

Olesen, J. E., Bocher, P. K., and Jensen, T. (2000), Comparison of scales of climate and soil data for aggregating simulated yields of winter wheat in Denmark, *Agric. Ecosys. Envir.* **82**: 213–228.

Palacios-Orueta, A. and Ustin, S. L. (1998), Remote sensing of soil properties in the santa monica mountains I. Spectral analysis, *Remote Sens. Envir.* **65**: 170–183.

Peñelas, J., Filella, I., Lloret, P., Munoz, F., and Vilajeliu, M. (1995), Reflectance assessment of mite effects on apple trees, *Int. J. Remote Sens.* **16**: 2727–2733.

Plummer, S. E. (2000), Perspectives on combining ecological process models and remotely sensed data, *Ecol. Model.* **129**: 169–186.

Price, J. C. (1979), Assessment of the urban heat island effect through the use of satellite data, *Monthly Weather Rev.* **107**: 1554–1557.

Prigent, C., Aires, F., Rossow, W., and Matthews, E. (2001), Joint characterization of vegetation by satellite observations from visible to microwave wavelengths: A sensitivity analysis, *J. Geophys. Res.* **106**: 20665–20685.

Prince, S. D. and Goward, S. N. (1995), Global primary production: A remote sensing approach, *J. Biogeogr.* **22**: 815–835.

Prince, S. D., Goetz, S. J., Dubayah, R. O., Czajkowski, K. P., and Thawley, M. (1998), Inference of surface and air temperature, atmospheric precipitable water and vapor pressure deficit using AVHRR satellite observations: Comparison with field observations, *J. Hydrol.* **213**: 230–249.

Priya, S. and Shibasaki, R. (2001), National spatial crop yield simulation using gis-based crop production model, *Ecol. Model.* **135**: 113–129.

Quarmby, M., Milnes, M., Hindle, T., and Silleos, N. (1993), The use of multitemporal NDVI measurements from AVHRR data for crop yield estimation and prediction, *Int. J. Remote Sens.* **14**: 199–210.

Quarmby, N. A., Townshend, J. R. G., Settle, J. J., White, K. H., Milnes, M., Hindle, T. L., and Silleos, N. (1992), Linear mixture modeling applied to AVHRR data for crop area estimation, *Int. J. Remote Sens.* **13**: 415–425.

Quattrochl, D. A., Luvall, J. C., Rickman, D. L., Estes, M. G., Laymon, C. A., and Howell, B. F. (2000), A decision support information system for urban landscape management using thermal infrared data, *Photogramm. Eng. Remote Sens.* **66**: 1195–1207.

Radhakrishnan, J., Liang, S., Teasdale, J. R., and Shuey, C. J. (2002), Remote sensing of weed canopies, in *From Laboratory Spectroscopy to Remote Sensed Spectra of Soils, Plants, Forests and the Earth*, R. S. Muttiah and R. N. Clark, eds., Kluwer Academic Publishers.

Rao, P. K. (1972), Remote sensing of urban "heat islands" from an environmental satellite, *Bull. Am. Meteorol. Soc.* **53**: 647–648.

Ray, S. S., Pokharna, S. S., and Ajai (1999), Cotton yield estimation using agrometeorological model and satellite-derived spectral profile, *Int. J. Remote Sens.* **20**: 2693–2702.

Reginato, R. J. and Howe, J. (1985), Irrigation scheduling using crop indicators, *J. Irrig. Drain. Eng.* **111**: 125–133.

Reynolds, C. A. (2001), Input data sources, climate normals, crop models and data extraction routines utilized by pecad, paper presented at 3rd Int, Conf. Geospatial Information in Agriculture and Forestry, Denver, CO.

Richtie, J. T., Singh, U., Godwin, D. C., and Bowen, W. T. (1998), Cereal growth, development, and yield, *Understanding Options for Agricultural Production*, G. Y. Tsuiji et al., eds., Lulisherswer Academic Publishers, pp. 79–98.

Riebsame, W. E., Changnon, S. A., and Karl, T. R. (1990), *Drought and Natural Resource Management in the United States*: *Impacts and Implications of the 1987–1989 Drought*, Westview Press.

Rosenzweig, C. and Abramopoulos, F. (1997), Land surface model development for the GISS GCM, *J. Climate* **10**: 2040–2054.

Roth, M., Oke, T. R., and Emery, W. J. (1989), Satellite-derived urban heat islands from three coastal cities and the utilization of such data in urban climatology, *Int. J. Remote Sens.* **10**: 1699–1720.

Rudorff, B. F. T. and Batista, G. T. (1990), Yield estimation of sugarcane based on agrometeorological-spectral models, *Remote Sens. Envir.* **33**: 183–192.

Running, S. W. and Coughlan, J. C. (1988), A general model of forest ecosystem processes for regional applications. I. Hydrologic balance, canopy gas exchange and primary production processes, *Ecol. Model.* **42**: 125–154.

Running, S. W., Collatz, G. J., Washburne, J., and Sorooshian, S. (1999), Land ecosystems and hydrology, in *EOS Science Plan—the State of Science in the EOS Program*, M. King, ed., NAS.

Sakamoto, C. M. (1978), The z-index as a variable for crop yield estimation, *Agric. Meteorol.* **19**: 305–313.

Schulze, R. (2000), Transcending scales of space and time in impact studies of climate and climate change on agrohydrological responses, *Agric. Ecosys. Envir.* **82**: 185–212.

Sellers, P. J., Dickinson, R. E., Randall, D. A., Betts, A. K., Hall, F. G., Berry, J. A., Collatz, G. J., Denning, A. S., Mooney, H. A., Nobre, C. A., Sato, N., Field, C. B., and Henderson-Sellers, A. (1997), Modeling the exchanges of energy, water, and carbon between continents and the atmosphere, *Science* **275**: 502–509.

Sellers, P., Hall, F., Kelly, R., Black, A., Baldocchi, D., Berry, J., Ryan, M., Ranson, J., Crill, P., Lettenmaier, D., Margolis, H., Cihlar, J., Newcomer, J., Fitzjarrald, D., Jarvis, P., Gower, S., Halliwell, D., Williams, D., Goodison, B., Wickland, D., and Guertin, F. (1997), BOREAS in 1997: Experiment overview, scientific results, and future directions, *J. Geophys. Res.* **102**: 28731–28769.

Sellers, P. J., Los, S. O., Tucker, C. J., Justice, C. O., Dazlich, D. A., Collatz, G. J., and Randall, D. A. (1996), A revised land surface parameterization (SiB2) for atmospheric gcms. Part II: The generation of global fields of terrestrial biophysical parameters from satellite data, *J. Climate* **9**: 706–737.

Shearer, S. A. and Jones, P. T. (1991), Selective application of post-emergence herbicides using photoelectrics, *Trans. ASAE* **34**: 1661–1666.

Shepherd, K. D. and Walsh, M. G. (2002), Development of reflectance spectral libraries for characterization of soil properties, *Soil Sci, Am, J.* **66**: 988–998.

Shonk, J. L., Gaultney, L. D., Schulze, D. G., and Van Scoyoc, G. E. (1991), Spectroscopic sensing of soil organic matter content, *Trans. ASAE* **34**: 1978–1984.

Sinclair, T. R., Kitani, S., Hinson, K., Bruniard, J., and Horie, T. (1991), Soybean flowering date: Linear and logistic models based on temperature and photoperiod, *Crop Sci.* **1**: 786–790.

Stafford, J. V. (2000), Implementing precision agriculture in the 21th century, *J. Agric. Eng. Res.* **76**: 267–275.

Streutker, D. R. (2002), A remote sensing study of the urban heat island of Houston, Texas, *Int. J. Remote Sens.* **23**: 2595–2608.

Supit, I. (1997), Predicting national wheat yields using a crop simulation and trend models, *Agric. Forest Meteorol.* **88**: 199–214.

Thompson, L. M. (1969), Weather and technology in the production of corn in the U.S. corn belt, *Agron. J.* **61**: 453–456.

Tucker, C. J., Holben, B., Elgin, G., and McMurtrey, J. (1980), Relationship of spectral data to grain yield variation, *Photogramm. Eng. Remote Sens.* **46**: 657–666.

Unganai, L. S. and Kogan, F. N. (1998), Drought monitoring and corn yield estimation in southern Africa from AVHRR data, *Remote Sens. Envir.* **63**: 219–232.

USDA (1991), *Agriculture and Environment: The 1991 Yearbook of Agriculture.* U.S. Govt. Printing Office, Washington, D.C.

Van Diepen, C. A., Wolf, J., and van Keuien, H. (1989), WOFOST: A simulation model of crop production, *Soil Use Manage.* **5**: 16–24.

Vogel, F. A. and Bange, G. A. (1999), *Understanding Crop Statistics*, USDA, Miscellaneous Publication 1554.

Watson, R. T., Noble, I. R., Bolin, B., Ravindranath, N. H., Verardo, D. J., and Dokken, D. J., eds. (2000), *Land Use, Land-Use Change, and Forestry. A Special Report of the IPCC (Intergovernmental Panel on Climate Change)*, Cambridge Univ. Press.

Weng, Q. (2001), A remote sensing-GIS evaluation of urban expansion and its impact on surface temperature in the Zhujiang Delta, China, *Int. J. Remote Sens.* **22**: 1999–2014.

Wiegand, C. L. and Richardson, A. J. (1990), Use of spectral vegetation indices to infer leaf area, evapotranspiration and yield, I. Rationale, *Agron. J.* **82**: 623–629.

Williams, J. R., Renard, K. G., and Dyke, P. T. (1983), EPIC: A new method for assessing erosion's effect onsoil productivity, *J. Soil Water Conserv.* **38**: 381–384.

Wilson, M. F., Henderson-Sellers, A., Dickinson, R. E., and Kennedy, P. J. (1987), Sensitivity of the biosphere-atmosphere transfer scheme (BATS) to the inclusion of variable soil characteristics, *J. Climate Appl. Meteorol.*, **26**: 341–362.

Xue, Y., Sellers, P., Kinter III, J., and Shukla, J. (1991), A simplified biosphere model for global climate studies, *J. Climate* **4**: 345–364.

Yazar, A., Howell, T. A., Dusek, D. A., and Copeland, K. S. (1999), Evaluation of crop water stress index for lepa irrigated corn, *Irrig. Sci.* **18**: 171–180.

Zeng, X. B., Shaikh, M., Dai, Y. J., Dickinson, R. E., and Myneni, R. (2002), Coupling of the common land model to the ncar community climate model, *J. Climate* **15**: 1832–1854.

Zwiggelaar, R. (1998), A review of spectral properties of plants and their potential use for crop/weed discrimination in row-crops, *Crop Protect* **17**: 189–206.

Appendix: CD-ROM Content

There are two directories (data and software) each arranged in the order of the chapters.

DATA DIRECTORY

../data/ch1/solar.irradiance: extraterrestrial solar irradiance data sets from MODTRAN4 (thkur.dat, cebchkur.dat, chkur.dat and newkur.dat)

../data/ch1/spectral.function: Sensor spectral response functions of a series of sensors (ALI, ASTER, AVIRIS, AVHRR, IKONOS, ETM + /TM, MISR, MODIS, POLDER, SPOT-VEGETATION)

../data/ch2/OPAC: Aerosol and cloud climatology on their optical properties developed by Hess et al. (see Section 2.3.5).

../data/ch2/gads: Aerosol climatology on the global distribution of microphysical and optical aerosol properties from the GRADS database developed by Koepke et al. (see Section 2.3.5).

../data/ch4/refractive.index: Refractive indices of pure water and ice developed by Warren, Flatau et al.

../data/ch12/spectra: 119 surface reflectance spectra (350–2500 nm) measured using ASD spectroradiometers over different cover types by Liang et al.

SOFTWARE DIRECTORY

../software/ch1/solpos.c: C codes for calculating the solar zenith and azimuth angles and the extraterrestrial solar irradiance for a given time at a specific geographic location.

../software/ch2/MIE: Mie code developed by Wiscombe for the spherical particles. Another Mie code is given in ch4 (sphere.f).

525

../software/ch2/disort: a discrete ordinate radiative transfer FOR-TRAN77 program developed by Stamnes et al. for a multi-layered plane-parallel medium (e.g., atmosphere). It can be also used for calculating bidirectional reflectance of soil and snow in Chapter 4.

../software/ch2/SBDART.readme: Information on SBDART (Santa Barbara DISORT Atmosphere Radiative Transfer), a free FORTRAN computer code designed for the analysis of a wide variety of radiative transfer problems encountered in satellite remote sensing and atmospheric energy budget studies developed by Ricchiazzi et al.

../software/ch3/kuusk/forest: Fortran codes (gzipped tar file for forward simulation and inversion) of a directional multispectral forest reflectance model developed by Kuusk et al.

../software/ch3/kuusk/MCRM: Fortran codes (gzipped tar file for forward simulation and inversion) of a homogeneous two-layer canopy reflectance model developed by Kuusk et al. The GUI version of the MCRM2 is also included (xmcrm20803.tar.gz).

../software/ch3/sail: pkzip file containing SAIL canopy radiative transfer model code and the simulation tools developed by Barnsley et al. (http://stress.swan.ac.uk/~mbarnsle/) where other tools are also available.

../software/ch3/GeoSAIL: The GeoSAIL forest model is a combination of the SAIL model and a geometric optical model (Huemmrich, K. F. (2001), The GeoSAIL model: a simple addition to the SAIL model to describe discontinuous canopy reflectance, *Remote Sens. Envir.*, 75:423–431). Tree shapes are described by cylinders or cones distributed over a plane. Spectral reflectance and transmittance of trees are calculated from the SAIL model to determine the reflectance of the three components used in the geometric model: illuminated canopy, illuminated background, shadowed canopy, and shadowed background. The model code is Fortran.

../software/ch3/NADIM: New Advanced DIscrete Model (NADIM) is a bidirectional reflectance factor model to simulate the radiative regime in the solar domain in the case of horizontally homogeneous canopies developed by Gobron et al. (http://www.enamors.org/Deli/software.htm) where other codes/tools are also available.

../software/ch3/prospect.f: Fortran code of PROSPECT leaf optics model developed by Jacquemoud et al. (see Section 3.2.1).

../software/ch3/FLIGHT: A gzipped tar file containing a forest Monte Carlo ray tracing code developed by Dr. North (see Section 3.5.1.3).

../software/ch4/brdf: FORTRAN codes developed by Mishchenko et al. to determine the bidirectional reflection function for a semi-infinite homogeneous slab composed of arbitrarily shaped, randomly oriented particles based on a rigorous numerical solution of the radiative transfer equation. It is suitable for soils and snow/ice. spher.f computes the Legendre expansion coefficients for polydisperse spherical particles using the standard Lorenz-Mie theory; refl.f computes Fourier components of the reflection function; interp.f computes the bidirectional reflection function for any Sun-viewing geometries.

../software/ch6/ac.dark: C codes of atmospheric correction based on the "dark object" technique (Liang et al., 1997) for retrieving surface reflectance from TM imagery (see Section 6.2.3).

../software/ch6/ac.cluster: Fortran codes of atmospheric correction based on the cluster matching technique (Liang et al., 2001) for retrieving surface reflectance from ETM + imagery (see Section 6.2.5).

../software/ch7/IPW.readme: information for downloading the free IPW (Image Processing workbench) software from the internet (see Section 7.3).

../software/ch8/SPECPR.readme: information for downloading Specpr, a free Interactive One Dimensional Array Processing System, with the tools needed for reflectance spectroscopy analysis.

../software/ch8/SPEX.readme: information for downloading SPEX, a free tool with IDL program for process and analysis of a single spectra.

../software/ch8/Tetracorder.readme: information for downloading the Tetracorder, a free software tool for creating "the maps of materials".

../software/ch12/fcover: Estimate canopy gap fraction/ground cover from nadir-viewing digital photographs of vegetation canopies (C source code, man page, test data and binary (compiled for Red Hat Linux 7.2)). Downloaded from http://stress.swan.ac.uk/~mbarnsle/#Software.

../software/ch12/hemiphot: Estimate canopy gap fraction/ground cover from nadir-viewing, hemispherical digital photographs of vegetation canopies (C source code, man page, test data and binary (compiled for Red Hat Linux 7.2)). Downloaded from http://stress.swan.ac.uk/~mbarnsle/#Software.

Index

Quantitative Remote Sensing of Land Surfaces. By Shunlin Liang
ISBN 0-471-28166-2 Copyright © 2004 John Wiley & Sons, Inc.